D0910173

Interpretation Theory
in Applied Geophysics

International Series in the Earth Sciences

Robert R. Shrock, *Consulting Editor*

Kenneth O. Emery, *Woods Hole Oceanographic Institution*
Fritz Koczy, *Institute of Marine Science, University of Miami*
Konrad Krauskopf, *Stanford University*
John Verhoogen, *University of California, Berkeley*
Sverre Petterssen, *University of Chicago*

Reversed refraction profile in sediments overlying granitic basement in Bolivia (cf. Fig. 4-18). (*Courtesy of The California Standard Company and Royal Dutch Shell Company.*)

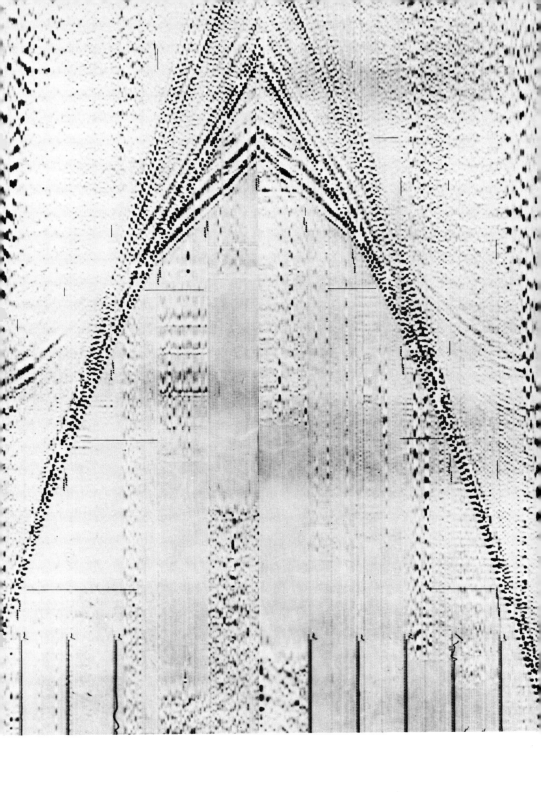

Interpretation Theory in Applied Geophysics

F. S. Grant

Associate Professor of Physics

G. F. West

Assistant Professor of Physics

University of Toronto

McGraw-Hill Book Company

New York   St. Louis   San Francisco
Toronto   London   Sydney

Interpretation Theory in Applied Geophysics

Library of Congress Catalog Card Number 64-8413

24100

2 3 4 5 6 – MP – 9 8 7 6

Preface

This is a book about applied geophysics written for two different audiences. It is meant for university students at the intermediate level in mathematics and physics who are preparing for future professional work in geophysics; it is also intended for persons working as professionals in geophysics or in related sciences who wish to brush up on the fundamentals of their specialty or broaden their acquaintance with others. The aim is to help build the mathematical and physical foundations for interpreting observational data in the areas of applied seismology, gravimetric and magnetometric surveying, and electrical (including electromagnetic) exploration. Wherever possible, we have tried to keep the two audiences impartially in view; but where their interests diverge—as inevitably they must—our attention is directed toward the student group rather more than to the professionals. However, the academic tone of this book should not obscure its essential function which is to try to relate theory to practice. Indeed if it were not so, the title would be difficult to justify.

Our purpose in writing this book has been to fill a gap in the literature of applied geophysics which has been in existence for many years and which today is widening inexorably. This is the gap in technical level between the introductory texts (written mostly for nonspecialists) and current research journals. There is not at the present time a single up-to-date book in the English language which even partially bridges it. In teaching applied geophysics to university students, this situation has for many years been an embarrassment to us and a handicap to students.   Moreover, some of our professional colleagues have suggested to us that the gap is also of concern to geophysicists outside the academic sphere, and that there is a corresponding need among professional people for a book of the type we proposed to write.   We hope that this book may go a part of the way toward satisfying the needs of both of these groups.

This is not a handbook for professionals, since it is devoted almost exclusively to theoretical principles and not to matters relating to instruments or to field practice.   Neither is it a reference book for advanced research workers, for it makes no serious attempt to review or to summarize current research activities in applied geophysics.   Nor does it contain the extensive lists of references and bibliographies that have today become a feature of such books.   Perhaps this is to be regretted; but, given the span of our topic, the contents could very easily have gotten out of hand.   And with some justification, we were anxious to keep the book to a tolerable size.

About the choice of material, it is obvious that the breadth of subject matter forces us to be selective.   In general, we have chosen those topics which in our opinion are best suited to university course work and yet are relevant to actual practice.   Probably the worst defect in this approach is that it sometimes tends to present the subject in an overly simplified, rigorous way.   We feel, however, that the introductory stage is not the time to stress the difficulties in interpretation nor to propagate doubts about the value of the theory; for as the professional reader well knows, the use of mathematics in real situations is tempered with a "feel" which comes only with experience.   Each section of the book is prefaced with a chapter which endeavours to provide the proper geological background for the theory that is to follow, and which we hope will prevent the inexperienced reader from taking an overly mathematical view of interpretation.   The problems and illustrations are for the most part concerned with geophysical prospecting for petroleum and minerals, although there is some material that relates more directly to regional exploration of the crust and upper mantle.   The difference is largely one of physical scale, not of basic approach.   Some topics such as engineering geophysics and groundwater exploration, which can be construed as branches of applied geophysics, have not been mentioned specifically in the text, although some of the theory that is developed for petroleum and mineral exploration problems is also relevant to these subjects.

To some extent, our choice of topics and our way of handling them reflects a personal viewpoint.   We do not feel that this is a matter for apology, since our selections are largely based upon direct experience and include methods which we believe are sound and practicable.   Occasionally, this criterion has compelled us to depart from what appear to be the more widely held views on certain topics, and to introduce a certain amount of new material.   This will almost certainly lead to divergences of opinion and we shall welcome criticism or comment which might be of benefit to future editions.   Also, if any errors appear in the text, we shall be grateful to have them pointed out to us.

F. S. Grant
G. F. West

Contents

xi

Chapter 3   *Plane Seismic Waves in Layered Media*   *51*

Chapter 4   *Analysis of Seismic Records*   *92*

Chapter 5   *Seismic Interpretation*   *127*

Chapter 6   *Reflection and Refraction of Spherical Waves*   *164*

Part II   **GRAVITY AND MAGNETIC METHODS**   **187**

Chapter 7   *Introduction to the Gravity and Magnetic Methods*   *189*

Chapter 8   *Potential Field Theory*   *210*

Chapter 15   *Electromagnetic Induction Methods*   *444*

Chapter 16   *Electromagnetic Theory*   *465*

Chapter 17   *Theory of Electromagnetic Induction*   *485*

Chapter 18   *Interpretation of Electromagnetic Surveys*   *547*

Appendix   *Coefficients for Downward Continuation*   *573*

Index   *577*

Interpretation Theory
in Applied Geophysics

Introduction

The cardinal aim in applied geophysics is to add a third dimension to geological maps. The trained eye of the field geologist is replaced with scientific instruments whose function is to detect changes in the physical properties of rocks which lie concealed beneath the surface. For instance, changes in rock density will cause minute variations in the earth's gravitational field; unusually magnetic rocks will produce disturbances in the earth's magnetic field; conductive rocks will affect the ground response to artificially stimulated electric or magnetic fields; and so on. Each of these phenomena involves the disturbance of a normal or predictable state of affairs by the addition of local perturbations, which in geophysics are called *anomalies*. Subsurface geology—the third dimension of the geological map—is unfolded somewhat obscurely to view in the pattern of anomalies that is observed on or above the ground. The geological picture is only vaguely adumbrated in lines of equal anomaly, and the primary job of the geophysicist is to interpret these observations in geological terms.

Measurements of geophysical phenomena are, for the most part, taken at the ground surface, and from these data the geophysicist must be able to outline the disturbing regions. This part of his work is closely controlled by well-established physical and mathematical laws, and although the deductions may often be ambiguous, the nature of the ambiguity at least is well understood. The next step, however, which is crucially important, is to translate these calculations into a reasonable geological picture. This part of his work is interpretative and speculative. It is surrounded with some of the mystery of a creative art. And not unlike other arts, success depends upon a proper appreciation and balancing of all the relevant factors—which in this case are physical and geological. For this, he will have to rely heavily upon past experience—his own and that of others. The whole operation of adducing a picture of the geology at depth from geophysical measurements is called *interpretation*, a word which aptly implies its indeterminate nature.

This book is very largely concerned with the first step in the interpretative process, i.e., with calculating the volumes of the disturbing regions from anomalies. It is not to be construed from this that greater importance is to be attached to performing these calculations than to understanding them in geological terms. The book has been written in its present form only because the second step—the translation of geophysics into geology—is based upon knowledge which can only be acquired through direct contact with real situations.

part I SEISMIC METHODS

chapter 1   Introduction to the Seismic Methods

1-1 Physical Basis of the Seismic Methods

Of all the physical methods used in geological exploration, the seismic methods are perhaps the most direct and—when applicable—give the least ambiguous results. These methods are used to find discontinuities in the *elastic* properties of rocks by detecting the faint "echoes" which return from these regions when a pulse of energy (usually a small explosion) is released near the surface of the ground. Rocks, being solid bodies, possess elasticity; because of this, stresses will be propagated away from the explosion as elastic waves. The velocities with which these waves will travel chiefly depend upon the elastic moduli of the rock. Since different kinds of rock have different elastic moduli, they will propagate stresses with different speeds. At interfaces where the velocities change abruptly, a part of the energy will be reflected and some of it will return eventually to the ground surface, where it can be picked up and recorded by sensitive instruments. If the times of flight of these small signals are accurately measured, we need know only the velocities in the formations which they have traversed in order to reconstruct the exact routes by which they have traveled from the source to the receiver. Generally these are worked out by using the ray-path methods of geometrical optics. Measurements of this kind are used in a systematic way to look for those variations in shape or in the properties of reflecting or refracting "horizons" which are known by association or experience to be favorable to the accumulation of petroleum, natural gas, water, or other minerals.

The electromechanical transducer which detects the ground motion and converts it into an electrical signal is called a *geophone* or a *seismometer*. The recording system which amplifies and displays these signals is called a *seismograph*. The output from the seismograph is a record of the ground motion at each seismometer location following the release of seismic energy, and it may be the photographic trace of a moving light spot, a signal impressed on magnetic tape, or a digital record of some kind. In petroleum prospecting the length of a seismic record seldom exceeds four or five seconds, and the events which appear on it are usually measured to the nearest two or three milliseconds. The "interpretation" of these events consists in identifying them as to their nature (i.e., reflection, refraction, surface wave, etc.) and in determining the depth and cause of their origin. Since these signals are accurately timed, the key to their interpretation lies in determining the elastic-wave velocities within the regions which they have traversed. The steps that one takes toward solving this problem are generally the most critical in the entire sequence of operations.

1-2 *Factors Influencing Seismic Velocity in Rocks*

As the result of extensive studies of the velocities of seismic waves in sedimentary formations in situ, Faust (1) concludes that the factors controlling this property are the depth of burial z and what he calls the "lithology" L. Included in "lithology" are porosity, lithification, chemical composition, texture, and fluid saturation. Geological age is implied in these properties in a general way and apparently need not be considered separately. Faust went further and endeavored to derive quantitative relationships involving these quantities. Because a large background of experience with well-logging methods had previously shown that electrical conductivity also changes with those factors included under "lithology," Faust chose to represent L with a quantity proportional to the *true* formation resistivity.[1] Thus he found that data from drilling cores having a wide geographical distribution over continental North America show that seismic velocities vary as the one-sixth power both of L and of z and show remarkably little

---

[1] For a definition of this term, see Chap. 13. Formation resistivity depends upon the conductivity of the saturating fluid (which the seismic-wave velocity does not) as well as upon the other factors included in L.

dispersion for so general a rule. These findings suggested a formula of the type

$$V = CL^{1/6}z^{1/6} \qquad (1\text{-}1)$$

where C turns out to be about 2×10^3 when the measurements are in feet, seconds, and ohm-feet. This is, of course, a purely empirical law, but it is remarkable that at about the same time a similar relationship was derived theoretically by Gassmann (2) for the velocity of compressional waves through a hexagonal (minimum volume) packing of spheres. The formula for a dry packing is

$$V = 800 \left[\frac{2\pi E^2 g}{(1 - \sigma^2)^2 n^3 \rho} \right]^{1/6} z^{1/6} \qquad (1\text{-}2)$$

where E, σ, and ρ are respectively Young's modulus, Poisson's ratio, and the density of the spheres, and n is the porosity of the medium. The expression in the brackets is, of course, modified when the pores are filled with fluid. The remarkable similarity between these two formulas strongly suggests that the increase of velocity with depth may be due largely to compression by the overburden.

Other workers have continued to study this problem. Acheson (3) in particular finds that longitudinal seismic-wave velocities in western Canada obey the rule

$$V(z) = Cz^{1/n} \qquad 8 < n < 20 \qquad (1\text{-}3)$$

although the value of n is prone to change within these limits with the geological environment.

The changes in seismic velocities that take place with depth are on the whole relatively minor, although they tend to be systematic and therefore ought to be considered. A much more important factor than the depth of burial is the gross textural composition of the rock. Seismically, we recognize five major categories of earth materials, which are, in order of increasing longitudinal-wave velocity, (1) alluvium, soil, and drift—relatively unconsolidated surface deposits that are generally recent in age; (2) sandstone and shale, no statistically significant difference being found between them; (3) limestone, dolomite, and related calcareous deposits; (4) granite, and other crystalline igneous and metamorphic rocks; and (5) salt. Velocity histograms for measurements which are given in Birch's "Handbook of Physical Constants" (4) are shown in Fig. 1-1. The 80 percent confidence ranges, also shown, indicate the ambiguity that attends any attempt to interpret lithology purely from seismic evidence.

In addition to the type of rock, several other factors also play a role in determining seismic-wave velocity. As indicated by the work of Faust, one of the more important of these is porosity. Since porosity tends to diminish with depth because of compression, the effect is strongest in the upper sedimentary layers. Laboratory measurements by Wyllie, Gregory,

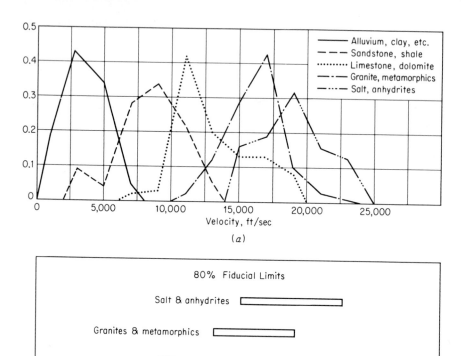

Fig. 1-1 (*a*) **Histograms of seismic-wave velocities for various classes of rocks;** (*b*) **80 percent fiducial limits. Data given by Birch (4).**

and Gardiner (5) on sandstone core specimens indicate that the velocity of longitudinal waves is roughly inversely proportional to porosity over a rather broad range (Fig. 1-2). There is little doubt that this accounts at least in part for the skewness in the frequency distribution of velocities in shale and sandstone toward lower values.

1-3 *Measurement of Seismic Velocity*

As an aid to geophysical interpretation, seismic-wave velocities are best measured in situ. Whenever possible the measurements are taken in deep boreholes with the aid of a logging device, so that a more or less continuous vertical record of longitudinal-wave velocity is obtained. An example of such a record is shown in Fig. 4-19. In point of fact, the measurements are not point values of the velocity but average values taken over the distance separating the electromechanical transducers in the logger—usually about 10 ft. Consequently the interfaces appearing on the log in transitional regions may be more numerous than is actually the case. It is usually advisable as a first step to study the log with a view to reducing the number of discontinuities, particularly in the transitional zones.

Major discontinuities, especially if they happen to coincide with abrupt changes in electrical resistivity (the two types of well log are often taken together), are generally noted as *markers*. Formation boundaries are frequently interpreted on the basis of such evidence. Between the marker horizons there should be no strong discontinuities, although the velocity may vary in a ragged manner. For many practical purposes the quasi-random part of this variation may be smoothed out, since it cannot appreciably affect the travel time of seismic waves whose wavelength is usually in excess of 100 ft. This leaves a smooth velocity function, which is called the *formation* velocity. In the case of flat-lying formations, the quantity which is defined by

$$\bar{V}_n = \frac{h_n - h_{n-1}}{\int_{h_{n-1}}^{h_n} dz/V_n(z)}$$

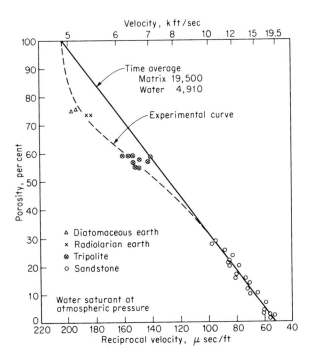

Fig. 1-2 Seismic-wave velocity as a function of porosity for silica rocks under directional pressure. [After Wyllie, Gregory, and Gardner (5).]

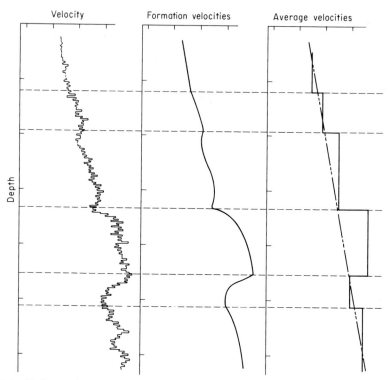

Fig. 1-3   Velocity log, formation velocities, and average velocities in a stratified section.

where $V_n(z)$ is the formation velocity and $h_n - h_{n-1}$ is the thickness of the nth formation in the sequence, is sometimes also called the *average* velocity in that formation.   The average velocity is calculated to yield the correct transit time for signals passing vertically through the formation, assuming that they travel at a constant speed.   In judging the degree of smoothing that is permissible in arriving at a formation velocity, it may be helpful to remember that velocity anomalies which occupy a distance less than one-quarter of a wavelength generally may be disregarded.   An exceptional circumstance in which this rule might fail is one in which there are several anomalies having a periodicity which might conceivably lead to Bragg reflection.

The profile of formation velocity can be integrated to yield an average velocity for waves traveling vertically through the entire sedimentary section.   For signals which reach a depth z, the average velocity is

$$\bar{V}(z) = z \left/ \int_0^z \frac{d\zeta}{V(\zeta)} \right. \tag{1-4}$$

where $V(z)$ is the total formation velocity including discontinuities, if any. Occasionally the actual formation velocity is replaced with a simple mathematical function which it roughly approximates. This is called the *velocity function*. It is used only for special purposes, such as the migration of reflection points (Chap. 5). From the average velocity it is an easy matter to draw a curve of two-way reflection time versus depth, assuming that there are only horizontal boundaries. From such a curve we can at once obtain the apparent depth of a reflecting zone from a measurement of the time of flight of its echo. To find the true depth, the point of reflection must be migrated to allow for the dip of the reflecting interface. Procedures for carrying out this step are given in Chap. 5.

Where deep boreholes are not available for logging or well surveys, the seismic velocities must be found from measurements which can be made at the surface of the ground. The theory of these procedures is discussed in Chap. 5. These methods are of course unable to predict subsurface velocities with nearly the same precision as that of the well-logging methods, and they give only a very coarse picture of the real velocity structure of the ground. What in fact it is possible to deduce from these methods are the average velocities between reflecting zones, from which a time versus depth curve can be constructed for making interpretations.

1-4 Hypotheses Used in the Interpretation of Seismic Data

In making interpretations we are confronted initially, not with the direct problem of calculating the times of echoes from the known velocities of the media through which they pass, but with the inverse problem of inferring the velocities and depths from the measured times. We shall find ourselves faced with an analogous situation in every method used in exploration geophysics. It is due to their "inverse" nature that geophysical interpretations are uncertain and ambiguous, that knowledge and experience in addition to the physical data are needed in order to achieve meaningful solutions, and that the task is rich in subtlety and challenge to the mind.

Any form of direct attack upon a problem of this kind requires in the first instance a flexible mathematical model whose predictions can readily be tested against the measurement provided by the field experiment. If these predictions do not agree well with the field observations, the model must be altered until the values do correspond at least to within satisfactory limits. The model succeeds if it is able to reproduce the more significant results of the physical experiment. However, it may be possible to find other models which predict just as accurately. At this point there is small logic in attempting to resolve the ambiguity by refining the calculations, since the limitations lie in the accuracy of the observations in any case. One must frequently accept a multiplicity of mathematical solutions and try to choose among them on other than purely physical grounds.

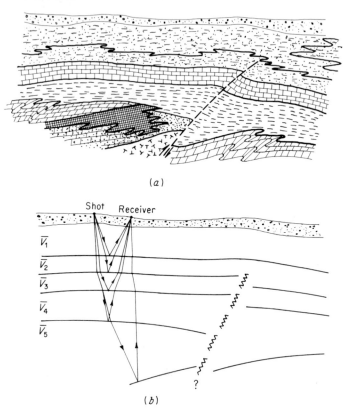

(a)

(b)

Fig. 1-4 (a) **A simplified stratigraphic section;** (b) **the geophysical interpretation which might be obtained by studying seismic reflections.**

At a later stage the model will be interpreted in geological terms that should be both meaningful and relevant. Here is where experience plays a most important part and where very little can be said in the way of general comment. If the results should later be proved incorrect, it is less likely that the mathematical solution is at fault than is the geological interpretation of this solution. Often this can be changed without affecting the calculations.

Now let us consider a few specific examples of mathematical models used in seismic interpretations. Since a great deal of seismic exploration is conducted in areas where the sediments are extensive and generally flat-lying, the horizontally stratified model consisting of a number of uniform layers is one at least that seems obvious. The typical geological situation to which this model corresponds is infinitely more complex, but the points of resemblance are usually apparent. The velocities within the layers correspond to average formation velocities, and the thicknesses, to the vertical distances between the centers of the reflecting or transitional zones.

In continuous-correlation reflection studies, for instance, we are frequently looking for changes in the shapes of boundaries which might indicate suitable petroleum reservoir structures. The geological section might look something like the sketch shown in Fig. 1-4a, or it might be much more complicated than this. The model likely to be used to represent this situation, together with the interpreted ray paths, is shown in Fig. 1-4b. Geophysics will yield values for the formation thicknesses and velocities. The reader may judge the subtleties of extracting meaningful geological information from these results.

A more striking example of the differences in detail between geology and geophysics is to be found in the model used to interpret seismic refraction data. On the hypothesis that the ground consists of parallel, uniform layers whose seismic-wave velocities without exception increase with depth,[1] the refraction time-distance measurements at the surface of the ground will lead to unique solutions for the formation velocities and thicknesses by assuming that the seismic energy has traveled along the ray paths shown in Fig. 1-5. This method, which is occasionally used in broad-scale reconnaissance, obviously lacks the resolving power of the continuous correlation

---

[1] Should this fail to happen in any layer, that layer will be missed in the interpretations and a cumulative error in the depth calculations would result. (See Sec. 5-7, under "blind zones.")

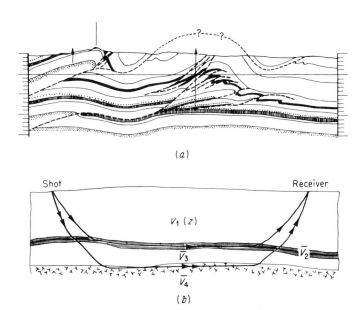

(a)

(b)

Fig. 1-5 (a) A stratigraphic section; (b) the geophysical interpretation which might be obtained by studying the arrival of refracted waves [Part (a) after T. C. Richards (6).]

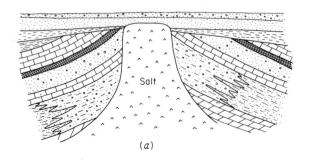

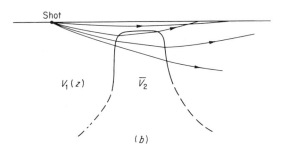

Fig. 1-6 (*a*) A salt dome; (*b*) the model used for geophysical interpretation based on the study of the times of arrival of direct waves.

reflection method insofar as details are concerned, but on the other hand it samples much greater volumes of the ground and on that account it is useful for such tasks as preliminary basin studies.

One of the early applications of the seismic method, and one which is still used to some extent, is in the direct search for large salt masses, particularly domes. Salt is characterized by, among other things, an unusually high seismic-wave velocity. Waves which spend an appreciable fraction of their travel time in salt will arrive at their destination anomalously early. A negative anomaly in the times of flight of direct waves is therefore taken to be a favorable indication. The geology is sketched in Fig. 1-6*a*, and the geophysical model, with ray paths, is shown in Fig. 1-6*b*.

There are obviously great differences between the mathematical models used in applied geophysics and the geological situations which they are supposed to represent. The differences lie essentially in the idealizations made in passing from the real to the representative situations. Geophysical interpretation consists in calculating the model and then in attempting in some sense to reverse the processes of idealization. Unfortunately, this is essentially an ambiguous undertaking; for, as Plato says, "The many are encountered without being defined, whereas ideas are defined although never encountered." The "many" to which Plato refers, of course, are commonplace things and phenomena of nature, while the Greek "idea" comes close to our notion of the abstract ideal. The difference between the abstract and the concrete worlds of thought could hardly be put more concisely.

In seeking to establish a two-way bridge between them, geophysical interpretation finds much of its enduring fascination and intellectual challenge.

References

1. L. Y. Faust, A Velocity Function Including Lithologic Variations, *Geophysics*, vol. 18, pp. 271–287, 1953.

2. F. Gassmann, Elastic Waves through a Packing of Spheres, *Geophysics*, vol. 16, pp. 673–685, 1951.

3. C. H. Acheson, Time-depth and Velocity-depth Relations in Western Canada, *Geophysics*, vol. 28, pp. 894–909, 1963.

4. F. Birch (ed.), "Handbook of Physical Constants," Geol. Soc. America Spec. Paper 36, 1942.

5. M. J. R. Wyllie, A. R. Gregory, and G. H. F. Gardiner, An Experimental Investigation of Factors Affecting Elastic Wave Velocities in Porous Media, *Geophysics*, vol. 23, pp. 459–493, 1958.

6. T. C. Richards, Wide Angle Reflections and Their Application to Finding Limestone Structures in the Foothills of Western Canada, *Geophysics*, vol. 25, pp. 385–407, 1960.

chapter 2   Elastic Waves in Unbounded Media

2-1   Fundamentals of Elasticity Theory

The theory of elasticity is concerned with the deformation of matter under stress. In this very brief introduction, we shall discuss only the most elementary topics prerequisite to the study of the propagation of seismic waves. Fuller treatments of the subject can be found in several books, to which references are given at the end of this chapter.

We shall take the point of view that, although recognizing the crystalline or granular nature of rocks, the phenomena in which we are interested are on such a scale that the local fluctuations of internal forces due to varying areas of contact between grains may be disregarded. Thus any force is the resultant of a very large number of small-scale effects which, taken as an ensemble, can be described by means of continuous variables. Thus the derivatives of force components, which would be difficult to describe on a microscopic scale, have a perfectly definite meaning in the macroscopic theory. The validity of this approach presupposes that the volumes throughout which these forces act contain a very large number of grains, so that in every case the "continuum" limit of the stresses is attained to well within the practical limits of observation. In the future, we shall refer to such macroscopic descriptions of the mechanical behavior of solid matter as the "continuum" theory.

2-2 The Analysis of Stress

Solid bodies are capable of propagating forces, for a load acting on one free surface produces a reaction at another. The elementary notion of *stress* within a body is concerned with the balance of internal action and reaction between different parts of the body at a given internal point. Physically, the concept of a stress derives from the analysis of tractions acting across surfaces of finite area, and therefore to speak of the "stress at a point" is somewhat ambiguous. It implies the limiting value of a traction or force acting over a given area as the area shrinks to zero. But while this idea of a limit provides us with a mathematical definition of stress, it is in no way concerned with the practical measurement of it. For in practice, the stress at any interior point of a solid is incapable of any form of direct measurement and can only be inferred from mechanical conditions at the outer surfaces, where internal reactions are balanced by external loads.

Let us consider then the force $\mathbf{F}$ acting at a point P within an elastic continuum. If ΔS is a small surface element which contains P and is cut from a plane whose orientation is specified by the unit normal vector $\mathbf{n}$, then the *stress* at P with respect to the direction $\mathbf{n}$ is given by the three components of the vector quantity $\lim_{\Delta S \to 0} \mathbf{F}/\Delta S$ (Fig. 2-1). One of these involves the component of $\mathbf{F}$, which is normal to ΔS and is called the *principal* stress component; the other two, which are derived from the tangential components of $\mathbf{F}$, are called *shearing* stress components. But since the direction of $\mathbf{n}$ is arbitrary, it is plain that other stress components can be derived from the same force $\mathbf{F}$ merely by changing the direction of $\mathbf{n}$. Thus the state of stress at P must in general be described in terms of the three force components acting across each of three mutually perpendicular surfaces, taken in the limit as each one of these surfaces shrinks to zero.

Let us surround the point P with the small volume ΔV in the form of a

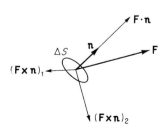

Fig. 2-1 Traction acting across a small surface.

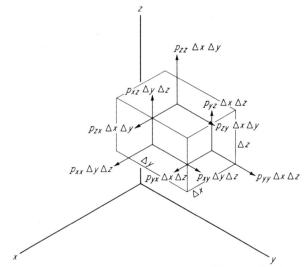

Fig. 2-2 Components of the stress at a point within an elastic continuum.

rectangular parallelepiped whose sides are Δx, Δy, and Δz (Fig. 2-2). The volume ΔV is assumed to be macroscopically small in the sense that all macroscopic variables converge to limiting values within it, but at the same time it is microscopically large in the sense that it contains a sufficient number of grains to give to these variables a recognizable meaning in terms of the continuum theory. For each of the six faces, there are three components of the stress, but opposite faces obviously involve identical quantities if the material is in equilibrium. Nine components are therefore required to specify the stress at P, which we denote with the symbol p_{mn}, each of the two subscripts being allowed independently to stand for x, y, or z. The first subscript specifies the direction of the normal to the area across which the associated force component acts, and the second specifies the direction of this force. A quantity having these directional properties is called a cartesian tensor of the second rank, and for this reason we shall speak in future of the *stress tensor*.

Not all of the nine components of the stress tensor are independent, however. We may show this by considering the total moment of all the forces acting across the six faces of ΔV about any axis through P. This moment, as we know, must equal the rate of change of angular momentum of ΔV about that axis. If I is the appropriate moment of inertia, then $\mathbf{h} = I\boldsymbol{\omega}$; but $I \approx \rho \Delta V (\frac{1}{2}\Delta x)^2$ and is therefore of the fifth order in the linear dimensions of ΔV. The moment of the surface forces on the other hand is $\approx p_{mn} \Delta S(\frac{1}{2}\Delta x)$ and is of only the third order. Therefore, if we are to avoid developing infinite angular accelerations as the dimensions of

ΔV become infinitesimal, the sum of the moments of all surface forces about any axis must vanish at all times.

Let us make ΔV so small that any changes in the stress from its value at P can be ignored. This is always permissible if p_{mn} is continuous. Then the moment about the axis through P in the direction Oz of the surface forces acting on ΔV is

$$-(p_{yx}\,\Delta x\,\Delta z)\,\Delta y + (p_{xy}\,\Delta y\,\Delta z)\,\Delta x = 0$$

i.e., $\qquad\qquad p_{yx}\,\Delta V = p_{xy}\,\Delta V \qquad \text{or} \qquad p_{yx} = p_{xy}$

Similarly, by considering moments about axes in the directions Ox and Oy, we find that $p_{xz} = p_{zx}$ and $p_{zy} = p_{yz}$. Thus we have in general $p_{nm} = p_{mn}$, and the stress tensor is therefore symmetrical. The number of independent components is thereby reduced from nine to six.

2-3 The Analysis of Strain

Consider two neighboring particles in an unstressed solid at $P(x,y,z)$ and at $Q(x + dx, y + dy, z + dz)$. Now suppose that the body deforms in some manner under an applied load, so that the particle at P is displaced by an amount $\mathbf{u}$ to a new position $P(x + u, y + v, z + w)$. Then if $\mathbf{u} + d\mathbf{u}$ is the displacement suffered by the particle at Q, we may write for the components of $d\mathbf{u}$ (du, dv, dw)

$$du \doteq \frac{\partial u}{\partial x}\,dx + \frac{\partial u}{\partial y}\,dy + \frac{\partial u}{\partial z}\,dz$$

plus similar expressions for dv and dw.

These three equations can be contracted into a single formula by the use of special indicial notation. Thus if we set

$$u = u_1 \qquad v = u_2 \qquad w = u_3$$

and $\qquad\qquad x = x_1 \qquad y = x_2 \qquad z = x_3$

we may condense the separate expressions for du, dv, and dw into the single form

$$du_i = \sum_{k=1}^{3} \frac{\partial u_i}{\partial x_k}\,dx_k \qquad i = 1, 2, 3$$

Fig. 2-3 Displacements of neighboring points within an elastic continuum.

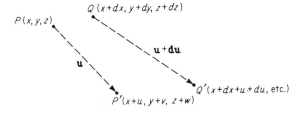

$Q\,(x+dx,\,y+dy,\,z+dz)$

$P\,(x,y,z)$

$\mathbf{u}+d\mathbf{u}$

$\mathbf{u}$

$Q'(x+dx+u+du,\,\text{etc.})$

$P'(x+u,\,y+v,\,z+w)$

which is more compact and a great deal more convenient for manipulations. The set of nine quantities $\partial u_i/\partial x_k$ constitutes a cartesian tensor of the second rank. Since it is nonsymmetrical, it can be divided into symmetrical and antisymmetrical components by introducing the two auxiliary tensors (1)

$$e_{ik} = \frac{1}{2}\left(\frac{\partial u_k}{\partial x_i} + \frac{\partial u_i}{\partial x_k}\right)$$

which is symmetrical, and

$$\xi_{ik} = \left(\frac{\partial u_k}{\partial x_i} - \frac{\partial u_i}{\partial x_k}\right)$$

which is antisymmetrical. Then we may write

$$\frac{\partial u_i}{\partial x_k} = e_{ik} - \frac{1}{2}\xi_{ik}$$

We first note some of the properties of the two tensors e_{ik} and ξ_{ik}. $\xi_{ik} = 0$ if $i = k$, and $\xi_{ik} = -\xi_{ki}$ if $i \neq k$. Therefore this tensor has just three independent components, ξ_{yz}, ξ_{zx}, and ξ_{xy}. Note that these happen also to be the components of the vector $\nabla \times \mathbf{u}$. We express this situation in vector notation as follows:

$$\boldsymbol{\xi} = \nabla \times \mathbf{u}$$

where $\boldsymbol{\xi}$ is the *vector* of the tensor ξ_{ik} and is defined as

$$\boldsymbol{\xi} = (\xi_{yz}, \xi_{zx}, \xi_{xy})$$

Since the curl of the displacement $\mathbf{u}$ is associated with a rigid rotation of the material about some axis through P, ξ_{ik} is called the *rotation tensor*.

The components of e_{ik} on the other hand are associated with actual deformation of the material. This is most easily demonstrated by considering the change produced in the distance PQ by loading. Reverting once again to indicial notation, we may write

$$d(PQ)^2 = (P'Q')^2 - (PQ)^2 = \sum_{i=1}^{3}[(dx_i + du_i)^2 - (dx_i)^2]$$

$$= \sum_{i=1}^{3}\left[\left(dx_i + \sum_{k=1}^{3}\frac{\partial u_i}{\partial x_k}dx_k\right)^2 - (dx_i)^2\right]$$

which, if we neglect the quadratic terms in $\partial u_i/\partial x_k$, becomes

$$d(PQ)^2 = \sum_{i=1}^{3}\sum_{k=1}^{3}\frac{\partial u_i}{\partial x_k}dx_k\, dx_i + \sum_{i=1}^{3}\sum_{j=1}^{3}\frac{\partial u_i}{\partial x_j}dx_j\, dx_i$$

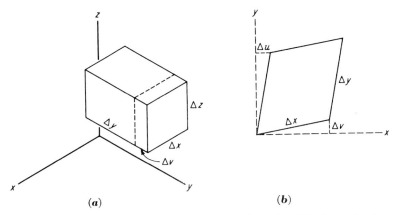

Fig. 2-4   Strain at a point. (*a*) **Principal strain;** (*b*) **shear strain.**

If in the second term we put $j = i$ and $i = k$ (which is permissible since i, j, and k are all dummy suffixes) we get

$$d(PQ)^2 = 2 \sum_{i=1}^{3} \sum_{k=1}^{3} e_{ik} \, dx_i \, dx_k$$

The physical meaning of the elements of the tensor e_{ik} is demonstrated in the two diagrams shown in Fig. 2-4. In the first, a deformation is produced in the y direction only. The original length of the parallelepiped in this direction is Δy. After the deformation, this length is changed to $\Delta y + \Delta v$. Since v is a function of y only, we may write

$$\Delta v = \frac{\partial v}{\partial y} \Delta y + \frac{\partial^2 v}{\partial y^2} \frac{(\Delta y)^2}{2!} + \cdots$$

The quantity

$$\lim_{\Delta V \to 0} \frac{\Delta v}{\Delta y} = \frac{\partial v}{\partial y} = e_{yy}$$

is called the *first-order principal strain* in the y direction and is seen to represent the fractional change in length of a small volume of the material, taken in the limit as the volume shrinks to zero. The implication of a limit in the definition of strain permits us to speak of the "strain at a point," although the concept is admittedly a little difficult to visualize. By similar arguments, the principal strains in the x and z directions are given by e_{xx} and e_{zz}, respectively.

In the second example of Fig. 2-4, the block is shown in plan in a state of shear. The face that originally was perpendicular to Ox has been rotated by an amount Δu in the positive x direction, while the face that was perpendicular to Oy has been rotated by an amount Δv in the positive y direction. Since the deformation Δu has a first-order dependence on y only, we may write

$$\Delta u = \frac{\partial u}{\partial y} \Delta y + \frac{\partial^2 u}{\partial y^2} \frac{(\Delta y)^2}{2!} + \frac{\partial^2 u}{\partial x \partial y} \Delta x \Delta y + \cdots$$

Similarly, the deformation Δv has a first-order dependence on x only; hence

$$\Delta v = \frac{\partial v}{\partial x} \Delta x + \frac{\partial^2 v}{\partial x^2} \frac{(\Delta x)^2}{2!} + \frac{\partial^2 v}{\partial x \partial y} \Delta x \, \Delta y + \cdots$$

The quantity

$$\lim_{\Delta V \to 0} \left(\frac{\Delta v}{\Delta x} + \frac{\Delta u}{\Delta y} \right) = \frac{\partial v}{\partial x} + \frac{\partial u}{\partial y} = 2e_{xy} = 2e_{yx}$$

is called the *first-order* (or infinitesimal) *shearing strain* perpendicular to Oz, and it is seen to represent a change in the shape of ΔV, but not a change in the volume. Notice that because $e_{xy} = e_{yx}$, a change of Δu in the direction of y may be accompanied by a change of Δv in the direction of x. The shearing-strain components in the other two principal directions are $2e_{yz}$ and $2e_{zx}$.

Cubical dilatation, or simply *dilatation,* is a term often used in elasticity theory to signify the fractional volumetric increase at P in the limiting case when the dimensions of the small volume at P shrink to zero. The symbol used is θ, and to a first order in the strains it is equal to the sum $e_{xx} + e_{yy} + e_{zz}$. In vector notation

$$\theta = \frac{\partial u}{\partial x} + \frac{\partial v}{\partial y} + \frac{\partial w}{\partial z} = \nabla \cdot \mathbf{u}$$

The dilatation is therefore just the divergence of the displacement vector $\mathbf{u}$. Being a scalar, it is invariant under a rotation of the axes. A negative dilatation is sometimes also called a compression.

The neglect of those terms in the expression for $d(PQ)^2$ which involve powers of $\partial u_i / \partial x_k$ of order two is the point of departure for what is called the *infinitesimal strain theory,* which represents the point of view commonly taken in seismology. This approximation allows us to represent the actual strains in the material with the components of the tensor e_{ik}, without modification. On the other hand, in certain other branches of elasticity theory where large deformations are the rule, and particularly in the theories of rubber elasticity and of plasticity, attempts have been made to include second-order terms in the representation of strain, and this leads to what has been termed *second-order strain theory* or *finite strain theory.* Addition of these terms leads to equations which are vastly more complicated and usually nonlinear. In seismology, however, the deformations are so small that the quadratic terms can be neglected.

2-4   *Stress-Strain Relations for a Perfectly Elastic Solid*

An elastic solid is one in which the strain at any point is determined by the stress at that point. The special case in which the components of the strain are homogeneous linear functions of the stress components (on the infinitesimal strain theory) is termed *perfect elasticity.* This is a generalization of what is commonly called Hooke's law, and it seems to agree with observations made in practice on a wide variety of materials and under a wide variety of conditions. It is not universally true, of course. Departures from Hooke's law are commonly observed in the behavior of materials which are called viscoelastic and plastic, and an adequate theoretical account of the mechanical behavior of these materials requires the use of stress-strain relations of a more complex nature. Fortunately, most of the properties of seismic waves are adequately described by Hooke's law, even in materials which are known not to be perfectly elastic under static loads.

By far the easiest route by which to proceed is to accept Hooke's law as an empirical truth about the elastic behavior of materials within a specified range of loading. To attempt to derive Hooke's law on purely a priori grounds is not an easy problem. And it is, after all, the simplest relationship between stress and strain that can be supposed to exist. We must remember, however, that for conclusions based upon Hooke's law to be valid, the stresses must not exceed the limits within which the material is approximately perfectly elastic.

The statement made by Hooke is that the deformation of a perfectly elastic body from its equilibrium configuration is directly proportional to the load applied to it. This is true only within a certain range of loading, whose upper limit is called the *proportional limit* of the material. Beyond this limit the stress-strain relation ceases to be linear, although the deformations may still be recoverable and hence the material may still be elastic. Under additional loading the deformations can then be described in terms of the nonlinear strain theory, but most materials soon reach a point thereafter at which they undergo plastic deformation which is irrecoverable. This limit, called the *elastic limit* of the material, usually depends upon the duration of the applied load. Thus it is very often much larger for rapidly changing loads, such as forced vibrations, than it is for static or slowly applied loads. This may be because dynamic stresses do not allow the material time to alter its microscopic structure or to "flow" nearly so readily as static stresses do. This time factor is one of many difficulties that arise in the descriptive theories of continuum mechanics.

Hooke's law in its most general form seems to apply even to the most aeolotropic materials. For the present, however, our interest is confined to isotropic substances, so that we may take a much more restricted interpretation of this law. One way of stating it is that in isotropic materials the stress tensor and the strain tensor always have the same principal axes and the relationship between them is linear. Thus when a bar is placed in simple tension, it extends longitudinally and contracts laterally, the transverse deformation being isotropic. If the stress tensor is reduced to a simple tension, let us say p_{xx}, and if the long axis of the bar is placed in the direction Ox, this state of affairs may be expressed in simple tensor notation as follows:

$$Ee_{xx} = p_{xx} \qquad Ee_{yy} = Ee_{zz} = -\sigma p_{xx}$$

where, according to convention, we regard tensile stresses and extensive strains as algebraically positive and compressive stresses and contractive strains as negative. In these equations E and σ, both of which are constants of the material, are called Young's modulus and Poisson's ratio, respectively. We note that inasmuch as the strain (as defined) is nondimensional, Young's modulus has the dimensions of stress and Poisson's ratio is a pure number.

Initially we had assumed that all stress components other than p_{xx} were zero. If we now suppose that p_{yy} and p_{zz} also exist, there will then be additional strains which must be added to those due to p_{xx}, giving

$$Ee_{xx} = p_{xx} - \sigma(p_{yy} + p_{zz})$$

with similar relations for e_{yy} and for e_{zz}. If we rewrite this equation as

$$Ee_{xx} = (1 + \sigma)p_{xx} - \sigma(p_{xx} + p_{yy} + p_{zz})$$

we see that the entire set of stress-strain relations may be written

$$Ee_{mm} = (1 + \sigma)p_{mm} - \sigma(p_{xx} + p_{yy} + p_{zz}) \qquad (2\text{-}1)$$

This tensor equation summarizes the stress-strain relationship when the coordinate axes coincide with the principal axes of stress.

In seismology the stresses are unknown, and we are required to solve differential equations of motion for the displacements. Clearly what is required is a relationship giving the stresses in terms of the displacements and therefore of the strains. This means that we must invert the formulas (2-1). To do so, we first sum up the three equations, getting

$$E\theta = (1 - 2\sigma)(p_{xx} + p_{yy} + p_{zz})$$

or

$$p_{xx} + p_{yy} + p_{zz} = \frac{E}{1 - 2\sigma} \theta$$

Putting this back into (2-1) gives

$$(1 + \sigma)p_{mm} = Ee_{mm} + \frac{\sigma E}{1 - 2\sigma} \theta$$

which, in indicial notation, may be written

$$p_{kk} = \lambda\theta + 2\mu e_{kk}$$

where $\qquad \lambda = \dfrac{\sigma E}{(1 + \sigma)(1 - 2\sigma)} \qquad$ and $\qquad \mu = \dfrac{E}{2(1 + \sigma)} \qquad (2\text{-}2)$

λ and μ are known as Lamé's constants for the solid.   Actually they have the dimensions of stress and are therefore elastic moduli.   Like E and σ, their values are independent of the stress, provided that it does not exceed the proportional limit of the material.

Let us now consider what happens under the action of a simple torque. If one face of a small block of the material is clamped in the plane $z = 0$ and the opposite face subjected to a tangential pressure in the Oy direction, only the p_{zy} and p_{yz} components of the stress tensor will exist.   The deformations caused by these stresses will be $2e_{yz}$ and $2e_{zy}$, and since these are the same and each is proportional to the stress, we may write

$$p_{yz} = 2\bar{\mu}e_{yz}$$

$\bar{\mu}$ is called the modulus of rigidity (or simply the rigidity) of the material. It is clear that the rigidity is a measure of the ability of the material to resist changing its shape under the application of a torque but that it is not concerned with changes in volume.   Since identical relations must exist also for p_{zz} and p_{xy}, we may put them into indicial notation by writing

$$p_{ik} = 2\bar{\mu}e_{ik} \qquad i \neq k \tag{2-3}$$

The equations (2-2) and (2-3) can be contracted into a single tensor equation if we set $\bar{\mu} = \mu$.   The proof of this identity requires an understanding of the transformation properties of cartesian tensors, and we must refer again to Jeffreys (1) for details.   The general stress-strain relationship, which applies when the coordinate axes no longer coincide with the principal axes of the stress tensor, is given by

$$p_{ik} = \lambda\theta\delta_{ik} + 2\mu e_{ik} \tag{2-4}$$

This is Hooke's law for isotropic elastic bodies, written in tensor form.   The term δ_{ik} in this expression is called the *unit tensor* or the *Kronecker delta*.   It is actually a second-rank tensor whose diagonal elements are all unity and whose off-diagonal elements are all zero.   It is readily apparent that the formulas (2-2) and (2-3) are both consistent with this expression.

We may note, parenthetically, that from Hooke's law it is possible to determine the incompressibility of an isotropic substance in terms of the Lamé constants.   The incompressibility, or bulk modulus, is defined as the ratio of hydrostatic stress to volumetric strain and is usually denoted with the symbol κ.   That is,

$$\kappa = -\frac{p}{\theta}$$

By putting $i = k$ in (2-4), we find that $\kappa = \lambda + \tfrac{2}{3}\mu$, with the dimensions of stress.   This modulus expresses the ability of a material to resist contraction under hydrostatic pressures.   It is of particular importance in the study of the propagation of seismic disturbances through fluids, where the stress tensor in (2-4) reduces to $p_{ik} = -p\delta_{ik}$ and $\mu = 0$.

2-5 Stress-Strain Relations in Aeolotropic (Transversely-isotropic) Media

The tensor formulation of Hooke's law (2-4) applies, as we have already noted, only to isotropic media. It is an extremely compact and convenient form for representing the entire set of stress-strain relations under those somewhat restricted conditions. The situation quickly deteriorates, however, as soon as aeolotropy is introduced. The elastic behavior of such materials can no longer be described in terms of just two elastic constants, and as a result, no simple relation of the kind (2-4) exists between the stress and the strain tensors. At a first glance, the mathematical situation for aeolotropic elasticity theory looks bleak indeed.

Aeolotropy is actually of little interest in seismology, except for one specific instance of it. That corresponds to what we might call—in the terminology used by Love—a medium having "hexagonal" symmetry, i.e., one which possesses two-dimensional isotropy. A homogeneous material of this kind, which we shall call *transversely-isotropic*, has elastic properties which are everywhere constant and which are the same in two directions but different in the third. Many earth materials possess a natural lamination or a foliation that is much too fine to permit a subdivision of the material into isotropic layers of macroscopic size, yet on a gross scale it produces a definite aeolotropy in the propagation of seismic waves.

A rigorous account of the dynamic elastic behavior of foliated materials consisting of finely interbedded isotropic layers has been given by Rytov (2), but it is somewhat beyond the scope of this book. We shall instead follow a more elementary approach, which is based essentially on the method used by Love (3).

In transversely-isotropic media, the stress and the strain tensors will have the same principal axes. Thus if the axis of symmetry is in the direction Oz, we may write

$$p_{xx} = Ae_{xx} + Ge_{yy} + Fe_{zz}$$
$$p_{yy} = Ge_{xx} + Ae_{yy} + Fe_{zz}$$
and
$$p_{zz} = Fe_{xx} + Fe_{yy} + Ce_{zz}$$

where A, C, F, and G are constants having the dimensions of stress. It is obvious that A and G must enjoy corresponding positions in the expressions for p_{xx} and p_{yy}. It is perhaps not quite so obvious why F should occur as it does in all three expressions. This is because the transformation properties of the symmetry group to which the transversely-isotropic medium belongs require that the coefficients appearing on the right-hand sides of these equations form a symmetrical matrix.[1] We may put these expressions into a form analogous to (2-4) by writing

$$p_{xx} = \lambda_\| \theta + 2\mu_\| e_{xx} + (\lambda_\perp - \lambda_\|)e_{zz}$$
$$p_{yy} = \lambda_\| \theta + 2\mu_\| e_{yy} + (\lambda_\perp - \lambda_\|)e_{zz} \qquad (2\text{-}5a)$$

and
$$p_{zz} = \lambda_\perp \theta + 2\mu_\perp e_{zz}$$

where
$$A = \lambda_\| + 2\mu_\|$$
$$G = \lambda_\|$$
$$F = \lambda_\perp$$
$$C = \lambda_\perp + 2\mu_\perp$$

The λ's and μ's occupy the same places in these equations as do Lamé's moduli for the isotropic solid.   Accordingly, we shall call them by that name, but we shall distinguish by the use of subscripts whether the stresses to which they correspond are parallel or perpendicular to the plane of isotropy.   Note that the expressions for other elastic moduli such as E or κ do not follow directly by substitution of either pair into the formulas derived in Sec. 2-4.

To Eqs. (2-5a) we may also add for the shearing stresses

$$p_{xy} = 2\mu_\| e_{xy}$$
$$p_{yz} = 2\nu e_{yz} \qquad (2\text{-}5b)$$

and
$$p_{zx} = 2\nu e_{zx}$$

where ν is a fifth elastic modulus which is not expressible in terms of the Lamé constants.   The six equations (2-5a) and (2-5b) describe Hooke's law for a transversely-isotropic elastic body.   Unfortunately, we cannot contract them further.

2-6 Equations of Motion

Let us now consider how time-dependent stresses may be transmitted through an unbounded elastic solid.   We shall assume the material to be macroscopically homogeneous and in equilibrium with respect to all body and surface forces in the undisturbed configuration.   If a disturbance passes through the material, the displacement of the point $P(x,y,z)$ at any instant t during its passage shall be specified by the vector $\mathbf{u}(x,y,z;t)$.

Let us first enumerate all the surface forces acting on the small parallelepiped ΔV in the z direction (Fig. 2-5).   We find that there is one component in this direction across each of the three pairs of opposite faces.   For example, from the faces perpendicular to Oy we obtain

$$\left(p_{yz} + \frac{1}{2}\frac{\partial p_{yz}}{\partial y}\,\Delta y\right)\Delta z\,\Delta x - \left(p_{yz} - \frac{1}{2}\frac{\partial p_{yz}}{\partial y}\,\Delta y\right)\Delta z\,\Delta x = \frac{\partial p_{yz}}{\partial y}\,\Delta V$$

[1] For further elucidation of this point, see Love (3), Chap. VI.

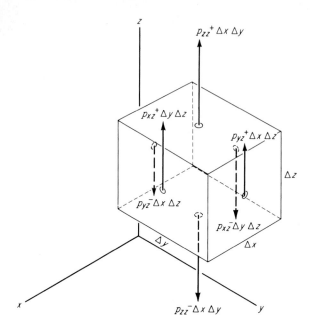

Fig. 2-5 Surface forces in the z direction that act on a volume element at P.

Similar expressions are obtained from the other pairs of faces. The net unbalanced surface force on ΔV in the direction Oz is therefore the sum of three terms, viz.,

$$F_z = \left(\frac{\partial p_{xz}}{\partial x} + \frac{\partial p_{yz}}{\partial y} + \frac{\partial p_{zz}}{\partial z} \right) \Delta V$$

or, in the indicial notation used in Sec. 2-2, this may also be written

$$F_3 = \sum_{k=1}^{3} \frac{\partial p_{k3}}{\partial x_k} \Delta V$$

Identical arguments show in general that

$$F_i = \sum_{k=1}^{3} \frac{\partial p_{ki}}{\partial x_k} \Delta V = \sum_{k=1}^{3} \frac{\partial p_{ik}}{\partial x_k} \Delta V$$

Thus the motion of the material at P, if we neglect gravitational forces, is given by the set of equations

$$\rho \frac{d^2 u_i}{dt^2} = \sum_{k=1}^{3} \frac{\partial p_{ik}}{\partial x_k}$$

The Lagrangian differential operator d/dt signifies differentiation following the motion of the material, and it is concerned with the total motion of a particular element or particle in space. The Eulerian operator $\partial/\partial t$ on the

other hand signifies differentiation at a point fixed in space, and it is the operator that is appropriate to the equations of motion of the material. The connection between the two operators is given by the relation

$$\frac{d}{dt} = \frac{\partial}{\partial t} + \sum_{i=1}^{3} \frac{\partial u_i}{\partial t} \frac{\partial}{\partial x_i} \qquad (2\text{-}6)$$

If we consider a sinusoidal disturbance having an amplitude a, a wavelength λ, and a period τ which passes the point P, the operator $\partial/\partial t$ reduces the amplitude by $O(1/\tau)$, while the operator $\partial/\partial x_i$ reduces it by $O(1/\lambda)$. Since

$$\frac{\partial u_i}{\partial t} \approx \frac{a}{\tau}$$

it follows that the second operator on the right-hand side of (2-6) introduces terms that are of $O(a/\lambda)$ times those introduced by the first. In seismology, amplitudes of ground motion are always almost infinitesimally small in relation to the wavelengths of the seismic disturbances. We are justified therefore in discarding the distinction between Eulerian and Lagrangian accelerations and in writing for the general equations of motion

$$\rho \frac{\partial^2 u_i}{\partial t^2} = \sum_{k=1}^{3} \frac{\partial p_{ik}}{\partial x_k} \qquad (2\text{-}7)$$

It may be noted in passing that stresses due to gravitational forces have been omitted from these derivations. Terms due to inhomogeneity in the earth's gravitational field may be added to the right-hand side of (2-7), but their effect upon the propagation of seismic waves is altogether negligible at the frequencies and within the distances encountered in applied seismology.

2-7 *Propagation in Isotropic Elastic Media*

If the medium is homogeneous, isotropic, and perfectly elastic, we may substitute for p_{ik} from (2-4), giving

$$\rho \frac{\partial^2 u_i}{\partial t^2} = \sum_{k=1}^{3} \frac{\partial}{\partial x_k} (\lambda \theta \delta_{ik} + 2\mu e_{ik})$$

$$= \lambda \frac{\partial \theta}{\partial x_i} + \mu \sum_{k=1}^{3} \frac{\partial}{\partial x_k} \left(\frac{\partial u_k}{\partial x_i} + \frac{\partial u_i}{\partial x_k} \right)$$

$$= (\lambda + \mu) \frac{\partial \theta}{\partial x_i} + \mu \nabla^2 u_i$$

These equations may also be written in vector form

$$\rho \ddot{\mathbf{u}} = (\lambda + \mu)\nabla\theta + \mu\nabla^2\mathbf{u} \qquad (2\text{-}8)$$

provided that the coordinate system is rectangular. If we take the divergence of both sides of this equation and rearrange the orders of differentiation, we get

$$\rho \frac{\partial^2 \theta}{\partial t^2} = (\lambda + 2\mu)\nabla^2 \theta$$

This is the standard wave equation for the propagation of cubical dilatation θ with a velocity given by $\alpha = \sqrt{(\lambda + 2\mu)/\rho}$. This may be interpreted as proof that compressional waves moving with this velocity can be propagated through any isotropic elastic solid.

Now if we take the curl of both sides of (2-8), we get

$$\rho \frac{\partial^2}{\partial t^2} (\nabla \times \mathbf{u}) = \mu \nabla^2 (\nabla \times \mathbf{u})$$

which once again is the standard wave equation, this time for the propagation of a pure rotational disturbance with velocity given by $\beta = \sqrt{\mu/\rho}$. This constitutes proof of the possibility of propagating transverse or rotational waves through an isotropic elastic solid, moving with the velocity β.

These two types of waves are usually referred to as P and S waves, respectively.[1] When considering the propagation of wavefronts at great distances from their source, where their initial curvature is lost and so for all practical purposes they may be considered as plane, it is clear that, as with electromagnetic waves, the rotational waves may be plane-polarized. Waves which are polarized such that the particle motion is in a horizontal plane are called SH waves; if in a vertical plane, they are called SV waves.

One small point that needs to be cleared up concerns the thermodynamics of elastic-wave propagation. We have associated the Lamé constants λ and μ (or the bulk modulus κ) with a stress tensor which is assumed to be isothermal since it contains no terms involving thermal expansion. The deformations which occur during the transmission of seismic waves, however, do so under conditions which seem a priori far more likely to be adiabatic, on account of the very low thermal conductivities of nearly all earth materials. The adiabatic and isothermal values of the elastic parameters of an isotropic solid are connected by the following relations [Jeffreys (1), chap. 3]:

$$\mu_A = \mu_I \qquad \lambda_A = \lambda_I + \frac{\kappa_I^2 \Gamma^2 T}{\rho c_v} \qquad \kappa_A = \kappa_I + \frac{\kappa_I^2 \Gamma^2 T}{\rho c_v}$$

which may be established from the second law of thermodynamics by including a term for temperature variation in the strain energy and assuming complete reversibility. Here the adiabatic and isothermal values are

---

[1] Other names by which these waves are known are: (1) P wave: longitudinal wave, dilatational wave, compressional wave; (2) S wave: rotational wave, transverse wave, shear wave. These names may at various times be used interchangeably.

denoted by the suffixes A and I, respectively; Γ is the coefficient of thermal cubical expansion, T is the temperature of the solid, and c_v is the specific heat at constant volume. Since most laboratory measurements on elastic properties are carried out under approximately isothermal conditions, one might expect a discrepancy between actual seismic velocities and those predicted from laboratory measurements. In fact, the relationship should be

$$\alpha_A{}^2 = \alpha_I{}^2 + \frac{\kappa_I{}^2\Gamma^2T}{\rho c_v} \qquad \beta_A = \beta_I$$

For ordinary solids and at ordinary temperatures, the discrepancy between the two values of α is less than 1 percent. There is therefore no need to distinguish between the two velocities, and we shall not do so in the future.

2-8 The Displacement Potentials

We have seen that the equations of motion separate, in homogeneous and isotropic media, into two different wave equations which imply the propagation of both dilatational (P) waves and rotational (S) waves. The former involve the propagation of principal stresses, and the latter, of shearing stresses. It is now clear that the two different types of stress are propagated through the solid at very different speeds.

It will be convenient for us to separate the displacement vector **u** into dilatational and rotational components. Using a method due to Helmholtz, we write **u** in terms of an arbitrary scalar ϕ and an arbitrary vector ψ; thus

$$\mathbf{u} = \nabla\phi - \nabla \times \psi \qquad (2\text{-}9)$$

Since we may impose one additional condition upon ϕ and ψ, let this be

$$\nabla \cdot \psi = 0$$

The two variables are now fully specified.

If we now take the divergence of (2-9), we find that

$$\nabla \cdot \mathbf{u} \equiv \theta = \nabla^2\phi$$

If instead we take the curl, we find that[1]

$$\nabla \times \mathbf{u} \equiv \xi = \nabla^2\psi$$

The connection of θ with ϕ and of ξ with ψ indicates that the two new variables represent a division between the dilatational and the rotational components of the displacement. The quantity ϕ is called the *dilatational displacement potential*, while ψ is called the *rotational displacement potential*. They have the dimensions $[L]^2$.

---

[1] The expression $\nabla^2\psi$ can be used only in rectangular coordinates. In curvilinear systems, we must return instead to the use of $\nabla \times (\nabla \times \psi)$.

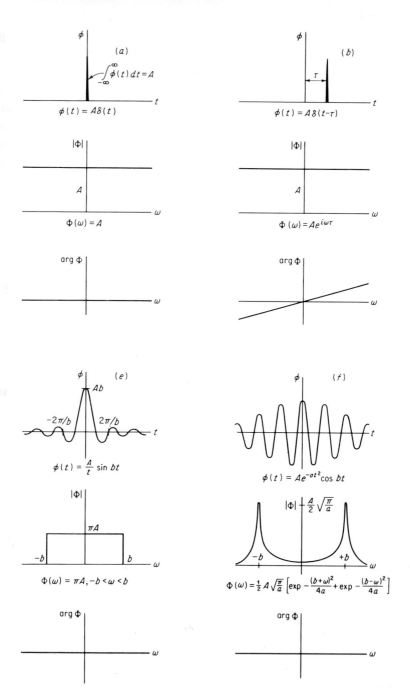

Fig. 2-6 Some simple transients and their Fourier spectra. (*a*) and (*b*) are ideal pulse transients (δ functions). The effect of a time delay is illustrated by the pairs (*a*) and (*b*), (*c*) and (*d*): The amplitude of the spectrum remains unchanged,

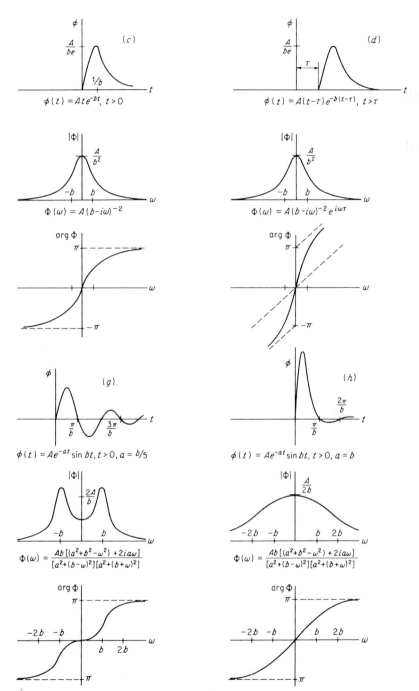

$$\phi(t) = Ate^{-bt}, \ t > 0$$

$$\Phi(\omega) = A(b - i\omega)^{-2}$$

$$\phi(t) = A(t - \tau)e^{-b(t-\tau)}, \ t > \tau$$

$$\Phi(\omega) = A(b - i\omega)^{-2}e^{i\omega\tau}$$

$$\phi(t) = Ae^{-at}\sin bt, t > 0, a = b/5$$

$$\Phi(\omega) = \frac{Ab[(a^2 + b^2 - \omega^2) + 2ia\omega]}{[a^2 + (b - \omega)^2][a^2 + (b + \omega)^2]}$$

$$\phi(t) = Ae^{-at}\sin bt, t > 0, a = b$$

$$\Phi(\omega) = \frac{Ab[(a^2 + b^2 - \omega^2) + 2ia\omega]}{[a^2 + (b - \omega)^2][a^2 + (b + \omega)^2]}$$

but the phase diagrams are rotated.   The pair (g) and (h) [and also the example (f)] shows that the more sinusoidal the pulse form becomes, the more sharply peaked is the spectrum.

If we introduce (2-9) into the equations of motion (2-8), we get

$$\rho\nabla\ddot\phi - (\lambda + 2\mu)\nabla(\nabla^2\phi) + \rho(\nabla \times \ddot\psi) - \mu\nabla^2(\nabla \times \psi) = 0$$

This rather complicated equation can be satisfied if we allow ϕ and ψ independently to satisfy the two wave equations

$$\rho\ddot\phi = (\lambda + 2\mu)\nabla^2\phi \tag{2-10}$$

and
$$\rho\ddot\psi = \mu\nabla^2\psi \tag{2-11}$$

which clearly represent P and S motions, respectively. It is advantageous in seismology to introduce these potentials since (2-10) and (2-11) are easier to solve than (2-8). By expressing stresses and displacements in terms of ϕ and ψ, the continuity conditions at elastic interfaces can be expressed in terms of the minimum number of quantities and the solution of boundary-value problems is greatly facilitated.

2-9   Plane-wave Solutions in Isotropic Media

In a homogeneous, isotropic elastic solid, the equations of motion are given by (2-10) and (2-11). In practical seismology the disturbances are generally short pulses, and we are concerned with the transient solutions of these equations. Since in most cases these are difficult to obtain directly, we make instead a synthesis of steady-state solutions by using Fourier's integral theorem. Thus we may write

$$\phi(\mathbf{r},t) = \frac{1}{2\pi}\int_{-\infty}^{\infty} \Phi(\mathbf{r},\omega)e^{-i\omega t}\,d\omega \tag{2-12}$$

where
$$\Phi(\mathbf{r},\omega) = \int_{-\infty}^{\infty} \phi(\mathbf{r},t)e^{i\omega t}\,dt$$

provided only that $\int_{-\infty}^{\infty} |\phi(\mathbf{r},t)|\,dt$ exists. The quantity Φ is the Fourier transform of ϕ. If we substitute (2-12) into Eq. (2-10), we find that Φ must be a solution of

$$\nabla^2\Phi + k_\alpha^2\Phi = 0 \tag{2-13}$$

where the wave number $k_\alpha = \omega/\alpha = \omega\sqrt{\rho/(\lambda + 2\mu)}$. Similarly, if we write

$$\psi(\mathbf{r},t) = \frac{1}{2\pi}\int_{-\infty}^{\infty} \boldsymbol\Psi(\mathbf{r},\omega)\,e^{-i\omega t}d\omega$$

where
$$\boldsymbol\Psi(\mathbf{r},\omega) = \int_{-\infty}^{\infty} \psi(\mathbf{r},t)e^{i\omega t}\,dt$$

then $\boldsymbol\Psi$ must be a solution of

$$\nabla^2\boldsymbol\Psi + k_\beta^2\boldsymbol\Psi = 0 \tag{2-14}$$

where $k_\beta = \omega/\beta = \omega\sqrt{\rho/\mu}$. k_α and k_β are known respectively as the P and S wave numbers of the medium.

Examples of transient signals and their Fourier transforms are shown in Fig. 2-6. Both the amplitudes and the phase angles are shown (since Fourier transforms are generally complex) in order to indicate how the *phase* of each Fourier element changes with the frequency.

The simplest solutions of (2-13) and (2-14) correspond to the propagation of plane waves. Thus if we may suppose that the displacements are independent of, let us say, the y coordinate, we may write

$$\frac{\partial^2\Phi}{\partial x^2} + \frac{\partial^2\Phi}{\partial z^2} + k_\alpha{}^2\Phi = 0$$

and a like equation for Ψ. Solving by the method of separation of variables, we get

$$\Phi(x,z;\omega) = A(\omega)\exp ik_\alpha(lx + nz) \qquad (2\text{-}15)$$

where $l^2 + n^2 = 1$, and similar expressions for the components of Ψ. This represents a plane wave whose front moves with the velocity α in the direction whose cosines are $(l, 0, n)$ (Fig. 2-7). The *apparent wave number* in the x direction is $k_\alpha l = k_\alpha \sin\vartheta$, and, correspondingly, the "apparent wavelength" and the "apparent velocity" are $2\pi \csc \vartheta/k_\alpha$ and $\alpha \csc \vartheta$, respectively.

It is of interest to remark that (2-15) is a solution of (2-13) provided only that $l^2 + n^2 = 1$. It is not required that both l and n have real values, and indeed as we shall see later there are physically realizable situations in which this will not be the case. If, for example, n were to become imaginary, then $l > 1$, and we would write

$$\Phi(x,z;\omega) = A(\omega)\exp(ik_\alpha lx \pm k_\alpha|n|z)$$

This corresponds to a wave which propagates in the x direction with the apparent velocity α/l, which of course is now $<\alpha$. The amplitude of the wave attenuates exponentially in the direction of positive or negative z. It

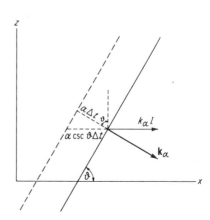

Fig. 2-7 A plane-wave front advancing in the direction $l,0,n$.

is equivalent to a plane wave whose front makes an angle of incidence $\vartheta = \pi/2 - is$ with the x axis, since

$$l = \sin \vartheta = \cosh s \qquad \text{and} \qquad n = \cos \vartheta = i \sinh s$$

Such waves are called *evanescent* or *inhomogeneous* waves, and we shall again discuss them in Chap. 3.

2-10   Conditions near a Source

Using Blake's example (4), the source that we shall consider is a symmetrical explosion which we represent as a radial pressure pulse, generated uniformly over the interior surface of a small spherical cavity of radius a.   If the actual form of this pulse is described by a function $p(t)$, then the dilatation at any point in the material may be determined by solving the equation

$$\frac{\partial^2 \theta}{\partial t^2} = \alpha^2 \nabla^2 \theta \tag{2-16}$$

subject to the boundary condition

$$-\left[(\lambda + 2\mu) \frac{\partial u_r}{\partial r} + 2\lambda \frac{u_r}{r} \right]_{r=a} = p(t)$$

which represents the continuity of stress at the cavity wall, and which we may rewrite as

$$\left(\frac{\partial u_r}{\partial r} + \frac{2\sigma}{1 - \sigma} \frac{u_r}{r} \right)_{r=a} = -\frac{p(t)}{\rho \alpha^2} \tag{2-17}$$

To express the equation of motion (2-16) and the boundary equation (2-17) in terms of a common variable, we must apply the Helmholtz separation method.   Since in this particular case the particle motion is purely dilatational and normal to the boundary, we may set $\psi = 0$.   Equation (2-9) then becomes

$$u_r(r) = \frac{\partial \phi}{\partial r}$$

since the motion is entirely radial.   Making the substitutions, we get a new equation of motion

$$\frac{\partial^2 \phi}{\partial t^2} = \alpha^2 \nabla^2 \phi \tag{2-18}$$

and a new boundary equation

$$\left(\frac{\partial^2 \phi}{\partial r^2} + \frac{2\sigma}{1 - \sigma} \frac{1}{r} \frac{\partial \phi}{\partial r} \right)_{r=a} = -\frac{p(t)}{\rho \alpha^2} \tag{2-19}$$

both of which are now given in terms of ϕ.

Let us begin by putting both $\phi(r,t)$ and $p(t)$ into spectrum form. If we substitute (2-12) for ϕ and if we write

$$p(t) = \frac{1}{2\pi} \int_{-\infty}^{\infty} P_\omega e^{-i\omega t}\, d\omega$$

where P_ω is the Fourier transform of $p(t)$, i.e.,

$$P_\omega = \int_{-\infty}^{\infty} p(t) e^{i\omega t}\, dt$$

then Eq. (2-18) reduces to the scalar Helmholtz equation

$$\nabla^2 \Phi + k_\alpha{}^2 \Phi = 0 \qquad k_\alpha = \frac{\omega}{\alpha} \tag{2-20}$$

while the boundary equation becomes

$$\left(\frac{d^2\Phi}{dr^2} + \frac{2\sigma}{1-\sigma}\frac{1}{r}\frac{d\Phi}{dr} \right)_{r=a} = -\frac{P_\omega}{\rho\alpha^2} \tag{2-21}$$

The solution of (2-20) which corresponds to outward-traveling spherical waves is

$$\Phi(r) = \frac{A_\omega e^{ik_\alpha r}}{r}$$

where A_ω is found by substituting this expression into (2-21). It is

$$A_\omega = -\frac{P_\omega a^3 e^{-ik_\alpha a}}{\rho\alpha^2 (Q - ik_\alpha a Q - k_\alpha{}^2 a^2)}$$

where $Q = 2(1 - 2\sigma)/(1 - \sigma)$. The final solution therefore becomes

$$\phi(r,t) = \frac{a^3}{2\pi r \rho \alpha^2} \int_{-\infty}^{\infty}\int_{-\infty}^{\infty} \frac{p(\tau) \exp\left(+i\omega\{\tau - [t - (r-a)/\alpha]\}\right)\, d\omega\, d\tau}{-k_\alpha{}^2 a^2 - ik_\alpha a Q + Q} \tag{2-22}$$

A suitable representation for the pressure pulse generated by a simple explosion is the following:

$$p(t) = \begin{cases} p_0 e^{-bt} & t > 0 \\ 0 & t \leq 0 \end{cases}$$

If we substitute this function into (2-22) and perform the integrations, we find that

$$\phi(r,t) = \frac{p_0 a}{\rho r [\chi^2 + (v - b)^2]^{1/2}} \left\{ \frac{e^{-bt_r}}{[\chi^2 + (v - b)^2]^{1/2}} \right.$$
$$\left. - \frac{e^{-vt_r}}{\chi} \cos\left(\chi t_r - \tan^{-1}\frac{v-b}{\chi}\right) \right\} \tag{2-23}$$

where
$$\chi = \frac{\alpha Q}{2a}\sqrt{\frac{4}{Q}-1} = \frac{\alpha}{a}\frac{\sqrt{1-2\sigma}}{1-\sigma}$$

$$v = \frac{\alpha Q}{2a} = \frac{\alpha}{a}\frac{1-2\sigma}{1-\sigma}$$

and
$$t_r = t - \frac{r-a}{\alpha}$$

Differentiation of (2-23) with respect to r gives the displacement pulse. In Fig. 2-8 we illustrate the form taken by this pulse when $\lambda = \mu$ and when $b = 2\alpha/3a = 10^3$ sec^{-1}. By setting $b = 0$, the solution may also be found for a pressure pulse represented by the Heaviside step function, and this in turn may be used to form the solution for any other pulse form by means of the convolution integral. We note that a pulsed disturbance generated inside a cavity within an elastic solid leads to the propagation of a strongly damped, oscillating (although nondispersive) wave train. This is because the elastic reactance (stiffness) of the material dominates the inertial forces, except possibly in materials of low rigidity, and then at very high frequencies.

The complete solution for the propagation of transient stresses in isotropic elastic media consists in finding the solution of the equation of motion which satisfies the conditions of continuity of stress and of displacement at all interfaces and which approaches a form such as (2-23) close to the source. A complete analysis of this kind is of importance only if we wish to follow the modifications in the shape of seismic pulses brought about by reflection or refraction at elastic boundaries. If on the other hand only

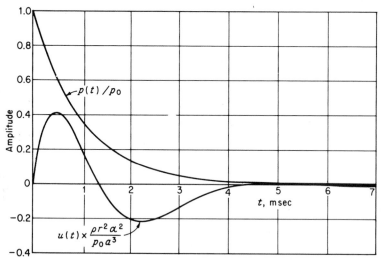

Fig. 2-8   Radial displacement u_r in an isotropic elastic solid generated by a simple pressure pulse within a spherical cavity.

the transit time and the change in amplitude of the pulse are held to be of interest, the precise description of conditions in the near vicinity of the source need not be considered and the analysis becomes very much simpler.

2-11 Waves in Imperfectly Elastic Media

It has been recognized for a very long time that a broad variety of earth materials, including silts, clays, sands, and shales, do not behave as perfectly elastic substances under any kind of loading. A very obvious defect in their elastic behavior is the phenomenon of gradual adjustment of elastic moduli to the duration of the load, an effect which is usually called elastic afterworking. The significant discrepancy that exists between the static and the dynamic moduli of these materials has naturally raised the question of whether important changes in elastic-wave propagation are brought about when the appropriate modifications are made in Hooke's law.

It has been ascertained that a stress-strain relationship that adequately describes the mechanical behavior of these materials is obtained by adding to the expression (2-4) a term that is proportional to rate of strain, i.e.,

$$p_{ik} = \lambda\theta\delta_{ik} + 2\mu e_{ik} + \eta(2\dot{e}_{ik} - \tfrac{2}{3}\dot{\theta}\delta_{ik}) \tag{2-24}$$

where η is called the "viscosity" of the solid. Materials which behave in this way are termed *viscoelastic* or Voigt solids; and if this modified form of the stress tensor p_{ik} is introduced into the equations of motion, we get

$$\rho\ddot{\mathbf{u}} = (\lambda + \mu)\nabla\theta + \mu\nabla^2\mathbf{u} + \tfrac{1}{3}\eta\nabla\dot{\theta} + \eta\nabla^2\dot{\mathbf{u}} \tag{2-25}$$

Taking the divergence of both sides of this equation, we find that

$$\rho\ddot{\theta} = \left[(\lambda + 2\mu) + \bar{\eta}\frac{\partial}{\partial t}\right]\nabla^2\theta \tag{2-26}$$

where
$$\bar{\eta} = \tfrac{4}{3}\eta$$

and taking the curl, we obtain

$$\rho\frac{\partial^2}{\partial t^2}(\nabla \times \mathbf{u}) = \left(\mu + \eta\frac{\partial}{\partial t}\right)\nabla^2(\nabla \times \mathbf{u}) \tag{2-27}$$

Thus both P and S waves are separately identifiable in viscoelastic media, but the form of the wave equation is altered in both cases by the addition of a damping term.

To investigate the influence of these additional terms, let us consider the propagation of compressional plane harmonic waves. For the displacement potential we shall write

$$\phi(x,t) = \phi(x) \cos \omega t$$

and if we substitute this expression into (2-26), we find that

$$\phi(x,t) = \phi_0 e^{-\gamma x} \cos (kx - \omega t) \tag{2-28}$$

where

$$\gamma = \left[\frac{\rho^2 \omega^4}{(\lambda + 2\mu)^2 + \bar{\eta}^2 \omega^2}\right]^{\frac{1}{4}} \sin \left(\frac{1}{2} \tan^{-1} \frac{\bar{\eta}\omega}{\lambda + 2\mu}\right)$$

and

$$k = \left[\frac{\rho^2 \omega^4}{(\lambda + 2\mu)^2 + \bar{\eta}^2 \omega^2}\right]^{\frac{1}{4}} \cos \left(\frac{1}{2} \tan^{-1} \frac{\bar{\eta}\omega}{\lambda + 2\mu}\right)$$

The first fact that emerges is that the waves are damped and that the damping depends upon frequency. The second is that the velocity also depends upon frequency, and the waves are therefore dispersive. However, if the frequency is small enough that $\bar{\eta}\omega \ll \lambda + 2\mu$, then $\gamma \doteq \dfrac{1}{2} \dfrac{\bar{\eta}\omega^2 \rho^{\frac{1}{2}}}{(\lambda + 2\mu)^{\frac{3}{2}}}$ approximately, and $k \doteq \omega/\alpha = k_\alpha$. Thus for low frequencies the absorption coefficient appears to be proportional to the square of the frequency, and the wave velocity is independent of frequency. In this approximation, therefore, the medium behaves as a filter but it is nondispersive. At high frequencies on the other hand, we find that both γ and $k \rightarrow (\rho\omega/2\bar{\eta})^{\frac{1}{2}}$, indicating that both the absorption coefficient and the wave velocity become proportional to the square root of the frequency.

In both instances, however, these conclusions are at variance with experimental observations. Although very few in number, they do strongly suggest that at least in some materials the absorption coefficient is linear in the frequency and the wave velocity is very largely frequency-independent. Figure 2-9, for instance, shows some results obtained by McDonal et al. (5) on the attenuation of compressional waves in the Pierre shale (Lower Cretaceous) in Colorado which seem to bear out these statements. A reconciliation of these facts can be brought about by putting $\bar{\eta} = C/\omega$, where C is a constant of the material having the dimensions of a stress. This form of dependence of the damping coefficient upon frequency, although

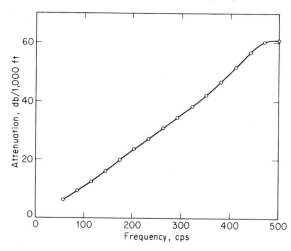

Fig. 2-9 **Attenuation of compressional waves in the Pierre Shale, Colorado.** [McDonal et al. (5).]

introduced here *ad hoc*, does actually correspond to what is predicted by theory for materials in which the loss mechanism is of a type called *solid friction* [Zener (6)].

When we let $\bar{\eta} = C/\omega$ in (2-28), we find that if $C \ll \lambda + 2\mu$

$$k = \frac{\omega}{\alpha} \qquad \gamma = \frac{\omega}{\alpha} \tan\left(\frac{1}{2} \tan^{-1} \frac{C}{\lambda + 2\mu}\right) \qquad \alpha = \left[\frac{(\lambda + 2\mu)^2 + C^2}{\rho^2}\right]^{1/4}$$

In the Pierre shale, which is the only rock formation thus far on which extensive field observations of this nature have been made, the wave velocity is approximately 7×10^3 ft/sec, and the attenuation constant is about $1.75 \times 10^{-5}\omega$ db/ft. From the relationships given above, this would imply a value $C/(\lambda + 2\mu) = 0.05$; and the effect of the internal damping upon the compressional-wave velocity is therefore only 0.06 percent. Thus although it would appear that the elastic behavior of rock materials may be more accurately described by the Voigt model rather than by the unmodified Hooke relationship, the effect of frictional damping on seismic-wave velocities is probably insignificant. Moreover, there seems to be a negligible amount of frequency dispersion caused by this effect, although we would again draw attention to the very small amount of experimental work that has been published on this problem.

2-12 Waves in Transversely-isotropic Media

As we mentioned earlier, some materials—mostly those which have formed through successive stages of compaction and consolidation—have an aeolotropy in their elastic properties. As a general rule, the form of this aeolotropy can be described as *transversely-isotropic*, meaning that the aeolotropy exists only in one direction. The stress-strain relations for such materials are given by the formulas (2-5), and so we may introduce these directly into (2-7) in order to find the general equations of motion.

To simplify matters, let us consider only the propagation of simple harmonic plane waves and let the direction of propagation be such that the wave normal lies parallel with the plane $y = 0$. Then if we consider only those motions which are independent of the y coordinate, we shall have

$$\rho \frac{\partial^2 u}{\partial t^2} = (\lambda_\| + 2\mu_\|) \frac{\partial^2 u}{\partial x^2} + \nu \frac{\partial^2 u}{\partial z^2} + (\lambda_\perp + \nu) \frac{\partial^2 w}{\partial z\,\partial x} \qquad (2\text{-}29)$$

and $\qquad \rho \dfrac{\partial^2 w}{\partial t^2} = (\lambda_\perp + \nu) \dfrac{\partial^2 u}{\partial z\,\partial x} + \nu \dfrac{\partial^2 w}{\partial x^2} + (\lambda_\perp + 2\mu_\perp) \dfrac{\partial^2 w}{\partial z^2}$

Into these expressions let us introduce the Helmholtz displacement potentials ϕ and ψ. For u and w, we get

$$u = \frac{\partial \phi}{\partial x} + \frac{\partial \psi}{\partial z} \qquad w = \frac{\partial \phi}{\partial z} - \frac{\partial \psi}{\partial x} \qquad (2\text{-}30)$$

respectively.   The quantity ψ in these relations is actually the y component of $\mathbf{\psi}$, but it will not be necessary for us to use a subscript because we shall have no need to use the other two components.   If these substitutions are made in (2-29), we arrive at the following pair of equations:

$$-\rho\omega^2(\phi_x + \psi_z) = (\lambda_{\|} + 2\mu_{\|})\phi_{xxx} + (\lambda + 2\nu)\phi_{xzz}$$
$$+ (\lambda_{\|} + 2\mu_{\|} - \lambda_{\perp} - \nu)\psi_{xxz} + \nu\psi_{zzz}$$

and $-\rho\omega^2(\phi_z - \psi_x) = (\lambda_{\perp} + 2\nu)\phi_{xxz} + (\lambda_{\perp} + 2\mu_{\perp})\phi_{zzz} - \nu\psi_{xxx}$
$$- (2\mu_{\perp} - \nu)\psi_{xzz}$$

where
$$\psi_{xzz} = \frac{\partial^3\psi}{\partial x\,\partial z^2} \qquad \text{etc.}$$

Now let us put

$$\phi = \phi_0 \exp ik(lx - nz) \qquad \psi = \psi_0 \exp ik(lx - nz)$$

corresponding to plane wavefronts advancing with a velocity $c = \omega/k$ in the direction $(l, 0, -n)$.   If ϑ is the angle of incidence between these wavefronts and the plane $z = 0$, then $l = \sin\vartheta$ and $n = \cos\vartheta$, and the wave advances in the direction of positive x and negative z.   We then get

$$\rho c^2(l\phi_0 - n\psi_0) = [(\lambda_{\|} + 2\mu_{\|})l^3 + (\lambda_{\perp} + 2\nu)ln^2]\phi_0$$
$$- [(\lambda_{\|} + 2\mu_{\|} - \lambda_{\perp} - \nu)l^2n + \nu n^3]\psi_0$$

and $\rho c^2(n\phi_0 + l\psi_0) = [(\lambda_{\perp} + 2\nu)l^2n + (\lambda_{\perp} + 2\mu_{\perp})n^3]\phi_0$
$$+ [\nu l^3 + (2\mu_{\perp} - \nu)ln^2]\psi_0$$

which, if we eliminate ϕ_0 and ψ_0, yields the following quadratic equation for c^2:

$$\begin{vmatrix} \rho c^2 l - (\lambda_{\|} + 2\mu_{\|})l^3 - (\lambda_{\perp} + 2\nu)ln^2 & -\rho c^2 n + (\lambda_{\|} + 2\mu_{\|} - \lambda_{\perp} - \nu)l^2 n + \nu n^3 \\ \rho c^2 n - (\lambda_{\perp} + 2\nu)l^2 n - (\lambda_{\perp} + 2\mu_{\perp})n^3 & \rho c^2 l - \nu l^3 - (2\mu_{\perp} - \nu)ln^2 \end{vmatrix}$$
$$= 0$$

i.e., $(\rho c^2)^2 - (\rho c^2)[(\lambda_{\|} + 2\mu_{\|})l^2 + (\lambda_{\perp} + 2\mu_{\perp})n^2 + \nu] + \nu[(\lambda_{\|} + 2\mu_{\|})l^4$
$$+ (\lambda_{\perp} + 2\mu_{\perp})n^4] + l^2n^2[(\lambda_{\|} + 2\mu_{\|})(\lambda_{\perp} + 2\mu_{\perp}) - \lambda_{\perp}(\lambda_{\perp} + 2\nu)] = 0$$

This equation in a slightly modified form was first given by Stoneley (7). It has two roots, viz.,

$$2\rho c^2 = [(\lambda_{\|} + 2\mu_{\|})l^2 + (\lambda_{\perp} + 2\mu_{\perp})n^2 + \nu]$$
$$\pm \{[(\lambda_{\|} + 2\mu_{\|} - \nu)l^2 - (\lambda_{\perp} + 2\mu_{\perp} - \nu)n^2]^2 + 4(\lambda_{\perp} + \nu)^2 l^2 n^2\}^{1/2} \quad \text{(2-31)}$$

Notice that beneath the square-root sign there is the sum of two perfect squares; consequently, both roots are real.   Since the various elastic moduli have values that are not greatly different from one another, the roots are also positive.   If the solid is nearly isotropic, the upper sign corresponds closely to a pure P wave, and the lower, to a pure S wave; but with increasing aeolotropy this distinction becomes less clear.

A *coefficient of aeolotropy* is occasionally used to describe the lack of symmetry of the wavefronts.   Usually it is taken as the ratio of the P-wave velocities transverse and parallel to the bedding.   Such a coefficient may be found from (2-31) by putting

1. $l = 0$, $n = 1$, corresponding to a vertically traveling wave.   This gives two roots for c, viz., $c = \sqrt{(\lambda_\perp + 2\mu_\perp)/\rho}$, $\sqrt{\nu/\rho}$.   The first is obviously a P wave transverse to the bedding, and the second is a vertically traveling SH wave.

2. $l = 1$, $n = 0$, corresponding to a horizontally traveling wave.   The two roots of (2-31) are $c = \sqrt{(\lambda_\parallel + 2\mu_\parallel)/\rho}$, $\sqrt{\nu/\rho}$.   The first is a P wave parallel with the bedding, and the second, a horizontally traveling SV wave. The coefficient of aeolotropy is equal to $\sqrt{(\lambda_\perp + 2\mu_\perp)/(\lambda_\parallel + 2\mu_\parallel)}$, which in practice seldom appears to exceed about 1.2.   For such small coefficients, we would expect the two roots of (2-31) to correspond to waves that distinctly resemble P and S waves in all directions.

The P and SV motions in a transversely-isotropic medium are necessarily mixed if the angle of incidence of the wavefronts to the plane of isotropy is other than 0 or 90°.   This is immediately evident from the fact that there is no solution of Eqs. (2-29) for either ϕ or ψ unless $\lambda_\perp = \lambda_\parallel$ and $\mu_\perp = \mu_\parallel = \nu$, in which case the material is isotropic.   Therefore pure P or SV waves cannot be propagated through a transversely-isotropic material except, as we have already noted, at normal or grazing incidence to the axis. SH motions, however, do have an independent existence.   The equation of motion for the displacement in the y direction is

$$\rho \frac{\partial^2 v}{\partial t^2} = \mu_\parallel \frac{\partial^2 v}{\partial x^2} + \nu \frac{\partial^2 v}{\partial z^2}$$

indicating that there is no coupling between the SH and the P-SV motions. In the horizontal (x) direction, SH waves will propagate with the velocity $\sqrt{\mu_\parallel/\rho}$.   In the vertical ($z$) direction, they will move with a speed $\sqrt{\nu/\rho}$, as we have already seen.   If the wave travels obliquely, its velocity will lie between these two values.   It will not, however, necessarily move with the same speed as the SV-type wave traveling in the same direction.

2-13   *Waves in Vertically-inhomogeneous Media*

Next we shall consider the implications of the third and last of the three assumptions imposed upon the wave theory, namely, that of homogeneity. In many respects this proves to be the most difficult restriction of all to remove, and without question it leads to the worst mathematical difficulties. Nevertheless, the existence of velocity gradients is so commonplace in geophysical problems that we cannot claim to have brought the theory to a practical stage without considering its effects upon the propagation of seismic waves.

Most occurrences of velocity gradients in otherwise uniform rock formations are related to the fact that materials tend to harden when they are subjected to compressive loads over long periods, and there is a resulting increase in the seismic-wave velocities. In the case of sandstones and also of shales, calcification and other cementing processes also play a part in the general hardening process, although the relative importance of pressure and solution on lithification is still an open question. In any case, the tendency for seismic-wave velocities within a given formation to increase with depth is almost universal. Horizontal velocity gradients, except in unusual circumstances, are generally very much smaller.

When gradients exist in the elastic parameters, the equations of motion require certain modifications. If we return to (2-7),

$$\rho \frac{\partial^2 u_i}{\partial t^2} = \sum_{k=1}^{3} \frac{\partial p_{ik}}{\partial x_k}$$

and substitute for p_{ik} from (2-4), we obtain, for variable λ and μ

$$\rho \ddot{\mathbf{u}} = (\lambda + \mu)\boldsymbol{\nabla}\theta + \theta\boldsymbol{\nabla}\lambda + \mu\nabla^2\mathbf{u} + \boldsymbol{\nabla}\mu \ \cdot\cdot \ e_{ik}$$

where the last term represents the double scalar product of the gradient of μ and the tensor e_{ik}.

If we disregard second and higher derivatives of the Lamé parameters λ and μ, the equation reduces to

$$\rho \ddot{\mathbf{u}} = (\lambda + \mu)\boldsymbol{\nabla}\theta + \theta\boldsymbol{\nabla}\lambda + \mu\nabla^2\mathbf{u} + (\boldsymbol{\nabla}\mu \cdot \boldsymbol{\nabla})\mathbf{u} + \boldsymbol{\nabla}(\boldsymbol{\nabla}\mu \cdot \mathbf{u}) \quad (2\text{-}32)$$

Let us take the divergence of both sides. Neglecting gradients in the density ρ, we find that

$$\rho \ddot{\theta} = (\lambda + 2\mu)\nabla^2\theta + 2\boldsymbol{\nabla}(\lambda + 2\mu) \cdot \boldsymbol{\nabla}\theta - 2\boldsymbol{\nabla}\mu \cdot (\boldsymbol{\nabla} \times \boldsymbol{\xi}) \quad (2\text{-}33)$$

where $\boldsymbol{\xi} = \text{curl } \mathbf{u}$. If we take the curl, we get

$$\rho \ddot{\boldsymbol{\xi}} = \mu\nabla^2\boldsymbol{\xi} + (\boldsymbol{\nabla}\mu \cdot \boldsymbol{\nabla})\boldsymbol{\xi} - \boldsymbol{\nabla}\mu \times (\boldsymbol{\nabla} \times \boldsymbol{\xi}) + 2\boldsymbol{\nabla}\mu \times \boldsymbol{\nabla}\theta \quad (2\text{-}34)$$

These two equations show that the compressional and rotational deformations are coupled to each other in a rather complicated way. The motion therefore does not separate distinctly into dilatational and rotational parts as it does when λ and μ are constants.[1]

Let us turn now to the special case in which λ and μ change in the vertical direction only. Such a medium we shall call *vertically-inhomogeneous*. We shall further simplify the problem by considering only the propagation of disturbances for which the wave normal has no y component. The displace-

[1] Except in a very few special cases where it is possible to separate the equation (2-32) by re-defining the potentials. The new potentials satisfy the P and S equations of motion, but they no longer correspond to pure dilatational or rotational deformations. For detailed discussion, we refer to Hook (8).

ment components u and w may then be replaced by the expressions (2-30) which, upon being substituted into (2-32), yield the following two equations in ϕ and ψ:

$$\left[\rho \frac{\partial^2}{\partial t^2} - (\lambda + 2\mu)\nabla^2 - 2\mu' \frac{\partial}{\partial z}\right]\frac{\partial \phi}{\partial x} + \left(\rho \frac{\partial^2}{\partial t^2} - \mu\nabla^2 - \mu' \frac{\partial}{\partial z}\right)\frac{\partial \psi}{\partial z}$$
$$+ \mu' \frac{\partial^2 \psi}{\partial x^2} = 0 \quad (2\text{-}35a)$$

and $$\left[\rho \frac{\partial^2}{\partial t^2} - (\lambda + 2\mu)\nabla^2 - 2\mu' \frac{\partial}{\partial z}\right]\frac{\partial \phi}{\partial z} - \lambda'\nabla^2\phi$$
$$- \rho\left(\frac{\partial^2}{\partial t^2} - \mu\nabla^2 - 2\mu' \frac{\partial}{\partial z}\right)\frac{\partial \psi}{\partial x} = 0 \quad (2\text{-}35b)$$

where $\lambda' = d\lambda/dz$ and $\mu' = d\mu/dz$. The third equation, corresponding to motion in the y direction, is

$$\rho \frac{\partial^2 v}{\partial t^2} = \mu\nabla^2 v + \mu' \frac{\partial v}{\partial z} \quad (2\text{-}36)$$

Let us first look at the two equations in ϕ and ψ. If we make the following substitutions:

$$P = \rho\ddot{\phi} - (\lambda + 2\mu)\nabla^2\phi \quad (2\text{-}37a)$$

and $$S = \rho\ddot{\psi} - \mu\nabla^2\psi \quad (2\text{-}37b)$$

we obtain

$$\frac{\partial P}{\partial x} + \frac{\partial S}{\partial z} = 2\mu' \frac{\partial w}{\partial x}$$

and $$\frac{\partial P}{\partial z} - \frac{\partial S}{\partial x} = -2\mu' \frac{\partial u}{\partial x}$$

provided that we may continue to disregard derivatives of ρ. Differentiating the first of these equations with respect to x and the second with respect to z, omitting second derivatives of λ and μ, and adding, we get

$$\nabla^2 P = -2\mu' \frac{\partial}{\partial x} \nabla^2\psi \quad (2\text{-}38a)$$

while performing the obverse set of operations and subtracting yields

$$\nabla^2 S = 2\mu' \frac{\partial}{\partial x} \nabla^2\phi \quad (2\text{-}38b)$$

The two equations will separate only if the inhomogeneous terms are small, which is tantamount to assuming either that $|\mu'|$ is very much smaller than the total changes in the elastic moduli within one wavelength or that the wave normal is nearly vertical. Thus we may conclude that in a vertically-inhomogeneous medium we might expect to encounter pure P or SV waves only if they are nearly normally incident to the bedding planes.

When this condition is fulfilled, we find from (2-37a) and (2-38a) that

$$\rho \frac{\partial^2 \theta}{\partial t^2} = (\lambda + 2\mu)\nabla^2\theta + 2(\lambda' + 2\mu')\frac{\partial \theta}{\partial z} \tag{2-39}$$

recalling that $\nabla^2\phi = \theta$. This equation describes the propagation of cubical dilatation and is therefore the wave equation for P waves. In the same way, we find from (2-37b) and (2-38b) that

$$\rho \frac{\partial^2 \xi}{\partial t^2} = \mu\nabla^2\xi + 2\mu'\frac{\partial \xi}{\partial z} \tag{2-40}$$

where $\xi = \nabla^2\psi$ is the y component of the "rotation" of the material. This is the equation for the propagation of SV waves.

The wave equations (2-36), (2-39), and (2-40) cannot be solved exactly, but they can be solved approximately. The solutions, called the WKBJ solutions, are accurate if the elastic moduli do not change greatly over distances comparable to the seismic wavelength (see Sec. 5-1). For simple harmonic waves having a frequency ω, they are

$$\theta \sim \sqrt{\frac{\alpha}{\omega}} \exp i\omega \int \frac{ds}{\alpha} \qquad \text{if } \omega \gg \left|\frac{d\alpha}{dz}\right|$$

where

$$\alpha = \sqrt{\frac{\lambda + 2\mu}{\rho}}$$

$$\xi \sim \sqrt{\frac{\beta}{\omega}} \exp i\omega \int \frac{ds}{\beta} \qquad \text{if } \omega \gg \left|\frac{d\beta}{dz}\right|$$

where

$$\beta = \sqrt{\frac{\alpha}{\rho}}$$

and

$$v \sim \sqrt{\frac{\beta}{\mu\omega}} \exp i\omega \int \frac{ds}{\beta} \qquad \text{if } \omega \gg \left|\frac{d\beta}{dz}\right|$$

For transient disturbances the solutions may be found by the method of Fourier transforms, provided that a way can be found of limiting the integrations to those frequencies for which the asymptotic method is valid. We shall have occasion to discuss this and other matters concerning the WKBJ method more fully in Chap. 5.

In each of the WKBJ solutions, the integral appearing in the exponent is evaluated in the direction of propagation, i.e., along the wave normal. Thus we have a true ray-path solution, in which the wave number changes continuously along the ray path. The velocities with which the wavefronts move also change, of course, so that the waves are continuously refracted during their travel. There are changes also in the *attenuation* of the waves, in spite of the fact that they are supposed to be plane. Note, in this respect, that the attenuation is not the same for SV and SH waves, despite the fact that the two waves are both propagated with the same velocity. This is a necessary condition if the total amount of energy in the wavefront is to be conserved.

To summarize, we have shown that in a continuously stratified medium which contains no very strong velocity gradients, waves can be propagated which we may recognize as predominantly *P*, *SV*, and *SH* types. The two former are coupled together, and will separate rigorously only if the direction of propagation happens to be normal to the bedding. The distinction between *P* and *SV* motions becomes less obvious at oblique incidence, however, and we shall later have to speak of "pseudo-*P*" and "pseudo-*SV*" waves in order to connote this ambiguity. Nevertheless, if the velocities do not change greatly within one wavelength, the two waves are very weakly coupled, and they will travel with velocities closely approximating the local values of α and β at each point in the wavefront. Thus their trajectories should not differ significantly from what we might predict for pure *P* and *SV* waves. We shall defer until Chap. 5 a fuller discussion of this point.

2-14 Propagation in Randomly Inhomogeneous Media

A second type of heterogeneity very common in geological settings is the uncorrelated or "random" type, in which the elastic properties change continuously but unsystematically from point to point. Rock strata classified as single geological formations may be many hundreds of feet thick; yet their textures usually vary considerably in the scale of tens of feet. Figure 4-19 shows an actual seismic velocity log taken from a deep borehole, and the random as well as the systematic nature of the velocity variation is evident. It is obvious that the random behavior of the velocity will scatter the seismic energy to some extent, but it is not obvious what the consequences of this scattering are apt to be.

To find an answer to this question, we begin, as always, by inventing a mathematical model to represent the ground. In this problem we shall suppose that the ground is uniform in the sense that the compressional-wave velocity has a mean value that is everywhere the same, but that it has in addition a small but random point-to-point variation. Thus we may write

$$\alpha(x,y,z) = \alpha_0[1 + \epsilon n(x,y,z)]$$

where $n(x,y,z)$ is a random variable whose rms value has been normalized to unity and ϵ is a number whose value is so chosen that $\epsilon\alpha_0$ is the rms value of the velocity variations. If $\epsilon \ll 1$, we may neglect the conversion between compressional and rotational motions which may be introduced by the inhomogeneities. Neglecting terms in ϵ^2, the equation for the propagation of a transient compressional stress in the medium is

$$\ddot{\phi} = \alpha_0^2[1 + 2\epsilon n(\mathbf{r})]\nabla^2\phi$$

If we take the Fourier transform of this equation and if we continue to neglect terms in ϵ^2 and higher powers, we get

$$\nabla^2\Phi + k_0^2\Phi = 2k_0^2\epsilon n(\mathbf{r})\Phi \tag{2-41}$$

where $k_0 = \omega/\alpha_0$. We shall treat this as an inhomogeneous equation and solve it by the use of the Green's function. The solution is

$$\Phi(\mathbf{r}) = \Phi_0(\mathbf{r}) - \frac{k_0^2 \epsilon}{2\pi} \int n(\mathbf{r}_0) \frac{\exp ik_0|\mathbf{r} - \mathbf{r}_0|}{|\mathbf{r} - \mathbf{r}_0|} \Phi(\mathbf{r}_0)\, d^3r_0$$

which is in the form of an inhomogeneous integral equation. This equation may be solved approximately by using the method of Born, which consists in substituting $\Phi_0(\mathbf{r}_0)$ in place of $\Phi(\mathbf{r}_0)$ under the integral sign. This will be valid only when ϵ is very small, but it reveals the nature of the solution clearly. Every point where $n \neq 0$ becomes an isotropic source of intensity $\Phi_0(\mathbf{r}_0)$. Thus a certain fraction of the incident wave energy is scattered incoherently in all directions.

Let us consider a plane wave passing normally through a randomly inhomogeneous layer of thickness L, and define a scattering or attenuation coefficient a by

$$a = \frac{\Delta I}{IL}$$

where I is the energy flux of the incident wave and ΔI is the scattered energy flow per unit cross-sectional area. In order to calculate a, it is necessary to specify the statistical properties of $n(\mathbf{r})$ more precisely. This is done with the autocorrelation function of n defined by

$$N(\xi,\eta,\zeta) = \iiint n(x,y,z)n(x + \xi,\, y + \eta,\, z + \zeta)\, dx\, dy\, dz$$

If the material is statistically isotropic, then the autocorrelation function will also be isotropic, i.e.,

$$N(\xi,\eta,\zeta) \equiv N(\rho) \qquad \rho^2 = \xi^2 + \eta^2 + \zeta^2$$

If in addition n has a Gaussian distribution, then $N(\rho) = \exp(-\rho^2/a^2)$, where a is to be interpreted as an average linear dimension of the inhomogeneities, which defines the standard deviation of n. The best justification that can be offered for using a Gaussian correlation is that the integrations needed to determine a may now be carried out, although they are somewhat involved. We shall refer the reader to Chernov (9) for the details. The result, however, is rather interesting

$$a = \pi^{1/2}\epsilon^2 k_0^2 a[1 - \exp(-k_0^2 a^2)]$$
$$\doteq \pi^{1/2}\epsilon^2 k_0^4 a^3$$

when the inhomogeneities are small in relation to the seismic wavelength (i.e., when $k_0 a \ll 1$).

Thus we see that the attenuation due to scattering increases rapidly with decreasing wavelength. There is therefore a greater tendency for the shorter wavelengths to be scattered from a pulse as it propagates forward, and this will result in a progressive change of form. This provides some explanation of why real earth materials almost invariably act as a high-cut filter to seismic signals, applying a strong attenuation at all frequencies higher than about 100 cps. If the medium also happens to be slightly dispersive, then the pulse velocity also changes progressively as the shorter wavelengths are eliminated. In most cases, this results in a gradual increase in velocity as the pulse moves outward.

2-15 Summary: Elastic versus Seismic Waves

In this chapter we have established the equations of motion of an ideal elastic solid, beginning from the elementary notions of stress and strain. We followed an altogether standard procedure (which the reader may find in many books), in the execution of which we were compelled to introduce three simplifying assumptions concerning the physical nature of the medium. These were isotropy, homogeneity, and perfectly elastic behavior, any or all of which may be violated when we come to deal with seismic-wave propagation through rocks. Our chief concern through all that followed was to find out what the effects might be on plane elastic waves if we were to relax these three assumptions.

1. In imperfectly elastic media which have internal viscosity or other loss mechanisms which depend upon the rate of deformation, the elastic waves are attenuated in their travel, and they are also dispersive. Present experimental evidence indicates, however, that the loss mechanism may be of the solid friction type, which largely eliminates dispersion. The attenuation (after allowing for geometrical spreading of the wavefront) and the changes in velocity appear to be unimportant at normal seismic wavelengths. Thus we conclude on the basis of present evidence that imperfect elastic behavior does not introduce any important modifications into the theory.

2. In aeolotropic media the velocities with which plane elastic waves will travel will depend, as we might have expected, upon their direction. There are always three waves, one of which is predominantly compressional. The particular case that we considered was that of a transversely-isotropic medium, i.e., one which possesses two-dimensional symmetry in its elastic properties. In such a medium a pure P wave can be propagated in the direction of the axis or at right angles to this direction, the velocities in the two directions having different values. In all other directions there is coupling between compressional and rotational motion. In most earth materials, however, the aeolotropy, if it is present at all, is not strongly pronounced. Under such circumstances the PS coupling is weak, and the equations of motion can effectively be separated. Thus there is generally one wave that is recognizably a P wave, another that is recognizably SV, and a third (in transversely-isotropic media) that is a pure SH wave. In geological formations, the P-wave velocity may vary by a few percent with changes in the direction of propagation.

3. In heterogeneous elastic media, P and S motions are again coupled in general. If the elastic properties change in the vertical direction only, the

SH motion separates from the coupled *P-SV* wave.   If the elastic moduli change only gradually within a seismic wavelength, however, the *P* and *SV* motions will effectively separate from each other (although there will always be a certain amount of energy transfer between them) to the extent that the wave equation divides into a dilatational and a rotational part.   Thus we may speak of "pseudo-*P*" and "pseudo-*SV*" waves as if each had an isolated existence.   The velocities of these waves at any point are given by the formulas for *P* and *S* velocities in homogeneous media, using local values of the elastic moduli and the density.

4. Inhomogeneity which is of a random rather than of a gradual kind leads to scattering which is stronger at the shorter wavelengths but does not, in the first approximation, lead to *PS* coupling.   It does, however, limit the frequencies that can be transmitted through any distance in the ground.

In the following chapters we shall be concerned with the effects on seismic waves caused by the presence of discontinuities.   Subject to the modest reservations discussed above, we may continue to make use of the wave equations for ideal elastic solids for the purpose of studying the nature of such effects.

References

1. H. Jeffreys, "Cartesian Tensors," Cambridge, London, 1931.
2. S. M. Rytov, The Acoustical Properties of a Finely-layered Medium, *Soviet Phys. Acoustics*, vol. 2, p. 67, 1956.
3. A. E. H. Love, "Mathematical Theory of Elasticity," Cambridge, London, 1927.
4. F. C. Blake, Spherical Wave Propagation in Solid Media, *Jour. Acoust. Soc. America*, vol. 24, pp. 211–215, 1952.
5. F. J. McDonal, F. A. Angona, R. L. Mills, R. L. Sengbush, R. G. van Nostrand, and J. E. White, Attenuation of Shear and Compressional Waves in Pierre Shale, *Geophysics*, vol. 23, pp. 421–439, 1958.
6. C. M. Zener, "Elasticity and Anelasticity of Metals," University of Chicago, Chicago, 1948.   See also L. Knopoff *Quart. Rev. Geophysics*, vol. 2. pp. 625–661, 1965.
7. R. Stoneley, The Seismological Implications of Aeolotropy in Continental Structure, *Royal Astron. Soc. Monthly Notices, Geophys. Supp.*, vol. 5, pp. 343–353, 1949.
8. J. F. Hook, Separation of the vector wave equation of elasticity for certain types of isotropic inhomogeneous media *Jour. Acoust. Soc. America*, vol. 33, pp. 302–313, 1961.
9. L. A. Chernov, "Wave Propagation in a Random Medium," transl. by R. A. Silverman, McGraw-Hill, New York, 1960.

Supplementary Reading

Bullen, K. E.: "Introduction to the Theory of Seismology," chaps. 2 and 4, Cambridge, London, 1947.
Jeffreys, H., and B. S. Jeffreys: "Mathematical Physics," Cambridge, London, 1951.
Kolsky, H.: "Stress Waves in Solids," chaps. 2 and 5, Oxford, London, 1953.
Landau, L. D., and E. M. Lifshitz: "Theory of Elasticity," chap. 1, Pergamon, New York, 1959.
Sokolnikoff, I.: "Mathematical Theory of Elasticity," chaps. 1–3, McGraw-Hill, New York, 1956.

chapter 3 — Plane Seismic Waves in Layered Media

Theoretical seismology is very largely concerned with the effects of velocity changes, and especially of discontinuities or boundaries, on the propagation of elastic waves. In solid media, the displacements and stresses must be continuous at all times and places, even across interfaces. Unless these continuity conditions are fulfilled, there is a direct violation of the laws of kinematics, and infinite forces or accelerations would result. Actually, the first of these conditions merely implies continuity of structure and pointedly assumes that no dislocation or slip occurs at any boundary. The second follows from Newton's third law of motion. In this chapter we shall be concerned with the effects of boundaries upon plane seismic waves, leaving aside most of the discussions of velocity gradients until Chap. 5 and those of curved wavefronts until Chap. 6.

3-1 Reflection and Refraction of Seismic Waves

Let us consider the propagation of a plane wave characterized by the displacement potentials ϕ and ψ as defined by (2-30). These correspond to the motion of a wavefront whose normal has no component in the direction Oy.   If the material is homogeneous and isotropic, the potentials ϕ and ψ and the displacement component v will satisfy the equations (2-10) and (2-11), which we rewrite as

$$\frac{\partial^2 \phi}{\partial t^2} = \alpha^2 \nabla^2 \phi$$

$$\frac{\partial^2 \psi}{\partial t^2} = \beta^2 \nabla^2 \psi$$

and $$\frac{\partial^2 v}{\partial t^2} = \beta^2 \nabla^2 v$$

We shall begin by writing ϕ, ψ, and v in their spectrum form.   The three equations of motion then reduce to

$$\nabla^2 \Phi + k_\alpha{}^2 \Phi = 0$$
$$\nabla^2 \Psi + k_\beta{}^2 \Psi = 0 \qquad\qquad (3\text{-}1)$$
$$\nabla^2 V + k_\beta{}^2 V = 0$$

where the capitalized variables denote, as usual, the Fourier transforms of their lowercase correlatives.   The solutions of Eqs. (3-1) which correspond to plane wavefronts advancing in the direction $(l,0,-n)$ (the minus sign signifying motion in the negative z direction) are

$$\Phi(x,z) = A(\omega)\exp ik_\alpha(lx - nz)$$
$$\Psi(x,z) = B(\omega)\exp ik_\beta(lx - nz) \qquad\qquad (3\text{-}2)$$
and $$V(x,z) = C(\omega)\exp ik_\beta(lx - nz)$$

If the wave is a pure P wave, then $B = C = 0$.   If it is an SV wave, $A = C = 0$.   If it is an SH wave, $A = B = 0$.

If a plane wave is incident upon a plane boundary separating two ideal elastic media as shown in Fig. 3-1, both reflected and transmitted plane waves may be expected to propagate away from the boundary.   If the direction of the displacement in the incident wavefront is oblique to the interface, shearing as well as compressive stresses will occur on account of the discontinuity in the elastic properties.   Thus both the transmitted and the reflected stress fields will contain, in general, both P and S components.

Since the media are uniform and the wavefronts are plane, we may without any restriction represent the progress of the wavefronts by means of ray paths.   These just represent the trajectories followed by points in the advancing wavefronts.   If the angle between the incident wavefront and the interface is ϑ, then ϑ will also be the *angle of incidence* of the incident ray path (Fig. 3-1).   Then for the incident wave, $l = \sin \vartheta$ and $n = \cos \vartheta$. Angles of reflection and refraction are similarly defined in relation to reflected and transmitted ray paths.

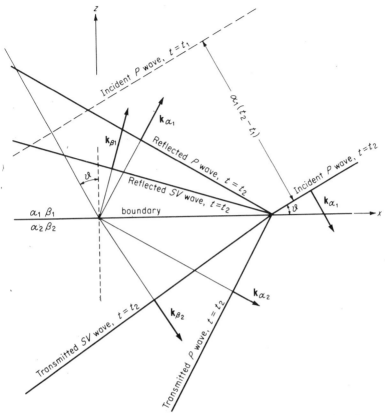

Fig. 3-1 Reflection and refraction of an incident plane compressional wave at a boundary.

The case that we shall consider in some detail is that of a compressional wave incident at an angle ϑ_P in medium 1. In this instance $B = C = 0$ in medium 1, and the incident disturbance is represented by the function

$$\Phi_i(x,z) = A_1(\omega) \exp ik_{\alpha_1}(l_{\alpha_1}x - n_{\alpha_1}z)$$

where

$$k_{\alpha_1} = \frac{\omega}{\alpha_1}$$
$$l_{\alpha_1} = \sin \vartheta_P$$
$$n_{\alpha_1} = \cos \vartheta_P$$

To characterize the reflected displacement field, we introduce the three wave functions

$$\Phi_r(x,z) = D_1(\omega) \exp ik_{\alpha_1}(l_{\alpha_1}x + n_{\alpha_1}z)$$
$$\Psi_r(x,z) = E_1(\omega) \exp ik_{\beta_1}(l_{\beta_1}x + n_{\beta_1}z)$$

and

$$V_r(x,z) = F_1(\omega) \exp ik_{\beta_1}(l_{\beta_1}x + n_{\beta_1}z)$$

and for the transmitted displacement field we shall use

$$\Phi_t(x,z) = A_2(\omega) \exp ik_{\alpha_2}(l_{\alpha_2}x - n_{\alpha_2}z)$$

$$\Psi_t(x,z) = B_2(\omega) \exp ik_{\beta_2}(l_{\beta_2}x - n_{\beta_2}z)$$

and

$$V_t(x,z) = C_2(\omega) \exp ik_{\beta_2}(l_{\beta_2}x - n_{\beta_2}z)$$

The directions taken by all these waves is given by Snell's law, according to which the apparent wave number in a direction parallel with the boundary is the same for all waves, i.e.,

$$k_{\alpha_1}l_{\alpha_1} = k_{\beta_1}l_{\beta_1} = k_{\alpha_2}l_{\alpha_2} = k_{\beta_2}l_{\beta_2}$$

while in each case $n = \sqrt{1 - l^2}$

To determine the coefficients of the six wave functions, we have six boundary conditions which specify the continuity of stresses and displacements at the interface. The stress components involved are

$$P_{zx} = 2\mu E_{zx} = \mu \left(2\frac{\partial^2 \Phi}{\partial z\, \partial x} - \frac{\partial^2 \Psi}{\partial x^2} + \frac{\partial^2 \Psi}{\partial z^2} \right) \tag{3-3a}$$

$$P_{zy} = 2\mu E_{zy} = \mu \frac{\partial V}{\partial z} \tag{3-3b}$$

$$P_{zz} = \lambda\Theta + 2\mu E_{zz} = \lambda\nabla^2\Phi + 2\mu \left(\frac{\partial^2 \Phi}{\partial z^2} - \frac{\partial^2 \Psi}{\partial z\, \partial x} \right) \tag{3-3c}$$

of which we have capitalized the Fourier transforms in order to conform to our general practice. The displacement components are

$$U = \frac{\partial \Phi}{\partial x} + \frac{\partial \Psi}{\partial z} \qquad V \qquad W = \frac{\partial \Phi}{\partial z} - \frac{\partial \Psi}{\partial x}$$

In medium 1 the wave potentials will be the sum of incident and reflected components, while in medium 2 there will be only the transmitted terms. The six conditions of continuity lead immediately to a set of equations which were first derived by Knott (1) and are known by his name. Another very similar set was derived by Zoeppritz (2) by a different method. They differ from the Knott equations only in that the unknowns are the displacement rather than the potential amplitudes. Both have been quoted frequently in seismology texts (3), (4), and we reproduce them again below, essentially following Knott's method but written in the present notation and in a somewhat compacted form.

For an incident P wave ($B_1 = C_1 = 0$) striking the boundary at an angle $\vartheta_P = \sin^{-1} l_{\alpha_1}$ or for an incident SV wave ($A_1 = C_1 = 0$) or SH wave ($A_1 = B_1 = 0$) incident at an angle $\vartheta_S = \sin^{-1} l_{\beta_1}$ (or in the unusual case of P and S waves incident at angles ϑ_P and ϑ_S when $\sin \vartheta_P/\alpha_1 = \sin \vartheta_S/\beta_1$), the Knott equations are

$$2\rho_1 l_{\beta_1}{}^2 \frac{n_{\alpha_1}}{l_{\alpha_1}} (A_1 - D_1) - \rho_1(n_{\beta_1}{}^2 - l_{\beta_1}{}^2)(B_1 + E_1)$$
$$= 2\rho_2 l_{\beta_2}{}^2 \frac{n_{\alpha_2}}{l_{\alpha_2}} A_2 - \rho_2(n_{\beta_2}{}^2 - l_{\beta_2}{}^2)B_2$$

$$\rho_1 n_{\beta_1} l_{\beta_1}(C_1 - F_1) = \rho_2 n_{\beta_2} l_{\beta_2} C_2$$

$$\rho_1(n_{\beta_1}{}^2 - l_{\beta_1}{}^2)(A_1 + D_1) + 2\rho_1 n_{\beta_1} l_{\beta_1}(B_1 - E_1)$$
$$= \rho_2(n_{\beta_2}{}^2 - l_{\beta_2}{}^2)A_2 + 2\rho_2 n_{\beta_2} l_{\beta_2}B_2 \quad (3\text{-}4)$$

$$(A_1 + D_1) - \frac{n_{\beta_1}}{l_{\beta_1}} (B_1 - E_1) = A_2 - \frac{n_{\beta_2}}{l_{\beta_2}} B_2$$

$$(C_1 + F_1) = C_2$$

$$\frac{n_{\alpha_1}}{l_{\alpha_1}} (A_1 - D_1) + (B_1 + E_1) = \frac{n_{\alpha_2}}{l_{\alpha_2}} A_2 + B_2$$

An additional relationship derived from the conservation of energy is useful for checking

$$\rho_1\left[\frac{n_{\alpha_1}}{l_{\alpha_1}} (A_1{}^2 - D_1{}^2) + \frac{n_{\beta_1}}{l_{\beta_1}} (B_1{}^2 - E_1{}^2 + D_1{}^2 - F_1{}^2)\right]$$
$$= \rho_2\left[\frac{n_{\alpha_2}}{l_{\alpha_2}} A_2{}^2 + \frac{n_{\beta_2}}{l_{\beta_2}} (B_2{}^2 + C_2{}^2)\right]$$

In calculating values of the various coefficients, it is usual to solve the equations directly on the computer for each set of parameters. Numerous authors have published tables and graphs of the refracted and reflected wave amplitudes which should arise under various conditions. Among them are McCamy, Meyer, and Smith (5); Koefoed (6); and Costain, Cook, and Algermissen (7). An algebraic solution has been obtained by Heelan (8).

As long as $\alpha_1/\alpha_2 > \sin \vartheta_P$[or $(\alpha_1/\beta_1) \sin \vartheta_S$], all the coefficients in (3-4) are real and solution is straightforward. If, however, the angle of incidence should exceed this "critical" value, then some coefficients become complex, indicating that changes in phase are taking place.

Figures 3-2 and 3-3, after Ergin (9), show the reflection and transmission coefficient of P and SV waves, respectively, when a plane compressional wave is incident at a fluid-solid boundary.[1] The fluid-solid interface is useful for the purpose of illustration, because the solutions are simpler and the curves are more easily interpretable than at the solid-solid interface. Some of the features of these curves are, at first sight, unexpected. We observe the very abrupt peak in the reflection coefficient at the "critical" angle of incidence for the P wave (Fig. 3-2). The sharp rise in the reflected energy at this angle is what we expect from the ray theory, since this is the direction at which total internal reflection of the P wave first takes place.

[1] A fluid can be taken as the limiting case of a solid when $\mu \to 0$ ($\beta \to 0$).

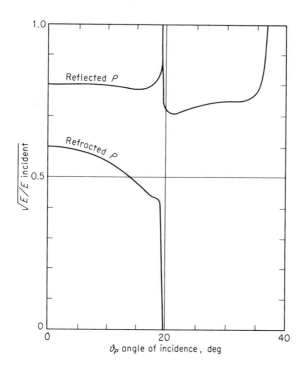

Fig. 3-2 (*a*) Square root of the energy ratio for the reflected *P* wave; (*b*) square root of the energy ratio for the transmitted *P* wave, for a plane *P* wave in water incident against a solid boundary. [After Ergin (9).]

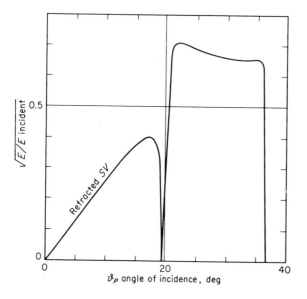

Fig. 3-3 Square root of the energy ratio for the transmitted *SV* wave for a plane *P* wave in water incident against a solid boundary. [After Ergin (9).]

What may seem surprising, however, is the equally sharp decline that follows as the critical angle is exceeded. The explanation is to be found in Fig. 3-3, which illustrates the transmission coefficient for SV waves. At the critical angle SV waves in the substratum abruptly disappear, but beyond this angle they reappear (with a change in phase) stronger than before. Beyond the critical angle, therefore, the absence of P-wave transmission is more than compensated for by the rise in SV transmission, and this accounts for the decline in the P-wave reflection coefficient. Eventually a second critical angle of incidence, at which the SV energy is also totally reflected, is reached, and beyond this angle no ray paths may enter the substratum. This curious behavior in the elastic reflection coefficients is quite real, and we shall have occasion to discuss actual examples of it later on.

At a solid-solid interface the situation is further complicated by the partition of the reflected, as well as the transmitted, energy between P and SV motions. A complete analysis of the reflection and transmission coefficients for incident P and SV waves has been given recently by McCamy, Meyer, and Smith[1] (5), and curves showing the behavior of the reflection coefficients with changes in the angle ϑ_P are shown in Figs. 3-4 and 3-5, for a variety of cases in which $\alpha_2 > \alpha_1$. In all the examples presented, it is assumed that $\lambda = \mu$ in both media.

In none of the cases described in Fig. 3-4 does a critical angle of incidence exist for the converted P-SV wave, so that in no case does total reflection of the seismic energy occur. The cusps are formed at the critical

[1] Cases 1–7 inclusive in this reference are incorrect.

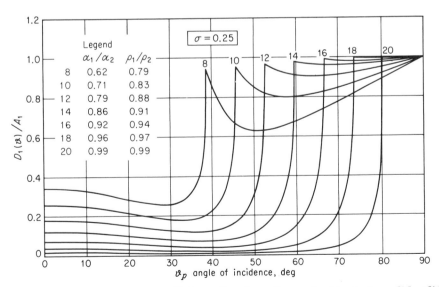

Fig. 3-4 **P-wave reflection coefficients for plane P wave incident at a solid-solid interface, $\alpha_2 > \alpha_1$.** [After McCamy, Meyer, and Smith (5).]

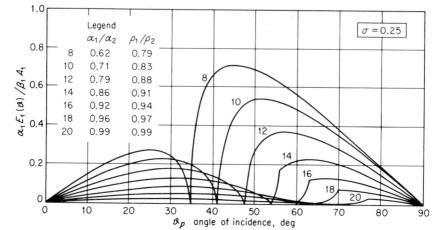

Fig. 3-5 **SV-wave reflection coefficients for plane _P_ wave incident at a solid-solid interface, $\alpha_2 > \alpha_1$.** [After McCamy, Meyer, and Smith (5).]

angle of incidence for the _P-P_ wave, where all the _P_ rays are turned back into the lower-velocity medium. Note that the reflected _SV_ wave suffers a phase reversal at this angle.

The behavior of the reflection coefficients when $\alpha_2 < \alpha_1$ is illustrated in Figs. 3-6 and 3-7. These curves are simple and do not seem to require special comment.

In the event that an incident _P_ wave strikes the boundary at normal incidence, no shear waves are generated. The reflection coefficient in this

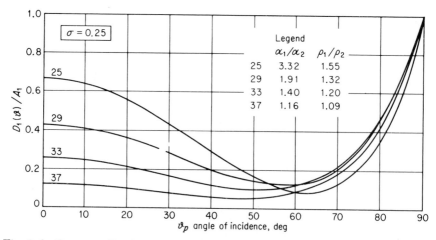

Fig. 3-6 **_P_-wave reflection coefficients for plane _P_ wave incident at a solid-solid interface, $\alpha_2 < \alpha_1$.** [After McCamy, Meyer, and Smith (5).]

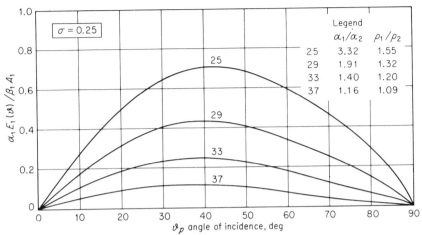

Fig. 3-7 *SV*-wave reflection coefficients for plane *P* wave incident at a solid-solid interface, $\alpha_2 < \alpha_1$. [After McCamy, Meyer, and Smith (5).]

case is

$$\frac{\rho_2\alpha_2 - \rho_1\alpha_1}{\rho_2\alpha_2 + \rho_1\alpha_1} \tag{3-5}$$

which is a particularly useful expression on account of its simplicity. This formula is frequently used to estimate rough magnitudes for the reflections from actual seismic boundaries even when the wavefronts are curved. It is noteworthy that in determining the effectiveness of a reflector, a contrast in seismic-wave velocity can be partly offset by an opposite contrast in density. Thus salt, for example, has a high velocity and a low density and is notoriously ineffective as a reflector of seismic energy.

3-2 Surface Elastic Waves

In addition to the plane-wave solutions (3-2), it is possible to write another type of solution for the equations of motion altogether. This is a form that corresponds to surface waves whose displacements are largely confined to the neighborhood of the boundary itself, and it was first proposed by Rayleigh (10). Let the displacement potentials ϕ and ψ be associated with a plane-wave disturbance containing mixed P and SV components and traveling with velocity c in the direction Ox. If Oz is perpendicular to the interface, then in the upper medium let us put

$$\Phi_1 = f_1(z)e^{ikx}$$
$$\Psi_1 = g_1(z)e^{ikx}$$
and
$$V_1 = h_1(z)e^{ikx}$$
where
$$k = \frac{\omega}{c}$$

and in the lower medium let us introduce a similar set of expressions for Φ_2, Ψ_2, and V_2. Substituting for Φ_1 into (3-1) gives the following ordinary differential equation

$$\frac{d^2 f_1}{dz^2} + k^2 \left(\frac{c^2}{\alpha_1{}^2} - 1 \right) f_1 = 0$$

of which the general solution is

$$f_1(z) = A_1 e^{ik r_1 z} + D_1 e^{-ik r_1 z} \qquad r_1 = \left(\frac{c^2}{\alpha_1{}^2} - 1 \right)^{\frac12}$$

where A_1 and D_1 are constants. The physical requirement on the solution is that $|f_1(z)| \to 0$ as $z \to +\infty$; therefore r_1 must be imaginary, and $D_1 = 0$. Similarly we find that

$$g_1(z) = B_1 e^{ik s_1 z} + E_1 e^{-ik s_1 z}$$

and

$$h_1(z) = C_1 e^{ik s_1 z} + F_1 e^{-ik s_1 z}$$

where

$$s_1 = \left(\frac{c^2}{\beta_1{}^2} - 1 \right)^{\frac12} \qquad \text{and} \qquad E_1 \text{ and } F_1 = 0$$

In the lower medium, the a priori requirement for energy conservation is that the solutions should vanish as $z \to -\infty$, so that $A_2 = B_2 = C_2 = 0$. Therefore, we shall have

$$\Phi_1 = A_1 \exp ik(x + r_1 z) \qquad \Phi_2 = D_2 \exp ik(x - r_2 z)$$
$$\Psi_1 = B_1 \exp ik(x + s_1 z) \qquad \Psi_2 = E_2 \exp ik(x - s_2 z) \qquad (3\text{-}6)$$
$$V_1 = C_1 \exp ik(x + s_1 z) \qquad V_2 = F_2 \exp ik(x - s_2 z)$$

As a case of particular interest, let us first examine the conditions at a free boundary by replacing the first medium with a vacuum. The boundary conditions now are that the components of stress (3-3) should vanish at $z = 0$. Thus, dropping the subscripts,

$$(c^2 - 2\beta^2)A + 2\beta^2 sB = 0$$

and

$$2\beta^2 r A - (c^2 - 2\beta^2)B = 0$$

Eliminating A and B, we get

$$(c^2 - 2\beta^2)^2 = 4\beta^4 rs$$

Rationalizing and dividing by c^2/β^2, we obtain therefore the following equation for c:

$$\frac{c^6}{\beta^6} - 8 \frac{c^4}{\beta^4} + c^2 \left(\frac{24}{\beta^2} - \frac{16}{\alpha^2} \right) - 16 \left(1 - \frac{\beta^2}{\alpha^2} \right) = 0 \qquad (3\text{-}7)$$

If we put $c = \beta$, the left-hand side of (3-7) $= +1$, while for $c = 0$ it becomes $-16(1 - \beta^2/\alpha^2)$, which is <0. Thus there is a real root of (3-7) lying between $c = 0$ and $c = \beta$. Note that such a value makes r and s both

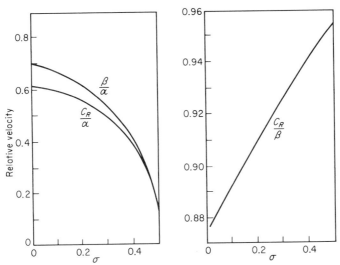

Fig. 3-8 Velocity of Rayleigh waves as a function of Poisson's ratio. [After Knopoff (11).]

imaginary so that the waves are damped as they penetrate the medium. This suggests that a bounded elastic solid may propagate a wave motion along its free surface which attenuates with the depth of penetration and travels with a velocity somewhat less than the shear-wave velocity of the solid. Such waves are called *Rayleigh waves*, and since their velocity does not depend upon the frequency of the disturbance, they are nondispersive. The velocity of these waves in an isotropic medium is shown in Fig. 3-8 as a function of Poisson's ratio.

It is interesting to study the nature of Rayleigh-wave motion. To simplify matters, we note that

$$\frac{B}{A} = \frac{2\beta^2 - c^2}{2\beta^2 s} = -\frac{2\beta^2 r}{2\beta^2 - c^2}$$

and so, taking real parts, it follows that

$$U = Ak\left[e^{-ikrz} - \left(1 - \frac{c^2}{2\beta^2}\right)e^{-iksz}\right]\sin kx$$

and

$$W = -Ak|r|\left[e^{-ikrz} + \left(1 - \frac{c^2}{2\beta^2}\right)^{-1}e^{-iksz}\right]\cos kx$$

For brevity, these may be written

$$U(x,z) = a(z)\sin kx \qquad W(x,z) = -b(z)\cos kx$$

Thus a particle at the free surface describes an elliptical path which is vertically polarized and in which the motion is retrograde (Fig. 3-9). Because

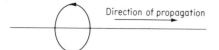

Direction of propagation

Fig. 3-9 Direction of the particle motion at the free surface of a solid, accompanying the passage of a Rayleigh wave.

of the nature of the particle motion, Rayleigh waves at the surface of the ground are often called *ground roll*.

The ratio of the vertical to horizontal motion at the free surface of an elastic solid is given as a function of Poisson's ratio in Fig. 3-10. We also observe that the horizontal motion has a zero at some depth. Below this level the motion changes sign and therefore transforms from elliptical retrograde to elliptical progressive. Because of this change in the nature of the particle motion, the level at which U vanishes is called the *nodal plane*. The depth of the nodal plane below the free surface of an isotropic solid is also shown in the figure.

Additional refinements to the theory are needed if the solid is aeolotropic or heterogeneous. For homogeneous, transversely-isotropic media, the equations of motion relevant to the problem will be (2-29), rather than (3-1). Then by taking ϕ and ψ as exponential functions of the depth and by forcing the stresses to vanish at the free surface, a rather complicated period equation can be obtained. Satô (12) has shown that the motion to which this equation corresponds is generally of the Rayleigh type. Details of the analyses, however, require rather specific assumptions about the precise nature of the aeolotropy, which in practice are difficult to make.

For isotropic, "vertically inhomogeneous" media on the other hand, the appropriate equations of motion are (2-39) and (2-40). The procedure here

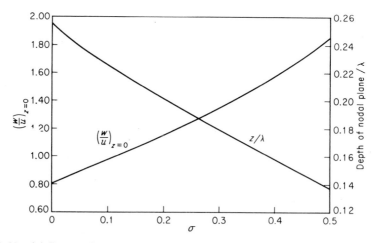

Fig. 3-10 (*a*) **Ratio of the vertical to horizontal amplitude at the free surface;** (*b*) **depth to the nodal plane of a Rayleigh wave in a semi-finite elastic solid, shown as a function of Poisson's ratio.**

is to substitute into these equations displacement potentials

$$\Phi = f(z) \exp ikx \qquad \Psi = g(z) \exp ikx$$

obtaining thereby a pair of simultaneous ordinary differential equations in the amplitude functions $f(z)$ and $g(z)$. These equations can be separated and integrated only under rather special circumstances, and then the application of the boundary conditions leads to some very laborious calculations. The result of these analyses, at least in those cases in which solutions have been found, appears to be a "rocking" sort of retrograde elliptical motion at the ground surface, which is highly dispersive although recognizably of the Rayleigh type.

Rayleigh waves do in fact exist in nature and are very easily observed. Because of their essentially two-dimensional nature, they attenuate less rapidly with distance than the P and S body waves do. Their amplitude diminishes roughly as $r^{-\frac{1}{2}}$ where r is the epicentral distance, as against r^{-1} for P and S waves. At great distances from the source of a seismic disturbance, they generally carry much greater energy than any other portion of the signal. As to the possible origin of Rayleigh waves, it seems at first to be a mystery, because in the plane-wave theory the elastic boundary conditions can be satisfied without them. The work of Lamb (13) has shown that in reality they arise from diffraction at the free boundary of the curved fronts of body waves. Thus Rayleigh waves do not appear as terms in the plane-wave solutions. We shall deal with this question at greater length in Chap. 6.

One further remark on Rayleigh waves may be made at this time with a view toward laying the foundation for a *rapprochement* between the plane-wave theory and the general theory. We have derived expressions for the displacement potentials of Rayleigh waves having the forms

$$\Phi(x,z) = A \exp ik(x - rz) \qquad \text{and} \qquad \Psi(x,z) = B \exp ik(x - sz)$$

These are strongly suggestive of plane-wave potentials, although the plane-wave interpretation in this case is somewhat obscure. Let us consider a plane P wave incident at the free boundary at the complex angle $\vartheta = \pi/2 - i\gamma$, where γ is real. Complex angles having this form correspond to perfectly real waves, as we saw in Chap. 2, although their trajectories cannot be represented by ray paths. The displacement potential of the reflected P wave will have the form

$$\Phi(x,z) = A \exp ik_\alpha(x \cosh \gamma - iz \sinh \gamma)$$

indicating that this wave is propagated in the x direction and is attenuated in the z direction away from the boundary. For the reflected SV wave, we shall have

$$\Psi(x,z) = B \exp ik_\beta(x \cosh \nu - iz \sinh \nu)$$

where (assuming that the laws of geometrical optics hold for complex as well as for real angles of incidence) the angle of reflection ν is given by

$$\frac{\cosh \gamma}{\alpha} = \frac{\cosh \nu}{\beta}$$

Let $c = \alpha/\cosh \gamma$. Then, since $\sinh \gamma = -i\alpha r/c$ and $\sinh \nu = -i\beta s/c$, where $r = \sqrt{c^2/\alpha^2 - 1}$ and $s = \sqrt{c^2/\beta^2 - 1}$, we may write

$$\Phi(x,z) = A \exp ik(x - rz)$$

and
$$\Psi(x,z) = B \exp ik(x - sz) \qquad \text{where } k = \frac{\omega}{c}$$

These expressions are identical in form with the Rayleigh-wave potentials. (The same result could also have been derived from a plane SV wave incident at the complex angle $\vartheta = \pi/2 - i\nu$.) *If the incident wave has zero amplitude*, then the application of the boundary conditions $p_{zz} = p_{zx} = 0$, $z = 0$, to the determination of the reflection coefficients A and B leads directly to (3-7). The generation of Rayleigh waves therefore appears to be associated with the internal reflection of plane body waves which seem to be entirely fictitious. The interpretation of this result, unfortunately, cannot be discussed in a meaningful way until we have had a chance to study the theory of curved wavefronts in Sec. 6-2. The concept of a complex wave number has been used extensively by Brekhovskikh (14) in the solution of various problems in the theory of wave propagation.

3-3   Interface Waves

We now consider the more general problem of two isotropic media and apply the boundary conditions to determine the constants A_1, B_1, C_1, D_2, E_2, F_2 in (3-6). Continuity of displacement components u, v, w gives

$$A_1 + B_1 s_1 = D_2 - E_2 s_2 \tag{3-8a}$$

$$A_1 r_1 - B_1 = D_2 r_2 - E_2 \tag{3-8b}$$

and
$$C_1 = F_2 \tag{3-8c}$$

Continuity of the stress components p_{zz}, p_{zx}, and p_{zy} yields the following equations, obtained by substituting the expressions (3-6) into the forms (3-3) and eliminating λ and μ:

$$\rho_1[(c^2 - 2\beta_1{}^2)A_1 - 2\beta_1{}^2 s_1 B_1] = \rho_2[(c^2 - 2\beta_2{}^2)D_2 + 2\beta_2{}^2 s_2 E_2] \tag{3-9a}$$

$$\rho_1[2r_1\beta_1{}^2 A_1 - (2\beta_1{}^2 - c^2)B_1] = \rho_2[-2\beta_2{}^2 r_2 D_2 + (c^2 - 2\beta_2{}^2)E_2] \tag{3-9b}$$

and
$$\rho_1\beta_1{}^2 s_1 C_1 = -\rho_2\beta_2{}^2 s_2 F_2 \tag{3-9c}$$

From (3-8c) and (3-9c) it follows at once that $C_1 = F_2 = 0$. Therefore there is no SH type of wave propagated along the interface between two isotropic media. The remaining equations (3-8) and (3-9) are four homogeneous equations in the four unknowns, A_1, B_1, D_2, E_2. If these equations have a real solution, the unknowns may be eliminated. This yields the following equation in c, which was first derived by Stoneley (15):

$$c^4[(\rho_2 - \rho_1)^2 + (\rho_1 r_2 + \rho_2 r_1)(\rho_1 s_2 + \rho_2 s_1)] + 4c^2(\rho_1\beta_1{}^2 - \rho_2\beta_2{}^2)(\rho_2 r_1 s_1$$
$$- \rho_1 r_2 s_2 + \rho_2 - \rho_1) + 4(\rho_1\beta_1{}^2 - \rho_2\beta_2{}^2)^2(1 - r_1 s_1)(1 - r_2 s_2) = 0 \quad (3\text{-}10)$$

When the second medium is a vacuum, this equation reduces to (3-7).

A solution of (3-10) has been shown by Scholte (16) to exist only under the stringent condition that β_2 differs only slightly from β_1, and the value is in that case intermediate to the Rayleigh- and shear-wave velocities of either medium. The wave to which this solution corresponds is in many respects similar to a Rayleigh wave but is usually called the Stoneley wave. Except in the immediate vicinity of the interface, its amplitude diminishes with distance from the boundary on both sides. Calculations of Stoneley-wave displacement amplitudes have been made by Yamaguchi and Satô (17).

In Fig. 3-11 the conditions under which the Stoneley wave may exist are illustrated graphically. Within the shaded region Eq. (3-10) has a real root, but outside this region Stoneley waves are not possible. Like

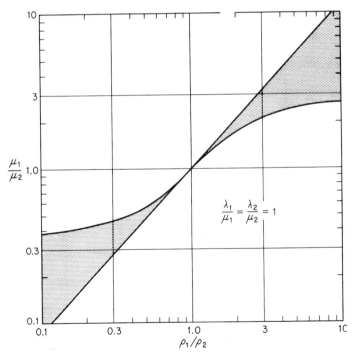

Fig. 3-11 Condition for the existence of the Stoneley wave. Solutions of the Stoneley wave equation exist within the shaded areas. [After Scholte (16).]

Rayleigh waves at a free surface, Stoneley waves are generated by the diffraction of curved fronts of body waves at a plane boundary. It can very easily be shown that the displacement potentials corresponding to the Stoneley wave can be derived from the reflection of a "fictitious" P wave incident in medium 1 at the complex angle $(\pi/2) - i \cosh^{-1}(\alpha_1/c)$ [or an equally fictitious SV wave incident at $\vartheta = (\pi/2) - i \cosh^{-1}(\beta_1/c)$], where c is a solution of (3-10). The amplitude of both Rayleigh and Stoneley waves has been shown by Cagniard (18) and others to diminish exponentially with the distance of the source from the boundary.

3-4   *Reflection of Plane Waves beyond the Critical Angle*

A few additional remarks about surface waves may be made without departing from the plane-wave theory. Let us consider what happens when a plane wave is incident at a plane boundary at an angle which exceeds one or both of the critical angles for the two media. Specifically, we shall suppose that the incident wave is a plane P wave, that the angle of incidence is real, and that it is $> \sin^{-1}(\alpha_1/\alpha_2)$, α_2 being the larger of the two P-wave velocities. Two different situations may then arise, according to whether $\beta_2 \gtrless \alpha_1$. We shall consider them separately.

1. $\beta_2 < \alpha_1$. According to Snell's law, the angle of refraction of the front of the transmitted P wave will be given by

$$\sin \rho_P = \frac{\alpha_2}{\alpha_1} \sin \vartheta$$

which is real and > 1. Therefore we must set $\rho_P = \pi/2 - i\gamma$, where $\gamma = \cosh^{-1}[(\alpha_2/\alpha_1)\sin \vartheta]$. If we write $c = \alpha_1/\sin \vartheta$ for the apparent phase velocity in the x direction, it follows from (3-2) that

$$\Phi_0(x,z) = A_1 \exp ik(x - r_1 z) \qquad \text{where} \qquad k = \frac{\omega}{c}$$

and

$$\Phi_t(x,z) = A_2 \exp ik(x - r_2 z)$$

where

$$r_1 = \cot \vartheta \qquad \text{and} \qquad r_2 = \sqrt{\frac{\alpha_1^2}{\alpha_2^2} \csc^2 \vartheta - 1}$$

We note that while r_1 is real, r_2 is imaginary. Thus the transmitted P wave is an "inhomogeneous" wave whose amplitude diminishes exponentially with distance from the boundary. The maximum rate of attenuation occurs when the incident wavefront just grazes the interface, i.e.,

$$|r_2|_{\max} = \sqrt{1 - \frac{\alpha_1^2}{\alpha_2^2}}$$

when $\vartheta = \pi/2$ and when $\gamma = \cosh^{-1}(\alpha_2/\alpha_1) = \gamma_{\max}$.

Further, we note that the P-wave reflection coefficient becomes complex, indicating that a phase change takes place upon reflection.

The situation with regard to SV waves is markedly different. The angle of refraction of the transmitted SV wavefront is real for all real values of ϑ; therefore the SV motion is able to propagate away from the interface. Consequently the transmitted SV wave can be represented by an ordinary ray path.

2. $\beta_2 > \alpha_1$. In this case there may be two critical angles. The first is reached when $\sin \vartheta = \alpha_1/\alpha_2$, and the second, when $\sin \vartheta = \alpha_1/\beta_2$. When $\sin^{-1}(\alpha_1/\beta_2) > \vartheta > \sin^{-1}(\alpha_1/\alpha_2)$, the situation is as described above. When $\sin \vartheta > \alpha_1/\beta_2$, the transmitted SV motion also becomes trapped at the interface. Thus according to Snell's law

$$\sin \rho_S = \frac{\beta_2}{\alpha_1} \sin \vartheta$$

which is real and > 1. Therefore $\rho_S = \pi/2 - i\nu$, where

$$\nu = \cosh^{-1} \frac{\beta_2}{\alpha_1} \sin \vartheta$$

According to (3-2)

$$\Psi_t(x,z) = B_2 \exp ik(x - s_2 z) \qquad \text{where} \qquad k = \frac{\omega \sin \vartheta}{\alpha_1}$$

and where $s_2 = \sqrt{(\alpha_1^2/\beta_2^2) \csc^2 \vartheta - 1}$. Since s_2 is now imaginary, the SV wave attenuates with distance from the boundary. The maximum rate of attenuation is

$$|s_2|_{\max} = \sqrt{1 - \frac{\alpha_1^2}{\beta_2^2}}$$

which is attained when the incident wave strikes the boundary at grazing incidence, and $\nu = \cosh^{-1}(\beta_2/\alpha_1) = \nu_{\max}$. Notice that the surface SV wave travels with the same phase velocity as the surface P wave.

The surface waves described above are not Stoneley waves. They are attached to the particle motions in the lower-velocity medium, and they propagate forward with a phase velocity $\alpha_1/\sin \vartheta$. Their amplitude and phase are determined by the plane-wave transmission coefficients (Sec. 3-1), and they arise as a result of the elastic compliance of the boundary. We shall discuss them again in relation to guided waves in Sec. 3-7.

3-5   The Critically Refracted Ray Path

Let the incident P wave strike the boundary at the critical angle for P-wave transmission, $\vartheta = \sin^{-1}(\alpha_1/\alpha_2)$. The transmitted P wave will move through the higher-velocity medium at grazing incidence, while the transmitted SV wave will propagate away from the boundary. In principle, each point on the interface traversed by the grazing P wave can transmit energy back into the low-velocity medium. According to Huygens' principle, the surfaces of constant phase for this reradiated energy must be parallel planes whose normals are directed away from the boundary at the critical angle. Thus the "critically refracted ray path" illustrated in Fig. 3-12 describes a possible trajectory of the seismic energy, since it conforms to the laws of geometrical optics.

The difficulty with this interpretation is that the plane-wave transmission coefficient predicts that zero energy ought to be carried along this trajectory. On the other hand, critically refracted signals are recorded daily by seismic instruments, and far from having vanishing amplitudes, many are very strong. The explanation of this apparent paradox is to be found in the theory for curved wavefronts, of which an outline is given in Chap. 6. It is now known that the energy is not actually carried along the critically refracted ray path but in a "head wave" which forms in the low-velocity medium between the reflected wavefront and the boundary (Fig. 3-13). This wave is in fact the physical manifestation of Huygens' principle as applied to points on the boundary ahead of the incident wavefront (in the range BC of Fig. 3-13). Thus it appears to emerge from the interface at the critical angle, and for this reason it will arrive at distant points at just the instant predicted by the ray-path theory. It should be remarked, however, that the ray-path description becomes asymptotically correct only at distances from the boundary which are sufficiently large (in terms of the wavelength) that the reflected and transmitted wavefronts have become fully decoupled from each other and from the interface. What occurs within the close vicinity of the source or of the boundary is somewhat conjectural, for the ray-path method tends to break down in these regions.

The head wave associated with the transmitted P wave is of course not the only one to form in elastic media. There are also a reflected and a transmitted S wave, and these will give rise to other head waves. In all, the possibility of as many as five different head waves exists, depending on the velocity relationships, each of which advances with the velocity of the body wave to which it remains attached and in a direction determined by the appropriate critical angle of incidence. These are illustrated in Fig. 6-5 for identification, but it is usually only the PPP head wave that is used in applied seismology. The dashed line in Fig. 3-13 defines a conical region

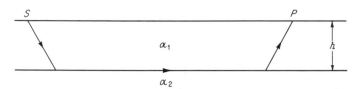

Fig. 3-12   Critical refraction at an interface.

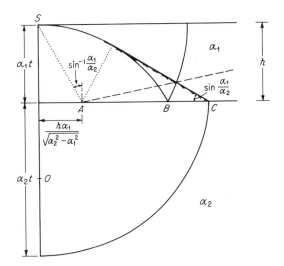

Fig. 3-13 The head wave.

above the interface, outside of which the PPP head wave arrives in front of the direct wave. The distance from the source to this locus is sometimes called the *crossover distance*.

3-6 Diffraction

The theory of reflection and refraction at a plane boundary applies in a limiting sense even to curved interfaces, provided that the radius of curvature is large compared with the seismic wavelength. When this is the case, local regions on the interface that are several wavelengths in areal extent may be considered as plane surfaces, and the theory for plane boundaries can be applied to them. If, however, the radius of curvature is not large, the phenomena of reflection and refraction can no longer be described by the simple theory for plane boundaries nor represented by the ordinary ray paths of geometrical optics. The wavefronts will suffer deviations which cannot readily be interpreted either as reflections or refractions and which are called *diffraction* effects. Diffraction will occur in the vicinity of any irregularity in an elastic interface, especially at a discontinuity. In geology, such irregularities are common. Diffraction phenomena associated with faulting are often recognizable in practice.

Because diffraction phenomena represent, in a certain sense, a breakdown of the geometrical method of describing wave propagation, the ordinary methods of geometrical optics fail to provide ray-path trajectories which represent the diffracted field. On the other hand, it is well known that at distances from a diffracting edge that are large in relation to the wavelength of the radiation, the diffracted wavefront can be found accurately by using the Huygens' construction. This suggests that there is a

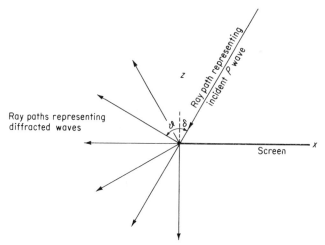

Fig. 3-14   Ray-path diagram for the diffraction of a plane wave by a rigid half-plane.   (Note that δ is negative in this diagram.)

geometrical interpretation of diffraction which is likely to improve toward the higher frequencies.   A bridge between the methods of geometrical optics and wave optics which permits a ray-path representation of the diffracted field of a discontinuity (at sufficiently small wavelengths) has been proposed by Keller (19).   We shall attempt to describe it briefly.

Let us consider at first a very simple problem.   Plane P waves are incident at an angle δ upon a thin, perfectly rigid sheet which terminates at an edge along Oy (Fig. 3-14).   If the wave is nonperiodic, we shall consider only a Fourier element of it which is sinusoidal.   Thus for the incident wave we may write

$$\Phi(x,z) = A \exp ik_\alpha(lx - nz)$$

and the boundary condition to be satisfied is

$$\frac{\partial \Phi}{\partial z} = 0 \qquad z = 0 \qquad x > 0$$

The solution of this boundary-value problem is too well known to be rederived here.   Sommerfeld (20) gives the exact solution, which for large values of $k_\alpha r$ (r being the distance from the edge) takes the form

$$\Phi_d(x,z) = D(\vartheta)r^{-\frac{1}{2}} \exp i\left(k_\alpha r + \frac{\pi}{4}\right) \tag{3-11}$$

where

$$D(\vartheta) = -A(8\pi k_\alpha)^{-\frac{1}{2}}[\sec \tfrac{1}{2}(\vartheta - \delta) - \csc \tfrac{1}{2}(\vartheta + \delta)]$$
$$\vartheta - \delta \neq \pi \qquad \vartheta + \delta \neq 0$$

In the above formula, ϑ is the polar angle measured from Oz, and $r = \sqrt{x^2 + z^2}$. Thus it is at once apparent that (3-11) describes a cylindrical wavefront diverging from the edge of the screen, whose amplitude falls away from a maximum at the geometrical shadow edge. It can be represented by a system of radial ray paths, the angular density of which is proportional to the *coefficient of diffraction* D/A.

At distances from the diffracting edge that are large in relation to the wavelength of the incident radiation, it is evident that the diffracted wavefront takes on a cylindrical form with its axis lying along the edge (Fig. 3-15). It follows also from (3-11) that the intensity of this wave diminishes with the frequency as $k_\alpha^{-\frac{1}{2}}$. Thus for nonperiodic disturbances the diffracted signal should be characterized by longer wavelengths than the incident waveform.

It is noteworthy that Sommerfeld's solution applies also to problems in which the incident wave normal does not lie in a plane perpendicular to the edge. If λ is the angle between the incident ray path and the axis Oy, the only change required is to multiply $D(\vartheta)$ by csc λ, provided that we now regard δ and ϑ as the angles of incidence and diffraction of the projections of the ray paths onto the plane normal to the edge. In this case the diffracted wavefront becomes a cone whose axis lies along the edge and whose apex contains the angle 2λ. This is just what is predicted by Huygens' principle.

From the statements made above, we can adduce the essential features of the geometrical theory of diffraction. The far-field diffraction pattern is calculated from the leading term in the asymptotic expansion, for large values of kr, of the exact solution for the diffracted field. This term has a

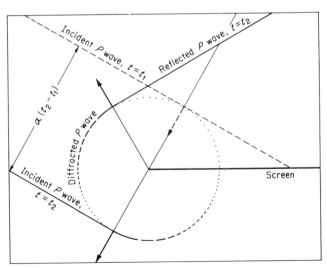

Fig. 3-15 Wave-front diagram for the diffraction of a plane wave by a rigid half-plane.

simple geometrical interpretation which corresponds to a curved wavefront describable by ray-path trajectories. The addition of further terms from the series is entirely possible in theory, but the ray interpretation of these terms is not altogether simple.

The geometrical method can of course be extended to diffracting surfaces other than the half-plane. In particular, if the half-plane is replaced by a rigid wedge whose interior angle is $(2 - n)\pi$, Sommerfeld's exact solution, according to Keller, gives

$$D(\vartheta) = \frac{1}{n} e^{i\pi/4} (2\pi k_\alpha)^{-\frac{1}{2}} \sin \frac{\pi}{n} \left[\left(\cos \frac{\pi}{n} - \cos \frac{\vartheta - \delta}{n} \right)^{-1} \right. $$
$$\left. + \left(\cos \frac{\pi}{n} - \cos \frac{\vartheta + \delta + \pi}{n} \right)^{-1} \right]$$

when $k_\alpha r \gg 1$. Burke and Keller (21) give a method for finding the diffraction coefficient of a rigid step (fault model) in terms of $D(\vartheta)$ above; however, the formula and its derivation are too lengthy to quote here.

3-7   *The Seismic Waveguide Effect in a Surface Layer*

Thus far we have discussed only the reflection, refraction, and diffraction of plane elastic waves at a single boundary. The upper part of the earth's crust is known to be made up of many different kinds of materials having a very wide range of elastic properties, so that a great many internal boundaries are apt to exist in nature. The effect of these boundaries will be to complicate the wave picture beyond any reasonable hope of mathematical analysis, and any further advancement of the theory will be achieved only through the use of simple models. Ideal assumptions must be imposed upon the elastic properties of the earth materials, upon the nature and shape of the internal boundaries, and upon the physical conditions at the seismic source. One such model, which has a wide range of application to geological problems, is the laminated half-space consisting of uniform, parallel layers. It roughly represents an undisturbed sedimentary environment, although it can also be applied to slightly disturbed regions.

Let us first consider a very simple multilayered problem; in fact, we shall choose the very simplest one possible. A layer of fluid resting upon a perfectly rigid substratum is excited by a source of acoustic energy. Since the energy cannot cross the interface into the substratum, it must in some fashion be propagated through the fluid. This very simple and somewhat artificial model thus provides an introduction to a phenomenon of great importance in seismology, namely, the guiding of seismic body waves by low-velocity surface layers. We shall at a later stage consider the effects of elastic properties upon this propagation, both within the layer and in the substratum, but many of the basic features of guided waves can be understood without this refinement.

Since the direction of energy propagation is horizontal, let us write for the Fourier element of the displacement potential

$$\Phi(x,z) = f(z) \exp ikx$$

Then, according to a result derived in Sec. 3-2, this may be written

$$\Phi(x,z) = A(\omega) \exp ik(x + rz) + D(\omega) \exp ik(x - rz)$$

where $r = \sqrt{c^2/\alpha^2 - 1}$, c being the phase velocity of the disturbance and α the acoustic wave velocity of the fluid. Notice that because the layer is bounded, the physical condition at infinity becomes irrelevant; thus both of the terms in the solution for $f(z)$ are needed to specify the wave function. The boundary conditions require that the normal stress must vanish at the free surface of the fluid and that the normal displacement must vanish at the interface. Thus if the interface lies at $z = 0$ and the free surface at $z = h$ (Fig. 3-16), the following two equations must be satisfied by the Fourier coefficients A and D:

$$A e^{ikrh} + D e^{-ikrh} = 0$$
$$A - D = 0$$

If we eliminate A and D from these two equations, we find that $\cos krh = 0$; hence

$$krh = (n + \tfrac{1}{2})\pi \tag{3-12}$$

where n may be any integer. Solving for c, we find that

$$c(\omega) = \left[\frac{1}{\alpha^2} - \frac{(n + \frac{1}{2})^2 \pi^2}{\omega^2 h^2} \right]^{-\frac{1}{2}} \tag{3-13}$$

showing that the phase velocity depends upon the frequency and that the propagation is therefore dispersive.

The appearance of the undetermined integer n in this expression suggests that more than one wave may be propagated through the fluid, which is indeed the case. Different values of n in (3-12) correspond to different "normal modes" of propagation. Normal modes in a waveguide arise in much the same way as do standing waves in an organ pipe. They are resonance effects brought about by the formation of pressure nodes at the free surface and pressure loops at the interface, and provided that these two

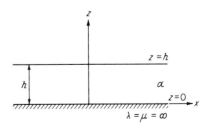

Fig. 3-16 Fluid layer resting upon a rigid substratum.

physical conditions are fulfilled continuously at both surfaces, unattenuated propagation of energy along the layer is possible. The lowest or "fundamental" mode ($n = 0$) corresponds to a pressure-depth profile that is a quarter of a sine wave; the second lowest mode ($n = 1$), with a pressure-depth profile that is three-quarters of a sine wave; and so on. Notice, however, that there is only a finite number of them; for in order to obtain a real root for c from (3-13), we must have

$$n + \frac{1}{2} \leq \frac{\omega h}{\pi}$$

Thus the number of normal modes is limited by the frequency ω of the source. A pulse will give rise to a large number of modes, but the amount of energy carried in the higher modes will normally be small.

The physical mechanism by which these modes are propagated can best be understood if we treat the same problem again from a slightly different point of view. Let us suppose that the source is in the fluid layer but that it is so far away that the wavefronts have become virtually plane. Then the Fourier element of the disturbance whose normal lies in the (x,z) plane may be represented by the displacement potential

$$\Phi(x,z) = A(\omega) \exp ik_\alpha(l_\omega x + n_\omega z) + D(\omega) \exp ik_\alpha(l_\omega x - n_\omega z)$$

corresponding to upgoing and downgoing wavefronts. If we apply the same set of boundary conditions to this function and eliminate the Fourier coefficients A and D, we find that $\cos ik_\alpha n_\omega h = 0$; hence

$$k_\alpha n_\omega h = (n + \tfrac{1}{2})\pi \tag{3-14}$$

Equation (3-14) corresponds identically with (3-13) if we set $c = \alpha/l_\omega$; hence we arrive at the following physical interpretation of normal mode propagation: Plane wavefronts, internally reflected at both the top and bottom of the fluid layer, will interfere with one another constructively if the angle of incidence satisfies (3-14); this interference gives rise to the propagation of energy through the fluid with a phase velocity given by (3-13). In Fig. 3-17 an advancing wavefront is shown at two consecutive positions GBE and LCM, which are separated by a time interval Δt. The ray path representing the route followed by this wavefront is the line $ABCDEF$, and it corresponds to total reflection at the bottom without change of phase and total reflection at the free surface with a phase change of π. If energy is to be propagated in the fluid layer, the wavefront in the

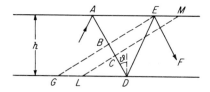

Fig. 3-17 Showing the condition for constructive interference among internally reflected plane waves in a fluid.

position GBE must be in phase with the coincident wavefront which has traversed the additional distance $BDE = 2h \cos \vartheta$. Therefore

$$2k_\alpha h \cos \vartheta - \pi = 2n\pi \qquad n = 0, 1, 2, \ldots$$

i.e.,

$$k_\alpha h \cos \vartheta = (n + \tfrac{1}{2})\pi$$

which is just (3-14). When this condition is fulfilled, the surfaces of constant phase will be the planes $x = $ const, and the plane which at time t is at the position G will during the subsequent interval Δt advance to the position L. Thus the velocity with which the surfaces of constant phase are propagated through the liquid will be

$$c = \frac{GL}{\Delta t} = \frac{BC}{\sin \vartheta \, \Delta t} = \frac{\alpha}{\sin \vartheta}$$

in agreement with (3-12) and (3-14).

Since the phase velocity exceeds the velocity of sound in the fluid, it is obvious that acoustic energy cannot be transported through the layer with the same speed as the surfaces of constant phase. It must travel with a velocity other than the phase velocity, which is also characteristic of the dimensions of the waveguide. If ϕ is nonperiodic, it may be written in spectrum form as follows:

$$\phi(x,z,t) = \frac{1}{2\pi} \int_{-\infty}^{\infty} \Phi(x,z;\omega) e^{-i\omega t} \, d\omega$$
$$= \frac{1}{2\pi} \cos \frac{(n + \tfrac{1}{2})\pi z}{h} \int_{-\infty}^{\infty} A(\omega) e^{i[k(\omega)x - \omega t]} \, d\omega$$

$A(\omega)$, which is the Fourier transform of the incident disturbance, will change slowly with ω in relation to the exponential term, especially at great distances from the source. The tendency will be for the exponential term to oscillate very rapidly when x and t are large, so that over a small range in ω it will produce a net effect which is very nearly zero. Exceptions to this rule will arise only in the vicinity of those frequencies at which the phase has a stationary value, i.e., where

$$\frac{d}{d\omega}[k(\omega)x - \omega t] = 0 \tag{3-15}$$

as the Fourier components whose frequencies lie close to these values will tend to travel coherently and will not destroy one another. At large distances, the signal may therefore be considered as a superposition of these frequency groups. The distance x which one of the groups will travel in the time t is called the *group velocity*, and according to (3-15) it will have the value

$$C(\omega) = \frac{x}{t} = \left[\frac{dk(\omega)}{d\omega}\right]^{-1} = c(\omega)\left[1 - \frac{\omega}{c(\omega)} \frac{dc}{d\omega}\right]^{-1} \tag{3-16}$$

Since the group of Fourier components of the nonperiodic signal which lie in the frequency range ω, $\omega + d\omega$ will presumably interfere constructively on surfaces moving with this speed, we identify the group velocity with the velocity of energy transport. A fuller discussion of this result and of certain exceptions to it can be found in Brillouin (22).

In the present instance $C(\omega) = \alpha^2/c(\omega)$, and since $c > \alpha$, it follows that $C < \alpha$, as indeed we should expect it to be. Curves of C/α, c/α for the two lowest modes plotted against the nondimensional quantity $k_\alpha h$ are shown in Fig. 3-18, showing the manner in which these two velocities converge toward each other at the higher frequencies, where the ray paths become very oblique. The diagram suggests that the signal due to a general disturbance in the fluid which arrives at a distant point will be a dispersed wave train, the higher frequencies coming in first. As the above analysis shows, however, the modes require a certain time to sort themselves out; for it is not until the wavefronts have become virtually plane that the conditions for normal mode propagation are fulfilled. Therefore normal mode behavior

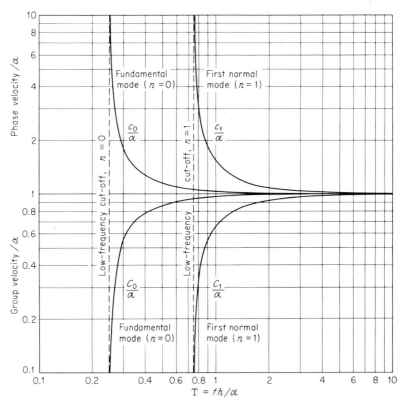

Fig. 3-18   Phase and group velocity curves for sound propagation in a fluid layer resting on a rigid substratum.

may be expected to occur only at distances from a point source which are large in terms of the layer thickness. At closer positions some dispersion may of course occur, but the normal mode solutions will not be valid.

3-8   *Normal Modes in a Liquid Layer on an Elastic Substratum*

Perhaps the most striking example of normal mode propagation to be found in a natural environment is in the oceans. The classic paper on this problem is by Pekeris (23). However, we can study it from a simpler viewpoint by extending the analysis of the previous section to include elastic properties within the substratum. Thus let us take as the Fourier transform of the displacement potential in the fluid the function

$$\Phi_1(x,z;\omega) = A_1(\omega) \exp ik(x + r_1z) + D_1(\omega) \exp ik(x - r_1z)$$

where $r_1 = \sqrt{(c^2/\alpha_1^2) - 1}$.

The pair of functions

$$\Phi_2(x,z;\omega) = A_2(\omega) \exp ik(x - r_2z) \qquad r_2 = \sqrt{\frac{c^2}{\alpha_2^2} - 1}$$

and $\qquad \Psi_2(x,z;\omega) = B_2(\omega) \exp ik(x - s_2z) \qquad s_2 = \sqrt{\frac{c^2}{\beta_2^2} - 1}$

will be introduced to represent the displacement potentials in the substratum. To satisfy the physical requirement of energy conservation, both r_2 and s_2 should be imaginary. The boundary conditions are

1. At $z = 0$, both p_{zz} and p_{zx} are continuous; hence

$$\rho_1\omega^2\Phi_1 = \rho_2\omega^2\Phi_2 - 2\mu_2\left(\frac{\partial^2}{\partial z^2} + \frac{\omega^2}{\alpha_2^2}\right)\Phi_2 + 2\mu_2\frac{\partial^2\Psi_2}{\partial z\,\partial x} \qquad z = 0$$

and $\qquad 0 = 2\dfrac{\partial^2\Phi_2}{\partial z\,\partial x} - \dfrac{\partial^2\Psi_2}{\partial x^2} + \dfrac{\partial^2\Psi_2}{\partial z^2} \qquad z = 0$

2. At $z = 0$ also, w is continuous, so that

$$\frac{\partial\Phi_1}{\partial z} = \frac{\partial\Phi_2}{\partial z} - \frac{\partial\Psi_2}{\partial x} \qquad z = 0$$

3. At $z = h$, p_{zz} vanishes, i.e.,

$$\Phi_1 = 0 \qquad z = h$$

From these four equations we may eliminate the Fourier coefficients A_1, D_1, A_2, and B_2 and so obtain the period equation for normal mode propagation in the water layer. When $\lambda_2 = \mu_2$, this equation is

$$\tan kr_1h = \frac{\rho_2}{\rho_1}\frac{\beta_2^4}{c^4}\frac{r_1}{r_2}[4r_2s_2 - (1 + s_2^2)^2]$$

It will be observed that this equation has a multiple-valued function on one side and a single-valued function on the other. Its solution is therefore ambiguous. To remove the ambiguity, we may write

$$kr_1h + n\pi = \tan^{-1} \frac{\rho_2}{\rho_1} \frac{\beta_2{}^4}{c^4} \frac{r_1}{r_2} [4r_2s_2 - (1 + s_2{}^2)^2] \qquad (3\text{-}17)$$

where n may be any integer. As in the previous problem, the value given to n defines the normal mode number. Thus $n = 0$ corresponds to the fundamental mode; $n = 1$, to the first normal mode; and so on. To keep c real, n must have an upper limit (which must be found from the root of a transcendental equation), and there can therefore be only a finite number of normal modes. These correspond to the formation of pressure nodes at the free surface and to the appropriate continuity conditions at the interface.

The fundamental mode ($n = 0$) is distinguished from the others in that it has no low-frequency cutoff. Thus it seems to be able to propagate energy at all wavelengths. If in Eq. (3-17) we allow $kr_1h \to 0$, we find that if $n = 0$, then

$$4r_2s_2 = (1 + s_2{}^2)^2$$

This is just the equation for the velocity of free Rayleigh waves along the interface, indicating that when the wavelength becomes much greater than the depth of the layer, the energy travels as a surface wave in the solid bottom, unmindful of the fluid overburden. This is the maximum phase velocity that can be attained in the lowest mode. The lowest phase velocity for which (3-17) has a real root when $n = 0$ is at the high-frequency limit, where the value becomes slightly less than α_1. This will correspond to an interface wave which is slightly damped with height above the boundary but strongly damped with the depth below. On this account it might be called a Stoneley wave. At intermediate frequencies, the phase velocity will shift gradually from one limit to the other.

In the first normal mode ($n = 1$), Eq. (3-17) does not have a real root for c unless both r_2 and s_2 are imaginary. Therefore the maximum value that c may have is β_2, which makes $s_2 = 0$. When $c = \beta_2$, the equation may be solved numerically to find ω. This corresponds to a low-frequency cutoff, below which unattenuated waves cannot be propagated through the fluid. The smallest value that c may have is α_1, which makes $r_1 = 0$. This value is attained only as a limiting solution of (3-17) as $\omega \to \infty$. Disturbances having very short wavelengths will travel approximately horizontally through the layer, their ray paths just grazing either boundary. Because $\alpha_1 \leq c \leq \beta_2$, the ray paths which represent energy propagation in the first (and also in all the higher) normal modes must strike the fluid-solid interface at angles of incidence which exceed the critical value for *PSV* refraction into the bottom $[\vartheta_c = \sin^{-1} (\alpha_1/\beta_2)]$. Thus, as we might have expected, there must be total reflection at the interface in order to prevent the loss of energy by radiation into the substratum either as ordinary P or SV waves. If such losses should occur, the waves in the layer will attenuate fairly rapidly. Accordingly, the motion in the substratum which accompanies the passage of the normal modes in the layer will be in the form of "inhomogeneous" or vertically attenuated waves, which exist on account of the elastic compliance of the boundary.

Curves showing the phase and group velocities in the fundamental and the first normal mode in a layer of fluid resting upon an elastic half-space are shown in Fig. 3-19. The calculations were given by Ewing, Jardetzky, and Press (24), and the two curves demonstrate the qualitative features that we have discussed up to this point. The main difference between these and the curves shown in Fig. 3-18 is that the group velocities now have minimum values. This is due entirely to the elastic properties of the bottom, and it has some important consequences which we shall discuss. In the meanwhile it is perhaps worth pointing out that as n increases, the penetration of energy into the bottom becomes progressively easier, with the result that the interface appears to become decreasingly "rigid" to the higher modes. This is due to the fact that, as Eq. (3-14) points out, if the frequency remains the same, a larger value of n corresponds to a steeper ray path, for which the coupling with the inhomogeneous wave is greater. This is illustrated in Fig. 3-20 by the "skin-depth" curve, which shows the distance below the interface (expressed in units of h) at which the amplitude of the vertical component w of the particle displacement in the two lowest modes is attenuated by the factor $1/e$. These curves give some indication of the volume

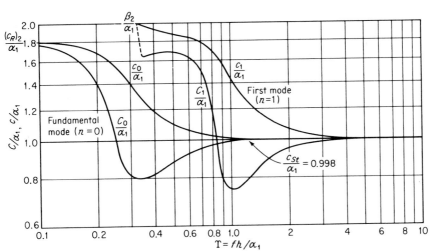

Fig. 3-19 Curves showing the phase velocity (c/α_1) and the group velocity (C/α_1) for sound propagation in a fluid layer resting on an elastic substratum, when $\alpha_2 = 2\sqrt{3}\,\alpha_1$, $\lambda_2 = \mu_2$, and $\rho_2 = 2.5\rho_1$. [After Ewing, Jardetzky, and Press (24).]

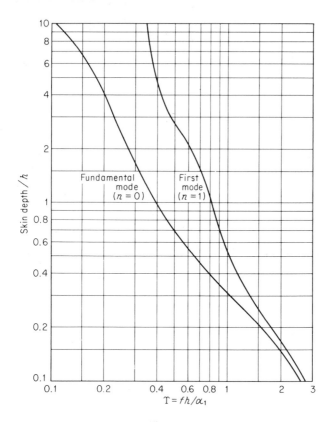

Fig. 3-20 Showing the depth in the substratum at which the amplitude becomes attenuated by a factor $1/e$, for the case illustrated in Fig. 3-19.

of the substratum that influences the wave structure in the fluid, and they also suggest one reason why the fundamental mode is almost invariably the strongest.

It is interesting to study, with reference to the dispersion curves in Fig. 3-19, the sequence of waves arriving at a seismometer placed in the water layer at some distance from a source of seismic energy. The first arrival will probably be the *PPP* head wave, which is nondispersive. Following this, a dispersive signal will come in from the left-hand sides of the group-velocity curves more or less simultaneously in all modes. Except under special conditions which we shall discuss presently, however, most of the energy will propagate in the fundamental mode, and so it will come in at first as a low-frequency Rayleigh wave. This low-frequency low-amplitude first arrival, often called the *ground wave*, increases steadily in both frequency and amplitude up to the moment of arrival of the high-frequency water wave, which is followed almost immediately by the Stoneley wave from the right-hand side. From this instant onward, contributions to the signal begin to come in simultaneously from both sides of the group-velocity curve, and the mixing of these two phases results for a time in a waveform consisting of a high-frequency rider superimposed upon the low-frequency ground wave. The two waves steadily approach each other in frequency until they merge to form a single wave pattern of relatively large amplitude, called the *Airy phase*, which terminates rather abruptly at a time corresponding to propagation at the minimum group velocity.

The tendency of wave energy to become enhanced within a restricted frequency band at the group velocity minimum is a phenomenon familiar in physics, and it can be understood in terms of classical dynamic principles. It comes about largely because the wave groups close to this frequency are strongly coherent, much more so than at other frequencies. Thus, we may expect the dominant mode of propagation to change from the fundamental to the first mode at frequencies close to $\omega = \alpha_1/h$. Since the group velocity attains a slightly lower value in the first mode than in the fundamental, a second Airy phase is likely, and so on. Each mode that possesses a group-velocity minimum will exhibit an Airy phase. To a receiving system in which noise may obscure some of the weaker parts of the dispersed wave train, these arrivals may appear as pulses.

3-9 The Solid Surface Layer: Love and Rayleigh Waves

Among the clearest pieces of seismological evidence of the layering of the earth's crust is the existence of surface SH waves. In Sec. 3-3 we learned that SH waves cannot exist at the free surface of a homogeneous half-space, and yet they are distinctly observed in nature. That this may be accepted as evidence of internal boundaries may be shown by considering SH normal mode propagation in a uniform solid surface layer.

Since normal mode propagation is a phenomenon that can be explained in terms of plane waves, it is sufficient in this problem to consider only the displacement component v, which separates from u and w in the plane-wave theory. Within the layer, we may write for the Fourier element of a nonperiodic disturbance

$$V_1(x,z;\omega) = C_1(\omega) \exp ik(x + s_1 z) + F_1(\omega) \exp ik(x - s_1 z)$$

and in the substratum

$$V_2(x,z;\omega) = C_2(\omega) \exp ik(x - s_2 z)$$

The condition at infinity requires that s_2 shall be imaginary. The boundary conditions that must be applied to these solutions are as follows:

1. At $z = 0$ the stress component p_{zy} is continuous; hence

$$\mu_1 \frac{\partial V_1}{\partial z} = \mu_2 \frac{\partial V_2}{\partial z} \qquad z = 0$$

2. At $z = 0$ the displacements are continuous also, so that

$$V_1 = V_2 \qquad z = 0$$

3. At $z = h$ the stresses vanish, i.e.,

$$\frac{\partial V_1}{\partial z} = 0 \qquad z = h$$

These conditions lead to the following three equations:

$$\mu_1 s_1 (C_1 - F_1) = \mu_2 s_2 C_2$$
$$C_1 + F_1 = C_2$$

and
$$C_1 e^{iks_1 h} = F_1 e^{-iks_1 h}$$

and if we eliminate the three Fourier coefficients, we obtain the following period equation:

$$\tan ks_1 h = -\frac{i\mu_2 s_2}{\mu_1 s_1}$$

i.e.,
$$ks_1 h + n\pi = \tan^{-1} \frac{\mu_2 s_2}{i\mu_1 s_1} \tag{3-18}$$

Surface *SH* waves can exist if (3-18) can be satisfied. This is possible if s is real and s_2 is imaginary, so that $\beta_1 \leq c \leq \beta_2$. If $\beta_1 > \beta_2$, such wave cannot exist. In earthquake seismology these are called *Love waves*, and since they appear as a normal-mode solution of the wave equation, they are evidently propagated by multiple internal reflections of *SH* waves within the surface layer. This interpretation is further fortified by observing that the high- and low-frequency limits of the phase velocity in all modes are, according to (3-18), β_1 and β_2, respectively. Thus the longer waves must travel along the steeper ray paths, whereas the shorter ones will travel more obliquely. Love waves are therefore *SH* waveguide modes which exist on account of the contrast in rigidity at the interface.

An important feature of Love-wave dispersion is that each mode except the fundamental mode has a low-frequency cutoff, below which unattenuated propagation through the layer cannot occur. Only in the fundamental mode is it theoretically possible for all wavelengths to propagate without horizontal attenuation. This low-frequency cutoff corresponds in each mode to propagation along the critically incident ray path. Once this limit is attained, further reducing the frequency makes it impossible to satisfy the conditions for constructive interference in the horizontal direction within a layer having the given thickness, and it therefore becomes impossible to find a real root for the period equation (3-18). In the simpler case in which a fluid layer rests on a rigid substratum this same condition has already been seen in Eq. (3-13), where the low-frequency cutoff in the nth mode is given by

$$\omega_n = \left(n + \frac{1}{2}\right)\frac{\pi\alpha}{h}$$

In seismology, the layer thickness and the elastic-wave velocities are seldom known in advance, but on the other hand, the group velocities can be measured by taking records of the ground motion at a number of locations. In this situation the dispersion curves may be used as a basis for geophysical interpretation, an actual example of which is given in Sec. 4-3. An example of theoretical dispersion curves for Love waves calculated by Stoneley (25) is shown in Fig. 3-21.

The dispersion of SV waves (including Rayleigh waves) within a surface layer may be calculated by applying the same mathematical procedures to the displacement components u and w. The essential features of the analysis are the same as those for Love waves; only the algebra is more tedious. The period equation for the phase velocity is a rather complicated transcendental equation which has several branches, indicating that the propagation of SV waves through the layer displays most of the usual normal-mode characteristics at large distances from the source. Examples of SV disper-

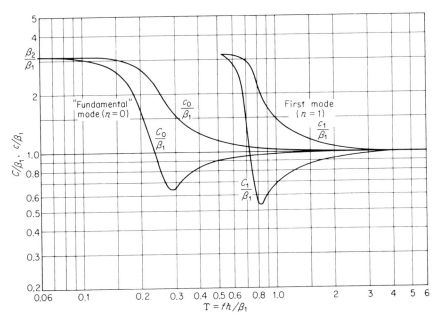

Fig. 3-21 **Dispersion curves for Love waves, when** $\beta_2 = 3.15\,\beta_1, \rho_2 = 1.39\rho_1$. [**After Stoneley (25).**]

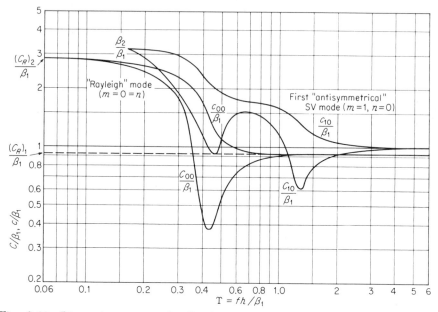

Fig. 3-22 Dispersion curves for Rayleigh and *SV* waves, when $\alpha_2 = 3.15\alpha_1$, $\beta_2 = 3.15\,\beta_1$, $\lambda_1 = \mu_1$, $\lambda_2 = \mu_2$, $\rho_2 = 1.39\,\rho_1$. **[After Tolstoy and Usdin (26).]**

sion curves due to Tolstoy and Usdin (26) are shown in Fig. 3-22, and as we can see, they are rather complicated in form. An interesting point about these solutions, however, is that in the lowest mode the high-frequency limit for the phase velocity is the solution of (3-7) using the wave velocities of the upper layer, while at the low frequencies the phase velocity approaches the solution of (3-7) in the substratum. Thus the fundamental mode at large distances corresponds to the Rayleigh wave, which now appears to be dispersive. We shall therefore call this the *Rayleigh mode* and refer to succeeding members as the first, second, etc., *SV* mode according to its branch number in the solution of the period equation. Thus the first *SV* mode, for example, has phase and group velocities which attain the value of the shear-wave velocity in the substratum at the low-frequency cutoff point and which approach the shear-wave velocity in the layer at very high frequencies. The existence of a minimum frequency (for a given value of h) below which unattenuated propagation in the layer cannot occur is a phenomenon common to all but the Rayleigh mode of *SV* guided waves.

The Rayleigh wave or "ground roll" which is observed at the surface of a layered ground at large distances from the source of seismic energy is accordingly identified with the lowest mode of *SV* propagation in the uppermost layer. Since this wave is generally strong, its dispersion characteristics are often used in earthquake seismology to study the structure and composition of the crustal layering. Since the group velocity attains a lower value in the Rayleigh mode than it does in the fundamental Love mode, the Rayleigh wave will generally arrive behind the Love wave.

3-10 Transmission of Plane Waves through Layered Media

A problem of some interest in practical seismology is the transmission of elastic waves through a succession of layers of different elastic media. The analysis is very often tied in with such questions as how much energy can be expected to return by reflection from a given interface and how the energy is divided during its travel. The boundary conditions make the detailed calculations extremely heavy if the number of layers is greater than two, but the problem has been set up by Thomson (27) in a matrix form that is easily handled by modern computing machines.

Let us suppose that a plane wave is incident on the first interface at an angle of incidence ϑ_0, and let us consider the nth layer in the sequence. For the Fourier components of the displacement potentials of the body waves within this layer, we may write

$$\Phi_n(x,z;\omega) = A_n(\omega) \exp ik_{\alpha_n}(l_{\alpha_n}x + n_{\alpha_n}z) + D_n(\omega) \exp ik_{\alpha_n}(l_{\alpha_n}x - n_{\alpha_n}z)$$
$$\Psi_n(x,z;\omega) = B_n(\omega) \exp ik_{\beta_n}(l_{\beta_n}x + n_{\beta_n}z) + E_n(\omega) \exp ik_{\beta_n}(l_{\beta_n}x - n_{\beta_n}z)$$

The first term in each expression represents the advancing wave, and the second, the reflected wave. Applying the continuity conditions for stresses and displacements gives the conditions on the Fourier coefficients shown on page 86.

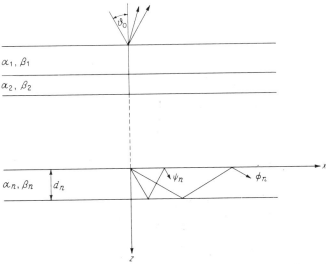

Fig. 3-23 Transmission of plane waves through a sequence of parallel layers.

at $x = 0 = z$, the continuity conditions give

$$
\begin{bmatrix} U^{(n-1)} \\ W^{(n-1)} \\ P_{zz}^{(n-1)} \\ P_{zz}^{(n-1)} \end{bmatrix} =
\begin{bmatrix}
ik_{\alpha_n} l_{\alpha_n} & 0 & -ik_{\beta_n} n_{\beta_n} & 0 \\
0 & -ik_{\alpha_n} n_{\alpha_n} & 0 & -ik_{\beta_n} l_{\beta_n} \\
-k_{\alpha_n}^2 \rho_n(\alpha_n^2 - 2\beta_n^2 l_{\alpha_n}^2) & 0 & -2k_{\beta_n}^2 \rho_n \beta_n^2 l_{\beta_n} n_{\beta_n} & 0 \\
0 & 2k_{\alpha_n}^2 \rho_n \beta_n^2 l_{\alpha_n} n_{\alpha_n} & 0 & k_{\beta_n}^2 \rho_n \beta_n^2 (l_{\beta_n}^2 - n_{\beta_n}^2)
\end{bmatrix}
\begin{bmatrix} A_n + D_n \\ A_n - D_n \\ B_n - E_n \\ B_n + E_n \end{bmatrix}
\tag{3-19}
$$

while at $x = 0, z = d_n$ we have

$$
\begin{bmatrix} U^{(n)} \\ W^{(n)} \\ P_{zz}^{(n)} \\ P_{xz}^{(n)} \end{bmatrix} =
\begin{bmatrix}
ik_{\alpha_n} l_{\alpha_n} \cos \mu_n & k_{\alpha_n} l_{\alpha_n} \sin \mu_n & -ik_{\beta_n} n_{\beta_n} \cos \nu_n & -k_{\beta_n} n_{\beta_n} \sin \nu_n \\
-k_{\alpha_n} n_{\alpha_n} \sin \mu_n & -ik_{\alpha_n} n_{\alpha_n} \cos \mu_n & -k_{\beta_n} l_{\beta_n} \sin \nu_n & -ik_{\beta_n} l_{\beta_n} \cos \nu_n \\
-k_{\alpha_n}^2 \rho_n(\alpha_n^2 - 2\beta_n^2 l_{\alpha_n}^2) \cos \mu_n & ik_{\alpha_n}^2 \rho_n(\alpha_n^2 - 2\beta_n^2 l_{\alpha_n}^2) \sin \mu_n & -2k_{\beta_n}^2 \rho_n \beta_n^2 l_{\beta_n} n_{\beta_n} \cos \nu_n & 2ik_{\beta_n}^2 \rho_n \beta_n^2 l_{\beta_n} n_{\beta_n} \sin \nu_n \\
-2ik_{\alpha_n}^2 \rho_n \beta_n^2 l_{\alpha_n} n_{\alpha_n} \sin \mu_n & 2k_{\alpha_n}^2 \rho_n \beta_n^2 l_{\alpha_n} n_{\alpha_n} \cos \mu_n & -ik_{\beta_n}^2 \rho_n \beta_n^2 (l_{\beta_n}^2 - n_{\beta_n}^2) \sin \nu_n & k_{\beta_n}^2 \rho_n \beta_n^2 (l_{\beta_n}^2 - n_{\beta_n}^2) \cos \nu_n
\end{bmatrix}
\begin{bmatrix} A_n + D_n \\ A_n - D_n \\ B_n - E_n \\ B_n + E_n \end{bmatrix}
\tag{3-20}
$$

where

$$\mu_n = k_{\alpha_n} n_{\alpha_n} d_n \qquad \nu_n = k_{\beta_n} n_{\beta_n} d_n$$

If we invert Eq. (3-19), we have

$$
\begin{bmatrix} A_n + D_n \\ A_n - D_n \\ B_n - E_n \\ B_n + E_n \end{bmatrix} = \begin{bmatrix} ik_{\alpha_n}l_{\alpha_n} & 0 & ik_{\beta_n}n_{\beta_n} & 0 \\ 0 & ik_{\alpha_n}n_{\alpha_n} & 0 & -ik_{\beta_n}l_{\beta_n} \\ & & \text{etc.} & \\ & & & \end{bmatrix}^{-1} \begin{bmatrix} U^{(n-1)} \\ W^{(n-1)} \\ P_{zz}^{(n-1)} \\ P_{zx}^{(n-1)} \end{bmatrix} \tag{3-21}
$$

Substituting this expression into (3-20) gives a recursion formula in stresses and displacements of the type

$$
\begin{bmatrix} U^{(n)} \\ W^{(n)} \\ P_{zz}^{(n)} \\ P_{zx}^{(n)} \end{bmatrix} = \begin{bmatrix} a_{11}^{(n)} & a_{12}^{(n)} & a_{13}^{(n)} & a_{14}^{(n)} \\ a_{21}^{(n)} & a_{22}^{(n)} & a_{23}^{(n)} & a_{24}^{(n)} \\ a_{31}^{(n)} & a_{32}^{(n)} & a_{33}^{(n)} & a_{34}^{(n)} \\ a_{41}^{(n)} & a_{42}^{(n)} & a_{43}^{(n)} & a_{44}^{(n)} \end{bmatrix} \begin{bmatrix} U^{(n-1)} \\ W^{(n-1)} \\ P_{zz}^{(n-1)} \\ P_{zx}^{(n-1)} \end{bmatrix} \tag{3-22}
$$

where the elements $a_{ik}^{(n)}$ of the 4×4 matrix are obtained by forming the product of the two 4×4 matrices appearing in (3-20) and (3-21).

The continuity of stresses and particle displacements enables us to express the quantities at the nth layer in terms of the quantities at the first layer as follows:

$$
\begin{bmatrix} U^{(n)} \\ W^{(n)} \\ P_{zz}^{(n)} \\ P_{zx}^{(n)} \end{bmatrix} = \begin{bmatrix} \\ a^{(n)} \\ \\ \end{bmatrix} \begin{bmatrix} \\ a^{(n-1)} \\ \\ \end{bmatrix} \begin{bmatrix} \\ a^{(n-2)} \\ \\ \end{bmatrix} \cdots \begin{bmatrix} \\ a^{(0)} \\ \\ \end{bmatrix} \begin{bmatrix} U^{(0)} \\ W^{(0)} \\ P_{zz}^{(0)} \\ P_{zx}^{(0)} \end{bmatrix}
$$

where, according to Snell's law,

$$
k_{\alpha_1}l_{\alpha_1} = k_{\beta_1}l_{\beta_1} = \cdots = k_{\alpha_n}l_{\alpha_n} = k_{\beta_n}l_{\beta_n}
$$

If at any boundary the incident angle of the advancing wave should exceed the critical angle for shear or compressional waves or both, then one or both of the angles of refraction will become complex. As we have already seen, the motions in this case will decay exponentially in passing through the plate. The analysis is equivalent to calculating the downward propagation of *SV* normal modes: If the plate in question is not too thick, enough energy may "leak" through it to yield a significant amount of transmission in the adjacent layers. These "leaking" modes can become reconverted into body waves if there is a subsequent velocity inversion.

This method of analysis is only concerned with the propagation of plane waves in the layered structure. It does not allow for the spreading of spherical wavefronts close to the source, for which special corrections will need to be made if an accurate account of the transmission characteristics is required. At nearly normal angles of incidence, however, the solutions will describe many of the qualitative features of the reflection and transmission of seismic waves in layered media.

Virtually the same methods have been used by Haskell (28) and several others to calculate surface-wave dispersion curves for multilayered media. Since normal mode theories are, as we have pointed out, closely related to plane-wave theory, Thomson's method is obviously appropriate to these studies. On the other hand, it does ignore head waves, which present an additional complicating factor at the wide angles of incidence where the normal-mode theory is valid.

3-11 Bragg Reflection

Now let us consider, as a particular instance of the problem described above, the reflection of a plane wave incident upon a region containing two different materials that are periodically interlayered. A phenomenon well known in physics is the reinforcement of partial reflections from a wave passing through a periodic structure at certain angles of incidence, leading to a returning wave having a large amplitude. The effect is called *Bragg reflection*. It is suspected to play a significant role in reflection seismology.

Let us look at Fig. 3-24. The incident wave is a plane simple harmonic P wave which strikes the topmost interface ($z = 0$) at an angle of incidence ϑ less than the smallest critical angle for the two media. The half-space $z < 0$ consists of alternating layers of medium 1 (wave velocities α_1, β_1 and density ρ_1), each having a thickness h_1, and of medium 2 (wave velocities α_2, β_2 and density ρ_2) with a thickness h_2. The entire half-space $z < 0$ thus possesses a periodic structure, the period being $h_1 + h_2$.

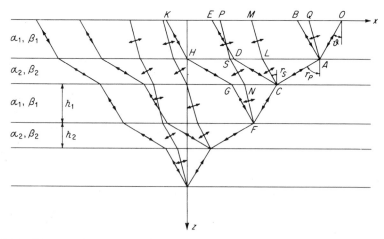

Fig. 3-24 Bragg reflection of plane waves in a periodically layered medium.

If we allow for all the possible conversions between P and SV motions at the interfaces, as well as for all the possible reflections and transmissions, the number of ray paths representing wavefronts in $z < 0$ becomes staggering. However, we shall assume here that the boundaries are rather weak; thus only single reflections need be considered. If the angle of incidence is small in relation to the various critical angles (which is certainly true for the weak boundaries and steep ray paths we are presently considering), it is quite in order for us to neglect altogether the conversion of energy upon transmission through any interface, since the conversion factor is practically zero in such cases. Thus the waves which lead to Bragg reflection are represented, in the first two layers, by the ray paths shown in Fig. 3-24.

Let us consider the case $\beta_2 < \beta_1 < \alpha_2 < \alpha_1$. Waves which have remained unconverted throughout their entire travel will lead to Bragg reflection at certain angles of incidence, which we shall call *Bragg angles of the first kind*. Waves which have suffered a P-SV conversion upon reflection will lead to *Bragg angles of the second kind*. Obviously there cannot be more than one such conversion on any ray path; hence there can be only two sets of Bragg angles.

Referring now to Fig. 3-24, the difference in phase at the points A and D on the rays $OACDE$ and OAB, respectively, will be $(2\omega h_2/\alpha_2) \sec r_P + \pi$. If the interference is to be constructive, we must set this equal to $2m\pi$, where m is an integer. Thus we get

$$\frac{2\omega h_2}{\alpha_2} \left(1 - \frac{\alpha_2^2}{\alpha_1^2} \sin^2 \vartheta \right)^{-\frac{1}{2}} = (2m - 1)\pi \qquad m = 1, 2, \ldots \qquad (3\text{-}23)$$

The difference in phase at D and H on the rays OCE and OFK, respectively, is $(2\omega h_1/\alpha_1) \sec \vartheta - \pi$. Again, we must set this equal to $2n\pi$, where n may be an integer or zero. Thus

$$\frac{2\omega h_1}{\alpha_1} \sec \vartheta = (2n + 1)\pi \qquad n = 0, 1, 2, \ldots \qquad (3\text{-}24)$$

From (3-23) and (3-24) we may calculate the Bragg angle of the first kind of order zero, for setting $m = 1$ and $n = 0$ gives

$$\frac{\alpha_2}{h_2} \sqrt{1 - \frac{\alpha_2^2}{\alpha_1^2} \sin^2 \vartheta} = \frac{\alpha_1}{h_1} \sin \vartheta$$

The condition for Bragg reflection of the first kind at normal incidence is

$$\frac{h_1}{h_2} = \frac{2m - 1}{2n + 1} \frac{\alpha_1}{\alpha_2}$$

For order zero, this reduces to the condition $h_1\alpha_2 = h_2\alpha_1$.

For Bragg reflection of the second kind, we must consider the ray paths $OACLM$ and OAQ. The difference in phase at the points A and L will be

$(\omega h_2/\beta_2) \sec r_S + (\omega h_2/\alpha_2) \sec r_P + \pi$, and setting this equal to $2m\pi$ gives

$$\frac{\omega h_2}{\beta_2}\left(1 - \frac{\beta_2{}^2}{\alpha_1{}^2}\sin^2\vartheta\right)^{-\frac{1}{2}} + \frac{\omega h_2}{\alpha_2}\left(1 - \frac{\alpha_2{}^2}{\alpha_1{}^2}\sin^2\vartheta\right)^{-\frac{1}{2}} = (2m-1)\pi$$

$$m = 1, 2, \ldots \quad (3\text{-}25)$$

At the points L and S on the ray paths OCM and OFP, respectively, the phase difference is $(\omega h_1/\beta_1) \sec \vartheta_S + (\omega h_1/\alpha_1) \sec \vartheta - \pi$, and setting this equal to $2n\pi$ gives

$$\frac{\omega h_1}{\beta_1}\left(1 - \frac{\beta_1{}^2}{\alpha_1{}^2}\sin^2\vartheta\right)^{-\frac{1}{2}} + \frac{\omega h_1}{\alpha_1}\sec\vartheta = (2n+1)\pi \qquad n = 0, 1, 2, \ldots$$

$$(3\text{-}26)$$

Thus putting $m = 1$ and $n = 0$ in (3-25) and (3-26) leads to an equation which must be solved numerically if we wish to find the Bragg angle of the second kind of order zero. There are, of course, no reflections of the second kind at normal incidence.

In practice, the signals used in seismic exploration have a transient nature, being either very short wave trains or pulses. They generally contain a rather wide range of frequencies, although in fact the scattering and absorptive properties of rocks generally limit the bulk of the seismic energy to a small band width extending from about 1 to 200 cps. If we imagine a plane pulse to be incident on a periodically layered half-space at a given angle, it is clear that certain frequencies will be selectively reflected. If, in addition, the angle of incidence is a Bragg angle, some of these frequencies will be particularly strongly reinforced. The spectrum of the returning wave train becomes highly modified as a result.

Rock strata, of course, never have a perfectly periodic structure, although interbedding is common. However, it is usually sufficient that a number of velocity reversals should occur within a relatively short depth range in order that partial reinforcement of the returning waves take place. The modifications in the pulse form brought about by this process will stamp upon the reflections a frequency "character" that is sometimes highly distinctive. When the stratification sequence is preserved over wide areas, this property often proves to be of considerable value in identifying and correlating reflecting zones.

References

1. C. G. Knott, On the Reflexion and Refraction of Elastic Waves, with Seismological Applications, *Phil. Mag.*, 5th ser., vol. 48, pp. 64–97, 1899.

2. K. Zoeppritz, Über Erdbebenwellen VII B. Über Reflexion und Durchgang Seismischer Wellen durch Ünstetigkerlsfläschen, Nachr. der Königlichen Gesell. Wiss., Gottingen, math.-phys. Kl., p. 66–84, 1919.

3. J. B. Macelwane, "Introduction to Theoretical Seismology," Part I, Geodynamics, Wiley, New York, 1936.

4. C. F. Richter, "Elementary Seismology," Freeman, San Francisco, 1958.

5. K. McCamy, R. Meyer, and T. J. Smith, Generally Applicable Solutions of Zoeppritz' Amplitude Equations, *Seism. Soc. America Bull.*, vol. 52, pp. 923–956, 1962.

6. O. Koefoed, Reflection and Refraction Coefficients for Plane Longitudinal Incident Waves, *Geophys. Prosp.*, vol. 10, no. 3, pp. 304–351, 1962.

7. J. K. Costain, K. L. Cook, and S. T. Algermissen, Amplitude, Energy, and Phase Angles of Plane SV Waves and Their Application to Earth Crustal Studies, *Seism. Soc. America Bull.*, vol. 53, pp. 1,039–1,074, 1963.

8. P. A. Heelan, On the Theory of Head Waves, *Geophysics*, vol. 18, pp. 871–893, 1953.

9. K. Ergin, Energy Ratio of Seismic Waves Reflected and Refracted at a Rock-Water Boundary, *Seism. Soc. America Bull.*, vol. 42, pp. 349–372, 1952.

10. Lord Rayleigh, On Waves Propagated along the Plane Surface of an Elastic Solid, *London Math. Soc. Proc.*, vol. 17, pp. 4–11, 1885.

11. L. Knopoff, On Rayleigh Wave Velocities, *Seism. Soc. America Bull.*, vol. 42, pp. 307–308, 1952.

12. Y. Satô, Rayleigh Waves Projected along the Plane Surface of a Horizontally Isotropic and Vertically Aeolotropic Elastic Body, *Seism. Soc. America Bull.*, vol. 28, pp. 23–30, 1950.

13. H. Lamb, On the Propagation of Tremors over the Surface of an Elastic Solid, *Royal Soc. Philos. Trans.*, ser. *A*, vol. 203, pp. 1–42, 1904.

14. L. M. Brekhovskikh, "Waves in Layered Media," Academic, New York, 1960.

15. R. Stoneley, Elastic Waves at the Surface of Separation of Two Solids, *Royal Soc. London Proc.*, ser. *A*, vol. 106, pp. 416–428, 1924.

16. J. C. Scholte, The Range of Existence of Rayleigh and Stoneley Waves, *Royal Astron. Soc. Monthly Notices Geophys. Supp.*, vol. 5, pp. 120–126, 1947.

17. R. Yamaguchi and Y. Satô, The Stoneley Wave—its Velocity, Orbit and the Distribution of Amplitude, *Tokyo Univ. Earthquake Research Inst. Bull.*, vol. 33, pp. 549–560, 1955.

18. L. Cagniard, "Reflection and Refraction of Progressive Seismic Waves," Gauthier-Villars, Paris, 1939 (transl. by E. A. Flinn and C. H. Dix, McGraw-Hill, New York, 1962).

19. J. B. Keller, Geometrical Theory of Diffraction, *Optic. Soc. America, Jour.*, vol. 52, pp. 116–130, 1961.

20. A. J. W. Sommerfeld, "Optics," Academic, New York, 1954.

21. J. E. Burke and J. B. Keller, Research Rept. EDL-E48, Electronic Defense Laboratories, Sylvania Electronic Systems, Mountain View, California, 1960.

22. L. Brillouin, "Wave Propagation and Group Velocity," Academic, New York, 1960.

23. C. L. Pekeris, Theory of Propagation of Explosive Sound in Shallow Water in M. Ewing and J. L. Worzel (eds.), "Propagation of Sound in the Ocean," Geol. Soc. America Memoir 27, 1948.

24. M. Ewing, W. Jardetzky, and F. Press, "Elastic Waves in Layered Media," McGraw-Hill, New York, 1957.

25. R. Stoneley, The Continental Layers of Europe, *Seism. Soc. America Bull.*, vol. 38, pp. 263–274, 1948.

26. I. Tolstoy and E. Usdin, Dispersive Properties of Stratified Elastic and Liquid Media: A Ray Theory, *Geophysics*, vol. 18, pp. 844–870, 1953.

27. W. Thomson, Transmission of Elastic Waves through a Stratified Medium, *Jour. Appl. Physics*, vol. 21, pp. 89–93, 1950.

28. N. A. Haskell, The Dispersion of Surface Waves in Multilayered Media, *Seism. Soc. America Bull.*, vol. 43, pp. 17–34, 1953.

In the two previous chapters the simpler parts of the theory of elastic-wave propagation in ideal solids have been discussed.   In this chapter we shall endeavor to show how the phenomena predicted by theory are observed in actual field experiments.   A single field record or *seismogram* may contain much information about subsurface conditions, but before it can be interpreted, the events which appear on the record must be clearly identified. Only then can measurements of the times, frequencies, and amplitudes of these events be used in interpreting the subsurface structure.   In this chapter we shall be concerned mainly with the recognition of events on seismograms.   The examples which follow will illustrate many of the phenomena which commonly occur in practice.

4-1   Seismograms

In studying a seismogram, it is necessary to take into careful consideration the method by which it is made.   The common practice in seismic surveying is to arrange the seismometers along one or more lines radiating from the shot point. Usually the seismic waves are created by the detonation of an explosive a few tens of feet below the surface of the ground. Each trace of the seismogram displays the output of a single seismometer or of a cluster of seismometers. The oscillatory signals which appear on the traces contain a record of the ground motion at the appropriate seismometer locations, although it is not necessarily a very simple record. A good vertical seismometer has an output which is more or less proportional to the vertical velocity of the ground motion for those Fourier components of the motion which have frequencies higher than the resonant frequency of the seismometer. Additional filtering is often introduced in the electronic system between the seismometer and the recorder which will modify this response. However, the instrumentation is usually arranged so that a reasonably true picture of the vertical velocity is obtained of those motions whose frequency components lie within a chosen band. Within this band the only important differences between the actual velocity and the form of the record will be in certain phase shifts or small time delays. Motions which have frequency components that lie outside of this band will not be

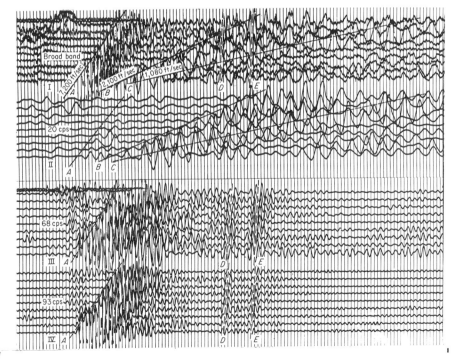

Fig. 4-1 A broadband seismic recording replayed through three different filters. The filters were centered about the frequencies noted on the records. [After Jakosky and Jakosky (1).]

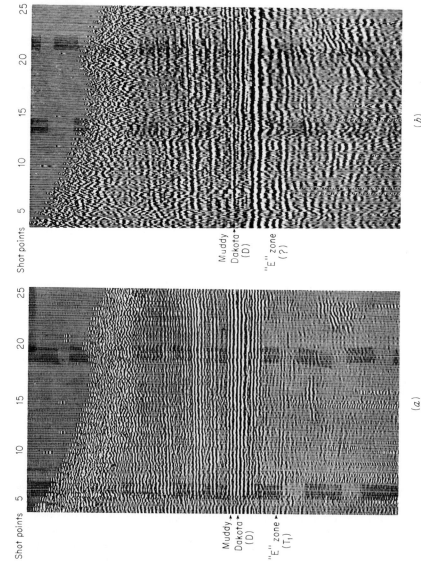

Shot points

Muddy
Dakota
(D)

"E" zone
(T_f)

(a)

Shot points

Muddy
Dakota
(D)

"E" zone
(?)

(b)

Fig. 4-2 Composite record sections presented in variable area form. Both are prepared from the same seismic recordings, but a was played back through a 40 to 134 cps very sharp cutoff filter and with considerable mixing, while b was played back through a 13 to 134 cps filter and had no mixing. [Sengbush (13)].

reproduced accurately, if at all. This must be kept in mind when actual seismograms are compared with theoretical calculations.

A study of the effects of filtering has been made by Jakosky and Jakosky (1), and one of their examples is shown in Fig. 4-1. Additional modification of the traces is frequently introduced either by grouping the seismometers or by mixing their outputs, so that an average of the ground motion over a certain region is recorded on the seismogram traces. The effects produced by these field procedures will depend, not only on the frequency composition of the ground motion, but also on the direction of propagation and apparent surface velocity of the waves.

Seismograms are not always presented in the form of "wiggly-trace" records. The use of recording systems in which the signal modulates the intensity or area of exposure along a thin strip of the recording film today is more common. These techniques are most frequently employed when large numbers of traces must be compiled into a single composite seismogram in order to facilitate the correlation of events from one trace to another. A pair of examples is shown in Fig. 4-2. These examples demonstrate how the effects of filtering and mixing reveal themselves with this type of record presentation and at the same time indicate how different events on a record may be emphasized by altering the recording techniques.

If we allow for the effects of instrumentation on the seismogram, we may compare the observed motions with predictions drawn from elastic-wave theory. However, we must keep in mind that the earth is very much more complicated even than the most elaborate model which can be handled theoretically. Moreover, the initial disturbance produced by a dynamite explosion is not, in general, the very simple pulse that we usually suppose it to be. We have seen in Sec. 2-10 that even if the explosion itself produces a simple pressure pulse, a decaying oscillation propagates through the medium.

4-2 Surface Waves

The strongest wave produced by a compressional source acting at or near the surface of a uniform elastic medium is the Rayleigh wave. This was shown originally by Lamb (2), but because of its mathematical complexity we did not discuss Lamb's analysis in detail in Chap. 3. However, we did show that a Rayleigh wave is propagated along a free surface with a wave velocity slightly less than the shear-wave velocity of the material. We should therefore expect that a pulse traveling with the Rayleigh-wave velocity will be a prominent feature on any seismic record unless the instrumentation has been designed specifically to reject it. We also found that if the elastic properties of the medium vary with depth, the Rayleigh wave is dispersive. Usually there are rapid changes in the elastic properties with depth near the ground surface owing to the presence of soil and the weathering of the surface bedrock; thus we should expect to find some dispersion in

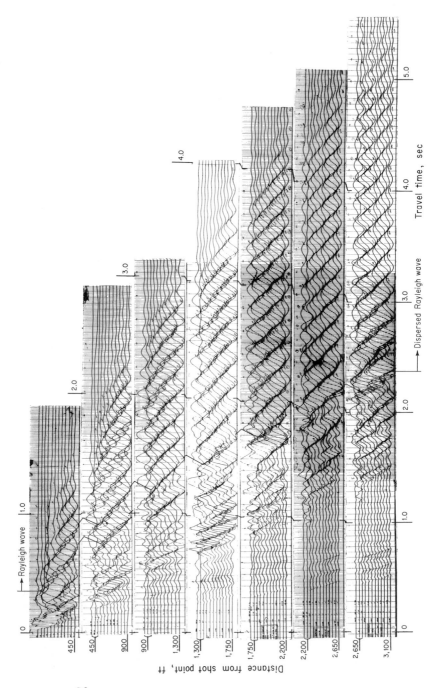

Rayleigh wave

1.0   2.0   3.0   4.0   5.0

Travel time, sec

→ Dispersed Rayleigh wave

Distance from shot point, ft

450
450 900
900 1,300
1,300 1,750
1,750 2,200
2,200 2,650
2,650 3,100

the observed surface wave.   In fact, we usually do.   Figure 4-3 shows a set of records obtained by Dobrin, Simon, and Lawrence (3).   They were recorded especially for the study of surface waves so that a broadband apparatus (5 to 200 cps) was employed, and horizontal seismometers as well as the usual vertical type were used.

By examination of the traces of the radial and vertical seismometers, the Rayleigh wave can be positively identified by its retrograde elliptical motion.   If we note that both upward vertical motion and outward radial motion will produce a downward deflection of the seismogram trace, we should expect oscillations on the radial trace to lead those on the vertical trace by 90° in phase.   This is seen to be approximately true for those events marked as Rayleigh waves on the record.   The earlier part of the records are caused by direct, reflected, and refracted body waves and do not show this phase shift.

The trace of the transverse seismometer is particularly noteworthy. If a purely compressional disturbance had been generated at the source and if the ground were perfectly horizontally stratified, no *SH* motion should have occurred; but the contrary appears to have happened, since strong deflections did take place.   The fact that the amplitude of this motion varies irregularly from record to record suggests that it may be caused by inhomogeneous conditions close to the shot.   The sinusoidal wave trains which arrive at about the same time as the dispersed Rayleigh waves are very probably Love waves.

To analyze the dispersion characteristics of the Rayleigh wave, we must measure the group or phase velocities of several of its various frequency components.   The group velocity can be measured by assuming that the wave originates as a sharp pulse at the instant of the shot and at the surface directly above the shot point.   On this assumption the travel time t for a particular cycle of period T observed at a horizontal distance x from the shot gives for the group velocity

$$C(T) = \frac{x}{t} \tag{4-1}$$

In fact, the assumption just stated has been shown by Lamb to be perfectly accurate if the medium is homogeneous and if there is no dispersion.   Equa-

Fig. 4-3   A composite seismogram showing the waves recorded by geophones on a line radiating 3,100 ft from the shot point.   The first three traces of each record show respectively the output of horizontal radial, horizontal transverse, and vertical motion geophones situated at a single point; the remaining traces show the output of vertical-motion geophones situated at 50-ft intervals along the line, the final trace of each record indicating motion at the same position as the first three traces of the next record.   Separate explosions were used to obtain each record of the composite.   [After Dobrin, Simon, and Lawrence (3).]

tion (4-1) is not strictly valid if the ground is inhomogeneous; but if there is horizontal homogeneity and if the distance between the shot and the seismometer is greater than the longest wavelengths recorded by the seismograph, then Eq. (4-1) will be sufficiently accurate.   Figure 4-4 shows values of the group velocity found by this method at various periods which were readily identifiable on the records.

In theory, the phase velocity can be found by numbering the peaks and troughs of the oscillations and by following them from one trace to another. Assuming that the period of a given cycle in the sequence does not change drastically from one trace to the next, we may write, approximately,

$$c(T) = \frac{\Delta x}{\Delta t} \tag{4-2}$$

where Δx is the distance between adjacent seismometers and Δt is the time that it apparently takes for the first, second, third, etc., cycle to travel this

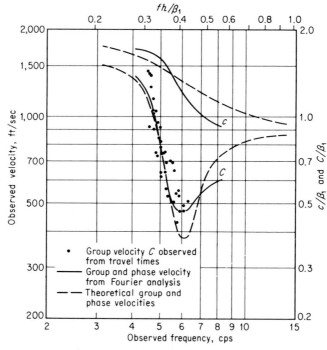

Fig. 4-4   Rayleigh-wave dispersion observed (3) on the seismogram of Fig. 4.3 compared with a theoretical dispersion curve for a single surface layer, $\beta_2/\beta_1 = 2.236$**,** $\alpha_1/\beta_1 = \alpha_2/\beta_2 = 1.732$**,** $\rho_1/\rho_2 = 1$**.   The best fit is obtained for a layer of thickness 68 ft and a shear velocity** $\beta_1 = 1,000$ **ft/sec.**

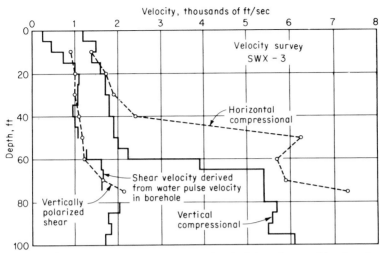

Fig. 4-5 Compressional and shear interval velocities observed in the shot hole. [After Dobrin et al. (3).]

distance. Unfortunately this is seldom an accurate enough method, and it is generally necessary to make a Fourier analysis of the pair of traces and measure Δt from the phase shift of the Fourier element whose period is T. Figure 4-4 also shows the results of such an operation performed on the two traces recorded at distances of 800 and 1,200 ft from the shot point. The group velocity can be calculated from the phase-velocity measurements by using the relation $C(T) = c(T) + T[dc(T)/dT]$, and these calculations are also shown. They compare well with the values obtained by direct measurement.

The actual stratification of the weathered layer was investigated by Dobrin, Simon, and Lawrence when the records were made. Figure 4-5 shows the variation in both P- and S-wave velocities down to a depth of 100 ft. Unfortunately the surface-wave dispersion curve for a structure as complicated as this cannot be calculated easily without the use of a computer programmed to make the analysis described in Sec. 3-10. However, the theoretical dispersion curve for a simple two-layer model which approximates the given situation is included in Fig. 4-4 for comparison. There is at least a qualitative agreement.

As a matter of further general interest, Dobrin et al. observed that the nodal plane of the Rayleigh motion lay at a depth of 0.136 wavelength at all the predominating frequencies. According to Fig. 3-10 the depth in a homogeneous solid ought to lie between 0.138 and 0.256 wavelength, depending upon the elastic properties. At the same time, the exponential decrease of the surface amplitude of the Rayleigh motion with the depth of burial of the source was also verified.

4-3   On the Use of Dispersion Curves

The dispersion measurements just described are frequently used as a basis for interpreting the earth's crustal thickness from the analysis of earthquake records. The process can best be illustrated by an example. The data shown in Fig. 4-6 are obtained from records of horizontal *SH* ground motion following an earthquake observed at a number of seismograph stations situated at various distances from the focus [see Ewing, Jardetzky, and Press (4), p. 212]. The records are divided into short time intervals, each one of which contains approximately one cycle and defines a *frequency group*. The mean frequency of the group is determined by measuring the period of the cycle. The group velocities are then obtained just as described in Sec. 4-2, by dividing the horizontal distance from the earthquake epicenter by the travel times of the various groups. The object then is to find a theoretical Love-wave dispersion curve which provides a match to these observational data. The model from which the theoretical curve is calculated may then be used to estimate the thickness and rigidity of the "waveguide."

Nowadays, the matching of dispersion curves is frequently carried out on a computer. However, automatic methods are generally effective only after approximate estimates of the model parameters (thicknesses, rigidities, and densities of the various layers) have been made. Here we shall concern ourselves only with the preliminary phase of the work, which will nevertheless demonstrate most of its essential features.

We shall consider only the single-layer model. According to Eq. (3-18), the dispersion is determined by five physical quantities, viz., the layer thickness h, the rigidities μ_1 and μ_2, and the rotational-wave velocities β_1 and β_2. Since only the ratio of the rigidities appears in the formula, there are actually four parameters to be chosen, viz., h, β_1, β_2, and μ_2/μ_1. The last can be converted into the ratio of the two densities ρ_2/ρ_1 by multiplying by β_1^2/β_2^2, since densities as a rule have a more direct geophysical significance than rigidities. If we now scale the group velocity in terms of β_1 and the frequency in terms of β_1/h, there remain just two "free" parameters, viz., the ratios β_2/β_1 and ρ_2/ρ_1. Thus for each value of β_2/β_1 there will be a family of theoretical dispersion curves (C/β_1 versus fh/β_1) corresponding to the allowable range of values of ρ_2/ρ_1. Each family has a common low-frequency asymptote ($C = \beta_2$) and a common high-frequency asymptote ($C = \beta_1$). Between these two limits the details of the curves will differ somewhat with the values of ρ_2/ρ_1.

The use of theoretical dispersion curves in interpreting a set of observational data is illustrated in Fig. 4-6. As the result of a systematic search through the complete portfolio of single-layer dispersion curves, the closest match to the measured group-velocity curve that could be found was provided by the set shown in the figure. The "fitting" process consists in plotting the observed dispersion data (C versus f) on logarithmic paper *of the*

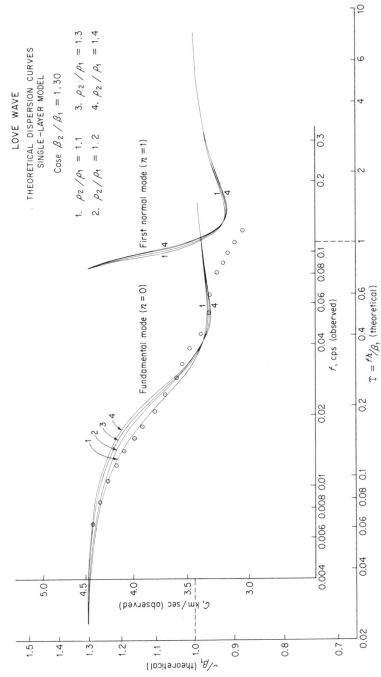

Fig. 4-6 Theoretical Love-wave dispersion curves fitted to data obtained from studies of earthquake seismic records.

same size as that on which the theoretical curves are drawn, and in moving this sheet horizontally and vertically over the theoretical curves until a satisfactory fit is found. Vertical displacements are equivalent to varying β_1, while horizontal displacements involve changes in β_1/h. The final position of the observational curve fixes both β_1 and β_1/h, and hence also h.

To illustrate, the best fit to the dispersion data shown on Fig. 4-6 appears to be given by the curve whose parameters are $\beta_2/\beta_1 = 1.3$ and $\rho_2/\rho_1 = 1.1$. At the low-frequency limit, the group velocity appears to be 4.49 km/sec, so that $\beta_1 = 3.45$ km/sec. Since the frequency at which $\Upsilon = 1$ appears to be 0.111 cps, it follows that

$$h = \frac{\beta_1}{(f)_{\Upsilon=1}} = 31 \text{ km}$$

This value of h suggests that the Love waves probably have been guided by the ground surface and the base of the earth's crust and that the entire crustal column has therefore participated in the motion. The mean crustal thickness across the continent, therefore, as interpreted from the single-layer model, turns out to be about 31 km. The average velocity of rotational waves in the upper mantle appears to be about 4.50 km/sec, and in the crust it is about 3.45 km/sec. The latter value agrees rather well with other investigations.[1] Furthermore, if we accept a mean density of 2.8 g/cm³ for the crust, a density of about 3.2 g/cm³ is implied for the upper mantle.

Notice that there is a shift in the main carrier from the fundamental mode toward the first normal mode as we move to higher frequencies. The reason is that the first normal mode has a minimum group velocity at these frequencies and near the minimum waves will travel more coherently in the first mode than in the fundamental. On the whole, however, the fit between the observational and theoretical dispersion data is not as close as we might wish, which suggests that the single-layer model is inadequate in some respects. An improvement might very possibly be brought about by introducing additional layers or else some special forms of heterogeneity into the model. This leads into somewhat specialized techniques of surface-wave dispersion analysis, and the calculations will almost certainly have to be done on a computing machine.

4-4   *Near-surface Transmission over Long Distances*

Next we shall study the earlier arrivals on the records shown in Fig. 4-3. These are enlarged and reproduced in Fig. 4-7. In Sec. 3-5 we described how a head wave may overtake the direct wave if the medium in which the source

---

[1] From the study of a large number of near earthquakes in California, Gutenberg gives values of 3.26 and 3.65 km/sec for rotational wave velocities in the two regions into which he divides the crust. From data taken in New England, Leet gives values of 3.45 and 3.93 km/sec.

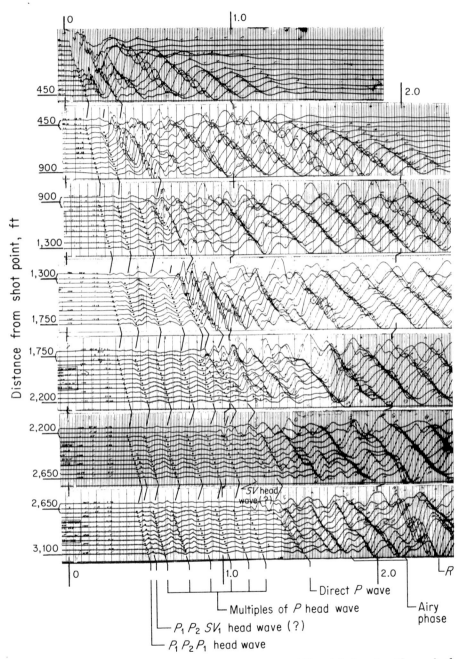

Fig. 4-7 An enlargement of a part of the records of Fig. 4-3 showing the arrival of various body waves.

is located is underlain by a material having a higher wave velocity.   Insofar as the situation under study fulfils this condition, we predict that the $P_1P_2P_1$ head wave will have a surface velocity of about 6,000 ft/sec, while the direct P_1 wave will have a velocity of about 1,900 ft/sec.   Both of these arrivals are clearly visible on the records.   If we assume the shear-wave velocity to be 1,000 ft/sec in the upper medium and 2,100 ft/sec in the lower, we predict the existence of three additional head waves, viz., the $P_1P_2SV_1$, $P_1SV_2P_1$, and $P_1SV_2SV_1$ waves.   The first should have the same surface velocity as the $P_1P_2P_1$ head wave and should follow it by about 0.03 sec.   The other two should have surface velocities of 2,000 ft/sec and should arrive only slightly ahead of the direct P_1 wave.   These waves can be identified only tentatively on the records and would therefore not be useful in the interpretation.   It is noteworthy that, although theoretically we would expect no SH waves, a definite first arrival which is probably the SH head wave is visible on the transverse seismometer trace on several of the records.   This further corroborates the identification of the SV_2 head waves.   The reason why these waves did not persist at distances beyond 2,500 ft is unknown, but may be due to lateral velocity changes in the two media.

The waves that we have discussed so far by no means account for all of the events to be found on the seismogram in front of the dispersed train of Rayleigh waves.   The most prominent feature of the early part of the records is the very regular wave train which follows the arrival of the first head wave.   This has a fundamental frequency of about 8 cps.   Another noticeable feature is that after the arrival of the direct P_1 wave, the frequency of the wave train drops sharply.   These two phenomena can be understood by considering multiple internal reflections of the $P_1P_2P_1$ head wave and of the P_1 wave in the surface layer.   From Fig. 4-8 it is easy to see why the P_1 wave train is composed of higher frequencies than the $P_1P_2P_1$ head wave.   This consideration of multiple reflections is qualitatively similar to the propagation of normal modes in a fluid layer, which was described in Sec. 3-8.   Indeed, there is such a close analogy between them that the

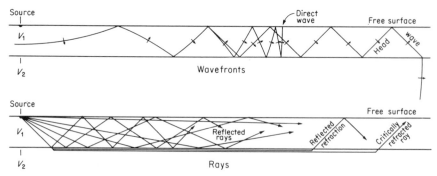

Fig. 4-8　Ray-path and wavefront diagrams showing some possible modes of interference.

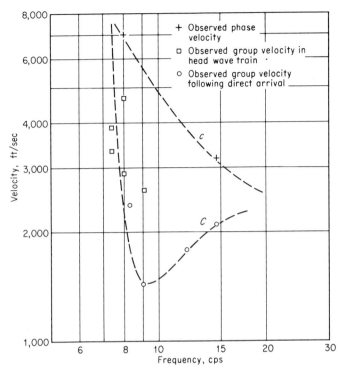

Fig. 4-9 Observed dispersion of the body waves on the seismogram of Fig. 4-7.

observed dispersion curve can be matched rather well to theoretical curves for the fundamental mode in a fluid layer overlying a fluid substratum [Officer (5)]. The data are shown in Figs. 4-9 and 4-10.

It is important to realize, however, that in a solid layer this multiply reflected wave train is not in fact a normal mode of propagation. Strictly speaking, a normal mode is an unattenuated horizontal wave, and this implies no downward transmission of energy. It is obvious that the head wave train will generate SV motion in the lower medium even though the critical angle for P transmission has been exceeded (Sec. 3-4). Moreover the theory predicts that the velocity of propagation in all the normal modes must always be less than β_2, a limit which is certainly exceeded in this case. Figure 4-11 shows the theoretical dispersion of the "Rayleigh" and of the first "SV" modes for a case roughly similar to the one we are studying. Unfortunately the β_2 value that had been used in the calculations is considerably higher than that which was found by measurement (Fig. 4-5), but since only a very few theoretical curves have been published, a better value was not available. Nonetheless it is obvious that the first SV mode propagates with velocities comparable to those of the Rayleigh mode, and it is therefore unlikely that one could separate it from the strong Rayleigh wave train except by careful Fourier analysis. At great distances from the source

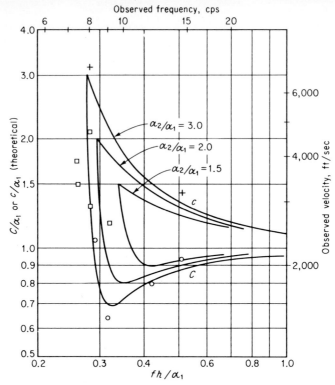

Fig. 4-10 Body-wave dispersion from Fig. 4-9, compared with first-mode dispersion curves for a fluid layer over a fluid half-space. The curve for $\alpha_2/\alpha_1 = 3.0$ fits very well for $\alpha_1 = 2,220$ ft/sec and $h = 78$ ft.

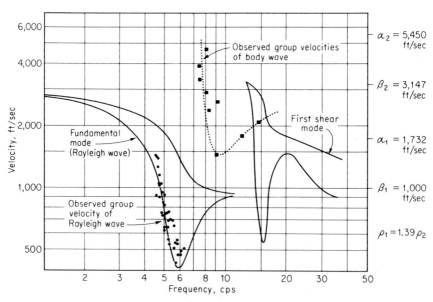

Fig. 4-11 Comparison of the observed dispersion (from Fig. 4-9) with the theoretical normal modes of a solid layer over a solid half-space. The Rayleigh wave is seen to be a true mode, while the earlier-arriving body wave is not. It is a leaky mode which disappears at large distances.

106

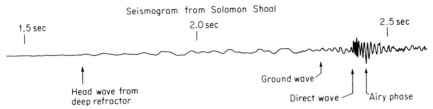

Fig. 4-12   A seismogram obtained at a range of 11,900 ft in 53 ft of water.   The ground wave and water wave approximately fit a dispersion curve for two liquids of velocity ratio 1.15.   The initiation of the ground wave is obscured by waves from deeper horizons.   [After Ewing, Worzel, Pekeris (6).]

we would expect the dispersive guided waves to become the predominant feature of the seismogram.

A simpler example of normal mode propagation is shown in Fig. 4-12. This is an example taken from experiments performed in the shallow ocean. Here the distance from the shot to the receiver is so large in comparison with the depth of the water that only the normal mode transmission is observed. The energy in the wave train is well dispersed over all wavelengths, and the entire dispersion curve for the fundamental mode can be deduced from it. Since the bottom is formed of unconsolidated material, it behaves very much like a liquid.   The observed and theoretical dispersion curves are shown in Fig. 4-13.

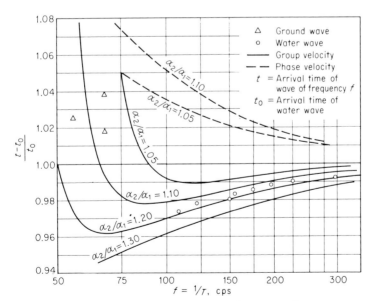

Fig. 4-13   Observed dispersion (6) of the ground wave and water wave of Fig. 4-12 fitted to liquid-media dispersion curves.   The theoretical curves are calculated for a velocity and depth of the layer of 4,925 ft/sec and 53 ft, respectively.

4-5 *Critical Refraction and Reflection*

Considerable amounts of seismic energy can penetrate into the ground to depths much greater than the separation between the shot and the detector, and some of it will be returned to the surface by deep-seated discontinuities in elastic properties. If a low-velocity layer exists at the surface, it is usually necessary to place the shot below this layer in order to ensure good penetration of the seismic energy into the substrata. Arrivals will then be produced on the seismogram by reflection and by critical refraction from the deeper horizons.

A remarkable set of seismic records on which a number of coherent arrivals can very easily be seen is shown in Fig. 4-14. These were obtained by T. C. Richards (7) in the Rocky Mountain Foothills of western Canada.

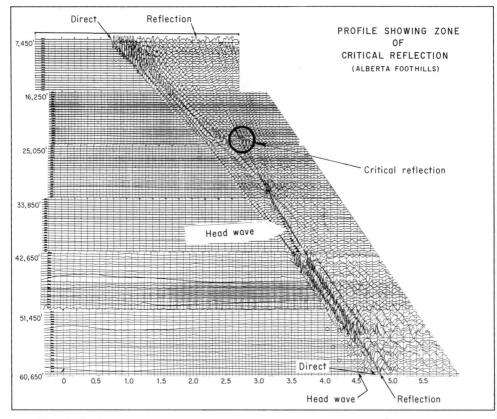

Fig. 4-14 A composite seismogram showing the output of geophones located at distances between 7,450 and 60,650 ft from shot point 1601. The seismometers were spaced at 400-ft intervals. Shot sizes for the individual records were 30, 50, 100, 150, and 175 lb, reading from the top. Amplifier gains were also increased by 6 db for each. [After T. C. Richards (7).]

The particular locality in which these records were made is comparatively simple geologically, and the seismogram is therefore easy to interpret. The profile was taken in the direction of the regional strike across a section of comparatively uniform rocks having a *P*-wave velocity of about 13,500 ft/sec. Underlying these at a depth of about 10,000 ft is a limestone formation roughly 1,000 ft thick having a *P*-wave velocity of 21,000 ft/sec. The equipment used should give a fairly accurate reproduction of ground velocity at frequencies above about 3 cps, since 2-cps seismometers and broadband amplifiers without automatic gain control were used. The various sections of the seismogram were obtained separately using increased dynamite charges and higher amplifier gain at the greater distances.

The main events visible on the records are (1) the direct wave traveling near the surface, (2) the head wave from the limestone, (3) the reflection from the limestone, and (4) weaker reflections from deeper horizons. The arrival times of these events are plotted on the time-distance graph of

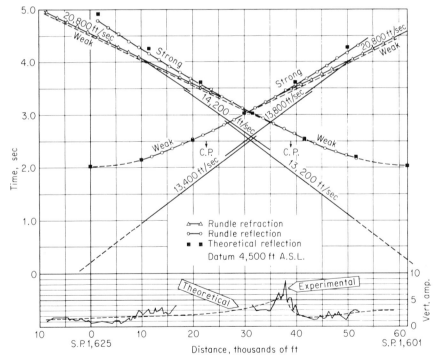

Fig. 4-15 **Time-distance plot of the arrivals seen on the composite seismogram of Fig. 4-14 (SP 1601) and on a similar seismogram obtained by shooting the line in the opposite direction from shot point 1625. A plot of the observed and theoretical amplitude of the reflected wave from the Rundle limestone (from SP 1601) is also shown. [After T. C. Richards (7).]**

Fig. 4-15 along with data from the same profile shot in the reverse direction. The fact that the two sets of curves are almost identical shows that the limestone surface has negligible dip in the direction of the seismometer spread. Examining first the "direct" wave, we find that the arrival times are only roughly predicted by a straight line through the origin having an inverse slope of 13,500 ft/sec, which is about what we would expect if the upper layer were perfectly homogeneous. The intercept of the observed time-distance curve is actually about 0.2 sec. This can be explained by accepting the not improbable hypothesis that the ground has a somewhat lower velocity near the surface. If this low-velocity zone is a distinct layer, then what we have called the direct wave is in fact a head wave. It is more difficult to account for the enechelon structure of the time-distance curve. One possible explanation is that the overburden above the limestone is made up of several strata each having a slightly different velocity.

At a distance of about 50,000 ft from the shot, a new wave becomes the first arrival. This is the head wave from the limestone. It is visible on the records even before it overtakes the direct wave. From Fig. 4-16 and

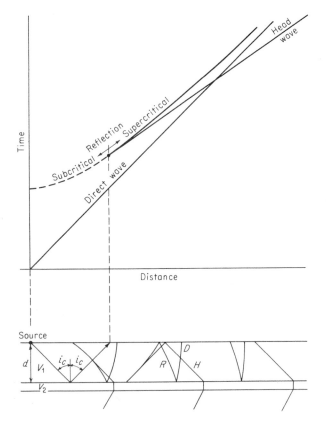

Fig. 4-16 The form of the *P* wavefront at three different instants after its reflection from the limestone, assuming the upper medium to be homogeneous and isotropic.

according to Sec. 3-5, we note that the head wave should be observed only beyond the point which corresponds to the arrival of the critically reflected ray, from which point the head wave and the reflected wave begin to separate. The distance and arrival time of the "critical" reflection can be predicted rather simply from the plane-wave theory using ray-path diagrams. The relationship between the arrival time of the reflection and the distance of the seismometer from the shot point should be hyperbolic, and the actual reflection times seem to follow this rather well. The critical angle of incidence is given by

$$i_c = \sin^{-1}\frac{V_1}{V_2} = \sin^{-1}\frac{13,400}{21,000} = 40°$$

and thus

$$x_c = 2d \tan 40° = 22,700 \text{ ft}$$

while

$$t_c = \frac{x_c}{V \sin 40°} = 2.65 \text{ sec}$$

The time-distance curves for the head wave and for the reflection from the limestone should separate from this point outward, the apparent surface velocity of the head wave remaining constant while that of the reflected wave diminishes steadily until it approaches asymptotically the P-wave velocity in the upper layer.

The head wave from the limestone actually does show a surface velocity of 20,800 ft/sec. On the other hand its time-distance line is not perfectly tangential to the reflection hyperbola but seems to lie somewhat above its theoretical position. This is probably due to the great difficulty in observing

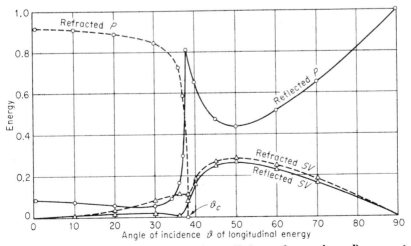

Fig. 4-17 Reflection and transmission coefficients for a plane P wave at an interface where $\alpha_1 = 12,800$ ft/sec, $\rho_1 = 2.40$ g/cm,^{3} $\alpha_2 = 21,000$ ft/sec, $\rho_2 = 2.65$ g/cm^3, $\alpha_1/\beta_1 = \alpha_2/\beta_2 = 7/4$. Amplitude ratios are the square root of the energy ratios. [After T. C. Richards (7).]

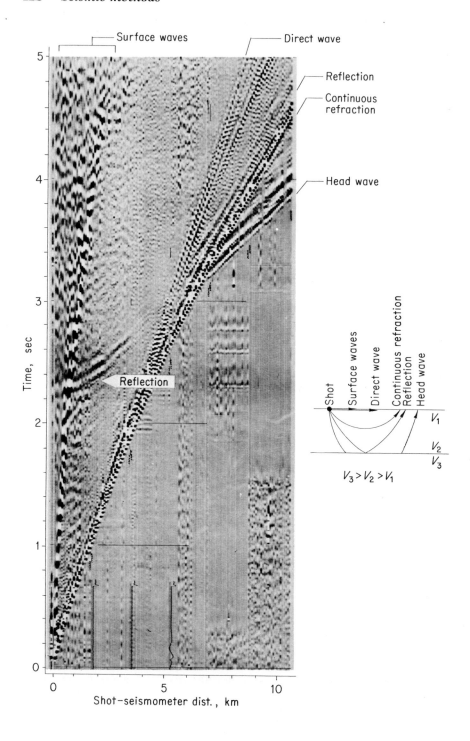

the exact onset of the head wave, which even in theory has a wide envelope with low initial amplitudes.[1]

Since the *P*-wave velocities in the limestone and the sediments above it are known and since the densities and the *S*-wave velocities can be estimated, it is possible to compare the theoretical amplitude of the reflected wave with what is actually observed. Assuming that the incident wavefront at the limestone surface is essentially a plane *P* wave, the transmission and reflection coefficients for the *P* and *SV* waves produced at this interface can be calculated for various angles of incidence from the formulas (3-4). The results of these calculations are shown in Fig. 4-17. The critical angle for *P-P* transmission is 38°, but since $\alpha_1 > \beta_2$, there is no critical angle for *P-SV* transmission and therefore no region of total reflection. To calculate the vertical displacement at the ground surface, the effects of the free boundary and the spreading of the wavefront must both be taken into account. Tables for the ground displacement due to incident plane waves have been computed by Jeffreys (8) and Knopoff et al. (9). A graph showing the vertical displacement to be expected from the reflected *P* wave is included in Fig. 4-15 along with the amplitudes actually recorded and corrected for the shot charge size and for amplifier gain. The agreement is remarkably good in spite of all the assumptions involved. The irregularities observed in these amplitudes are likely to have been caused by some heterogeneity in the ground close to the seismometers. In one region the reflected wave was almost totally obscured by interference from other arrivals, but elsewhere it was relatively well isolated. The effect of the rapid drop in transmitted *P*-wave energy as the critical angle is reached can be seen in the disappearance of reflections from deeper interfaces with increasing angles of incidence.

Figure 4-18 shows a second example of a composite seismogram on which a prominent head wave is displayed. The records are in variable density format this time. The more prominent events have been identified in the figure.

4-6 *Vertical Reflections and Synthetic Seismograms*

Several reflections from deep strata were apparent on the seismogram shown in Fig. 4-14 and were even discernable on the traces from the seismometers

---

[1] In his interpretation of these records, Richards assumed some aeolotropy of the material overlying the limestone layer. He found that a horizontal velocity of 13,360 ft/sec and a vertical velocity of 12,380 ft/sec improved the value found for the depth of the limestone and corresponded to a coefficient of aeolotropy close to that found elsewhere in a similar geological environment.

Fig. 4-18 Composite seismogram (variable density) from a refraction survey in Bolivia. A relatively uniform sedimentary sequence overlies the crystalline basement. The velocity in the sedimentary overburden seems to increase more or less uniformly with depth. At the basement surface there is a sudden large increase. (*California Standard Oil Co.*)

nearest the shot point.   In that example the amplitudes of the surface waves and of the near-surface multiple reflections were rather weak and caused no appreciable interference.   Thus, even though the reflection coefficients at nearly normal incidence were small, other conditions were favorable to the appearance of reflections.   This is not usually the case.   If suitable precautions are taken to reduce the amplitudes of near-surface phenomena as they are recorded on the seismogram, however, reflections from deep horizons can very frequently be observed.   The use of nearly vertical raypaths to map reflecting horizons is essentially the working basis of the reflection method of seismic prospecting.   From the standpoint of procuring significant subsurface geological information, the reflection method has proved to be more fruitful than any other.

A typical record on which there are several strong reflections and very little interference or noise is shown in Fig. 4-19.   This seismogram shows traces from two groups of seismometers, one of which has been set out on a line running 1,800 ft north from the shot point and the other on a line running 1,800 ft east.   The modification of the signals by the recording instruments is very substantial, since seismometer grouping, filtering, mixing, and automatic gain control have all been employed.   However, the traces give a fairly accurate picture of the vertical ground motion produced by those waves which have arrived at the surface at nearly normal incidence and which have frequencies lying within the passband of the filters.   Trace deflections for other ground displacements are highly attenuated.

Beside the seismogram in Fig. 4-19 is shown the geological log and the short-interval P-wave velocity log of a well situated very close to the shot point.   The interval velocities have been integrated to obtain a curve of vertical travel time versus depth, and by the use of this curve both the geology and the velocity log have been plotted on a linear time scale rather than on a linear depth scale.   This makes it possible to correlate events appearing on the records with specific zones within the ground.   Even a casual examination of the velocity log shows that the elastic properties of the ground are highly complex and that the relation which the log bears to the seismogram is not at all obvious.   It is clear, however, that energy must be reflected almost continuously during the penetration of the wavefront and that the oscillations which occur on the seismogram are therefore likely to be interference phenomena.

In order to try to identify more precisely the origins of "reflections" in terms of the subsurface lithology, a "synthetic" seismogram can be constructed from the velocity log.   This is intended to be a record of the theoretical response of the actual ground to a given input signal.   To calculate the synthetic seismogram, a simplified representation of the true geological section is used.   It is supposed that the ground is made up of a large number of thin horizontal layers, each having a constant P-wave velocity whose value is taken from the corresponding point on the velocity

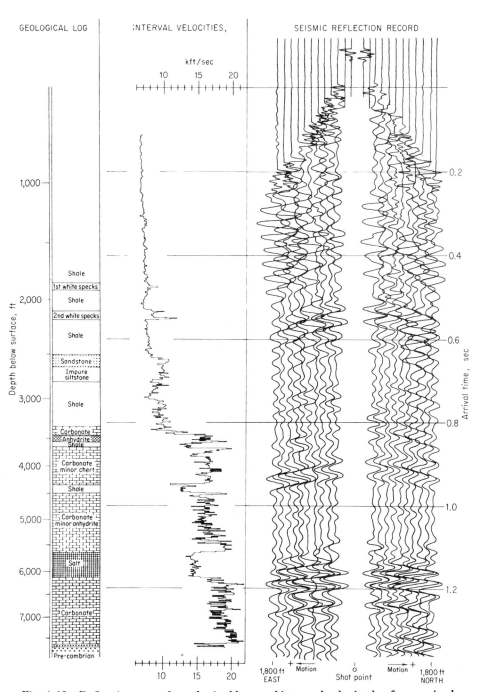

Fig. 4-19   Reflection record, geological log, and interval velocity log from a single location.   The logs are scaled to reflection times for ease of correlation.   (*Imperial Oil Ltd.*)

log. Since the densities of the strata are generally unknown, they are assumed to be the same. They vary much less than the velocity in any case. Below the bottom of the borehole in which the velocity has been measured, the ground is assumed to be uniform. We can then calculate the motion of the upper surface which would follow if a plane-wave pulse of arbitrary form were to begin traveling vertically downward from the top layer. The analysis is based entirely on plane-wave theory, although a geometrical correction for the spherical divergence of the wavefronts is sometimes included in the calculations. The effects of instrumental filtering are included in the form postulated for the incident wave.

There are many ways of obtaining the desired result for the model described above. A first approximation may be found rather readily by noting that the fraction of the incident energy reflected from any interface is given by

$$\left(\frac{\alpha_1 - \alpha_2}{\alpha_1 + \alpha_2}\right)^2 \tag{4-3}$$

when differences in density can be neglected. Since this is usually a small quantity, only single reflections are considered in this approximation and the attenuation of the primary wave by reflection losses is neglected. The response of the section to an incident unit impulse is then just a series of impulses of which the jth member is delayed by $2\Delta z_j/\alpha_j$ sec, and has an amplitude

$$\left(\frac{\alpha_j - \alpha_{j+1}}{\alpha_j + \alpha_{j+1}}\right) \tag{4-4}$$

The response $R_w(t)$ to a postulated incident wavelet $I(t)$ is then obtained by the superposition of the unit impulse response $R_i(t)$ and the incident wavelet in the convolution integral

$$R_w(t) = \int_0^\infty I(t - \lambda) R_i(\lambda) \, d\lambda \tag{4-5}$$

Calculations are greatly simplified by assuming layer thicknesses which give constant transit times, i.e., $\Delta z_j/\alpha_j = \text{const}$.

To obtain a more exact expression which takes into account all the possible reflections arising from a plane-wave input pulse is a much more difficult problem. It can be solved in several different ways. A convenient method is described in a paper by P. C. Wuenschel (10), in which he has given a number of interesting examples. It is important to note that the exact plane-wave solution, in contrast to the first approximation, is profoundly affected by the presence of the free surface at the top of the section. This is because all upcoming waves are totally reflected by it and the number of multiple reflections that have significant amplitude is thereby greatly increased.

The calculation of the synthetic seismogram is readily carried out on an electronic digital computer using the method outlined above. The impulse solution is just the solution by the Laplace transform method. The problem can also be solved by use of the Fourier transform. This is somewhat simpler analytically, but the computation takes much longer, since at the end a Fourier integral must be computed numerically instead of the simpler convolution integral of Eq. (4-5). Numerous analogue methods have also been devised to yield the solution in the first approximation, and some are also available which compute the exact plane-wave response.

Figure 4-20 shows a velocity log and a reflection seismogram taken at the same site, together with synthetic seismograms calculated with and without multiple reflections. The results demonstrate—not altogether to our surprise perhaps—that seismic energy is received in appreciable quantity at all times after the transmittal of the incident pulse and that oscillations having large amplitude are as apt to be the result of constructive interference as of prominent individual velocity contrasts. The large amount of energy being reflected at approximately 0.72 sec. does not produce noticeably large oscillations on the seismogram; yet the sequence of smaller impulses at about 0.8 sec. produces the strongest "reflection" recorded. This is because of the similarity of their separation and polarity to the shape of the input pulse.

The differences between the single-reflection and the multiple-reflection synthetic seismograms are really rather small, especially at the earlier arrival times. At later times the two begin to become appreciably different, although much of this is due to phase shifts which are not easily noticed when comparing the records visually. Both records in fact predict events on the actual seismogram very well. A particularly striking point is that the multiple synthesis continues to predict events which lie outside the time-depth range of the velocity information, demonstrating very convincingly that many of the later events on actual records may really be shallow multiple reflections.

A second example is shown in Fig. 4-21. The continuous velocity log shows much smaller velocity contrasts than that of the previous example. The amplitudes of the oscillations on the seismograms nonetheless remain rather high. The essential difference between the synthetic seismograms in this and in the previous example is in the *envelope* of the oscillations. In the first example the oscillations show rapid changes of amplitude, while in the second they vary smoothly as if the entire record had been played through a very narrow band filter. The multiple-reflection and single-reflection synthetic seismograms still appear to be very similar, although a subtraction of one from the other in fact shows an increasing divergence at later times, most of which is due to phase shift. The prediction of events appearing on the observed seismogram is not quite as convincing as that in the previous example. In this instance, however, the multiple synthesis is definitely an improvement over the single one. This is particularly evident at the times 1.15, 1.4, and 1.65 sec.

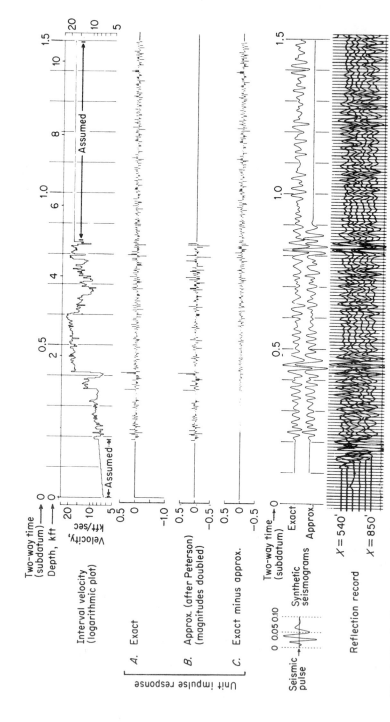

Fig. 4-20 A plane-wave synthetic seismogram computed by the exact and first-approximation methods. The resemblance to the actual seismogram is striking. Note in particular the accuracy of the prediction by the exact seismogram beyond the bottom of the velocity information. [After P. C. Wuenschel (10).]

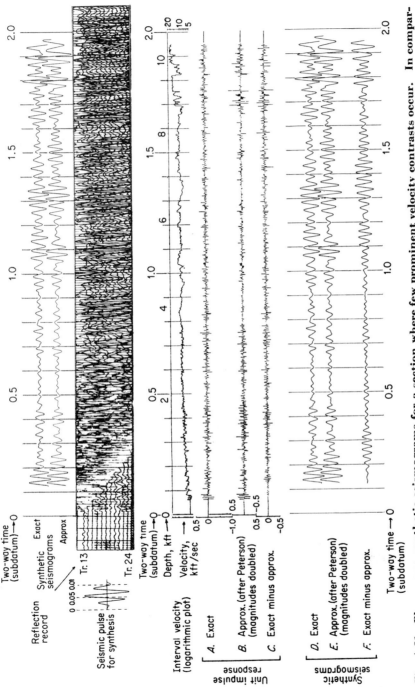

Fig. 4-21 Plane-wave synthetic seismograms for a section where few prominent velocity contrasts occur. In comparing this with Fig. 4-19, note that the scale of the velocity log is smaller. A rather oscillatory seismogram is obtained. [After P. C. Wuenschel (10).]

119

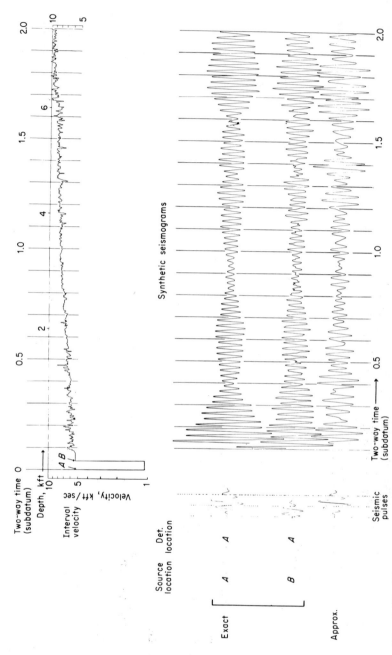

Fig. 4-22 Plane-wave synthetic seismograms when a surface low-velocity layer is included. To compensate for the inability of the computer program to allow for a changed density, a very low velocity has been assigned to the layer. The approximate seismogram does not include any multiple reflection in this layer and actually predicts the observed results the best. The exact seismograms are very similar to ringing records obtained in water-covered areas. [After P. C. Wuenschel (10).]

In both of these examples the origin of the incident wave and the detector of the reflected wave were assumed to be located on a free surface at the top of the geological section. No low-velocity surface layer was included. Allowance was made for the "ghosting" of the incident wave by the low-velocity layer in postulating the incident pulse form. Note that different pulse forms were needed in the two examples in order to predict the observed seismograms.

A third example (Fig. 4-22) shows a pair of multiple-reflection synthetic seismograms, in which the effects of the low-velocity surface layer are included, and a single-reflection synthetic seismogram, which does not include this feature. The latter is almost identical in appearance with a multiple-reflection seismogram without the weathered layer, which is not shown. Curiously enough it is this calculation which resembles the observed record the most closely. Apparently, therefore, multiple reflections of the upcoming waves within the weathered layer do not occur. Perhaps this is not surprising since the unconsolidated and heterogeneous materials in the weathered layer have a high attenuation factor. On the other hand, records of the kind predicted by the multiple-reflection synthesis are frequently obtained in water-covered areas where attenuation in the surface layer would be due only to scattering caused by the small roughness of the top and bottom surfaces.

4-7 Diffraction at Discontinuities

So far all the examples which we have examined have demonstrated wave propagation in surroundings which were approximately uniform horizontally but vertically inhomogeneous. Often, however, sharp breaks in the strata may exist, such as those at the offset produced by a fault. A wavefront incident at such a discontinuity will be diffracted by it. If the energy carried by the wave is made up predominantly of short wavelengths, the wave will tend to form distinct shadows and reflections. The longer wavelengths, however, will show no such sharp distinctions, and they will be spread out by the discontinuity. The very long wavelengths will be almost unaffected.

The diffraction of elastic body waves in solids by obstacles other than spheres or thick screens has not yet been fully studied, although approximate solutions in the wave zone have been found for some special objects such as plane wedges and cylinders. However, the theory of scalar (sound) wave diffraction and the use of Huygens' principle seem to be adequate to explain the general nature of the effects. T. Krey (11) has worked out the diffraction pattern to be observed from a perfectly rigid half-plane reflector for a pulse from a point source of a predominantly 60-m wavelength, on the basis of the Fresnel diffraction theory (which is virtually identical with that described in Sec. 3-6). The wavefront pattern which occurs has previously

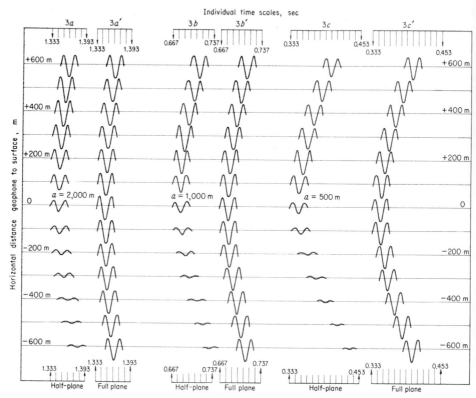

Fig. 4-23   **The reflection expected from a horizontal half-plane reflector situated with its edge at a depth _a_ directly below a point source. The predicted reflection is compared with that from a full plane at the same depth. [After T. Krey (11).]**

been shown diagrammatically in Fig. 3-17. The pattern of events we might expect to find on a seismic reflection record is shown in Fig. 4-23. If the energy is concentrated within a fairly narrow frequency band as in this example, the pulse is not strongly dispersed, but a reflection of considerable strength will be found beyond the limit of the ordinary geometrical reflection. To confirm the applicability of this method to problems involving elastic waves, Krey's predictions can be compared with the seismogram given by a two-dimensional model apparatus (Fig. 4-24). It was obtained by F. A. Angona (12) by using high-frequency pulse techniques, involving piezoelectric transducers and sheet-metal models. The reflection from the top of the step has very much the same form as the theoretical wavefront shown in Fig. 3-17, even though the latter refers to a spherical wave incident on a rigid half-plane while the model makes use of cylindrical waves.

Figure 4-25 shows a field reflection seismogram which displays a diffraction pattern due to an offset bed. The pattern of the reflections is essentially

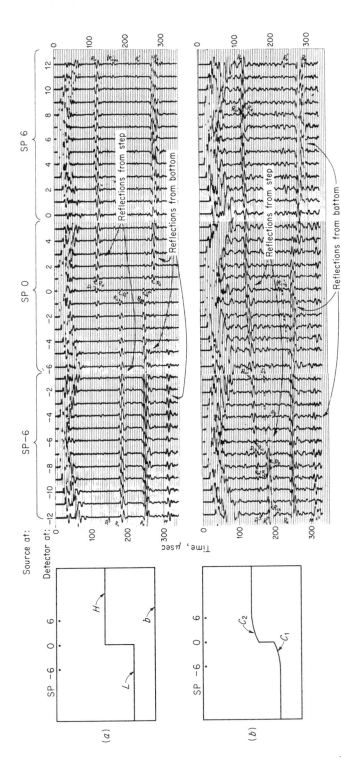

Fig. 4-24 Two-dimensional model seismograms over a step reflector. Shot point 0 is directly above the step. (*a*) Flat reflector surfaces; (*b*) curved reflector surfaces. [After F. A. Angona (12).]

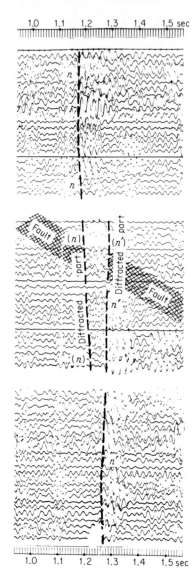

Fig. 4-25   Three reflection seismic records from which a faulted horizon has been interpreted.   [After T. Krey (11).]

what would be predicted for two parallel half-planes which are such weak reflectors that reflection of the diffracted wave can be neglected.

A second model seismogram which demonstrates the effect of curvature of the faulted surfaces is also shown in Fig. 4-24.   It can be seen that accurate interpretation of the form of features which are only a few wavelengths in size is almost an impossible task.   A field example showing a large number of diffracted events is displayed in Fig. 4-26.   The diagnostic hyperbolic time-distance curve indicating the arrival of a spherical wave-

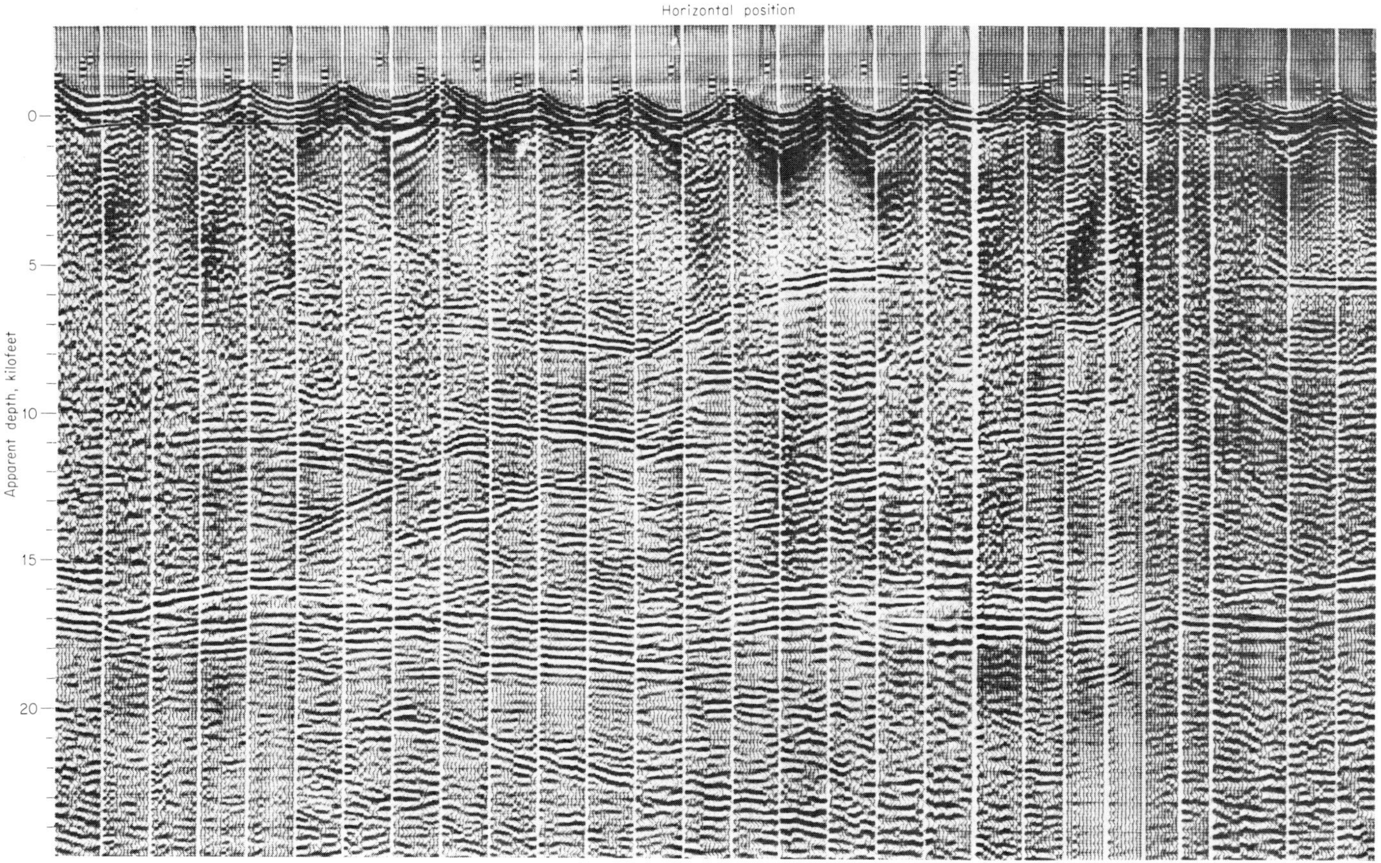
Horizontal position
Apparent depth, kilofeet
0
5
10
15
20

Fig. 4-26 Composite corrected reflection seismograms (variable intensity) which show a large number of diffracted events. (*Pan American Oil Co. Ltd.*)

front is clearly visible in several places. Thus diffraction phenomena place a very definite limit on the resolving power of seismic techniques.

References

1. J. J. Jakosky and J. Jakosky, Frequency Analysis of Seismic Waves, *Geophysics*, vol. 17, pp. 721–738, 1952.
2. H. Lamb, On the Propagation of Tremors over the Surface of an Elastic Solid, *Royal Soc. Philos. Trans.*, vol. 203, pp. 1–42, 1904.
3. M. B. Dobrin, R. F. Simon, and P. L. Lawrence, Rayleigh Waves from Small Explosions, *Am. Geophys. Union Trans.*, vol. 32, pp. 822–832, 1951. See also M. B. Dobrin, P. L. Lawrence, and R. L. Sengbush, Surface and Near-Surface Waves in the Delaware Basin, *Geophysics*, vol. 19, pp. 695–715, 1954.
4. M. Ewing, W. S. Jardetzky, and F. Press, "Elastic Waves in Layered Media," McGraw-Hill, New York, 1957.
5. C. B. Officer, Jr., "Introduction to the Theory of Sound Transmission," McGraw-Hill, New York, 1958.
6. M. Ewing, J. L. Worzel, and C. L. Pekeris, "Propagation of Sound in the Ocean," Geol. Soc. America Memoir 27, 1948.
7. T. C. Richards, Wide-angle Reflections and Their Application to Finding Limestone Structures in the Foothills of Western Canada, *Geophysics*, vol. 25, pp. 385–407, 1960.
8. H. Jeffreys, The Reflection and Refraction of Elastic Waves, *Royal Astron. Soc. Monthly Notices Geophys. Supp.*, vol. 1, pp. 321–334, 1926.
9. L. Knopoff, R. W. Fredericks, A. F. Gangi, and L. P. Porter, Surface Amplitudes of Reflected Body Waves, *Geophysics*, vol. 22, pp. 842–847, 1957.
10. P. C. Wuenschel, Seismogram Synthesis Including Multiples and Transmission Coefficients, *Geophysics*, vol. 25, pp. 106–129, 1960.
11. T. Krey, The Significance of Diffraction in the Investigation of Faults, *Geophysics*, vol. 17, pp. 843–858, 1952.
12. F. A. Angona, Two-dimensional Modelling and its Application to Seismic Problems, *Geophysics*, vol. 25, pp. 468–482, 1960.
13. R. L. Sengbush, Stratigraphic Trap Study in Cottonwood Creek Field, Big Horn Basin, Wyoming, *Geophysics*, vol. 27, pp. 427–444, 1962.

chapter 5   Seismic Interpretation

In applied seismology, interpretation usually consists in calculating the positions of, and identifying geologically, concealed interfaces or sharp transition zones from seismic pulses returned to the ground surface by either reflection or refraction. Rays may be used as an aid in following the route taken by seismic body waves, provided that the seismic wavelength is reasonably short in relation to the velocity gradients. If the velocity is known, the ray path can be computed, or vice versa. However, the use of ray-path methods is not always valid. They fail, for example, to predict diffraction effects, including surface waves. They do not provide an adequate theory of the interference of waves, a phenomenon of central importance to an understanding of the origin of reflections and guided waves. They also fail to give an accurate account of the amplitude of wave motion.

Yet in spite of these shortcomings, the laws of geometrical optics form the foundation upon which the great majority of seismic interpretation methods rest. In such a situation, it is often surprising to observe how uncritically their validity in seismological problems is accepted.   In Chaps. 5 and 6 we shall study this question more closely, with particular reference to velocity gradients and curved wavefronts.   Our purpose in Chap. 5 is to investigate the range of validity of ray-path methods in relation to the plane-wave theory and to indicate how these methods may be used in the geophysical interpretation of seismograms.

5-1   The Approximations of Geometrical Optics

We shall start with the equations derived in Sec. 2-13 for the propagation, in a vertically inhomogeneous medium, of waves whose direction of propagation has no y component.   The SH motion is propagated according to the equation

$$\rho \ddot{v} = \mu \, \nabla^2 v + \mu' \frac{\partial v}{\partial z} \tag{5-1}$$

and the P and SV motions, at small angles of incidence, according to the equations

$$\rho \ddot{\theta} = (\lambda + 2\mu) \, \nabla^2 \theta + 2(\lambda' + 2\mu') \frac{\partial \theta}{\partial z} \tag{5-2a}$$

and
$$\rho \ddot{\xi} = \mu \, \nabla^2 \xi + 2\mu' \frac{\partial \xi}{\partial z} \tag{5-2b}$$

respectively.   Since λ, μ, and ρ are nonmathematical functions, we cannot solve these equations exactly.   Therefore we shall attempt to find approximate solutions from which the behavior of the wavefronts can be ascertained.

Let us take as an example Eq. (5-1) for the propagation of SH displacements.   We shall begin by writing the displacement component v in spectrum form, i.e.,

$$v(x,z,t) = \frac{1}{2\pi} \int_{-\infty}^{\infty} V(x,z,\omega) e^{-i\omega t} \, d\omega$$

so that Eq. (5-1) reduces to the form

$$\nabla^2 V + \frac{\mu'}{\mu} \frac{\partial V}{\partial z} + k_\beta^2 V = 0 \qquad k_\beta = \omega \sqrt{\frac{\rho}{\mu}} = \frac{\omega}{\beta}$$

Using the method of Jeffreys, we first get rid of the term in $\partial V/\partial z$ by employing the substitution $V = \mu^{-\frac{1}{2}}\mathcal{v}$, which further reduces the equation to

$$\frac{\partial^2 \mathcal{v}}{\partial x^2} + \frac{\partial^2 \mathcal{v}}{\partial z^2} + [k_\beta^2(z) - f(z)]\mathcal{v} = 0 \tag{5-3}$$

in which $(f)z = -\dfrac{1}{4}\left(\dfrac{\mu'}{\mu}\right)^2$.   In this approximation, we have neglected the term in f which involves the second derivative of μ.   If we also disregard derivatives of the density ρ, we may put $f(z) = (\beta'/\beta)^2$. Equation (5-3) bears an obvious resemblance to the Helmholtz equation. It has an asymptotic solution which is valid when the variation of the function $q(z) = k_\beta{}^2(z) - f(z)$ over one wavelength is small and which is usually called the WKBJ solution.[1] We shall show that this solution is formally equivalent to propagation along the ray-path trajectories determined by Snell's law.

Let us first make the attempt to solve (5-3) by the method of separation of variables. This leads us to write $\mathcal{U}(x,z) = e^{ikx} Z(z)$, where k is the "apparent wave number" in the direction Ox. Formally, of course, k is just the separation constant, and it may have any positive value. The equation in $Z(z)$ now becomes

$$\frac{d^2Z}{dz^2} + [q(z) - k^2]Z = 0 \tag{5-4}$$

where $q(z) = k_\beta{}^2(z) - f(z)$.

If q were constant, the solution of (5-4) would be $\exp i \sqrt{q - k^2}\, z$ in the absence of any boundaries. This suggests a solution of the form

$$Z(z) = \exp iw(z)$$

when $q(z)$ is slowly varying. Substitution of this expression into (5-4) gives the following equation for $w(z)$

$$\left(\frac{dw}{dz}\right)^2 - i\frac{d^2w}{dz^2} = q(z) - k^2 = \chi^2(z) \tag{5-5}$$

This equation is simple in appearance, but it is nonlinear. If $q(z)$ is indeed a slowly varying function, however, the second term on the left-hand side will be smaller than the first, so that we may solve (5-5) approximately by iteration.

In the lowest approximation we may write

$$\frac{dw}{dz} = \chi(z)$$

and if we use the derivative of this solution as a substitute for d^2w/dz^2 in Eq. (5-5), we get

$$\left(\frac{dw}{dz}\right)^2 = \chi^2 + i\frac{d\chi}{dz}$$

Since the second term on the right-hand side is smaller than the first, we may approximate the square root by the first two terms of the binomial

---

[1] The initials are taken from the surnames of G. Wentzel, H. Kramers, L. Brillouin, and H. Jeffreys, who more or less independently rediscovered the procedure in connection with the solution of different problems. The method was in use much earlier, however, and was probably due to Liouville.

expansion, giving

$$\frac{dw}{dz} \cong \chi + \frac{i}{2\chi} \frac{d\chi}{dz} \tag{5-6}$$

Integrating, we obtain finally

$$w(z) \cong \int^z \chi(\zeta) \, d\zeta + \frac{i}{2} \ln \chi(z)$$

and thus

$$Z(z) = \chi^{-\frac{1}{2}}(z) \exp i \int^z \chi(\zeta) \, d\zeta$$
$$= [q(z) - k^2]^{-\frac{1}{4}} \exp i \int^z \sqrt{q(\zeta) - k^2} \, d\zeta \tag{5-7}$$

This is the WKBJ solution of Eq. (5-4). Its validity is limited by the accuracy of the binomial approximation (5-6) for the square root of $(dw/dz)^2$, and thus the essential requirement for its good behavior is that

$$\left| \frac{d\chi}{dz} \right| \ll \chi^2(z) \qquad \text{or} \qquad \left| \frac{dq}{dz} \right| \ll 2[q(z) - k^2]^{\frac{3}{2}}$$

Since k is undetermined, the *minimum* requirement that must be met is when $k = 0$, viz.,

$$\left| \frac{dq}{dz} \right| \ll 2q^{\frac{3}{2}}$$

and if, as on other occasions, we disregard second derivatives of μ, this is equivalent to the statement that

$$\omega \gg \left| \frac{d\beta}{dz} \right|$$

Thus it develops that the WKBJ solution is valid only at frequencies that are large in relation to the velocity gradients. If we make this approximation, we may replace $q(z)$ with $k_\beta^2(z)$ and write

$$Z(z) \cong [k_\beta^2(z) - k^2]^{-\frac{1}{4}} \exp i \int^z \sqrt{k_\beta^2(\zeta) - k^2} \, d\zeta$$

so that the solution for V becomes

$$V(x,z) \cong \mu^{-\frac{1}{2}}(k_\beta^2 - k^2)^{-\frac{1}{4}} \exp i \left[kx + \int^z \sqrt{k_\beta^2(\zeta) - k^2} \, d\zeta \right] \tag{5-8}$$

In the WKBJ approximation, the surfaces of constant phase, or the *wavefronts*, are defined by the relation

$$kx + \int^z \sqrt{k_\beta^2(\zeta) - k^2} \, d\zeta = \text{const}$$

and thus

$$k \, dx + \sqrt{k_\beta^2(z) - k^2} \, dz = 0 \qquad \text{or} \qquad \frac{dx}{dz} = -\frac{\sqrt{k_\beta^2(z) - k^2}}{k}$$

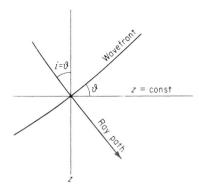

Fig. 5-1

If ϑ is the angle which the wavefront makes with the plane $z =$ const (Fig. 5-1), then

$$\sin \vartheta = \frac{1}{\sqrt{1 + (dx/dz)^2}} = \frac{k}{k_\beta}$$

or

$$k_\beta \sin \vartheta = k \tag{5-9}$$

But k is an arbitrary constant. Its value does not depend in any way upon the position of the wavefront. Consequently we recognize in (5-9) the formal statement of Snell's law, which asserts that the sine of the angle of incidence of each wavefront to the surfaces of constant velocity bears a constant proportionality to the wave velocity along the entire trajectory of that wavefront. The trajectories of all points in the wavefront will therefore follow "ray" paths whose geometrical properties are fully determined by (5-9). The WKBJ solution is thus formally equivalent to propagation of energy along the ray paths of geometrical optics; accordingly it is often called the *geometrical optics approximation*. The final solution for V in this approximation is

$$V(x,z;\omega) \cong (\omega\rho\beta \cos \vartheta)^{-\frac{1}{2}} \exp i k_\beta \left[x \sin \vartheta + \int^z \cos \vartheta(\zeta) \, d\zeta \right]$$

The amplitude factor is required in order to conserve the total amount of energy in the wavefront.

The easiest way to examine the accuracy of the WKBJ solution is to calculate the next higher approximation. This may be done by differentiating (5-6) and reinserting it into (5-5). The result is

$$\left(\frac{dw}{dz}\right)^2 = \chi^2 + i\chi' - \frac{1}{2} \frac{d}{dz} \left(\frac{\chi'}{\chi}\right)$$

from which we obtain

$$Z(z) = \chi^{-\frac{1}{2}}(z) \exp i \int^z \left(\chi - \frac{1}{8} \frac{\chi'^2}{\chi^3}\right) d\zeta$$

The wavefronts are now given by the relation

$$kx + \int^z \left\{ \sqrt{q(\zeta) - k^2} - \frac{[q'(\zeta)]^2}{32[q(\zeta) - k^2]^{5/2}} \right\} d\zeta = \text{const}$$

The appearance of the new term under the integral sign indicates that Snell's law is no longer valid and that a different ray path exists for every frequency. According to Fermat's principle, however, the path predicted by Snell's law is a path of minimum time, and so the higher frequencies ought always to arrive first at any position. Some dispersion may therefore be introduced by inhomogeneity, but at the frequencies commonly used in applied seismology it is likely to be very weak. This dispersion is opposite to that associated with normal mode propagation, in the sense that the longer wavelengths in this case will appear to travel the more slowly. Thus Snell's law, which at first sight seems to be independent of frequency, is actually a geometrical property of the wavefronts that holds true only at the shorter wavelengths.

By following essentially the same analytical byways, WKBJ solutions can also be found for the P and SV waves. In media whose properties change little within one seismic wavelength, we can proceed from Eqs. (5-2) and, without repeating steps already demonstrated, write down their first-order WKBJ solutions directly

$$\Theta(x,z;\omega) \cong (\omega\rho^2\alpha^3 \cos \vartheta)^{-1/2} \exp ik_\alpha \left[x \sin \vartheta + \int^z \cos \vartheta(\zeta) \, d\zeta \right]$$

$$\Xi(x,z;\omega) \cong (\omega\rho^2\beta^3 \cos \vartheta)^{-1/2} \exp ik_\beta \left[x \sin \vartheta + \int^z \cos \vartheta(\zeta) \, d\zeta \right]$$

Here Θ and Ξ are the Fourier transforms of the cubical dilatation θ and the SV rotation ξ, respectively.

5-2   Solution Near a Turning Point

The WKBJ method obviously fails when $k = k_\beta$ (or k_α, in the case of P waves). According to (5-9), this situation arises when $\vartheta = \pi/2$, which means that the ray path is just on the point of being turned back again toward the ground surface. This implies that the point of maximum penetration has been reached by the ray path whose initial angle of incidence is $\sin^{-1} (\beta_0/\beta)$, β_0 being the wave velocity at the source. Accordingly, the value $k = k_\beta$ is called the *turning point* of Eq. (5-4).

In the vicinity of the turning point, special precautions must be observed in order to avoid singularities of the kind found in the solution (5-8). This singularity does not really exist, of course, since the turning point is a perfectly regular point of the differential equation (5-4). It is introduced solely in the approximation which led to (5-6) and which is clearly not permissible in this region. Therefore, in order to find a solution of (5-4) which is valid when $k^2 = q(z)$, let us identify the maximum depth of penetration of a particular ray path as z_0. Within the WKBJ approximation and according to Eq. (5-9), we may set $k = k_\beta(z_0)$. Close to this point, we may write

$$k_\beta^2(z) \doteq k_\beta^2(z_0) + (z - z_0) \left(\frac{dk_\beta^2}{dz} \right)_{z=z_0}$$

assuming that the derivative exists.[1] Replacing the value of the derivative at the turning point with the constant $a^2\omega^2$, we may set

$$q(z) - k^2 \cong a^2\omega^2(z - z_0)$$

and Eq. (5-4) therefore becomes

$$\frac{d^2Z}{dz^2} + a^2\omega^2(z - z_0)Z = 0 \tag{5-10}$$

This is called Airy's equation, and its exact solutions are well known [see, for example, Jeffreys and Jeffreys (2), Chap. 17]. They are linear combinations of the quantities

$$\sqrt{z - z_0} \, J_{\pm\frac{1}{3}}[\tfrac{2}{3}a\omega(z - z_0)^{\frac{3}{2}}]$$

which, as ω becomes large, tend asymptotically to the values

$$\left(\frac{2}{\pi a\omega} \right)^{\frac{1}{2}} (z - z_0)^{-\frac{1}{4}} \cos \left[\frac{2}{3} a\omega(z - z_0)^{\frac{3}{2}} - \frac{m\pi}{12} \right]$$

where $m = 1$ or 5 according as the $J_{-\frac{1}{3}}$ or the $J_{\frac{1}{3}}$ term is taken. Notice that these asymptotic forms are exactly what is obtained by making the same approximation as above for $q(z) - k^2$ in Eq. (5-8) except for the phase angle $m\pi/12$. Thus the solution at the turning point corresponds to propagation along the ray paths of geometrical optics. It is interesting that the asymptotic solution by the exact method should include such a phase factor, whereas the WKBJ method does not. This suggests that, in close analogy with the problem of the reflection of plane waves at an interface beyond the critical angle, a change in phase normally takes place when the ray path is turned back toward the surface of the ground.

5-3 *Applications of Ray-path Theory*

In many areas the P-wave or pseudo-P-wave velocity can be represented adequately, at least down to some particular horizon, by a continuous function $V(z)$ of depth only. By assuming Snell's law in its usual form

$$\frac{\sin i}{V(z)} = p(\text{const})$$

[1] The derivative will not exist if a velocity reversal takes place at $z = z_0$, and in that case we must go to the next higher term. For the solution of this and related problems, we refer to Morse and Feshbach (1), sec. 9-3.

we can immediately write down expressions for the increments of time and horizontal distance along a particular ray path (Fig. 5-1)

$$dt = \frac{1}{V(z)} \frac{dz}{\cos i} = \frac{1}{V(z)} \frac{dz}{\sqrt{1 - p^2 V^2(z)}}$$

$$dx = \tan i \, dz = \frac{pV(z) \, dz}{\sqrt{1 - p^2 V^2(z)}}$$

For a ray traversing a path from depth $z = z_1$ to $z = z_2$ (Fig. 5-2)

$$t = \int_{z_1}^{z_2} \frac{d\zeta}{V(\zeta) \sqrt{1 - p^2 V^2(\zeta)}}$$

and

$$x = p \int_{z_1}^{z_2} \frac{V(\zeta) \, d\zeta}{\sqrt{1 - p^2 V^2(\zeta)}}$$

(5-11)

This pair of integrals constitutes an implied relationship between x and t through the *ray-path parameter p*. If $V(z)$ is known for $z_1 \leq z \leq z_2$, then in theory at least, we may perform the integrations and eliminate p to obtain a direct relationship between t and x for given values of z_1 and z_2. If x is given, this result may be carried one stage further, and a table may be set up for $t(z_2)$ when $z_1 = 0$. Such a table or graph could be used directly in the interpretation of reflection times. In many cases, however, $V(z)$ is unknown and the problem is how to derive it from measurements of t, x, and z.

5-4 Velocity Measurements in Boreholes

If deep boreholes are available, the most direct way to measure seismic velocities "in situ" is by timing the passage of a pulse from an explosion at

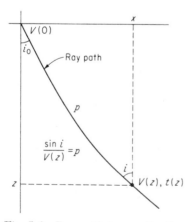

Fig. 5-2 Ray path in a vertically-inhomogeneous medium.

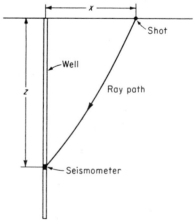

Fig. 5-3 Schematic diagram of a "well-shoot."

the surface of the ground to a transducer lowered to a given depth in the borehole. The arrangement is illustrated schematically in Fig. 5-3. The experiment is repeated for various seismometer positions until a satisfactory graph is obtained of travel time versus seismometer depth.

To make a continuous log of velocity from these measurements, we proceed on the two assumptions that Snell's law is obeyed and that within the short lateral distances from the borehole reached by the Fermat ray paths there are no significant horizontal velocity gradients. Thus the ray-path integrals (5-11) may be used. We further notice that inasmuch as each seismometer position implies a different Fermat path from the shot, the ray-path parameter p will be a continuous function of z although it is independent of ζ. Thus

$$t(z) = \int_0^z \frac{d\zeta}{V(\zeta)\sqrt{1 - p^2(z)V^2(\zeta)}} \tag{5-12a}$$

and

$$x = p(z) \int_0^z \frac{V(\zeta)\,d\zeta}{\sqrt{1 - p^2(z)V^2(\zeta)}} \tag{5-12b}$$

are the parametric equations for the rays. The problem is now to eliminate $p(z)$ between these two equations and so obtain a relationship between t and V.

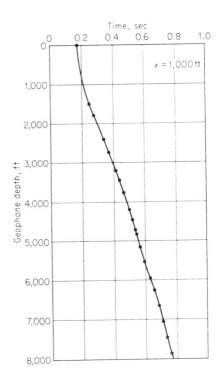

Fig. 5-4 Curve of travel time versus geophone depth obtained from a Louisiana well.

First of all, we note that since x is constant,

$$dx = \frac{p(z)\,V(z)\,dz}{\sqrt{1 - p^2 V^2(z)}} + \frac{dp}{dz}\,dz \int_0^z \frac{V(\zeta)\,d\zeta}{[1 - p^2 V^2(\zeta)]^{3/2}} = 0$$

Therefore, if we differentiate the t equation

$$\frac{dt}{dz} = \frac{1}{V(z)\,\sqrt{1 - p^2 V^2(z)}} + p\,\frac{dp}{dz} \int_0^z \frac{V(\zeta)\,d\zeta}{[1 - p^2 V^2(\zeta)]^{3/2}}$$

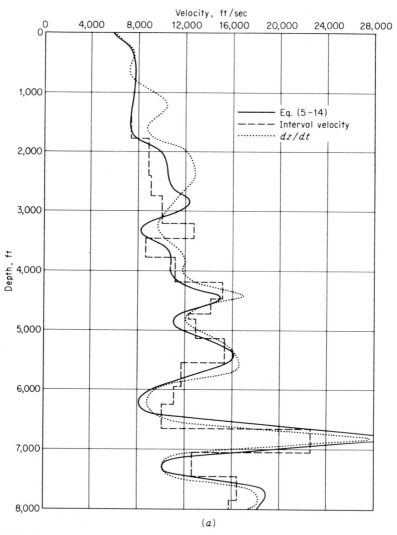

Fig. 5-5 (*a*) *dz/dt* measured from Fig. 5-4 and the velocity *V(z)* derived from it by using Eq. (5-14). For comparison, interval velocities have been computed from the same data by using the standard straight ray-path method.

and substitute for dp/dz, we find that

$$\frac{dt}{dz} = \frac{\sqrt{1 - p^2 V^2(z)}}{V(z)}$$

and hence

$$p^2(z) = \frac{1}{V^2(z)} - \left(\frac{dt}{dz}\right)^2 \qquad (5\text{-}13)$$

By introducing (5-13) into (5-12a), we eliminate p and obtain

$$t(z) = \int_0^z \left\{ 1 - V^2(\zeta) \left[\frac{1}{V^2(z)} - \left(\frac{dt}{dz}\right)^2 \right] \right\}^{-\frac{1}{2}} \frac{d\zeta}{V(\zeta)} \qquad (5\text{-}14)$$

This is an integral equation for $V(z)$ which must be solved numerically. To do so, it is necessary to make a table of t and of dt/dz and then to apply

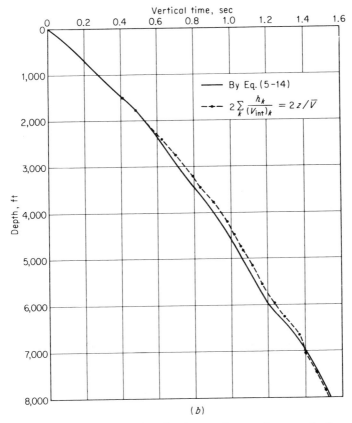

(b) Vertical time versus depth by the integral-equation method and by the straight ray-path method.

formulas for numerical quadratures to find values of V. If x is small enough, dz/dt by itself may give a sufficiently good estimate.

Most of the standard references on well-shooting methods disregard the curvature of the ray paths and employ very much simpler computing techniques based upon rectilinear propagation between shot and receiver [see, for example, Dix (3), chap. 7]. This gives an "average velocity" at each depth, which is always less than the true velocity since the assumed path is shorter than the actual one. In some cases the discrepancies between the two methods are not large enough to justify the more general treatment, but in others they are significant. In Fig. 5-4 we show the time-depth curve obtained from a well in Louisiana, where the offset of the shot point from the wellhead was 1,000 ft. The velocity-depth relation as computed by (5-14) and also the interval velocities computed by the rectilinear ray-path method are shown for comparison in Fig. 5-5a, and the vertical two-way times obtained from these functions by integration are shown in Fig. 5-5b. Although the difference between them does not seem large, it is nonetheless equivalent to a discrepancy in the depth of as much as 400 ft. We must balance this against other uncertainties in reflection interpretations in order to decide whether or not it is necessary to compute the integral (5-14).

5-5 *Velocity Measurements by Surface-to-surface Refraction*

In areas where no boreholes are available, the velocities must be found by using only measurements which can be made at the surface of the ground. One such method makes use of the refractions that occur when the velocity increases continuously with depth. The situation is illustrated in Fig. 5-6, where we shall assume to start with that the velocity increases smoothly. Under these conditions any ray path from S that obeys Snell's law will return to the surface of the ground at a distance Δ, given by

$$\Delta = 2p(z_m) \int_0^{z_m} \frac{V(\zeta)\, d\zeta}{\sqrt{1 - p^2(z_m)V^2(\zeta)}} \tag{5-15}$$

where z_m is the maximum depth of penetration of the ray. If we make the following change of variables: $y(z_m) = p^2(z_m)$, $\eta(\zeta) = 1/V^2(\zeta)$, and if we

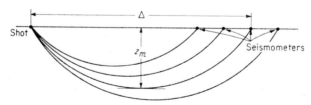

Fig. 5-6 Schematic diagram of surface-to-surface refraction when velocity increases continuously with depth.

write $f(y) = \Delta/2p$, then (5-15) becomes

$$f(y) = \int_{\eta_0}^{y} \frac{\zeta'(\eta)\, d\eta}{\sqrt{\eta - y}} \qquad (5\text{-}16)$$

where $\zeta' = d\zeta/d\eta$ and $\eta_0 = 1/V^2(0)$. This is now in the standard form of Abel's integral equation

$$f(\xi) = \int_a^{\xi} \frac{u(x)\, dx}{(\xi - x)^n} \qquad 0 < n < 1$$

of which the solution is

$$u(x) = \frac{\sin n\pi}{\pi} \frac{d}{dx} \int_a^{x} \frac{f(\xi)\, d\xi}{(x - \xi)^{1-n}}$$

Making the appropriate substitutions, therefore, we have for the solution of (5-16)

$$\frac{1}{i} \frac{dz}{d\eta} = \frac{1}{\pi} \frac{d}{d\eta} \int_{\eta_0}^{\eta} \frac{f(y)\, dy}{\sqrt{\eta - y}}$$

or

$$z(\eta) = -\frac{1}{\pi} \int_{\eta_0}^{\eta} \frac{f(y)\, dy}{\sqrt{y - \eta}}$$

Next let p_1 be the parameter of the ray path whose maximum penetration is z. Then $\eta = 1/p_1^2$, since η is now a function of z. Therefore

$$z(p_1) = -\frac{1}{\pi} \int_{1/V(0)}^{p_1} \frac{\Delta(p)\, dp}{\sqrt{p^2 - p_1^2}}$$

After integrating by parts, this reduces to

$$z(p_1) = \frac{1}{\pi} \int_0^{\Delta_1} \cosh^{-1} \frac{p(\Delta)}{p_1}\, d\Delta \qquad (5\text{-}17)$$

Now by differentiating (5-11), we find that

$$\left(\frac{\partial x}{\partial p}\right)_z = \frac{1}{p}\left(\frac{\partial t}{\partial p}\right)_z \qquad \text{and therefore} \qquad \left(\frac{\partial t}{\partial x}\right)_z = p \qquad (5\text{-}18)$$

so that $p(\Delta)$ is given by the slope of the time-distance graph. It also transpires that, since $\sin i/V = p$, which is constant along each ray path, the velocity at depth z must be $1/p_1$ since at this point $i = 90°$. Consequently, the integral (5-17) is in a form suitable for the direct computation of z as a function of V, since

$$z(V) = \frac{1}{\pi} \int_0^{\Delta} \cosh^{-1}\left(V \frac{dt}{dx}\right)_z dx \qquad (5\text{-}19)$$

where

$$\frac{1}{V} = \left(\frac{dt}{dx}\right)_{x=\Delta}$$

This is known as the Wiechert-Herglotz-Bateman (or WHB) integral. [Slichter (4)].

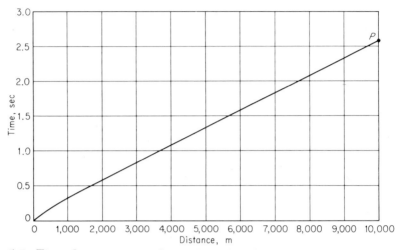

Fig. 5-7 **Time-distance curve observed on the Greenland Ice Cap by Joset and Holtzscherer (5). The effects of variable velocity can be seen in the first 2,000 meters.**

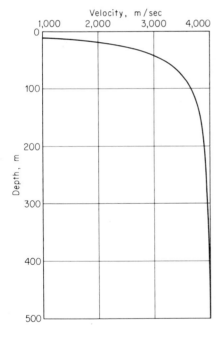

Fig. 5-8 **Outcome of the application of the WHB integral to the time-distance curve of Fig. 5-7. [After Joset and Holtzscherer (5).]**

Figure 5-8 shows an example of the use of this integral. The plot of arrival times versus distance shown in Fig. 5-7 is the outcome of some experiments made on thick inland Arctic ice. The velocity-depth relationship calculated from (5-19) shows clearly the approximate thickness of the firn and the depth at which the transition into ice takes place.

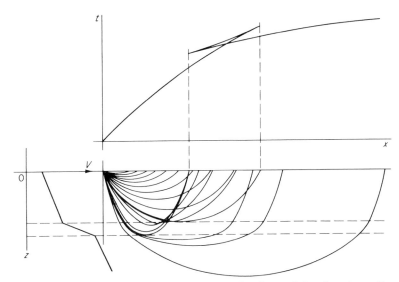

Fig. 5-9 Example showing how a loop may be formed in the time-distance curve if the velocity increases suddenly. [After Goguel (8).]

The form of the time-distance curve is not always as simple as that in Fig. 5-7, however. If the velocity suddenly increases at some depth, Δ will decrease with increasing penetration of the rays below that depth. One such instance is shown in Fig. 5-9, in which a loop is formed in the time-distance curve. If the WHB integral is to give the correct velocity-depth function, the entire curve including any loops must be used in the integration. Failure to do this through not recognizing the later arrivals will produce some serious errors. This creates a task of peculiar difficulty in seismogram analysis, since even if a loop is recognized, it is often troublesome to locate the second and third arrivals accurately. Yet without this information it is not possible to make unambiguous interpretations, because the velocity at each depth is determined from the apparent surface velocity of the ray that just penetrates to that depth. If the loop is unobservable, a blind zone exists in which the velocity is undetermined. The conditions on $V(z)$ for the formation of a loop are rather complicated and have been discussed in detail by Bullen (6).

5-6 *Velocity Measurements Using Reflections*

It should be noted that under most circumstances the method of surface-to-surface refraction requires very great distances between shot and receiver in order to achieve even moderate depths of penetration by the rays. It also breaks down in the presence of seismic discontinuities. There are other methods, however, which make direct use of discontinuities as a means

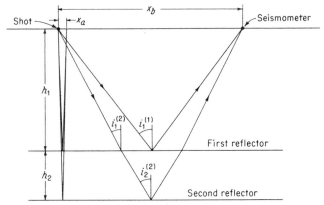

Fig. 5-10   Schematic diagram of reflections from homogeneous strata.

of estimating the velocities of underground formations by employing the differential reflection times between two seismometers located at different positions.   These are the so-called "t^2,x^2" and "Δt" methods.

The simplest possible example is one in which there is a single horizontal interface and the velocity down to the reflector is constant.   Here we can see immediately that

$$t^2 = \frac{x^2 + 4z^2}{V^2} \tag{5-20}$$

where x is the horizontal distance from the source and z is the thickness of the layer.   Thus for a pair of measurements at x_a and x_b

$$t_b{}^2 - t_a{}^2 = \frac{x_b{}^2 - x_a{}^2}{V^2}$$

or, if we write $t_b = t_a + \Delta t$,

$$\Delta t(2t_a + \Delta t) = \frac{x_b{}^2 - x_a{}^2}{V^2} \tag{5-21}$$

If it happens that $x_b \ll z$, then $\Delta t \ll t_a$ and (5-21) may be further reduced to the simple expression

$$2t_a \, \Delta t \doteq \frac{x_b{}^2 - x_a{}^2}{V^2}$$

Virtually the same idea may be extended to any number of horizontal and homogeneous layers.   Thus if we are presented with seismograms which exhibit a number of reflections which can clearly be identified and correlated over a sufficiently large area, we may remove most of the effects of dip by averaging the reflection times in pairs from shots made in opposite directions and the effect of variable surface conditions by averaging the reflection times over the area as a whole.   We may then suppose that the reflections have originated from horizontal interfaces separating a number of homogeneous strata, as indicated in Fig. 5-10. Beginning with the first reflection and assuming for convenience that $x_a \ll x_b$ or h_1, we have

$$t_1 = \frac{2h_1}{V_1}$$

and

$$\Delta t_1(2t_1 + \Delta t_1) = \frac{x_b{}^2}{V_1{}^2}$$

from which we may calculate h_1 and V_1.

For the second reflection we shall have

$$t_2 = \frac{2h_1}{V_1} + \frac{2h_2}{V_2} \tag{5-22}$$

and

$$t_2 + \Delta t_2 = \frac{2h_1 \sec i_1{}^{(2)}}{V_1} + \frac{2h_2 \sec i_2{}^{(2)}}{V_2} \tag{5-23}$$

plus two auxiliary equations

$$x_b = 2h_1 \tan i_1{}^{(2)} + 2h_2 \tan i_2{}^{(2)} \tag{5-24}$$

and

$$\frac{\sin i_1{}^{(2)}}{V_1} = \frac{\sin i_2{}^{(2)}}{V_2} = p_2 \tag{5-25}$$

where $i_k{}^{(n)}$ is the angle of incidence at the base of the kth layer of the ray path which is subsequently reflected at the base of the nth layer. Expanding in powers of p_2, we get from (5-24) and (5-25)

$$x_b = 2p_2(h_1V_1 + h_2V_2) + O(p_2{}^3) + \cdots \tag{5-26}$$

and from (5-22) and (5-23)

$$\Delta t_2 = \frac{2h_1}{V_1}\left(\frac{1}{\sqrt{1 - p_2{}^2V_1{}^2}} - 1\right) + \frac{2h_2}{V_2}\left(\frac{1}{\sqrt{1 - p_2{}^2V_2{}^2}} - 1\right)$$

$$= p_2{}^2(h_1V_1 + h_2V_2) + O(p_2{}^4) + \cdots \tag{5-27}$$

Combining (5-26) and (5-27), we find that

$$\Delta t_2 = \tfrac{1}{2}p_2 x_b$$

$$= \frac{x_b{}^2}{4(h_1V_1 + h_2V_2)} + O(p_2{}^3) + \cdots$$

This gives us the following two equations to estimate h_2 and V_2:

$$t_2 - t_1 = \frac{2h_2}{V_2}$$

and

$$\frac{x_b{}^2}{4\,\Delta t_2} = h_1V_1 + h_2V_2$$

from which we get

$$V_2{}^2 = \frac{x_b{}^2}{2\,\Delta t_2(t_2 - t_1)} - \frac{2V_1{}^2 t_1}{t_2 - t_1}$$

and

$$h_2 = \tfrac{1}{2}V_2(t_2 - t_1)$$

In a similar way, for the nth reflection we have

$$t_n = 2 \sum_{k=1}^{n} \frac{h_k}{V_k}$$

and $t_n + \Delta t_n = 2 \displaystyle\sum_{k=1}^{n} \frac{h_k \sec i_k{}^{(n)}}{V_k} = 2 \displaystyle\sum_{k=1}^{n} \frac{h_k}{V_k}\left(1 + \tfrac{1}{2}p_n{}^2 V_k{}^2 + O(p_n{}^4) + \cdots\right)$

where

$$x_b = 2 \sum_{k=1}^{n} h_k \tan i_k{}^{(n)}$$

$$= 2p_n \sum_{k=1}^{n} h_k V_k + O(p_n{}^3) + \cdots$$

Thus $\Delta t_n \doteq p_n{}^2 \displaystyle\sum_{k=1}^{n} h_k V_k$, and consequently by eliminating p_n, we find that

$$\sum_{k=1}^{n} h_k V_k \doteq \frac{x_b{}^2}{4\,\Delta t_n}$$

If we now get rid of all but h_n and V_n, we get finally

$$V_n{}^2 = \frac{x_b{}^2}{2(t_n - t_{n-1})}\left(\frac{1}{\Delta t_n} - \frac{1}{\Delta t_{n-1}}\right) \tag{5-28}$$

and

$$h_n = \tfrac{1}{2}V_n(t_n - t_{n-1}) \tag{5-29}$$

This is a very simple calculation to make, once the reflection data are given. From the values of V_n and h_n so obtained, it is at once possible to make a graph of the two-way vertical reflection time versus the depth, since

$$z_n = \sum_{k=1}^{n} h_k = \frac{1}{2}\sum_{k=1}^{n} V_k(t_k - t_{k-1}) \tag{5-30}$$

A practical difficulty with the "Δt" method arises when the seismometers are placed too close to the shot point, for then the Δt values become small. Since the errors in making the measurements will remain about the same, the dispersion in the values of V_n becomes correspondingly large. The difficulty is relieved somewhat by extending the distance x_b, but beyond a certain point the convergence of the various binomial expansions becomes doubtful.

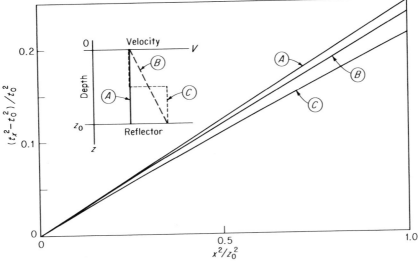

Fig. 5-11 Theoretical t^2, x^2 curves for three different velocity distributions.

An alternative expression to (5-28), which can be used whenever such doubts arise, is the following:

$$V_n{}^2 = \frac{1}{(t_n - t_{n-1})} \left(\frac{t_n}{b_n} - \frac{t_{n-1}}{b_{n-1}} \right) \qquad (5\text{-}31)$$

where

$$b_n = \frac{d[t_n(x)]^2}{d(x^2)}$$

Equation (5-31) is in fact identical with (5-28) within the approximation $2t\,\Delta t_n/x_b{}^2 = b_n$, but it was deduced on the assumption that the travel-time curve of the nth reflection is perfectly described by the equation

$$t_n{}^2(x) = t_n{}^2(o) + b_n x^2$$

Its utility rests on the fact that in virtually all realizable situations the t^2, x^2 loci for reflections from any depth are—to within the limits of practical measurement—perfectly straight lines.   This fact is illustrated in Fig. 5-11 for two different artificial models of an inhomogeneous ground, and it is further demonstrated in Fig. 5-12 for the pair of actual seismograms, parts of which were shown in Fig. 4-19.   Thus to compute V_n from the formula (5-31), all that is required in practice is to measure the slopes of the t^2, x^2 lines which are obtained from the reflection times observed at several well-separated distances from the shot point.

We illustrate the two methods by using the seismograms mentioned previously.   Ten distinct reflections in all could be identified on these records. Accompanying these data is a continuous velocity log, shown in Fig. 5-13,

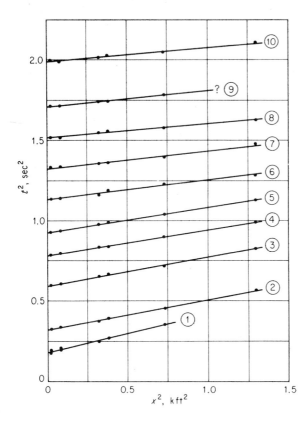

Fig. 5-12 A plot of t^2 versus x^2 for ten reflections visible on the records shown in Fig. 4-19. Data obtained from geophones placed at distances greater than 1,800 ft from the shot are also included in this diagram.

which was taken from a well close to the site of the shot point. A curve of the two-way vertical reflection time versus the depth was constructed from this log by integration, and should provide a fairly rigorous check on the calculations.

The results are given in Table 5-1 in three columns. One column gives the depth calculated by the "Δt" method using (5-28) and (5-30) at each of the ten reflecting horizons. Another gives their depths as calculated by the "t^2" method, using (5-31) and (5-30). A third column shows the depths obtained by integrating Fig. 5-13. The agreement among all these calculations is quite satisfying and appears to lie well within the experimental error of any one of the sets of values. Less satisfying, however, is the agreement between the calculations and the velocity log (see Fig. 5-13), for not one of the calculated depths lies close to a major velocity discontinuity. On closer examination, however, it appears that the "Δt" and the "t^2" velocity profiles can be brought into closer coincidence with the velocity log by an upward displacement of about 300 ft, or else by subtracting approximately 40 msec from each of the vertical reflection times. That such a delay is actually introduced both in the initial buildup of the incident

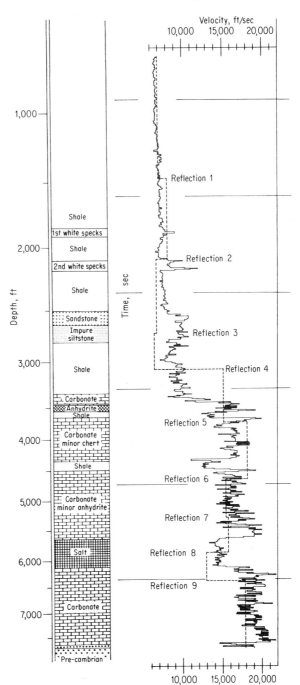

Fig. 5-13 A continuous velocity log (plotted on a time scale) of a well located near the shot point used to obtain the records shown in Fig. 4-18. On it is shown the velocity structure interpreted by the Δt method.

Table 5-1

$$x = 1,800 \text{ ft}$$

Reflection no. n	$t_n,$ sec	$\Delta t_n,$ sec	"Δt" z_n, ft	"t^2" z_n, ft	Depth from velocity log, ft
1	0.422	0.072	1,480	1,350	1,380
2	0.565	0.049	2,090	2,030	1,900
3	0.768	0.040	2,710	2,700	2,710
4	0.877	0.037	3,070	3,400	3,210
5	0.961	0.026	3,700	3,850	3,780
6	1.063	0.017	4,610	4,550	4,610
7	1.150	0.014	5,280	5,200	5,300
8	1.232	0.012	5,910	6,020	5,960
9	1.307	0.011	6,390	6,470	6,600
10	1.410	0.009	7,310	7,200	7,550

pressure pulse and in the formation of "reflections" by interference among the returning wavefronts is demonstrated by the examples in Sec. 4-6. It is intrinsic to the propagation of seismic waves and should be allowed for in the interpretations.

5-7 On the Use of Head Waves

The head wave discussed in Sec. 3-5 forms the basis of the refraction technique of seismic prospecting. For clear, unambiguous identification it is generally desirable to observe the head wave beyond the "crossover distance" from the source, where it reaches the receiver in front of the direct wave. The head wave certainly exists as a later arrival within the crossover distance, but it is normally difficult to identify because of its very low amplitude. Thus, to apply the refraction technique to exploration for deep interfaces, strong sources are needed in order that distant receiver positions may be used. The same methods are also used on a smaller scale to locate shallow velocity discontinuities.

The interpretation of head-wave arrival times is generally based upon a simple layered model of the ground. If all geological formations are considered to be uniform both in thickness and velocity, then in the one-layer model where a single discontinuity occurs at the depth h the problem is analytically very simple. A picture of the advancing wavefront is shown in Fig. 5-14, together with the equivalent ray-path diagram and a plot of the arrival time versus the shot-to-receiver separation. Within the distance $x < x_c$, the first signal to reach the geophone has traveled directly through the surface layer with the compressional-wave velocity V_1; hence the initial

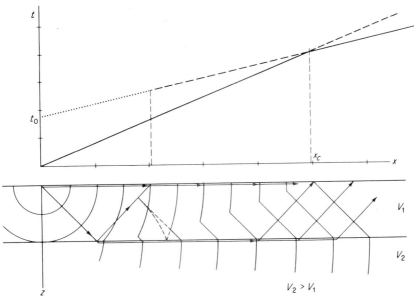

Fig. 5-14 Schematic diagram of the advancing wavefront which transmits the first arrival of energy to the geophone.

segment of the time-distance curve is given by the formula

$$t = \frac{x}{V_1}$$

Beyond x_c, however, the head wave comes in first, and the time-distance relation thereafter is

$$t = \frac{x}{V_2} + \frac{2h \sqrt{V_2^2 - V_1^2}}{V_1 V_2}$$

according to the ray-path theory. Two methods are available for determining h: The first makes use of the time intercept of the head wave

$$t_0 = \frac{2h \sqrt{V_2^2 - V_1^2}}{V_1 V_2} \tag{5-32}$$

and the second employs the "crossover distance"

$$x_c = 2h \sqrt{\frac{V_2 + V_1}{V_2 - V_1}} \tag{5-33}$$

In both cases the velocities V_1 and V_2 are determined from the inverse slopes of the two line segments. Preference between the two formulas will depend entirely on the circumstances. If h is small, there will be very few points on the first segment, and t_0 probably will be more accurately measurable

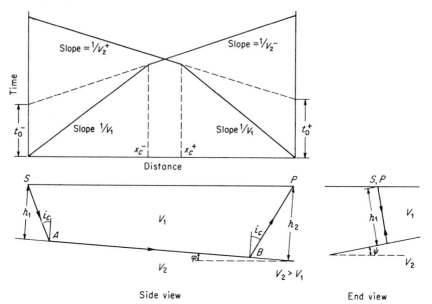

Side view   End view

Fig. 5-15   Critical refraction at a sloping interface underlying a uniform layer.

than x_c.   If h is large or if the velocity contrast is small, the contrary may
be true.   In either case the difference is likely to be small enough that the
one calculation may profitably be used to provide a check on the other.

Now let us consider what occurs when the interface is sloping.   A ray-
path diagram representing this situation is shown in Fig. 5-15.   The sec-
tional view of the ray path is not taken in the vertical plane but in the plane
which contains both the shot point and receiver and which is perpendicular
to the interface.   The end view of the ray-path trajectory is also shown in
order to clarify the meaning of this statement.   The interpretation requires
the determination of two quantities: the dip of the interface and the depth
below the shot point.   Since only the component of the dip that lies in the
direction of the profile can be obtained from a single line, two intersecting
profiles are needed to calculate the true dip.

If in Figure 5-15 the shot point is at S, the arrival time of the head wave
will be given by the formula

$$t^- = \frac{x \sin (i_c + \varphi)}{V_1} + \frac{2h_1 \cos i_c}{V_1}$$

where φ is the component of the dip in the direction of the profile and
$i_c = \sin^{-1} (V_1/V_2)$.   The apparent velocity in the substratum obtained by
shooting downdip is therefore

$$V_2^- = \frac{V_1}{\sin (i_c + \varphi)}$$

Now let the position of the shot be moved to P and the line be repeated in the reverse direction. The arrival time of the head wave in this case will be given by

$$t^+ = \frac{x \sin (i_c - \varphi)}{V_1} + \frac{2h_2 \cos i_c}{V_1}$$

and the apparent velocity in the substratum obtained by shooting updip is

$$V_2^+ = \frac{V_1}{\sin (i_c - \varphi)}$$

From these formulas, we find that

$$\varphi = \frac{1}{2} \left(\sin^{-1} \frac{V_1}{V_2^-} - \sin^{-1} \frac{V_1}{V_2^+} \right)$$

and

$$\frac{1}{V_2^+} + \frac{1}{V_2^-} = \frac{2 \cos \varphi}{V_2}$$

which permits us to calculate both V_2 and φ. Once we have these two quantities, we obtain h_1 from the head-wave intercept time shooting from S by means of the formula

$$t_0^- = 2h_1 \frac{\sqrt{V_2^2 - V_1^2}}{V_1 V_2}$$

To calculate the actual depth of the interface below S, the true dip must first be found, and this requires shooting a transverse profile from S. X- or L-shaped spreads are usually the most convenient to use for this purpose. In refraction surveys, some form of reverse shooting is the only means of determining whether or not dip is present.

There is no difficulty in principle in extending this analysis to any number of interfaces. The major problem that arises is the entirely practical one of recognizing the small changes in the slope of the time-distance curve which might indicate the presence of additional layers.

Some modification in the theory is required when the velocity in the overburden increases continuously with depth. A ray-path diagram representing this situation is shown in Fig. 5-16. To approach this problem, we should first know the behavior of V_1 down to the depth h, and this we can determine by applying the WHB integral (5-19) to the direct wave out to the distance at which the ray path just grazes the interface. At this point the direct wave stops abruptly and the time-distance curve forms a cusp (point A in Fig. 5-16). Long before we reach this point, however, the head wave has overtaken the direct wave, and so the arrivals that we are now discussing will appear on the records as second events; but they will

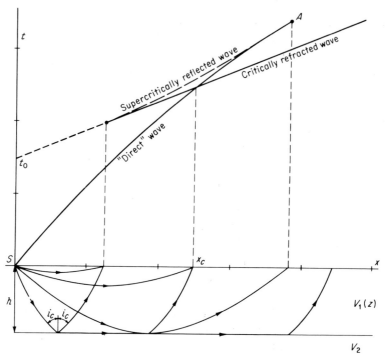

Fig. 5-16   The effect on the time-distance curve of an increasing velocity in the region above the refracting horizon.

carry relatively large amounts of energy, and they should be recognizable. The head wave itself will arrive at a time which is given by

$$t = \frac{x}{V_2} - \frac{2\Delta_1}{V_2} + 2t_1 \qquad (5\text{-}34)$$

where Δ_1 is the horizontal displacement of the ray path in the overburden and t_1 is the time taken by the energy to traverse the overburden.   According to (5-11) we may write

$$\Delta_1 = p \int_0^h \frac{V_1(z)\,dz}{\sqrt{1 - p^2 V_1^2(z)}} \qquad t_1 = \int_0^h \frac{dz}{V_1(z)\,\sqrt{1 - p^2 V_1^2(z)}}$$

but since $\sin i/V = p$ is constant along the entire ray path, it follows that $p = 1/V_2$, and it is therefore equal to the slope of the straight line beyond the crossover distance.   Consequently

$$\Delta_1 = \int_0^h \frac{V_1(z)\,dz}{\sqrt{V_2^2 - V_1^2(z)}} \qquad \text{and} \qquad t_1 = V_2 \int_0^h \frac{dz}{V_1(z)\,\sqrt{V_2^2 - V_1^2(z)}}$$

$$(5\text{-}35)$$

hence

$$t = \frac{x}{V_2} + \frac{2}{V_2} \int_0^h \frac{\sqrt{V_2^2 - V_1^2(z)}}{V_1(z)} \, dz \qquad (5\text{-}36)$$

In principle, the depth h can be found by solving Eq. (5-36) for the head-wave intercept time, but in practice this method is too cumbersome. The following approximate procedure may be used instead: Choose first an average value or estimate for V_1 and use (5-32) to find h. Let the observed time intercept of the head wave be $(t_0)_{\text{obs}}$ and let $(t_0)_{\text{calc}}$ be the calculated value for t according to (5-36). Then let us form the difference of these two quantities and write, to a first order,

$$(t_0)_{\text{obs}} - (t_0)_{\text{calc}} = \frac{\partial}{\partial h} (t_0)_{\text{calc}} \, \Delta h = \frac{2 \sqrt{V_2^2 - V_1^2(h)}}{V_2 V_1(h)} \, \Delta h$$

and solve for Δh. The adjusted value $h + \Delta h$ is a first improvement on the initial estimate; if the change is not negligible, the process can be carried through a second iteration, and a third, until the calculations converge satisfactorily. Ordinarily, only a very few iterations should be needed.

Difficulties arise in refraction interpretation when the velocity does not increase monotonically with the depth. Since no rays can be returned to the surface from within a low-velocity zone, there is no way to measure $V(z)$ within such zones. The WHB integral method is therefore not applicable; in fact it is frequently impossible even to recognize the existence of such a zone from an examination of the arrival-time curve. Figure 5-17 shows two such examples. In the first, two layers having uniform velocities produce a time-distance graph for first arrivals similar to that of a single layer. Application of the formulas (5-32) and (5-33) will yield an erroneous depth to the high-velocity medium. In the second example, we find that ray-path theory predicts a shadow zone where no refracted arrival occurs. However, this is not likely to be a useful diagnostic feature since normal field procedures provide only a limited number of points on the time-distance curve; consequently a small shadow zone could very easily be missed.

Velocity minima are not the only sources of difficulty. Since it is common practice in refraction methods to observe only the first arrival at each seismometer, a direct measurement of $V(z)$ is usually possible only over a restricted range of z. The remainder is known as the blind zone. For a structure made up of uniform layers it actually includes the entire section except for those interfaces where the velocity increases. Where $dV/dz > 0$, however, the blind zone is more limited. Figure 5-18 shows an example where initially the velocity increases linearly with depth from a value of 4,500 ft/sec. Underlying this region is a uniform medium which has a velocity of 12,500 ft/sec. If only first-arrival times are available, a blind

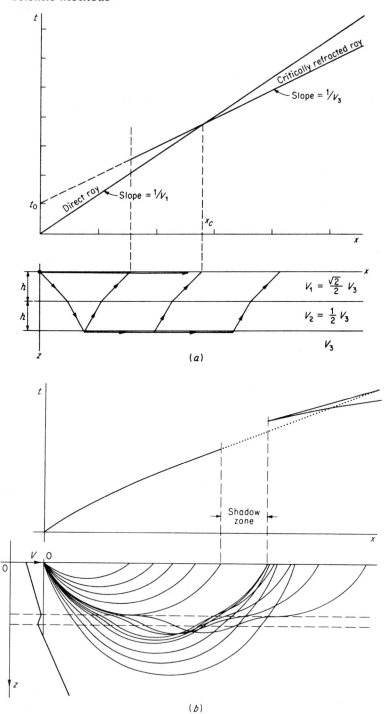

Fig. 5-17   (*a*) **The effect of a homogeneous low-velocity layer;** (*b*) **the formation of a shadow zone by a region of decreasing velocities.**   [After Goguel (8).]

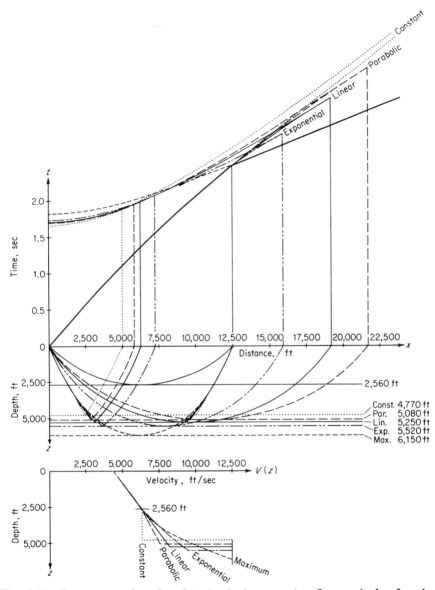

Fig. 5-18 **Some examples of ambiguity in interpreting first-arrival refraction data.** [After Hagedoorn (7b).]

zone exists from 2,625 ft to the depth where $V(z)$ reaches 12,500 ft/sec. Between these limits $V(z)$ may be assumed to have any value, provided only that the crossover distance remains the same. Hagedoorn (7b) has worked out interpretations by making five different assumptions, and he finds depths to the high-velocity medium ranging from 4,770 to 6,150 ft.

5-8   *On the Interpretation of Reflections*

Barring unfavorable circumstances, reflections may be used to map the topography of the zones or interfaces from which they originate. One of the most common field techniques is the "split-spread," or "continuous-profiling," method. By laying out the seismometers as shown in Fig. 5-19, it is possible to correlate events from record to record and so build up a continuous profile of reflections. To obtain a three-dimensional picture of the reflector, the lines are laid out in a network on the ground surface. For details of field procedures, see Dix (3), chap. 2.

Reflections from the deeper horizons arrive almost simultaneously at all the seismometers. This helps to distinguish them from the slow surface waves which will cross the seismic record more obliquely. The arrival time of a reflection is ordinarily chosen as close as is practicable to the time of onset, so that the energy may be interpreted as having followed the geometrical ray path, or the path of least time. Sometimes small allowances must be made for the blunting or dispersion of the pulse and for the delay introduced by the electronic filters in order to refine the estimates of geometrical ray-path times. Reflection coefficients and amplitudes, whenever they are used, are calculated from the plane-wave theory.

To translate ray-path reflection times into depths, it is the usual practice to introduce two simplifications. Firstly, times are taken from the trace of the seismometer nearest the shot point, so that the upward ray path virtually retraces the downward one. Secondly, the local velocity function is assumed to depend only upon the depth. When the velocity function is determined, it is possible by making an integration to draw up a curve of the two-way

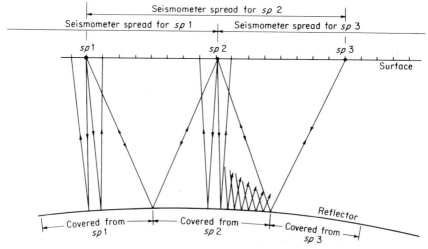

Fig. 5-19   The field procedure for continuous-reflection profiling.

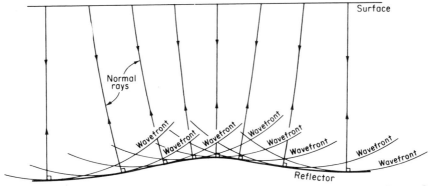

Fig. 5-20   The reflector surface as the envelope of wavefronts drawn from the shot points.

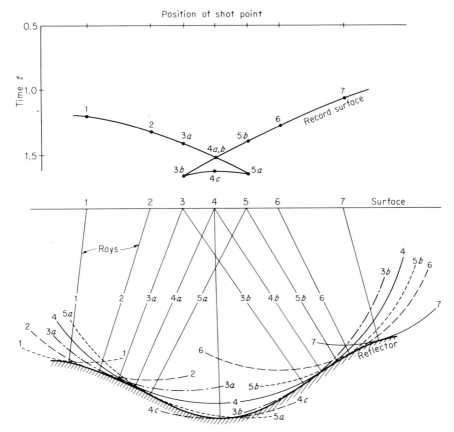

Fig. 5-21   Double reflections and cusps observed over a reflector more concave than the incident wavefront.

reflection time versus the depth for vertical ray paths. This curve enables us then to convert record times into equivalent depths, which are true depths only if the ray paths are vertical. The surface defined in this way will be called the *record surface*, and the actual lithologic interface to be derived from it will be called the *reflector surface*. The two surfaces coincide if they are horizontal.

Let us now consider the nature of the modifications that must be introduced into this picture when the reflector surface is sloping or uneven. Since the ray paths with which we are concerned are at all times normally incident on the reflector, the reflector surface is tangent to the incident wavefront, and it is therefore an osculating surface. The wavefronts, whose shape depends upon the velocity function, can be drawn at each shot point corresponding to half the measured reflection time at the nearest seismometer. The reflector surface will then be the common envelope of these wavefronts (Fig. 5-20). For an envelope to have meaning, however, its curvature must always be less than that of the generating surfaces. This statement may not always be true of the reflector surface, although it is likely to fail only in regions where the structures are highly disturbed. The meaning of this statement can easily be seen from Fig. 5-21. If the curvature of the reflector exceeds that of the wavefront, reflections can originate

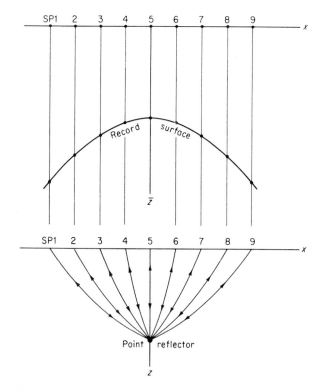

Fig. 5-22 Record surface of maximum convexity formed by reflections from a point.

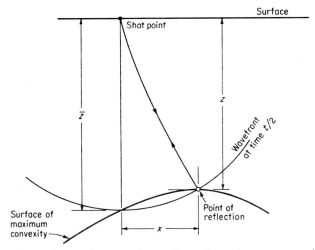

Fig. 5-23   A wavefront and a surface of maximum convexity.

simultaneously from more than one point, and an ambiguity arises in locating the point of reflection.

The record surface can also be defined as an envelope. The generating surface is called a *surface of maximum convexity*, and it is in some sense analogous to a wavefront. It is in fact the record surface which corresponds to the reflections originating from a point reflector (Fig. 5-22). Like the wavefronts, the surfaces of maximum convexity can be derived from the velocity-depth function. In particular, if the velocity is constant, they are a family of hyperboloids of revolution, while the wavefronts are a family of hemispheres. The record surface, then, is the envelope of the surfaces of maximum convexity corresponding to each of the points on the actual reflector.

To draw the family of wavefronts, we require a table of x and z values which satisfy the known relation $z = f(x;\bar{z})$, and for the surfaces of maximum convexity we require x and $\bar{z}$ values which satisfy the reciprocal relation $\bar{z} = g(x;z)$, where $\bar{z}$ is the equivalent vertical depth (the maximum depth reached by a wavefront or the vertical distance to a surface of maximum convexity). The f and g functions themselves are defined from the ray-path integrals as follows (see Fig. 5-23):

$$2 \int_0^z \frac{d\zeta}{V(\zeta)\sqrt{1 - p^2 V^2(\zeta)}} = t = 2 \int_0^{\bar{z}} \frac{d\zeta}{V(\zeta)}$$

and

$$p \int_0^z \frac{V(\zeta)\,d\zeta}{\sqrt{1 - p^2 V^2(\zeta)}} = x$$

Except for the simplest forms of $V(z)$ it is impractical to try to perform the integrations and to eliminate p in order to obtain $z = f(x;\bar{z})$ and

$\bar{z} = g(x;z)$ in explicit form. Consequently it is necessary, in practice, to tabulate t and x for a large number of p and z values (including $p = 0$, for which $z = \bar{z}$) and then to obtain the necessary pairs of values of x and z and of x and $\bar{z}$ by an interpolation procedure.

If the record surface is single-valued, it is a straightforward matter to deduce the reflector surface from it. The process, known as migration, is shown in two dimensions in Fig. 5-24. The position of the point of reflection corresponding to the point Q on the record surface is found by fitting one of the family of surfaces of maximum convexity to the slope of the record surface at Q. The point of reflection must then be situated at the position of minimum depth of that surface (marked P in the figure). The slope of the reflector surface at P is given by the slope of the wavefront through that point.

The process is shown in three dimensions in Fig. 5-25. Since the surfaces of maximum convexity and the wavefronts are both axially symmetrical, their slopes are radial. The plane in which the migration takes place is therefore a vertical plane through P and Q, and on contour maps the line along which the migration takes place must always be perpendicular to the contours of the record surface at Q or to those of the reflector surface at P.

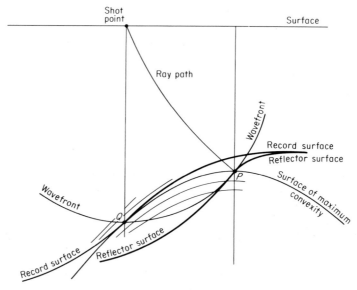

Fig. 5-24 Two-dimensional migration. A family of surfaces of maximum convexity is moved laterally (keeping its origin at the surface and its z axis vertical) until one of the curves matches the slope of the record surface at Q. The point of reflection must then lie at P, the position of the "point reflector" which generates that surface. The attitude of the reflector may be determined by drawing the wavefront through P.

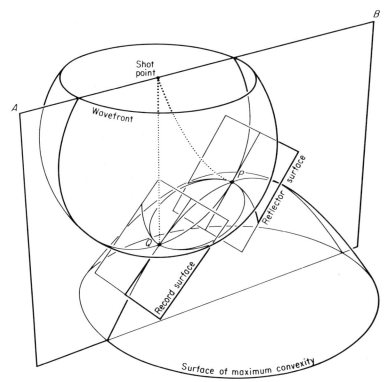

Fig. 5-25 **The migration process in three dimensions. Notice that the trace of the migration is perpendicular to the contours of the record surface at *Q* and to the contours of the reflector surface at *P*. [After Hagedoorn (7*a*).]**

The actual procedure for performing the migration of a record cross section or of a time contour map may be carried out in a variety of ways. The most direct is a graphical process described by Hagedoorn (7*a*), from which the foregoing outline is drawn. Other methods both graphical and analytical are available, but for the most part they deal only with special velocity-depth relationships.

The concept of a wavefront chart may be extended to situations where there is a large separation between the shot point and seismometer. The equivalent of a family of wavefronts is a family of surfaces of constant reflection time. These can be drawn up from a pair of wavefront diagrams as is shown in Fig. 5-26. If the velocity is uniform, the surfaces are prolate spheroids with the shot point and seismometer located at the foci. The times measured at distant geophones are used to find the dip and migration of the reflector only if there are no adjacent records with which correlations can be made. The methods of deriving the migration are usually approximate and based upon special velocity-depth relationships.

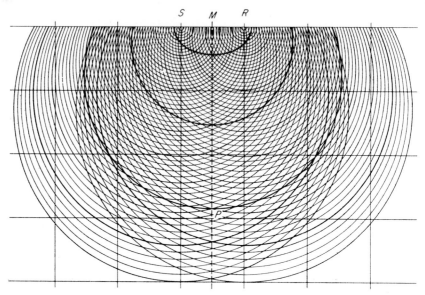

Fig. 5-26 Derivation of surfaces of equal reflection time from two wavefront charts by joining crossing points where the sum of the two reflection times is a constant. [After Hagedoorn (7a).]

5-9 *Velocity Functions*

In many interpretation procedures (e.g., in the mapping of reflector surfaces) the assumption is made that the seismic velocity varies with the depth only. Since velocities are measured in practice only at a very few locations within a given area, there is little to be gained by using a detailed velocity-depth function such as a velocity log, since a smooth function in general will predict local values just as reliably as the more elaborate one will. At the same time it is convenient to approximate the velocity-depth relationship with a simple mathematical form $V = V(z)$ for the purpose of drawing templates. Two commonly used functions are

$$V = V_0\left(1 + \frac{z}{l}\right) \qquad \text{i.e.,} \qquad t = V_0 l \ln\left(1 + \frac{z}{l}\right)$$

$$\text{and} \qquad V = V_0\left(1 + \frac{z}{l}\right)^{\frac{1}{2}} \qquad \text{i.e.,} \qquad t = 2V_0 l\left[\left(1 + \frac{z}{l}\right)^{\frac{1}{2}} - 1\right]$$

To find a smooth velocity-depth relationship, an obvious procedure would be to fit a simple function such as the one above to the velocity data. However, it must be remembered that in almost all cases it is the relationship between time and depth which is required for interpretation. The fitting process should therefore be carried out on the time-depth graph whenever possible. Plotting on double logarithmic scales can facilitate the

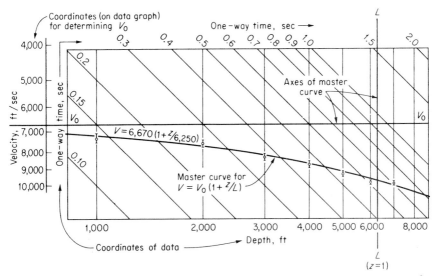

Fig. 5-27 Fitting a velocity function to observed time-depth data with a logarithmic master curve having oblique coordinates. [After Hagedoorn (7b).]

fitting of certain types of velocity functions since the free parameters can then be obtained by fitting a single master curve to the data. An example of this is shown in Fig. 5-27. However, this method has one important disadvantage; viz., the near-surface part of the data is strongly emphasized on the graph, and consequently great care must be taken in the fitting to allow for this. If confidence limits are plotted around each of the observed values of time and depth, more meaningful results can be obtained.

References

1. P. Morse and H. Feshbach, "Methods of Theoretical Physics," McGraw-Hill, New York, 1954.

2. H. Jeffreys and B. S. Jeffreys, "Mathematical Physics," Cambridge, London, 1951.

3. C. H. Dix, "Seismic Prospecting for Oil," Harper, New York, 1956.

4. L. B. Slichter, Interpretation of Seismic Travel-time Curves in Horizontal Structures, *Physics*, vol. 3, pp. 273–295, 1932.

5. A. Joset and J. J. Holtzscherer, Étude des vitesses de propagation des ondes seismiques sur l'inlandsis du groenland, *Annales Géophysique*, vol. 9, pp. 330–344, 1953.

6. K. E. Bullen, Seismic Ray Theory, *Royal Astron. Soc. Geophys. Jour.*, vol. 4, pp. 93–105, 1961.

7. J. G. Hagedoorn, (a) A Process of Seismic Reflection Interpretation, *Geophys. Prosp.*, vol. 2, pp. 85–127, 1954. (b) Templates for Fitting Smooth Velocity Functions to Seismic Refraction and Reflection Data, *Geophys. Prosp.*, vol. 3, pp. 325–338, 1955.

8. J. M. Goguel, Seismic Refraction with Variable Velocity, *Geophysics*, vol. 16, pp. 81–101, 1951.

chapter 6   The Reflection and Refraction of Spherical Waves

Up to this point, we have dealt with the propagation of seismic waves in layered media within the framework of the plane-wave theory. So far it has been possible to discuss most of the phenomena that take place near boundaries without having to abandon the simplicity of this limited viewpoint. However, the shortcomings of the plane-wave theory have already been felt. It does not, for instance, account for the existence of head waves nor, for that matter, of surface waves, except by the use of physically unimaginable complex angles of incidence. It has been suggested earlier that these phenomena are in some way related to the curvature of the incident wavefront, and in this chapter we shall show that this is indeed the case.

It is obvious that curved wavefronts tend to become plane as they propagate outward to great distances from the source, and so we should expect that in the "far-field" zone the plane-wave theory will provide a good approximation to the exact theory. But what is the exact theory and how good is this approximation, and for that matter where does the far-field zone begin?

When we remember that in applied seismology the source of the seismic waves is a point source, and that in terms of the seismic wavelength it is seldom far removed from elastic boundaries, these questions are obviously germane. In seeking to answer them we shall find out more about the reliability of the methods of geometrical optics, upon which the great majority of seismic interpretations are based.

6-1 The Weyl Solution

The problem of adapting the plane-wave theory to radiation from a point source is truly forbidding if the medium is aeolotropic, but it can be worked out with few formal difficulties if the medium is isotropic and homogeneous. The solution was obtained by H. Weyl in 1919 (1) by the use of an integral representation of the spherical Hankel function. Thus, as an example of a more general result [see, for example, Stratton (2), p. 577] we may write

$$\frac{e^{ikR}}{R} = ikh_0^{(1)}(kR) = ik \int_0^{\pi/2 - i\infty} e^{ikR \cos \gamma} \sin \gamma \, d\gamma$$

where the quantities R and γ are illustrated in Fig. 6-1. The left-hand side of this equation may be recognized as the solution of the scalar Helmholtz equation with a point singularity at Q. Accordingly it represents the isotropic steady-state radiation from a point source located at Q with the time factor $e^{-i\omega t}$ omitted. Using the treatment of the problem given by Stratton, we may introduce an unspecified equatorial angle τ, whose axis coincides with R, and write

$$\frac{e^{ikR}}{R} = \frac{ik}{2\pi} \int_0^{2\pi} \int_0^{\pi/2 - i\infty} e^{ikR \cos \gamma} \sin \gamma \, d\gamma \, d\tau$$

in order to give it the general appearance of an integration over the surface of a sphere. Since the integral (assuming it exists) is invariant under a

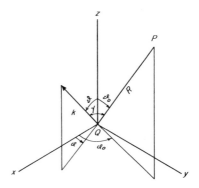

Fig. 6-1 Coordinates used in the Weyl solution.

rotation of the reference axes, we may write it as

$$\frac{e^{ikR}}{R} = \frac{ik}{2\pi} \int_0^{2\pi} \int_0^{\pi/2 - i\infty} \exp ik(x \sin \vartheta \cos \varphi + y \sin \vartheta \sin \varphi + z \cos \varphi)$$
$$\sin \vartheta \, d\vartheta \, d\varphi$$

where

$$x = R \sin \vartheta_0 \cos \varphi_0 \qquad y = R \sin \vartheta_0 \sin \varphi_0 \qquad z = R \cos \vartheta_0 \quad (6\text{-}1)$$

thus bringing out the plane-wave nature of the solution. In this way, the radiation from a point source in an unbounded isotropic medium may be represented as a bundle of plane waves traveling outward in all directions, complex as well as real. Where boundaries exist, additional waves are generated as a consequence of the boundary conditions. These new waves may be given a general integral representation, in the spirit of Weyl's method, as follows:

$$\Psi_R(\mathbf{r}) = \frac{ik}{2\pi} \int_0^{2\pi} \int_0^{\pi/2 - i\infty} f(\vartheta, \varphi) \exp (ikR \cos \gamma) \sin \vartheta \, d\vartheta \, d\varphi$$

where the amplitude function $f(\vartheta, \varphi)$ must be determined from the boundary conditions as described in Sec. 3-1.

If we follow the method of Weyl to its conclusion, it would seem that the complete solution of the problem of propagation in bounded elastic media of a disturbance originating from a point source can be set up in terms of plane harmonic waves. The block diagram shown in Fig. 6-2 illustrates the sequence to be followed. Thus it would seem that the solutions obtained in earlier chapters for plane waves could be adapted to the study of curved wavefronts through a direct synthesis.

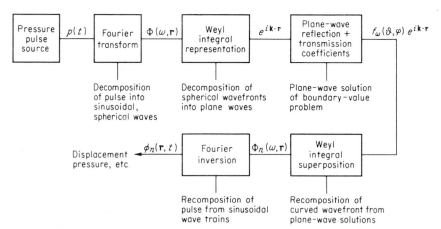

Fig. 6-2 The sequence used in solving transient problems involving a point source.

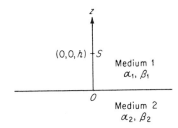

Fig. 6-3 Point source near a plane interface.

For example, let us consider a point explosion near a plane interface separating two different isotropic elastic half-spaces. Let the distance of the source from the interface be h, and let us assume that only a compressional strain is generated. The Fourier element of the incident pulse that corresponds to the wave number k_{α_1} may then be represented by an integral such as (6-1). If we choose an origin at O (Fig. 6-3) and write for the displacement potential of the incident wave

$$\Phi(\omega,\mathbf{r}) = \frac{\exp ik_{\alpha_1}|R - h|}{|R - h|}$$

$$= \frac{ik_{\alpha_1}}{2\pi} \int_0^{2\pi} \int_0^{\pi/2 - i\infty} \exp ik_{\alpha_1}\, [x \sin \vartheta \cos \varphi + y \sin \vartheta \sin \varphi$$
$$+ (h - z) \cos \vartheta]\, d\Omega$$

where $d\Omega = \sin \vartheta\, d\vartheta\, d\varphi$, then for the displacement potentials of the reflected waves we introduce integral representations as follows:

$$\Phi_1(\omega,\mathbf{r}) = \int_0^{2\pi} \int_0^{\pi/2 - i\infty} f_1(\vartheta) \exp ik_{\alpha_1}\, [x \sin \vartheta \cos \varphi + y \sin \vartheta \sin \varphi$$
$$+ (z + h) \cos \vartheta]\, d\Omega$$

$$\Psi_1(\omega,\mathbf{r}) = \int_0^{2\pi} \int_0^{\pi/2 - i\infty} g_1(\vartheta) \exp ik_{\beta_1}\, [x \sin \vartheta \cos \varphi + y \sin \vartheta \sin \varphi$$
$$+ (z + h) \cos \vartheta]\, d\Omega$$

(6-2a)

and for the transmitted waves

$$\Phi_2(\omega,\mathbf{r}) = \int_0^{2\pi} \int_0^{\pi/2 - i\infty} f_2(\vartheta) \exp ik_{\alpha_2}\, \left(x \sin \vartheta \cos \varphi + y \sin \vartheta \sin \varphi \right.$$
$$\left. - z \cos \vartheta + h \sqrt{\frac{\alpha_2{}^2}{\alpha_1{}^2} - \sin^2 \vartheta} \right) d\Omega$$

$$\Psi_2(\omega,\mathbf{r}) = \int_0^{2\pi} \int_0^{\pi/2 - i\infty} g_2(\vartheta) \exp ik_{\beta_2}\, \left(x \sin \vartheta \cos \varphi + y \sin \vartheta \sin \varphi \right.$$
$$\left. - z \cos \vartheta + h \sqrt{\frac{\beta_2{}^2}{\alpha_1{}^2} - \sin^2 \vartheta} \right) d\Omega$$

(6-2b)

where the reflection and transmission coefficients, which obviously are independent of φ, are given by the solutions of Eqs. (3-4). In this manner, all the wave functions are represented as superpositions of elementary plane waves which make complex as well as real angles of incidence with the boundary. But although it is an easy matter to set up the formal solutions in this way, the evaluation of these integrals proves to be forbiddingly difficult because of the complexity of the coefficients, and it has been accomplished only under rather restricted conditions.

6-2 The Sommerfeld Solution

A method of solution due to Sommerfeld makes direct use of the physical symmetry of the problem of the point source by employing cylindrical polar coordinates. The displacement vector u will have components u_r, u_{ϑ}, u_z in this system. However, if the incident disturbance is a pure P wave—an explosion, for example—no SH components of displacement are generated at the boundary, so that $u_{\vartheta} = 0$ everywhere. The displacement vector at any point is therefore fully specified by the two components u_r and u_z, and neither of these depends in any way upon ϑ. Consequently, the only component of the rotational displacement potential ψ that is needed is ψ_{ϑ}, and the problem reduces essentially from three dimensions to two.

In terms of the potentials ϕ and ψ the two components of the displacement are, in cylindrical coordinates,

$$u_r = \frac{\partial \phi}{\partial r} + \frac{\partial \psi_{\vartheta}}{\partial z} \quad \text{and} \quad u_z = \frac{\partial \phi}{\partial z} - \frac{1}{r}\frac{\partial}{\partial r}(r\psi_{\vartheta}) \tag{6-3}$$

According to (2-8), the two potentials are solutions of the scalar equations

$$\frac{\partial^2 \phi}{\partial t^2} = \alpha^2 \nabla^2 \phi \tag{6-4a}$$

and

$$\frac{\partial^2 \psi_{\vartheta}}{\partial t^2} = \beta^2 \left(\nabla^2 \psi_{\vartheta} - \frac{\psi_{\vartheta}}{r^2} \right) \tag{6-4b}$$

or, if we write $\psi_{\vartheta} = \partial \chi / \partial r$, we have instead of (6-4b)

$$\frac{\partial^2 \chi}{\partial t^2} = \beta^2 \nabla^2 \chi \tag{6-4c}$$

The axially symmetrical solutions of (6-4a) and (6-4c) for sinusoidal waves of frequency ω are, respectively, functions of the type

$$\Phi(\omega, \mathbf{r}) \sim e^{\pm \nu z} J_0(\sqrt{\nu^2 + k_{\alpha}^2}\, r)$$

and

$$X(\omega, \mathbf{r}) \sim e^{\pm \nu z} J_0(\sqrt{\nu^2 + k_{\beta}^2}\, r)$$

where ν may have any real value. Consequently we may build up general solutions for the incident, reflected, and transmitted waves as follows. Using the identity

$$\Phi_0(\omega, \mathbf{r}) = \frac{\exp\, ik_{\alpha_1} R_1}{R_1} = \int_0^{\infty} \exp\left(-|z - h| \sqrt{k^2 - k_{\alpha_1}^2}\right) J_0(kr) \frac{k\, dk}{\sqrt{k^2 - k_{\alpha_1}^2}}$$

where $k = \sqrt{\nu^2 + k_{\alpha_1}^2}$ $\qquad R_1 = \sqrt{r^2 + (z - h)^2}$

to represent the radiation from a point source of P waves at $z = h$ in medium 1, we may write for the reflection terms

$$\Phi_1(\omega, \mathbf{r}) = \int_0^\infty f_1(k) \exp\left(- \sqrt{k^2 - k_{\alpha_1}{}^2}\, z\right) J_0(kr)\, dk$$

and
$$X_1(\omega, \mathbf{r}) = \int_0^\infty g_1(k) \exp\left(- \sqrt{k^2 - k_{\beta_1}{}^2}\, z\right) J_0(kr)\, dk$$

and a similar pair of integrals for the potentials of the transmitted wave. The stress components at the boundary, written in terms of the displacement potential functions, are

$$p_{zz} = \lambda \nabla^2 \phi + 2\mu \left[\frac{\partial^2 \phi}{\partial z^2} - \frac{\partial}{\partial z}\left(\frac{\partial^2 \chi}{\partial r^2} + \frac{1}{r}\frac{\partial \chi}{\partial r} \right) \right]$$

$$p_{zr} = \mu \left[2\frac{\partial^2 \phi}{\partial r\, \partial z} - \frac{\partial}{\partial r}\left(\frac{\partial^2 \chi}{\partial r^2} + \frac{1}{r}\frac{\partial \chi}{\partial r} - \frac{\partial^2 \chi}{\partial z^2} \right) \right]$$

and these must be continuous. The displacement components (6-3) must also satisfy the condition of continuity. Application of these boundary conditions gives the reflection and transmission coefficients $f_1(k)$, $g_1(k)$, $f_2(k)$, and $g_2(k)$, which have forms as complicated as the plane-wave coefficients given in Sec. 3-1. To evaluate the integrals, it is best to employ Cauchy's method and to make use of asymptotic approximations. This requires us to identify all the singularities in the integrands and to attach the correct physical interpretation to the residue terms from each.

There are four branch points, which lie at $k = k_{\alpha_1}$, k_{β_1}, k_{α_2}, and k_{β_2}. These contribute waves of a new type, called *head waves*, which we have already mentioned in Chap. 3 and about which we shall have much more to say. The derivation of these waveforms from the Sommerfeld solutions is very difficult, and we shall not make the attempt. A simple pole exists where

$$k^2[2k^2(\mu_2 - \mu_1) - \omega^2(\rho_2 - \rho_1)]^2$$
$$- \sqrt{k^2 - k_{\alpha_1}{}^2}\, \sqrt{k^2 - k_{\alpha_2}{}^2}\, [2k^2(\mu_2 - \mu_1) - \rho_2\omega^2]^2$$
$$- \sqrt{k^2 - k_{\beta_1}{}^2}\, \sqrt{k^2 - k_{\beta_2}{}^2}\, [2k^2(\mu_2 - \mu_1) + \rho_1\omega^2]^2 - (\sqrt{k^2 - k_{\alpha_1}{}^2}\, \sqrt{k^2 - k_{\beta_2}{}^2}$$
$$+ \sqrt{k^2 - k_{\beta_1}{}^2}\, \sqrt{k^2 - k_{\alpha_2}{}^2})\omega^4\rho_1\rho_2 + 4(\mu_2 - \mu_1)^2 k^2 \sqrt{k^2 - k_{\alpha_1}{}^2}\, \sqrt{k^2 - k_{\beta_1}{}^2}$$
$$\times \sqrt{k^2 - k_{\alpha_2}{}^2}\, \sqrt{k^2 - k_{\beta_2}{}^2} = 0$$

This is exactly the equation for the Stoneley wave velocity c given by (3-10), if we write $k = \omega/c$. The integrands will thus have poles on the real axis under the conditions that Stoneley waves may exist. From this we may surmise that the residue term contributes the interface wave.

This approach has been developed extensively by Ewing, Jardetzky, and Press (4). It is formally equivalent to the method of Weyl, since the one may be transformed into the other by a change of coordinates. The demonstration of this statement is given by Stratton and is also to be found in several other textbooks. The numerical evaluation of the integrals has been carried out by the use of approximate methods, which become invalid if the receiver is close to the source or to the boundary but which are satisfactory in regions where the various wavefronts are fully formed.

6-3   The Work of L. Cagniard

In a memoir published in 1939, L. Cagniard (5) detailed the most thorough yet general theoretical investigation of the problem of the point source of seismic energy in the vicinity of a plane interface between two elastic media that has yet appeared. In many respects, it may even be said that the work of Cagniard has completed this problem, if it were not for the fact that practical difficulties in calculating the integrals have left possibilities open for the later publication of a number of less rigorous solutions, which can be more easily computed. Cagniard's solutions, however, being completely general, are well worth studying.

It is impossible within a brief space, and at the mathematical level we are entitled to assume, to do more than summarize the main results. Cagniard set about the problem in the following way: He employed the Sommerfeld formula for the steady-state source e^{ikR}/R in cylindrical coordinates, and the reflection and transmission coefficients appeared in the usual way as a result of applying the boundary conditions (6-3). To synthesize solutions for pulse forms, however, Cagniard employed the Laplace transform in preference to the Fourier transform. This allowed him to investigate solutions for pulses [in particular for the Heaviside step function] which in the strict sense have no Fourier transform. Then by the use of the convolution integral, the solution corresponding to any transient disturbance can be determined. This approach also has a heuristic advantage in that it is free of having to assume a priori the existence of steady-state solutions out of which the pulse problem must be synthesized. For, while the synthesis of steady-state solutions to form transients has a fairly rigorous mathematical basis, it is not always physically obvious that steady-state solutions should exist; for this reason the method of Fourier transforms is sometimes unsatisfying to physicists.

The pivotal feature of Cagniard's method consists in expressing Laplace-transformed quantities in the form of Laplace integrals through Carson's integral equation [Carson (6), Chap. 1], dispensing altogether with the need to perform some extremely complicated inversions from transform space back into the time domain. The method in outline is as follows: Let us denote functions in the time domain by lowercase symbols and the Laplace transforms of these quantities by the corresponding uppercase symbols. Then let us begin with a function $f(t)$, which describes the displacement potential of the source, which is represented as a point. Actually, although this departs from Cagniard's treatise, there is an arguable case for taking as an excitation function the displacement potential at the surface of a small spherical cavity whose radius is much smaller than the distance from the source to the interface, rather than at the very center, where the behavior must be highly conjectural. The incident wave can then be represented by (2-22), and we shall avoid some troublesome questions concerning differentiations at the singularity itself. With or without this minor modification, however, we can represent the Laplace transform of the excitation function by a function $F(s)$, known or calculable, of a real, positive variable s. If the source happens to be a simple explosion, then at a distance $r > a$ from its center

$$F(s) = \int_0^\infty e^{-st} f\left(t - \frac{r - a}{\alpha}\right) dt$$

$$= \frac{e^{-s(r-a)/\alpha}}{r - a} \int_0^\infty e^{-st} f(t) \, dt = \frac{e^{-s(r-a)/\alpha}}{r - a} F_0(s)$$

By applying the boundary conditions, we shall obtain in the usual way additional displacement potentials for the reflected and transmitted waves, whose Laplace transforms are given in each of the two media by formulas of the kind

$$\Phi(\mathbf{r},s) = X(\mathbf{r},s) F(s)$$

and
$$\Psi(\mathbf{r},s) = Y(\mathbf{r},s) F(s)$$

If we write

$$X(\mathbf{r},s) = sP(\mathbf{r},s) \qquad Y(\mathbf{r},s) = sQ(\mathbf{r},s)$$

the functions $p(\mathbf{r},t)$ and $q(\mathbf{r},t)$ to which the Laplace transforms P and Q correspond will be solutions of Carson's equation. The solutions for the displacement potentials in the time domain can then be calculated by means of the convolution theorem for Laplace integrals

$$\phi(\mathbf{r},t) = \int_0^t f(\tau) p(\mathbf{r},\, t - \tau) \, d\tau$$

$$\psi(\mathbf{r},t) = \int_0^t f(\tau) q(\mathbf{r},\, t - \tau) \, d\tau$$

By using the Sommerfeld formula in cylindrical coordinates and applying the boundary conditions, the quantities $X(s)$ and $Y(s)$ are obtained without much difficulty. They are in the form of Sommerfeld integrals which contain rather complicated reflection and transmission coefficients. The formal transition from X and Y to p and q is effected through some rather skillful manipulations of the independent variable, and at this stage the theoretical solution is essentially complete. Practically, however, there remain considerable difficulties in evaluating the quantities p and q, which are in the form of complex integrals. These integrals must be evaluated by the Cauchy residue theorem, and it is therefore imperative to attach the correct physical interpretations to the terms corresponding to the various singularities of the integrands.

To illustrate the nature of the solution, the form for the transmission term $q_2(\mathbf{r},t)$ is

$$q_2(\mathbf{r},t) = \frac{4}{\pi r} \oint \frac{\nu - a_1h - b_1z}{\sqrt{\lambda^2r^2 + (\nu - a_1h - b_1z)^2}} f(\lambda)\lambda \, d\lambda$$

where ν is the equivalent of time and f is the transmission coefficient given by

$$f(\lambda) = \frac{(\lambda^2 + u^2 + v^2)(\lambda^2 + v^2) - a_2b_2(\lambda^2 + u^2)}{\Delta(\lambda)} \qquad (6\text{-}5)$$

where $\quad \Delta(\lambda) = \lambda^2(\lambda^2 + u^2 + v^2)^2 + \lambda^2 a_1a_2b_1b_2 - a_1b_1(\lambda^2 + v^2)^2$

$$- a_2b_2(\lambda^2 + u^2)^2 + u^2v^2(a_1b_2 + a_2b_1)$$

in which
$$a_1 = \sqrt{\lambda^2 + \frac{1}{\alpha_1^2}} \qquad b_1 = \sqrt{\lambda^2 + \frac{1}{\beta_1^2}}$$

$$a_2 = \sqrt{\lambda^2 + \frac{1}{\alpha_2^2}} \qquad b_2 = \sqrt{\lambda^2 + \frac{1}{\beta_2^2}}$$

and
$$u^2 = \frac{\rho_1}{2(\rho_1\beta_1^2 - \rho_2\beta_2^2)} \qquad v^2 = \frac{\rho_2}{2(\rho_2\beta_2^2 - \rho_1\beta_1^2)}$$

The path of integration in the complex λ plane is indicated in Fig. 6-4.

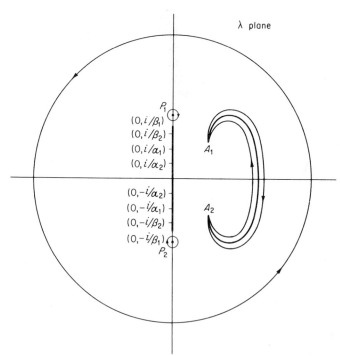

Fig. 6-4   Path of integration used in evaluating the transmission coefficient $q_2(\mathbf{r},t)$. [According to Cagniard (5).]

The condition for a simple pole is that $\Delta(\lambda) = 0$, and this is exactly equivalent to Eq. (3-10), which specifies the condition for the existence of Stoneley waves. Thus we may anticipate that the residue terms will correspond to waves propagated along the interface.

The integrand also contains two branch points which lie at $\lambda = \pm i/\alpha_2$. Thus there will be a contribution from the branch line

$$\frac{r}{\alpha_2} + a_1 h + b_1 z = \nu \qquad r > \frac{\alpha_2 \nu}{a_2{}^2 b_1{}^2 + 1}$$

since the other branch line lies outside of the path of integration. The above equation may be rewritten as

$$\sqrt{\alpha_2{}^2 - \beta_1{}^2}\, z + \frac{h\beta_1 \sqrt{\alpha_2{}^2 - \alpha_1{}^2}}{\alpha_1} + \beta_1(r - \alpha_2 t) = 0 \qquad r > \frac{\beta_1{}^2}{\alpha_2} t \quad (6\text{-}6)$$

Equation (6-6) is easily identified as the frustum of a right circular cone which is coaxial with the z axis and whose side makes an angle $\sin^{-1}(\beta_1/\alpha_2)$ with the interface $z = 0$. It is truncated at the upper end at the line of tangency with the reflected SV wavefront, and propagates with the shear-wave velocity β_1. This conical wavefront therefore corresponds to the $P_1 S_2 S_1$ head wave, which must exist if the coefficients of (6-6) are real. The existence of $P_1 P_2 P_1$, $P_1 P_2 S_1$, and $P_1 S_2 P_1$ head waves may be demonstrated under the appropriate velocity conditions in a similar manner. Cagniard thus provided, for the first time it appears, a rigorous mathematical theory of head waves, and in doing so finally laid to rest arguments and debates concerning the problem of critical refraction which had arisen repeatedly throughout the previous half-century.[1] Since then, the reality of head waves has been demonstrated many times in the laboratory by means of high-speed photography.

[1] In fairness to earlier writers, it ought to be emphasized that the "Cagniard technique," as described above, did not originate altogether in the work of Cagniard. What it amounts to essentially is working the boundary and initial value solutions in (Laplace) transform space into forms which can be recognized as Laplace integrals. The solution of the transient problem can then be written down more or less by inspection. (It ought to be pointed out that this is equivalent to *constructing* a plausible solution, not to deducing it; however, Cagniard was able to establish the uniqueness of such a solution.) According to this definition, the Cagniard technique (or the Laplace transform method) was in use much earlier. In particular, the methods used by Sir Horace Lamb (7) to study the causes of Rayleigh waves bear some resemblance to Cagniard's work on certain points, as do the further developments of Lamb's results by Lapwood (8). Considering the power of the method of Laplace transforms, this is not surprising. However, Cagniard has made the most complete analysis of the problem involving two different elastic media that is known to the authors.

It may be of further interest to note that Cagniard's problem has also been solved with a somewhat different approach by Zvolinskii (9).

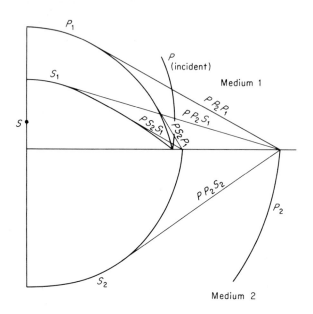

Fig. 6-5 The head waves at an interface where $\alpha_1 < \beta_2$. [After Cagniard (5).]

The accomplishment of the Cagniard theory does not end quite here. Because the solutions are exact and asymptotic methods are not used, the method is capable of determining the forms of the wavefronts from the instant that the incident pulse reaches the boundary. Asymptotic formulas, such as the method of steepest descents, require sufficient time to elapse for all the wavefronts to become fully formed before they are able to give a reasonably accurate picture of events. The results are interesting and to some degree unexpected. If the incident disturbance is a pure P pulse and the wavefront is spherical, the reflected P wave has a spherical front, but that of the reflected S wave is hyperboloidal. The fronts of the transmitted P and S waves are both spheroidal, although they approach the spherical form fairly rapidly. If both compressional and shear-wave velocities in medium 2 exceed the P-wave velocity in medium 1, there will be five head waves altogether, of which one will occur in medium 2 and the remaining four in medium 1. This is the maximum number possible. A wavefront diagram for such a case at an instant shortly after incidence is shown in Fig. 6-5. If the shear-wave velocity in medium 2 is less than the compressional-wave velocity in medium 1, the transmitted shear wave has a hyperboloidal front, and there will be three head waves, of which two will lie in medium 1. This case is illustrated shortly after incidence in Fig. 6-6.

The Cagniard theory has been verified experimentally in a very striking manner. Using a spark source and a small piezoelectric detector, Roever and Vining (10) measured the pressure response to a very sharp pulse emitted in a deep layer of water lying on a thick substratum of pitch. The solution of this problem was also computed from Cagniard's formulas by

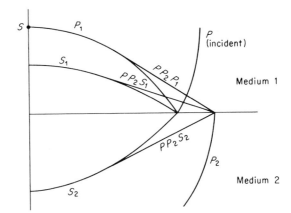

Fig. 6-6 The head waves at an interface where $\alpha_1 > \beta_2$; $\beta_1 < \beta_2$. [After Cagniard (5).]

Strick (10), who used a Dirac delta function as an input signal. The two response curves are shown together in Fig. 6-7. The PPP head wave t_p, the direct wave t_D, and the Stoneley wave t_{st} have been predicted by the calculations with remarkable accuracy both in their phase velocities and in their frequency spectra. The inability of the apparatus to detect the reflected wave close behind the direct wave is undoubtedly due to shock excitation ("ringing") in the transducer.

One final remark concerns the interface waves. As Cagniard pointed out and as Strick has elaborated, a special situation arises when the medium containing the source is a fluid whose wave velocity is less than that of the Rayleigh wave in the substratum. In such a case a new head wave appears

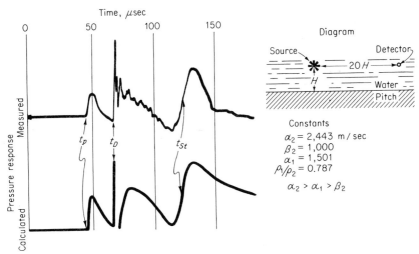

Fig. 6-7 Experimental and theoretical pressure response in water due to a step pulse emitted from a transducer close to a water-pitch interface. [After Roever, Vining, and Strick (10).]

in the fluid, which is coupled with the Rayleigh wave[1] on the interface. Such waves are called *Rayleigh-coupled head waves* by seismologists. Near the boundary, they are more intense than the *P*-coupled head wave, but being more oblique, their critical distance from the source is greater. These waves have been observed in laboratory models by many experimenters.

6-4   The Work of L. M. Brekhovskikh

The great success of Cagniard's method lay in producing for the first time a rigorous solution of the problem of the reflection and transmission of spherical waves at a plane elastic boundary. These solutions appear to be valid everywhere, even in those regions near the boundary where ray-path solutions fail. However, it cannot be denied that the computations require a great deal of labor to perform, particularly as the time of propagation increases. Another type of solution, which is more easily calculated but less general than Cagniard's since it is based upon the far-field approximation, is found in a contribution by L. M. Brekhovskikh (11). The main body of this work has recently been included in a general treatise on wave propagation by the same author (12), but because of its great importance to our curriculum we shall summarize the essential points here.

Brekhovskikh's analysis is based on the solution of Weyl, given in (6-2). If we first perform the integrations with respect to φ, we shall get four expressions of the kind

$$\Phi_1(\omega, \mathbf{r}) = ik_{\alpha_1} \int_0^{\pi/2 - i\infty} f_1(\vartheta) \exp\left[ik_{\alpha_1}(z + h) \cos \vartheta\right] J_0(k_{\alpha_1} r \sin \vartheta) \sin \vartheta \, d\vartheta$$

for the Fourier transforms of the displacement potentials corresponding to the reflected and transmitted waves. By making use of the identity

$$J_0(u) = \tfrac{1}{2}[H_0^{(1)}(u) + H_0^{(2)}(u)]$$

we may rewrite the integrals in the form

$$\Phi_1(\omega, \mathbf{r}) = \frac{ik_{\alpha_1}}{2} \int_{-\pi/2 + i\infty}^{\pi/2 - i\infty} f_1(\vartheta) \exp\left[ik_{\alpha_1}(z + h) \cos \vartheta\right] H_0^{(1)}(k_{\alpha_1} r \sin \vartheta) \sin \vartheta \, d\vartheta$$

by virtue of the fact that $H_0^{(2)}(-u) = -H_0^{(1)}(u)$, while $f_1(-\vartheta) = f_1(\vartheta)$.

In the wave zone, where $kr \gg 1$, the Hankel function may be replaced by its asymptotic expression. Thus

$$H_0^{(1)}(k_{\alpha_1} r \sin \vartheta) \cong \sqrt{\frac{2}{\pi k_{\alpha_1} r \sin \vartheta}} \left(1 + \frac{1}{8ik_{\alpha_1} r \sin \vartheta} + \cdots\right)$$
$$\times \exp i\left(k_{\alpha_1} r \sin \vartheta - \frac{\pi}{4}\right)$$

---

[1] What we refer to in this paragraph as the "Rayleigh wave" is in reality a fluid-coupled surface wave of the Rayleigh type, which tends to a pure Rayleigh wave as the fluid density diminishes. Some writers, including Cagniard, prefer to apply the term "Rayleigh wave" to the disturbance which travels along only a perfectly free boundary and instead describe the fluid-coupled motion by the term "pseudo-Rayleigh wave."

and the integrals take the form

$$\Phi_1(\omega,\mathbf{r}) \cong e^{i\pi/4}\sqrt{\frac{k_{\alpha_1}}{2\pi r}}\int_{-\pi/2+i\infty}^{\pi/2-i\infty} f_1(\vartheta)\left(1+\frac{1}{8ik_{\alpha_1}r\sin\vartheta}+\cdots\right)$$
$$\times \exp[ik_{\alpha_1}R_1\cos(\vartheta-\vartheta_0)]\sin^{1/2}\vartheta\,d\vartheta \quad (6\text{-}7)$$

where $r = R_1\sin\vartheta_0$ and $z + h = R_1\cos\vartheta_0$ (Fig. 6-8).

The integral (6-7) and its three correlatives are now to be evaluated by the method of steepest descent. The application of this method requires deforming the path of integration in the complex ϑ plane into one such that the real part of the integrand passes through its maximum value within the shortest possible distance along it. It is shown in several textbooks why such a contour must be a path of constant phase for the integrand and how the principal contribution to the integral arises from within the near vicinity of the saddle points on this line. In this instance the saddle points lie at $\vartheta = \gamma$ where

$$\frac{d}{d\vartheta}\,\text{Im}\,[\cos(\vartheta-\vartheta_0)] = 0$$

There is only one such point and that is at $\vartheta = \vartheta_0$. Therefore the line of steepest descent is determined by the condition that $\text{Re}\,[\cos(\vartheta-\vartheta_0)] = 1$; it must begin at $\vartheta = -\pi/2 + \vartheta_0 + i\infty$, cross the real axis at $\vartheta = \vartheta_0$ at an angle of 45°, and finish at $\vartheta = \pi/2 + \vartheta_0 - i\infty$. The original path of integration and the line of steepest descent are indicated as C_1 and C, respectively, in Fig. 6-9.

In the far-field approximation, where the method of steepest descent is valid, the integral (6-7) becomes

$$\Phi_1(\omega,\mathbf{r}) \cong \frac{\exp ik_{\alpha_1}R_1}{R_1}\left[f_1(\vartheta_0) - \frac{iN(\vartheta_0)}{2k_{\alpha_1}R_1} + \cdots\right] \quad (6\text{-}8)$$

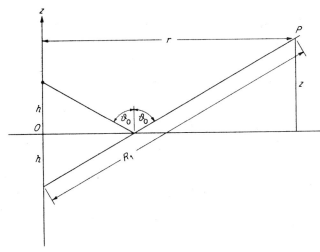

Fig. 6-8 Coordinates used in the asymptotic solutions of the Weyl integrals.

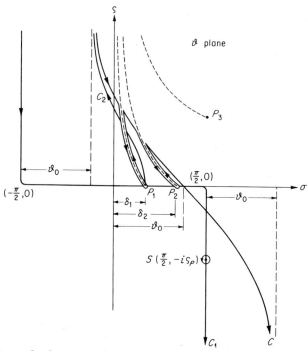

Fig. 6-9 **The path of steepest descent for the integrals representing the reflected and transmitted waves.**

where $N(\vartheta_0) = f_1''(\vartheta_0) + \cot \vartheta_0 f_1'(\vartheta_0)$. This clearly represents a spherical wave propagating away from the image formed by the source in the boundary and having an amplitude which is a complex function of ϑ. The first term, which is the most important, is the reflection predicted by geometrical optics using the reflection coefficient derived from the plane-wave theory. The interpretation of this result is that the elementary plane waves reflected in directions close to $\vartheta = \vartheta_0$ will travel more or less in phase with one another, whereas cancellations will occur in all other directions. The second term gives a minor correction to this result when kr is not too large, although it is rarely used in practice. For most purposes, the reflection coefficients derived in Sec. 3-1 are also used when the wavefronts are curved.

The integral (6-7) has a residue term if the integrand contains a pole. Poles will occur where the determinant of (3-4) $= 0$. The condition for this is given by Eq. (3-10) if we make the identification $c = \alpha_1/\sin \vartheta$, so that the residue term contributes the Stoneley wave. We have already noted that (3-10) has a real root only when $\beta_1 \approx \beta_2$ and that the value lies close to either shear-wave velocity. Thus the pole of the integrand must lie close to the point $\vartheta = \vartheta_P$, where $\sin \vartheta_P = \alpha_1/\beta_1$. Since $\sin \vartheta_P > 1$, we may put

$\vartheta_P = \pi/2 - i s_P$, where $s_P = \cosh^{-1}(\alpha_1/\beta_1)$. This point is well removed from the saddle point and therefore will not affect the reflected wave.

In addition to the pole, the reflection coefficient $f_1(\vartheta)$ has branch points at $\sin \vartheta = \alpha_1/\alpha_2$, α_1/β_2, and α_1/β_1. These points are indicated in Fig. 6-8 at P_1, P_2, and P_3, respectively. At each of these points, the ϑ plane must be cut along the line that causes the imaginary part of n_{α_2}, n_{β_2}, or n_{β_1} in Eq. (3-4) to vanish. Thus, for example, the cut which terminates at P_1 will form the junction of two Riemann sheets, on the upper one of which $\mathrm{Im} \sqrt{\alpha_1{}^2/\alpha_2{}^2 - \sin^2 \vartheta} > 0$ and on the lower, $\mathrm{Im} \sqrt{\alpha_1{}^2/\alpha_2{}^2 - \sin^2 \vartheta} < 0$. Since a negative imaginary part would, according to Sec. 3-5, give rise to a surface wave which would amplify rather than attenuate in the lower medium with increasing distance from the boundary, it is quite clear that the line C must in all cases remain on the *upper* Riemann surface. The equation of the branch cut is in this case

$$\frac{\alpha_1{}^2}{\alpha_2{}^2} - \sin^2 \vartheta = \lambda \qquad 0 \leq \lambda < \infty$$

where λ is real. The three branch lines are shown in Fig. 6-9. The line of steepest descent may intersect two of them, but not the third.

It now appears that in order to follow the line of steepest descent, it may be necessary to evaluate certain branch-line integrals. The number of these integrals may be two, one, or zero, depending upon the value of ϑ_0 (i.e., upon the distance of the receiver from the source). We shall now look at these in some detail. Let us first consider the branch cut terminating at the point P_1 where $\vartheta = \delta_1 = \sin^{-1}(\alpha_1/\alpha_2)$. Neglecting the correction term of $O(1/kR_1)$, we may write for the branch-line integral issuing from (6-7)

$$\Phi_1^*(\omega,\mathbf{r}) \cong e^{i\pi/4} \sqrt{\frac{k_{\alpha_1}}{2\pi r}} \int_{\delta_1}^{i\infty} f_1^*(\vartheta) \exp[ik_{\alpha_1} R_1 \cos(\vartheta - \vartheta_0)] \sin^{1/2} \vartheta \, d\vartheta$$

where $f_1^*(\vartheta)$ is the difference between the values of $f_1(\vartheta)$ taken on the upper and on the lower Riemann sheets at the same value of ϑ. We shall evaluate this integral approximately by the method of stationary phase, according to which we replace the branch line with the contour along which

$$\mathrm{Re}[\cos(\vartheta - \vartheta_0)] = \mathrm{const} = \cos(\delta_1 - \vartheta_0)$$

Let us write $\vartheta = \sigma + is$. Then

$$\cos(\vartheta - \vartheta_0) = \cos(\delta_1 - \vartheta_0) - i \sin(\sigma - \vartheta_0) \sinh s$$

along the path of stationary phase, and since the major contribution to the integral will occur in the near vicinity of $\vartheta = \delta_1$ (where the real part of the integrand decreases the most rapidly), we may set $\sigma \approx \delta_1$ and allow s to remain small.

According to a general property of analytic functions, we may allow

$$f_1^*(\vartheta) = \sqrt{\frac{\alpha_1{}^2}{\alpha_2{}^2} - \sin^2 \vartheta}\, F_1(\vartheta)$$

and assume that $F_1(\vartheta)$ does not possess an essential singularity at $\vartheta = \delta_1$. In the approximation discussed above

$$f_1^*(\vartheta) \cong e^{-i\pi/4} \sqrt{2 \sin \delta_1 \cos \delta_1}\, F_1(\delta_1)\, s^{\frac{1}{2}}$$

Hence the integral becomes

$$\Phi_1^*(\omega,\mathbf{r}) \cong \frac{i\alpha_1}{\alpha_2} \sqrt{\frac{k_{\alpha_1} \cos \delta_1}{\pi r}}\, F_1(\delta_1)\, \exp\, i k_{\alpha_1} R_1 \cos(\delta_1 - \vartheta_0)$$

$$\times \int_0^\infty \exp\,[-k_{\alpha_1} R_1 \sin(\delta_1 - \vartheta_0)s] s^{\frac{1}{2}}\, ds$$

$$= -\frac{1}{2} \frac{\alpha_1}{\alpha_2} \frac{\sqrt{\cos \delta_1}}{i k_{\alpha_1} \sqrt{r}}\, F_1(\delta_1)\, \frac{\exp\, i k_{\alpha_1} R_1 \cos(\delta_1 - \vartheta_0)}{[R_1 \sin(\delta_1 - \vartheta_0)]^{\frac{3}{2}}} \qquad (6\text{-}9)$$

Referring now to Fig. 6-10, we have $r = L + (z + h) \tan \delta_1$; hence

$$R_1 \cos(\delta_1 - \vartheta_0) = L_0 + L_1 + L \sin \delta_1$$

and

$$R_1 \sin(\delta_1 - \vartheta_0) = L \cos \delta_1$$

Consequently (6-9) may be rewritten as

$$\Phi_1^*(\omega,\mathbf{r}) \cong -\frac{\tan \delta_1 F_1(\delta_1)}{2 i k_{\alpha_1} r^{\frac{1}{2}} L^{\frac{3}{2}}} \exp\,[i k_{\alpha_1}(L_0 + L_1) + i k_{\alpha_2} L] \qquad (6\text{-}10)$$

which is clearly equivalent to propagation along the "critically refracted" ray path shown in Fig. 6-10. The solutions along the branch lines thus represent head waves. Unless the angle of incidence exceeds the critical angle, the line of steepest descent will traverse the branch line at two points, and there will then be no branch-line integral and hence no head wave. In the example shown in Fig. 6-9, it is assumed that $\alpha_1 < \beta_2$, so that two of the branch points lie on the real axis. If $\vartheta_0 > \delta_2 = \sin^{-1}(\alpha_1/\beta_2)$, there will thus be two head waves, both of which remain attached to the reflected P wavefront. At the boundary, the two head waves will terminate at the fronts of the transmitted P and SV waves. If $\vartheta_0 < \delta_1 = \sin^{-1}(\alpha_1/\alpha_2)$, no head waves will form from the reflected P wave. There may also, of course, be other head waves attached to the reflected SV wave, the complete system of these waves in medium 1 being represented in the case of incident P waves by the four possibilities $P_1P_2P_1$, $P_1S_2P_1$, $P_1P_2S_1$, and $P_1S_2S_1$. A more detailed account of the head-wave theory as applied to seismology has been given by Heelan (13).

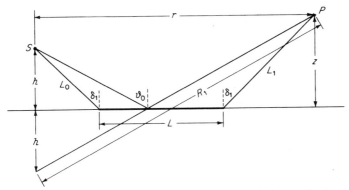

Fig. 6-10 The ray-path interpretation of the branch-line solutions. [After Brekhovskikh (12).]

The expression (6-10) is of course only the leading term in an asymptotic series, and as such it represents the solution yielded by geometrical optics. Further terms need to be added as kr diminishes. However, three facts of importance to applied seismology are obvious: (1) as r increases, $L \to r$ and the head-wave amplitudes diminish as r^{-2}. This rate of change is much faster than that for body waves. (2) On account of the factor k^{-1}, head-wave amplitudes also diminish with the frequency. Thus the head wave in seismology is generally characterized by longer wavelengths than either the direct or the reflected waves. (3) The amplitudes of the head waves increase as L diminishes, even if r remains the same. This suggests that the energy carried in the head wave increases along the wavefront from the interface (where it is a minimum) to the point where it joins the reflected wavefront (at "critical incidence"). One further point worth remarking is that $F_1(\delta_1)$ is generally an increasing function of δ_1. Thus, all other things being the same, headwave energies will rise with the critical angle, indicating that weak boundaries will propagate more intense head waves than strong ones. This conjecture has been corroborated by Heelan, although systematic tables of head-wave coefficients are not yet available. The present situation appears to be that the work of Cagniard and Brekhovskikh has put the head-wave theory on a sound mathematical basis, and a more detailed study of amplitudes would seem to be in order.

6-5 Reflection near the Critical Angle

As the angle of incidence approaches the "critical" value, something obviously goes wrong with the solutions (6-8) and (6-10) for the reflected wave and the head wave, since both become singular—the first because of the correction term involving $N(\vartheta)$ and the second because L vanishes. The cause of the difficulty is that the saddle point of the integrand in (6-7) has now also become a branch point, and the method of steepest descent cannot therefore be used. The exact theory, in the far-field approximation, for the reflection and refraction of spherical waves at a plane boundary in acoustics as well as in optics has been worked out by Brekhovskikh (11), and the results are given in terms of Weber functions (parabolic cylindrical functions) of order $-\frac{3}{2}$. It turns out that the exact solution in the wave zone for the $P_1P_2P_1$ head wave is found by multiplying the "geometrical optics" approximation (6-10) by the function

$$M(\eta) = e^{-i5\pi/8}(2\eta^2)^{\frac{3}{4}}e^{i\eta^2/2}D_{-\frac{3}{2}}(\eta + i\eta)$$

where
$$\eta = \sqrt{2k_{\alpha_1}R_1}\,\sin\frac{\vartheta_0 - \delta_1}{2}$$

and where $D_n(z)$ is Weber's function of order n.

In the limit, as $\eta \to 0$, we find that

$$\Phi_1(\omega,\mathbf{r}) \cong \frac{\exp ik_{\alpha_1}R_1}{R_1}\left[1 + \frac{M'(\delta_1)}{(k_{\alpha_1}R_1)^{\frac{1}{4}}}\right] \tag{6-11}$$

where $M'(\vartheta)$ is not the same quantity as $N(\vartheta)$, and

$$\Phi_1^*(\omega,\mathbf{r}) \cong M(\delta_1)[R_1(k_{\alpha_1}R_1)^{\frac{1}{4}}]^{-1}\exp ik_{\alpha_1}R_1 \tag{6-12}$$

$M(\vartheta)$ being different from either $f_1(\vartheta)$ or $F_1(\vartheta)$, and in general differing only by a numerical factor from $M'(\vartheta)$. Now we observe that the head wave attenuates less rapidly with distance at points close to the critical distance than elsewhere. The reflected wave is similarly enhanced. We have seen already in some examples used in Chap. 4 distinct evidence of the increased amount of energy carried by both waves in this direction as compared with other directions. In fact, as we can practically surmise by comparing (6-11) with (6-12), the head wave and the reflected wave merge at this point into a single disturbance of unusual intensity. Note, incidentally, that in directions close to critical incidence the higher frequencies in the head wave are less strongly attenuated than elsewhere—as $k^{-\frac{1}{4}}$ rather than as k^{-1}.

It is clear from the results given in Secs. 6-4 and 6-5 that by the use of ray paths (the method of geometrical optics) the direction and phase of both the reflected and the head waves are correctly predicted *in the wave zone* of the source and the boundary, where $kR_1 \gg 1$. There are certain limitations on the validity of geometrical optics even in those regions, however, and we shall discuss some of these in the following section. On the other hand, ray-path methods fail to give any account of the head-wave amplitudes, even in the far-field zone. In fact, since they represent essentially a high-frequency limit, they predict that the amplitudes of these waves should be zero. In the near-field regions ($kR_1 \sim 1$), the asymptotic methods described above cannot be used, and the solutions must be calculated from the exact integrals given by Cagniard.

6-6   Ray-path Diagrams and the Wave Theory

The work of Cagniard, Brekhovskikh, and others again raises the question of the validity of ray-path diagrams as a means of interpreting seismic data. Granted that the use of ray-path representations is a shorthand method that does not take into account *all* of the effects of elastic boundaries (surface waves in particular being a conspicuous omission), if we wish to consider the net value of these diagrams, it is important to form some idea of their limitations.   This we can do only with reference to the exact solutions given by the wave theory.

It is appropriate in this connection to reconsider the asymptotic solutions given in the two preceding sections.   In the solution for the reflected wave potential (6-8), it is clear that the first term is predicted by geometrical optics and that the second is a correction of the first order in $(kR_1)^{-1}$ provided by the wave theory.   The condition for the validity of the ray-path solution is that this correction (and succeeding terms, which form a series having asymptotic convergence except at critical incidence) should be negligible.   This condition is fulfilled when

$$k_{\alpha_1} R_1 \gg \left| \frac{N(\vartheta_0)}{f_1(\vartheta_0)} \right| = \left| \frac{f_1''(\vartheta_0) + \cot \vartheta_0 f'(\vartheta_0)}{f_1(\vartheta_0)} \right|$$

If we exclude values of ϑ_0 close to the critical angle, where $f_1'(\vartheta_0)$ becomes singular, we may write the above condition as

$$k_{\alpha_1} R_1 \gg \cot \vartheta_0 \left| \frac{d}{d\vartheta} \ln f_1(\vartheta) \right|_{\vartheta=\vartheta_0}$$

Difficulties in satisfying this condition will obviously arise as $\vartheta_0 \to \pi/2$, i.e., as the angle of incidence approaches grazing.   Under these conditions the source appears to be almost at the boundary, so that $h \sim 0$ and $R_1 \cos \vartheta_0 \approx z$.   Thus

$$k_{\alpha_1} z \gg \left| \frac{d}{d\vartheta} \ln f_1(\vartheta) \right|_{\vartheta=\vartheta_0} \tag{6-13}$$

The obvious implication of this result is that if ray-path methods are to be used, the receiver must not be too close to the boundary.

The ray-path solution for the refracted wave similarly requires certain corrections, according to the wave theory.   These are mainly of two different kinds, viz., (1) a correction based on the second term of the asymptotic series, which is of $O(1/kR)$, and (2) the addition of new waves which are not predicted by ordinary geometrical optics and which result from plane waves incident on the boundary at complex angles (Sec. 3-3).   If the refracting medium has the lower wave velocity, the latter disturbances will propagate away from the boundary as ordinary plane waves.   With regard to the first correction, all that is required to render it negligible is that the receiver be sufficiently far from the boundary. With regard to the second, it is clear that it will diminish as the source moves away from the boundary and as the incident wavefront becomes less curved. Indeed, as the asymptotic solutions for the transmitted wave potentials show, the amplitudes of these surface effects attenuate approximately exponentially as $kh \cos \vartheta_0$, and so it appears that the *source* must be positioned several seismic wavelengths from the boundary before the ray-path diagrams will become really representative.

The above comments are indicative of the requirements that must be met if ray diagrams are to be used to describe reflected and refracted events. Using virtually the same approximations, Keller (14) has extended the use of ray diagrams to include descriptions of the diffracted fields from edges, etc. (Sec. 3-6). Like the solutions (6-8) and (6-10), the potential of the diffracted wave in the far-field (Fresnel) approximation is the leading term in an asymptotic expansion. It will have the form

$$\Phi_d(\omega, \mathbf{r}) = \frac{e^{ik_\alpha R}}{k_\alpha R} \left[\bar{f}_d(\bar{\vartheta}_0) + \frac{N_d(\bar{\vartheta}_0)}{k_\alpha R} \right]$$

where $\bar{f}_d(\bar{\vartheta}_0)$ is the diffraction coefficient predicted by the plane-wave theory and R is the distance from the edge. $\bar{\vartheta}_0$ in this case is the angle of diffraction of ray paths projected into the plane which contains the source and which is perpendicular to the edge. The method of finding $\Phi_d(\omega, \mathbf{r})$ is analogous to the asymptotic methods described above, and it leads to a convergent solution if $k_\alpha R \gg 1$. Thus it becomes clear that the Fresnel theory does not apply within distances of a few wavelengths of any obstacle or boundary and that more exact methods must be used within these regions.

References

1. H. Weyl, Ausbreitung elektromagnetischer Wellen über einem ebenen Leiter, *Annalen Physik*, vol. 60, pp. 481–500, 1919.

2. J. Stratton, "Electromagnetic Theory," McGraw-Hill, New York, 1941.

3. A. Sommerfeld, Über die Ausbreitung der Wellen in der drahtlosen Telegraphie, *Annalen Physik*, vol. 28, pp. 665–736, 1909.

4. M. Ewing, W. Jardetzky, and F. Press, "Elastic Waves in Layered Media," McGraw-Hill, New York, 1957.

5. L. Cagniard, "Reflection et refraction des ondes progressives seismiques," Gauthier-Villars, Paris, 1939 (transl. by E. A. Flinn and C. H. Dix, McGraw-Hill, New York, 1962).

6. J. R. Carson, "Electrical Circuit Theory and the Operational Calculus," Chelsea, New York, 1953.

7. H. Lamb, On the Propagation of Tremors over the Surface of an Elastic Solid, *Royal Soc. Philos. Trans.*, ser. A, vol. 203, pp. 1–42, 1904.

8. E. R. Lapwood, The Disturbance Due to a Line Source in a Semi-infinite Elastic Medium, *Royal Soc. Philos. Trans.*, ser. A, vol. 242, pp. 63–100, 1949.

9. N. V. Zvolinskii, Reflected Waves and Head Waves Arising at a Plane Boundary between Two Elastic Media, I, II, III, *Izv. Akad. Nauk S.S.S.R., Ser. Geofiz.*, 1957–1958.

10. W. L. Roever, T. F. Vining, and E. Strick, Propagation of Elastic Wave Motion

from an Impulsive Source along a Fluid-Solid Interface, *Royal Soc. Philos. Trans.,* ser. *A*, vol. 251, pp. 455–523, 1959.

11. L. M. Brekhovskikh, The Reflection of Spherical Waves at a Plane Interface between Two Media, *Jour. Tech. Physics* [U.S.S.R.], vol. 18, pp. 455–482, 1948. (in Russian).

12. L. M. Brekhovskikh, "Waves in Layered Media" (transl. by D. Lieberman, ed. by R. T. Beyer), Academic, New York, 1960.

13. P. A. Heelan, On the Theory of Head Waves, *Geophysics,* vol. 18, pp. 871–893, 1953.

14. J. B. Keller, The Geometrical Theory of Diffraction, *Jour. Optical Soc. America,* vol. 52, pp. 116–130, 1962.

part II GRAVITY AND
MAGNETIC METHODS

chapter 7   Introduction to the Gravity and Magnetic Methods

Density is a physical property that changes significantly from one rock type to another. Another such property is ferromagnetic susceptibility, and a third is natural magnetization. A knowledge of the distribution of any of these properties within the ground would presumably convey information of great potential value about the subsurface geology, even if the rocks themselves could not be otherwise identified.

Each of the properties mentioned above is the source of a potential field which is intrinsic to the body possessing that property and which acts at a distance from it. The strength of the gravitational field of a body is in proportion to its density, and that of its magnetic field is in proportion to its magnetization. Thus the means exist, at least in principle, of studying the distribution of density or of magnetization under the ground from the gravitational or magnetic fields at the surface, provided that the appropriate physical measurements can be made. We shall discuss some of the geological implications of this fact in this chapter and again in Chap. 12. Chapters 8 to 11 will be concerned mainly with the problem of determining the configurations of density and magnetization from observations of potential fields.

7-1   The Physical Basis of Gravity Surveying

All materials in the earth influence gravity, but because of the inverse-square law of behavior, rocks that lie close to the point of observation will have a much greater effect than those farther away.   The bulk of the gravitational pull of the earth (i.e., of the weight of a unit mass), however, has little to do with the rocks of the earth's crust.   It is caused by the enormous mass of the mantle and core, and since these are regular in shape and smoothly varying in density, the earth's gravitational field is, in the main, regular and smoothly varying also.   Only about three parts in one thousand (0.3%) of g are due to the materials contained within the earth's crust, and of this small amount roughly 15 percent (0.05% g) is accounted for by the uppermost 5 km of rock—that region of the crust which we generally think of as being the seat of "geological" phenomena.   Changes in the densities of rocks within this region will produce variations in g which generally do not exceed 0.01% of its value anywhere.   Fluctuations in the value of g which may be associated with bodies that have a commercial mineral value are unlikely to exceed even a small fraction of this minute amount— perhaps 10^{-5} g altogether.   Thus geological structures contribute very little to the earth's gravity, but the importance of that small contribution lies in the fact that it has a point-to-point variation which can be mapped. From these maps we can, by applying potential theory, adduce a shadowy and ambiguous picture of the density distribution underneath the ground. From such impressions we can hazard an interpretation of the subsurface geology.   Two imperatives that must be fulfilled in order to produce meaningful maps are (1) the measuring apparatus must be sufficiently sensitive to detect the effects of geology on g and (2) effective methods must be used to compensate the data for all sources of variation other than the local geology.   These corrections will include chiefly the effects of changing elevations and of crustal heterogeneity on a broad scale.

When the data are finally reduced to a form meaningful in terms of local geology, they must be interpreted.   This is done chiefly in two parts: First, a solution is sought to the "inverse" potential problem, which consists in deducing the shape of the source from its potential field.   These solutions are never unique, but neither are they wholly ambiguous, since they are limited by physical factors other than gravity potentials.   Second, the solutions deduced from potential field theory are interpreted in geological terms.   This requires some knowledge of the factors which determine the densities of rocks.   It is to the second task that we shall direct our attention in this chapter.

7-2 The Precise Measurement of g

The gravitational field of the earth has a world-wide average value of about 980 gals,[1] with a total range of variation from equator to pole of about 5 gals, or $\pm 0.5\%$. Mineral ore bodies and geological structures of interest in the search for petroleum and other minerals seldom produce fluctuations in g exceeding a few milligals, or about 1 part in 10^6 of g. To calculate details of the size and shape of these structures, we require a reasonably precise knowledge of these fluctuations, so that for practical purposes a reading sensitivity of 0.01 mgal would be desirable. In fact, about one-half of this sensitivity is readily obtained by present-day instruments under average operating conditions. This figure represents about 1 part in 10^8 of the gravitational field of the earth.

No instrument yet devised can measure g absolutely to this accuracy. The most accurate method for absolute measurement employs the time of free fall of an object in vacuo measured by means of light pulses [Preston-Thomas et al., (1)]; however, the apparatus from the viewpoint of field use is extremely complex, and the highest accuracy so far obtained is about ± 1.5 mgals. Modern gravimetry instead makes use of a device which responds to variations in g, although it cannot measure g itself. This is accomplished by measuring the minute changes in the weight of a small mass that occur when it is moved from place to place. The device is essentially an extremely sensitive spring balance, called a _gravity meter_. Modern field gravity meters have already achieved reading sensitivities of 0.01 mgal when operating under good conditions, and the small variations in g that may be due to geological causes fall well within their range of detectability. However, since g is not measured by these instruments, the values necessarily refer to an arbitrary datum. Ordinarily this is not a disadvantage, because the gravity method is used to search for changes in geological conditions and only the variations in g are of interest. But if we wish to make a map of g itself, at least one point in the survey must be tied to a place where g has been measured absolutely.

The development of gravity meters has by now practically attained the point at which the limitation on the accuracy of interpretations is determined, not by the reading of data, but by the corrections which must be applied to them. These will be discussed more fully in Chap. 9, but we may make the remark here that the full utilization of the sensitivity of modern gravity meters often demands the use of precise leveling, which is painstaking and costly.

Gravity measurements are also carried out on ships at sea and on the sea bottom itself, and efforts are also under way to make measurements in

[1] The cgs unit employed for measuring gravitational field strength is the _gal_, after Galileo. Dimensionally, it is equivalent to an acceleration of 1 cm/sec².

aircraft. In moving vehicles such as these, the sensitivity needed for geological or geodetic purposes can be achieved only by using instruments and instrument platforms designed to average out the fluctuating accelerations of the vehicle, and then a number of new corrections enter on account of the general motion of the instrument in space. The equivalence between gravitational fields and accelerations indeed suggests that matters might be improved by measuring the gradient of g rather than g itself; for the gradient will contain essentially the same geological information as g, and at the same time it will be free of objectionable interference from linear accelerations of the instrument. As yet (1965), however, although the principle is much discussed, no successful gradiometer has been invented.

7-3 *Factors Which Influence Rock Density*

The following statements are necessarily of a very general kind and cannot be applied rigorously to all rock materials. They describe certain widely observed tendencies, but there will of course be numerous exceptions.

1. *Sedimentary rocks*

a. Composition. Gravitationally, sedimentary rocks fall into four major groupings. In order of increasing density, these are soils and alluvia, sandstones and conglomerates, shales and clays, and calcareous rock including limestones and dolomites. Here we must distinguish between the *bulk density*, which is the volume density of specimens of macroscopic size, and the *grain density*, which is the density of the rock-forming mineral in powder form. With respect to the latter, the above ordering is generally quite valid, but with respect to the former, exceptions may occur in cases of unusual porosities.

A further distinction is usually made between *dry* and *wet* bulk density. Dry bulk density refers to macroscopic specimens in a completely dessicated state, whereas wet bulk density implies that the specimen is fully impregnated with water. In the case of highly porous materials, the difference between dry and wet densities may be as much as 30 to 40 percent. The proper value to use in gravity interpretations will depend upon the depth of the formation in relation to the water table, which in turn will depend upon whether the climatic region is arid or moist. In most cases, including as a rule those in which the depth of the water table is unknown, the wet bulk density (being the more conservative) is used.

b. Age and depth of burial. The effects of these two factors are usually inseparable. Sediments that remain buried for long periods usually consolidate and lithify, resulting in a reduction in porosity and an increase in density. In sandstones and limestones this is accomplished as a general rule by the infiltration of cements unaccompanied by volumetric change, whereas in clays and shales the dominant process appears to be compaction.

According to Hedberg (2) the compaction of shales takes place in four distinct stages: First, there is the initial settling and rearrangement of the grains. This is followed by the dehydration of the colloids and the gradual expulsion of water adsorbed on the grains. Additional gravitational pressure leads to granulation and deformation of the soft grains, followed ulti-

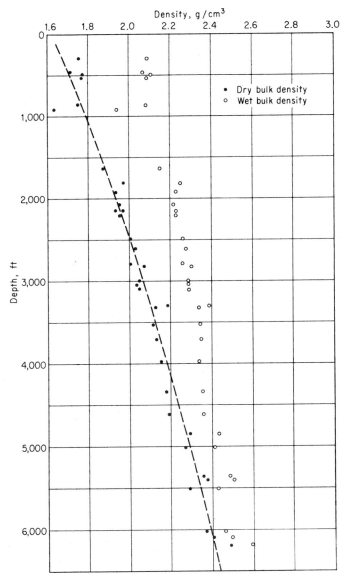

Fig. 7-1 Dry-bulk and wet-bulk densities of a number of samples taken from a thick section of Tertiary shale in Venezuela. [After Hedberg (2).]

mately by the recrystallization of minerals into denser forms. Since duration and depth of burial are both involved in these processes and the processes themselves overlap at various stages in the compaction, it is not surprising that no simple mathematical theory is able to express the density-depth relation for shales over a broad range of depths. Figure 7-1 shows the densities of samples taken from a thick Tertiary section in Venezuela. These strata were horizontal and apparently undisturbed, and most important, they were complete, i.e., none of the overburden had been removed at any time. The grain densities show no significant changes, but the bulk densities, both wet and dry, show definite gradients due to gravitational compaction. A second and equally striking example is shown in Fig. 7-2. These measurements were made by Athy (3) on core samples taken from wells in western Oklahoma, and wet bulk densities are shown plotted against stratigraphic depth. The curve has been extrapolated back to the average density of surface clay, and the negative depth intercept has been interpreted as the amount of material that has been removed by erosion. Although neither of these curves provides a sufficiently general basis for predicting the increase in the density of shales with depth, they indicate that a significant variation is to be expected.

Shales and clays are the most highly compressible of all sedimentary rocks, and they therefore show the greatest amount of compaction. Sand-

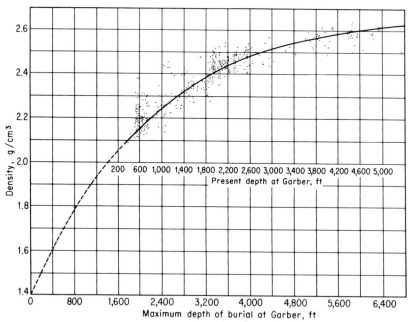

Fig. 7-2 Wet-bulk densities of core samples taken from wells in Garber County, Oklahoma. [After Athy (3).]

stones and limestones subjected to the same environment may be expected to experience much smaller density changes.

2. *Igneous rocks*

a. Composition. The density of igneous rocks generally rises as the silica content falls. Thus we usually find that a steady degradation in rock density more or less follows the acidity line, whether the rocks are plutonic or volcanic. For example, gabbro is ordinarily more dense than granite, diabase than syenite, and basalt than rhyolite. The extreme variation seldom exceeds 0.5 g/cm³, or about 20 percent, and this can be partly offset by differences in texture.

b. Texture. Holocrystalline rocks which have solidified at some depth are generally more dense than effusive rocks having the same chemical composition. Intrusive rocks are consequently somewhat heavier than lavas, even though they have the same constitution. For example, we would expect aphanitic rocks to be lighter than phanerites, volcanic glasses to be lighter than granitoid rocks, and so on. The effect of texture on density is not large, being as a rule less than 10 percent, and it can be highly variable. Even among the effusive rocks, density is lowered by the presence of amorphous material. This can be observed, for example, in obsidian and basalt, or in vitrophyre and rhyolite.

c. Mineralization. Because metallic mineral densities are on the whole very high, it is easy to overestimate the effect of mineralization upon rock density. The net contribution from this source is usually quite small because of the low fractional concentration of mineral. Massive sulfide ore specimens may have densities twice as large as that of the host rock, but bulk densities higher than 4 g/cm³ are highly localized and quite rare.

d. Porosity and fracturing. Igneous rocks are seldom porous, so that the distinction between dry and wet bulk density tends to lose significance. However, they may be fractured, with a marked effect upon their bulk density. Even small cracks, if sufficiently numerous, may have an influence as great as all the factors we have mentioned.

3. *Metamorphic rocks*

As a class, these rocks are gravitationally the most heterogeneous and the least susceptible to rules of behavior. Generally speaking, density shows a tendency to increase with the degree of metamorphism, because of increased filling of the pore spaces and recrystallization of the material into denser mineral forms. Slates, for example, being more fine-grained and less porous than shales, are also more dense. Quartzites are often heavier than sandstones, and marbles than dolomites. This general tendency seems to extend even to the igneous-metamorphic series, although certainly it is not so marked. We find that greenstones generally exceed basalts in density, and orthogneisses are slightly more dense than granites.

Among the schists and gneisses, density appears to run counter to the line of acidity, as it does among igneous rocks. The heavier schists usually contain amphiboles and pyroxenes, while mica and biotite species tend to be lighter. The total effect of composition on the bulk density, however, is not large, probably not more than a few percent.

A strong characteristic of metamorphic rock is the irregularity of their density behavior. Distinct zoning and chemical differentiation are features often found within a single formation. The migration of certain minerals into concentrated zones may produce intense irregularity within a distance of a few feet, and this often makes the sampling of metamorphic rocks for density a difficult task. Pods and dikelets of quartz, amphibolite, or serpentinite are common examples of this kind of heterogeneity.

7-4 The Sampling of Rock Densities

A major difficulty with the gravity method is that there is not yet any very satisfactory way to measure the densities of rocks in situ. Borehole density logging devices, although much discussed, have not yet attained very widespread use. Consequently we are forced to rely largely upon laboratory measurements for quantitative information on rock densities, which means that the best that one can do is to establish formation densities within the rather broad fiducial limits provided by a few small specimens. Much too frequently, however, authors who publish density measurements fail to quote their sample statistics, so that the value of these data to others is vitiated.

An extreme example of the difficulty of sampling occurs in a sulfide ore body. Since the relative proportions of mineral and host rock within small volumes change rapidly from place to place, the density measurements that are provided by diamond-drill core samples may be expected to be highly erratic. A set of measurements made on small fragments (of about 30 g each) taken from a typical suite of cores is illustrated in Fig. 7-3. Over short distances, small-specimen bulk densities will fluctuate widely; yet over longer intervals they appear to follow a trend. If we were to smooth out the fluctuations—by taking group averages, for instance—we would find in general that the variation in the bulk density within a formation can be reduced to a smooth quantity which we may call the *formation density*. In special applications of interpretation theory, the formation density is sometimes approximated with a simple mathematical form, which we call the *density function*. Density functions are only rarely used, however, in view of all the other uncertainties in making interpretations of gravity measurements. Since the total gravitational effect of the body is given by a volume integral, local variations in the formation density will not affect the field unless they are sustained over distances of several tens of feet.

An illustration of this problem in a slightly different context is shown in Fig. 7-4. Here we attempt to show a log of wet bulk density (*a*) as it

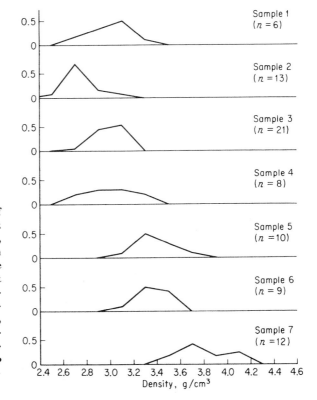

Fig. 7-3 Histograms of densities of small rock specimens (average size, about 3 cms) taken from the cores of a single diamond drill hole at Lake Mattagami, Quebec. Each sample consists of n specimens, which collectively occupy about 1 ft of diamond drill core. (*Rio Tinto Mining Company.*)

might appear if observations were taken every few inches down the hole, (b) if observations were taken a few feet apart, and (c) as it might affect a gravity meter placed on the surface of the ground at a short distance from the wellhead. Clearly the information relevant to the interpretation of gravity effects (and, conversely, the information that can be adduced from such interpretations) is fully represented by the *formation densities*. Without extensive prior sampling, however, the details of the formation densities will be lacking, and the formation density is then usually considered to be constant. We shall call this the *average* or *bulk formation density*. Even this gross approximation does not seem to affect interpretations seriously.

Choosing an average formation density without first sampling the formation is always a risky venture. Yet in many cases the gravity meter is the first instrument of exploration used in a new area, far in advance of any subsurface investigations. It is in fact typical of the gravity method that speculation often runs far ahead of geological control, for this situation is forced upon it by its role as a method of reconnaissance. Just how problematical the choice of average formation density may be is illustrated in Fig. 7-5. This shows a histogram of the densities of sedimentary rock samples quoted in the tables of Birch (4). The rocks are divided into four major

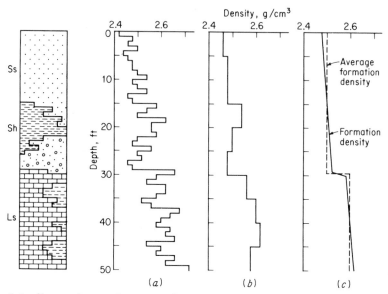

Fig. 7-4   **Formation and average densities within a stratigraphic sequence.**

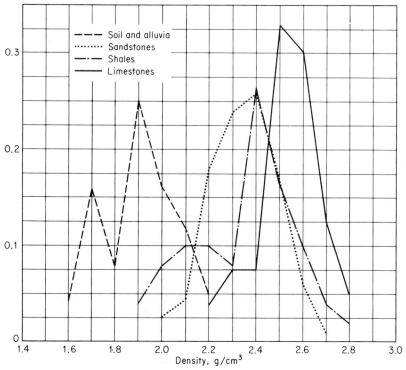

Fig. 7-5   **Histograms of small-specimen bulk densities of various kinds of sedimentary rocks.**   [From Birch (4).]

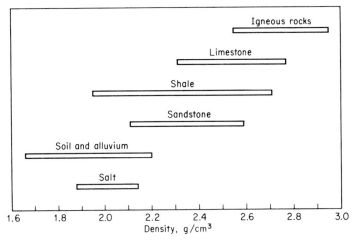

Fig. 7-6 80 percent fiducial limits of small-specimen bulk densities of various kinds of rocks. [From Birch (4).]

groupings and are plotted without regard to age or depth of burial, and because sampling statistics were not provided with the tables, each entry was accorded the same weight.   Two things are clear: there are extensive overlaps among all groups, and in most cases the frequency distributions are noticeably skewed.   The overlapping is best demonstrated by the 80 percent fiducial limits derived from these data and shown in Fig. 7-6.   The skewness probably can be explained for shales by the nonlinearity of the porosity-depth relation, and for limestones, by the sampling of reefal materials having anomalously high porosity.   In any event, even if the lithological aspects of two contiguous formations are known, it is evidently impossible to predict with confidence the difference in average density between them, without having access to other geological information.

Should seismic information be available within the area, however, then the situation is less bleak.   As we have already indicated that both seismic-wave velocities and bulk densities are affected by lithification accompanying age and compaction, it should not be surprising to discover a tendency for these two physical properties to increase together.   Figure 7-7 shows a plot of seismic velocity measurements versus bulk density determinations which shows that they do.   This graph, prepared by Nafe and Drake (5), contains data from a large number of core samples taken throughout a considerable range of depths and from widely distributed points.   The mean curve, fitted by least squares, can be used for predicting the bulk density of a formation from its seismic velocity when only the latter is known.   Although there is a considerable scatter about this curve, it is a useful device for gravity interpretations when no other information about the densities is available.

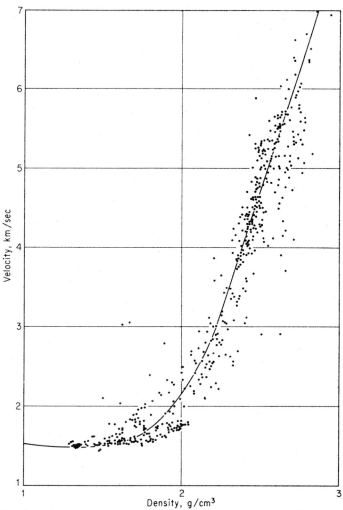

Fig. 7-7   A plot of seismic P-wave velocity versus bulk density.   [Courtesy of C. Drake.]

7-5   *Some Typical Hypotheses Used in Gravity Interpretations*

Because of chronic inability to measure the relevant physical property, the interpretation of gravity data tends to become somewhat theoretical in its function.   Usually the aim is to calculate the shape of a supposed discontinuity in average density from surface gravity measurements, after *assuming* a value for the average density difference.   Very often several such interpretations are made from a single set of data by changing the density contrast.   Since all these solutions are equally correct insofar as the geophysical data are concerned, one is free to choose *post hoc* whichever best fits all the available evidence.

We should like to demonstrate a few typical examples of gravity-interpretation problems. To begin with, since an almost infinite variety of suppositions concerning subsurface density structure is possible, it is usually desirable to keep the hypotheses simple. Since this undoubtedly will oversimplify the geology very greatly, we must not be misled into believing that a geophysical interpretation accurately represents a set of geological facts. What it endeavors to do, rather, is to schematize them, leaving out many of the details and merely adumbrating their essential features with a few simple lines. The information given by geophysics is of a general nature. The details can only be guessed.

One of the oldest applications of the gravity method is in the location of salt domes. Salt rises through the heavier overlying strata as a result of differential hydrostatic pressure. Being a lighter substance than the rock which it displaces, salt will diminish the earth's gravity field locally; thus the target of the survey is a negative anomaly. Ancillary features which can assist in identifying salt domes on gravity maps sometimes occur. One of the most common, and certainly one of the most characteristic, is a heavily compacted, and hence a relatively dense cap rock which rests on top of the salt and which may produce a very local positive gravity effect in the middle of the "low." Away from the piercement structure the sediments are usually rather flat, so that the regional gravity field tends to be smooth. Typical of the sort of density distribution assumed in this kind of interpretation problem is the one shown in Fig. 7-8. The usual object of the interpretation is to calculate the depth, size, and shape of the dome from the map of its gravity field.

Another application in which the gravity method has proved its usefulness is in mapping structural uplifts and depressions in areas where the stratification tends to be regular. If there are prominent density contrasts between reasonably thick sections of the sedimentary column, flexures in the strata will yield observable gravity anomalies. As suggested by Fig. 7-9a, structures may be brought about by folding, differential compaction over buried hills, or any of several other agencies. If density contrasts are found among the materials involved in these displacements, positive gravity effects will be observed over those regions where the heavier material is uplifted, and negative effects, where it is depressed and replaced by lighter material. For the purpose of geophysical interpretation, the situation illustrated in Fig. 7-9a is idealized in Fig. 7-9b. The object of the interpretation will probably be to find the amplitude and shape of structural relief at a given interface. In this type of problem there is often a double source of ambiguity, however, because, not only are the density contrasts uncertain, but the source of the gravity anomaly may even be in doubt. Within the crystalline basement rocks inhomogeneity is commonplace, and heavy intru-

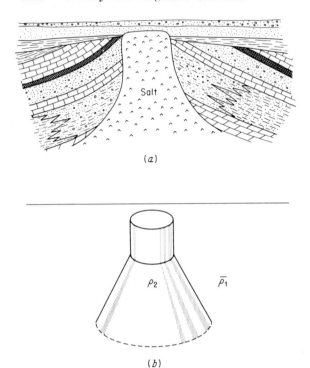

Salt

(a)

ρ_2 $\bar{\rho}_1$

(b)

Fig. 7-8 A salt dome and its interpretational model.

sives at great depths may resemble gravitationally a gentle flexure under relatively shallow overburden. Postdepositional vulcanism is another source of ambiguity, but it is usually more easily recognizable (e.g., with the aid of magnetics).

Faults are structures which often show up strongly and unmistakably

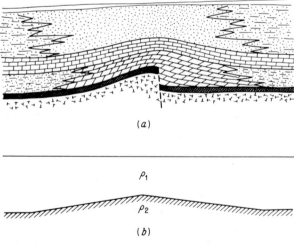

(a)

ρ_1

ρ_2

(b)

Fig. 7-9 A structural uplift and its interpretational model.

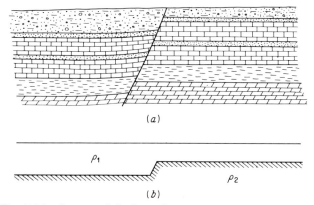

(a)

(b)

Fig. 7-10 A normal fault and its interpretational model.

on gravity maps.   The situation shown in Fig. 7-10a will produce a marked gravity gradient in the vicinity of the fault strike line.   The geology is abstracted in Fig. 7-10b into a particularly simple form suited to geophysical interpretation, where the object is to determine the depth, dip, and throw of the fault.   For practical purposes, the figure shown in Fig. 7-10b is usually assumed to extend to an infinite distance along its strike, even though the fault itself obviously must end somewhere.

Gravity-meter surveys are sometimes used to map bedrock topography under the overburden, especially for the purpose of locating buried river valleys for water supplies.   A typical geological situation is illustrated in Fig. 7-11a and the geophysical representation of it is shown in Fig. 7-11b. Here we have greatly modified the geology by replacing several interfaces with two.   Consequently we do not anticipate very great accuracy in

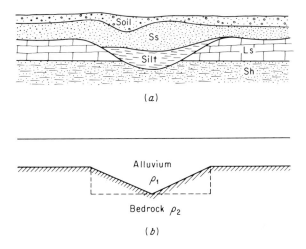

(a)

(b)

Fig. 7-11 A buried stream valley and its interpretational model.

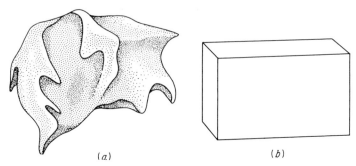

(a)   (b)

Fig. 7-12   A massive ore body and its interpretational model.

locating the sides or depth of the valley, but we may hope for results that are at least in rough agreement.

Many ore bodies, especially veins and heavy sulfide masses, have one small dimension and two large ones; hence they are generally thought to be dikelike in structure. The ore body represented in Fig. 7-12a may have a very irregular and complex shape in three dimensions, but for geophysical purposes it is often represented by the model shown in Fig. 7-12b. There is usually assumed to be a constant density difference between the prism and its surroundings and a sharp, well-defined boundary separating them. Neither of these assumptions seems very likely to be correct, since most real ore zones are banded or disseminated to some extent and tend to be rather ragged in outline. The gravity method will determine the general form of the massive part of the ore.

In each of the instances cited above, a figure having a very simple geometry is chosen to represent the anomalous formation. We call this figure an *interpretational model*. Its dimensions are to be calculated from the gravity field after assuming a value for the average formation density contrast between the model and its surroundings. This procedure is followed most commonly in the interpretation of gravity data. Methods will differ only in so far as trial-and-error or else more direct techniques are used to calculate the model dimensions. The difference between the average formation densities is almost always assumed beforehand, as it is rarely ever measured. For this, tables of density measurements on rock samples may be helpful in a general, but seldom in a specific, way, since little is known about the rock properties before interpretation. The fiducial limits given in Fig. 7-6 provide adequate information on these values for a general understanding of their trend. Interpretations usually retain some flexibility with regard to density contrast, so that different values, or at least upper and lower limits, may be used as desired. The final task is the development of the solution into a picture which can be understood in geological terms.

7-6   The Physical Principles of Magnetic Surveying

The earth, as is commonly known, is a weak magnet, and as far as we can tell, it has always possessed a magnetic field. The present geomagnetic field is almost dipolar; but unlike the gravitational field, which is very smooth and regular on the whole, the earth's main magnetic field is too complicated to be expressed by a simple mathematical function. Nevertheless it is smooth when measured on a scale of hundreds of miles, and against this background geological effects are easily distinguishable.

Many kinds of rocks possess a ferromagnetic susceptibility, so that when placed in a magnetic field such as the earth's, they acquire an appreciable magnetization of their own. These induced effects will show up in the main magnetic field as *anomalies*, which may be used to outline the zones of high susceptibility within the ground. Some rocks, on the other hand, have a permanent or intrinsic magnetization known as a *natural remanent magnetization* (NRM). This is independent of the earth's magnetic field, and it may lie in a quite different direction (Chap. 12). Rocks possessing this property create their own magnetic field, and in sufficient volume their NRM will also produce magnetic anomalies. If the NRM is appreciable in relation to the induced magnetic moment, the rock formation may be magnetized in a direction quite different from that of the earth's main magnetic field; and since anomaly patterns change very markedly with the direction of magnetization, this causes additional uncertainties in interpretation.

The magnetic properties of rocks are subject to sudden and violent fluctuation over short distances. We therefore consider large volumes of rock to possess an *average formation susceptibility*, or an *average NRM*. The object of geophysical interpretation is to determine the shape and magnetic properties of the source of an anomaly, and it would be a great help in the geological interpretation of these results to be able to relate magnetic properties to rock type. Tables of the susceptibilities or of the remanent magnetizations of scattered rock specimens are, like tables of densities, generally informative but not of much practical help. Indeed, in this respect the problem in magnetics is somewhat worse than that in gravity, because magnetic properties depend, in the main, upon trace rather than on bulk constituents. Consequently the fiducial limits on tabulated susceptibilities and NRMs of rock specimens are so broad as to be of little practical value in rock identification.

7-7   The Measurement of Magnetic Fields

The main magnetic field of the earth (H_0) is roughly equal to that of a dipole of moment 8×10^{25} emu, which is situated close to the center of the earth and is inclined at about 11° to the geographical axis. The maximum (polar) intensity of the field is about 0.65 oersted, and its minimum (equatorial) intensity is about 0.35 oersted.   Magnetic anomalies due to concentrations of ferromagnetic minerals in the upper parts of the earth's crust may reach amplitudes as large as H_0, or even larger.   For the most part, however, they lie in the range $0.001H_0$ to $0.1H_0$.   The practical unit for measuring field strength in geophysical surveys is the gamma, which is 10^{-5} oersted.   The geomagnetic field thus has a strength between 30,000 and 60,000 gammas, while local anomalies may have amplitudes of several hundreds of gammas, and sometimes of a few thousands.   In a number of cases anomalies having amplitude higher than 100,000 gammas have been recorded on the ground over magnetite deposits.

Usually the sensitivity needed to measure magnetic anomalies in sufficient detail for geophysical interpretation is about 5 gammas, or about one part in 10^4 of H_0.   This accuracy can be achieved with a variety of different types of field instruments, called *magnetometers*.   Broadly speaking, magnetometers fall into two classes: (1) those that measure only the total intensity of the combined (geomagnetic + anomalous) fields and (2) those that resolve the total field into vertical and horizontal components. Instruments in the first category can be operated either from aircraft or on the ground, and at present they are either the *flux-gate* or the *nuclear-precession* type [for details, see Dobrin (6), Chap. 16].   Instruments in the second category are generally of the *balanced-torque* type, that is, the mechanical torque acting on a freely pivoted magnetic needle in the earth's field is balanced by an opposing moment supplied either by gravity or by a torsion fiber, or else by canceling the field with a current-carrying coil or permanent magnet.[1]   Instruments of this kind operate in a leveled position, so that they may measure either the horizontal or the vertical component of the magnetic field.   The geophysical interpretation of field data recorded by these two different kinds of magnetometer differs in certain essential details.   A point worth mentioning is that many magnetometers are designed to measure absolute (rather than relative) values of the total field. This is especially valuable in extending old surveys or in tying together new ones.

A difficulty that arises in the planning and execution of magnetic surveys is that the intensity of the earth's magnetic field does not remain constant with time but suffers small, temporal changes.   Most of the time the changes are smooth and they tend to be more or less cyclical with a 24-hr period and an amplitude whose world-wide average is of the order of 100 gammas, or about $0.002H_0$.   This is not a large amount, but it is enough to distort the magnetic-field pattern.   These small fluctuations are related to large-scale motions in the ionosphere which are probably due to radiations

---

[1] Since writing these paragraphs, several new types of magnetometers have appeared. These include the portable flux-gate magnetometer for measuring the vertical component of the field, and a total-field instrument that measures the Zeeman effect caused by H_0. The second instrument appears to be capable of very high sensitivity (about 0.01 gamma).

from the sun. While their intensities will vary with the sun's azimuth, their horizontal gradients are such that they may be considered uniform over distances of up to a few tens of miles. Thus their effects may be removed from field observations made within a limited area either by monitoring H_0 continuously at a fixed point or else by repeating observations at certain stations at regular intervals of about 1 hour during the day so that all observations may be brought back to a common time. Occasionally the fluctuations become sudden and violent, indicating the onset of a *magnetic storm*. During a magnetic storm, the value of H_0 at a fixed point may change suddenly and unpredictably by several hundreds of gammas, making the continuance of magnetic surveys altogether unprofitable. The storm may last for several hours, during which time all magnetic surveying activity is normally discontinued. These occurrences generally follow strong corpuscular outbursts from the sun by approximately 30 hr, and they can sometimes be predicted by solar observatories.

The aims of magnetometer surveys are often very much the same as those of gravity surveys, viz., to find from surface measurement the shapes or the volumes of formations whose susceptibilities or NRMs differ markedly from those of their surroundings. The "models" used to represent geological structures are similar to those used in gravity interpretations, but unfortunately the selection of suitable values for the susceptibility contrasts or NRMs is highly problematical on account of the wide variability of these properties, even within a single rock type. The factors which affect magnetic properties of rocks will be the subject of a separate study in Chap. 12. One way or another, susceptibility or magnetization is mainly a function of the quantity of magnetic minerals in the rock, and inasmuch as these are usually present in small amounts in any case, it is plain that the concentrations may vary widely. Therefore methods of interpretation based upon the shapes, but not upon the amplitudes, of anomaly patterns are generally used, so that the question of physical property contrast may be deferred to as late a stage in the calculations as possible.

In practice there are three well-recognized uses of magnetic methods. The first is in direct exploration. By creating local anomalies in the natural field of the earth, magnetic ores reveal their presence and provide us with a simple and effective tool for their discovery. The methods used in interpretation in this case bear a strong resemblance to those used in gravity. The second application is in geological mapping, in which the magnetization of a rock may give some clue to its composition and provide a method for tracing the extent and shape of the formation. The third application is less direct, there being no intrinsic interest in the magnetized rocks whatsoever. A typical example is in the reconnaissance of sedimentary basins with the airborne magnetometer, in which the depths of the magnetic rocks below the aircraft are calculated from the anomaly patterns. If these rocks are assumed to lie in the crystalline complex of the Precambrian "basement," and if a sufficient number of interpretable anomalies can be found, the limiting depth of the sedimentary basin can be roughly delineated.

Further evidence for the widespread use of the magnetometer as an indirect aid to exploration will be found in the following examples from mineral-exploration case histories:

Asbestos. Asbestos deposits usually occur in ultrabasic intrusive rocks, very frequently in association with peridotite. Many of these rocks contain significant amounts of magnetite, both as a consequence of serpentinization and as an original constituent. Peridotites therefore tend to be much more magnetic than the adjacent formations, and the magnetometer may be used to locate them and outline their contact zones. At the same time, it may sometimes be used to distinguish between serpentinized peridotite and usually more weakly magnetized pyroxenites, and even to locate replacement or altered zones within the ultrabasic body, such as zones of carbonates or of talc carbonates, which are often characterized by a scarcity of magnetite and correspondingly low magnetic values.

Gold. In the Val d'Or area in Quebec, gold ore deposits are closely controlled by large masses of granodiorite which have intruded into ancient volcanic tuffs. The ore deposition appears to be associated with relatively late intrusive activity, and nearly all the known deposits are in quartz veins along the edges of these masses. The targets for magnetometer surveys in this area are the granodiorite intrusives, which are less magnetic on the whole than the displaced volcanics. Favorable indications are rounded areas of low magnetic values exhibiting characteristic crisscrossing patterns, which reflect fracturing and subsequent mineralization in the fracture zones. These masses commonly have a characteristic anomaly pattern around their border, which is produced by displaced banded volcanic rocks.

Columbium-uranium. One of the more common columbium ores is pyrochlore, a metasomatic mineral usually found in the aureoles surrounding alkaline intrusives. Because of the alkaline environment, magnetite is formed in these zones whenever iron is present in the host rock. The magnetite aureoles have a characteristic ring anomaly pattern which serves to reveal and identify the intrusive rock and thus a potential zone of mineralization.

These examples give some indication of the variety of purposes to which the magnetometer may be put in the exploration for minerals. In some cases a fairly precise interpretation of the data is required, while in others a characteristic anomaly pattern is sufficient. While only experience can be the judge, the second approach is often the most rewarding as far as positive results are concerned. There obviously cannot be a wholly systematic attitude to magnetic interpretations, and we must bear this in mind when we come to Chap. 11, which deals with the rather specialized problems of making quantitative interpretations.

References

1. H. Preston-Thomas, L. G. Turnbull, E. Green, T. M. Dauphinée, and S. N. Kalra, An Absolute Measurement of the Acceleration Due to Gravity at Ottawa, *Canadian Jour. Physics*, vol. 38, pp. 824–852, 1960.

2. H. Hedberg, The Gravitational Compaction of Clays and Shales, *Am. Jour. Sci.*, vol. 31, pp. 241–287, 1936.

3. L. F. Athy, Density and Porosity of Sedimentary Rocks, *Am. Assoc. Petroleum Geologists Bull.*, vol. 14, pp. 1–25, 1930.

4. F. Birch (ed.), "Handbook of Physical Constants," *Geol. Soc. America Spec. Paper* 36, 1942.

5. C. Drake, Private communication, 1962.

6. M. B. Dobrin, "Introduction to Geophysical Prospecting," 2d ed., McGraw-Hill, New York, 1960.

chapter 8   Potential Field Theory

The mathematical problem of gravity interpretation may be said to consist in finding the mass distribution whose gravitational field is given on a plane surface. In the interpretation of magnetics, the problem is to find the distribution of magnetized material whose magnetic field is given on a plane surface. In either instance, we are confronted with the "inverse" problem of potential field theory, i.e., of determining the source from its potential. This problem, unfortunately, does not have a unique solution, for there is insufficient information to determine the size and shape of the source completely and unambiguously from its potential field. The indeterminacy is subtle and makes its entry in many different forms. The essential difficulty arises in trying to separate the physical size from the density of the gravitating mass, or from the magnetization of the magnetic mass. Potential fields are inherently ambiguous in this respect.

The geophysical interpretation of potential field measurements consists essentially in finding a source whose size and shape may be adjusted to fit the observations. This is an exercise that requires a fair degree of skill in the manipulation and use of potential field theory, and the object of this chapter is to provide the necessary background for this task. The standard treatises on potential theory are of little direct help, since they rarely deal with the "inverse" problem. We therefore restrict Chap. 8 to some formal exercises, the results of which we shall later use in Chaps. 9, 10, and 11.

8-1 The Potentials

1. The gravitational potential. Newton's law of universal gravitation may be stated as follows: The mutual gravitational attraction between two particles having masses m_1 and m_2 which are separated by a distance r has the value

$$\mathfrak{F}(r) = G\frac{m_1 m_2}{r^2}$$

and acts in the direction of r. The force per unit mass on a particle at any point P at a distance $\mathbf{r}$ from m_1 is defined as the *gravitational field* of the particle m_1. This is written

$$\mathbf{F(r)} = -G\frac{m_1}{r^3}\mathbf{r}$$

Such a field is conservative, and it is therefore derivable from a scalar potential function $U(\mathbf{r})$ as follows:

$$\mathbf{F(r)} = -\boldsymbol{\nabla}U(\mathbf{r})$$

where $U(\mathbf{r}) = -Gm_1/r$ is the gravitational potential of the mass m_1. All other field quantities may be expressed in terms of U.

Since potentials in free space are additive, the gravitational potential due to continuous distributions of matter may be calculated at external points by integration. Thus if mass is distributed continuously with the density $\rho(\mathbf{r}_0)$ throughout the volume V, the gravitational potential at an exterior point P is (see Fig. 8-1)

$$U_P(\mathbf{r}) = -\int_V \frac{G\,dm}{|\mathbf{r} - \mathbf{r}_0|}$$

$$= -G\int_V \frac{\rho(\mathbf{r}_0)d^3r_0}{|\mathbf{r} - \mathbf{r}_0|} \tag{8-1}$$

where
$$|\mathbf{r} - \mathbf{r}_0| = \sqrt{r^2 + r_0{}^2 - 2rr_0\cos\gamma}$$

If the volume integration is carried out over the entire earth, one obtains the earth's gravitational potential in free space, from which the gravitational field may be found by differentiation. If P is on the earth's surface, the gravitational field there is denoted by the symbol g. Actually, since g is usually taken with a positive sign, it is really the *negative* field

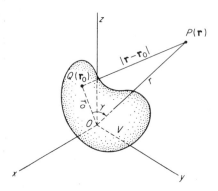

Fig. 8-1

intensity, i.e., $g = |-\mathbf{F}(\mathbf{r})| = |\nabla U_P|$. It is called variously the gravitational field, the gravitational acceleration, or the acceleration of free fall, although in fact it is a force per unit mass that is usually measured, not an acceleration.   The unit of g is the *gal* ($= 1 \text{ cm/sec}^2$), and the direction is by definition everywhere vertical.   The *weight* of a mass m at rest in the gravitational field is the force vector whose magnitude is mg and whose direction is vertically downward.

The value of g depends upon (*a*) the volume distribution of the mass of the earth, as expressed by the density function $\rho(\mathbf{r}_0)$, and (*b*) the actual shape of the earth, as specified by the integral limits.   Variations in $\rho(\mathbf{r}_0)$ near P will of course have a much greater effect upon g than those occurring farther away, in accordance with the inverse-square law, and in fact, it would appear that the integral even becomes singular near $|\mathbf{r} - \mathbf{r}_0| = 0$.   Thus even minor density changes in the very shallow subsurface materials will often cause variations in g which can be measured.

2. The magnetostatic potential.   Magnetic dipoles can also interact with one another at a distance, by means of their magnetic fields.   The force per unit pole strength exerted by a magnetic dipole of moment $\mathbf{m}$ upon a single magnetic pole (if such a thing can be imagined) at a distance r from the dipole is

$$\mathbf{H}(\mathbf{r}) = \frac{2m \cos \vartheta}{r^3} \mathbf{r}_1 + \frac{m \sin \vartheta}{r^3} \boldsymbol{\vartheta}_1$$

where ϑ is the angle measured from $\mathbf{m}$ to $\mathbf{r}$, and where $\mathbf{r}_1$ and $\boldsymbol{\vartheta}_1$ are unit vectors in the directions of increasing r and of increasing ϑ, respectively. $\mathbf{H}$ is called the *magnetic field intensity* at P.

From the structure of this force field, it is evident that it too is derivable from a scalar potential function.   We may write

$$\mathbf{H}(\mathbf{r}) = -\nabla A(\mathbf{r})$$

where
$$A(\mathbf{r}) = \frac{m \cos \vartheta}{r^2}$$

and it is evident that

$$A(\mathbf{r}) = -\mathbf{m} \cdot \nabla \left(\frac{1}{r} \right)$$

$A(\mathbf{r})$ is called the scalar magnetic potential of the dipole whose moment is $\mathbf{m}$.

Permeable materials, placed in an external magnetic field such as the earth's, acquire an induced magnetism whose strength is determined by the magnetic susceptibility of the material. Other substances exhibit a permanent magnetism unrelated to environment. In either case we may suppose that the material that fills a volume V has a continuously distributed magnetic dipole moment per unit volume which is given by $\mathbf{M}(\mathbf{r}_0)$. Then, by superposition, the magnetic scalar potential at a point P outside V is

$$A(\mathbf{r}) = -\int_V \mathbf{M}(\mathbf{r}_0) \cdot \nabla \frac{1}{|\mathbf{r} - \mathbf{r}_0|} \, d^3 r_0 \qquad (8\text{-}2)$$

The total magnetic field intensity at P is $\mathbf{H}(\mathbf{r}) = -\nabla A(\mathbf{r})$. Substituting for $A(\mathbf{r})$ in this equation, we get

$$\mathbf{H}(\mathbf{r}) = \nabla \int_V (\mathbf{M} \cdot \nabla) \frac{1}{|\mathbf{r} - \mathbf{r}_0|} \, d^3 r_0$$

If we assume that the direction of magnetization is the same throughout V and denote it with the symbol a, so that $\mathbf{M} \cdot \nabla = M(\partial/\partial a)$, this expression becomes

$$\mathbf{H}(\mathbf{r}) = \nabla \frac{\partial}{\partial a} \int_V M(\mathbf{r}_0) \frac{1}{|\mathbf{r} - \mathbf{r}_0|} \, d^3 r_0 \qquad (8\text{-}3)$$

An interesting relationship between the gravitational and magnetic fields due to a homogeneous mass distribution develops when we make ρ and M in (8-1) and (8-3) both constant. Then it turns out that

$$\mathbf{H}(\mathbf{r}) = \frac{M}{G\rho} \frac{\partial}{\partial a} \mathbf{F}(\mathbf{r}) \qquad (8\text{-}4)$$

This connection was first noted by Poisson. By means of it, all properties of the magnetic field due to a homogeneous body are derivable from the gravitational field. This fact becomes extremely useful to us in our later development of magnetic interpretation theory.

8-2 The Field Equations

At points outside the volume V, the volume integrals representing the potential functions $U(\mathbf{r})$ and $A(\mathbf{r})$ are nonsingular and clearly satisfy Laplace's equation. Thus the field equations in free space are

$$\nabla^2 U = 0 \qquad \text{for the gravitational field}$$

$$\nabla^2 A = 0 \qquad \text{for the magnetic field}$$

Within V on the other hand, the integrals become singular at $\mathbf{r} = \mathbf{r}_0$. To isolate this singularity, we enclose it within a small sphere of radius ϵ and volume v. The potential U may then be written

$$U(\mathbf{r}) = -G \int_{V-v} \frac{\rho(\mathbf{r}_0)\, d^3 r_0}{|\mathbf{r} - \mathbf{r}_0|} - G \int_v \frac{\rho(\mathbf{r}_0)\, d^3 r_0}{|\mathbf{r} - \mathbf{r}_0|}$$

The first term is now nonsingular, and so it is harmonic everywhere. In the second, if we make ϵ small enough, we may regard $\rho(\mathbf{r}_0)$ as constant. This allows us to write

$$\nabla^2 U(\mathbf{r}_0) = -G\rho(\mathbf{r}_0) \int_v \boldsymbol{\nabla} \cdot \boldsymbol{\nabla} \frac{1}{|\mathbf{r} - \mathbf{r}_0|}\, d^3 r_0$$

which according to Gauss' theorem is

$$\nabla^2 U(\mathbf{r}_0) = -G\rho(\mathbf{r}_0) \int_s \mathbf{n} \cdot \boldsymbol{\nabla} \frac{1}{|\mathbf{r} - \mathbf{r}_0|}\, d^2 r_0$$

Now s is the surface of the small sphere whose radius is ϵ, and therefore in the above integral $|\mathbf{r} - \mathbf{r}_0| = \epsilon$ and $\mathbf{n} \cdot \boldsymbol{\nabla} = \partial/\partial\epsilon$. Thus in the limit as $\epsilon \to 0$,

$$\nabla^2 U(\mathbf{r}_0) = -G\rho(\mathbf{r}_0) \left(\frac{d}{d\epsilon} \frac{1}{\epsilon} \right) 4\pi\epsilon^2$$

$$= 4\pi G \rho(\mathbf{r}_0) \tag{8-5}$$

Similarly, $$\nabla^2 A(\mathbf{r}_0) = 4\pi \boldsymbol{\nabla}_0 \cdot \mathbf{M}(\mathbf{r}_0) \tag{8-5a}$$

These equations, known as Poisson's equations, are satisfied by the potentials everywhere within regions occupied by massive or magnetic bodies.

8-3   *The Equivalent Stratum*

We now show how a gravitational field whose normal component is given on a horizontal plane can be related to a surface distribution of density on that plane. Consider a material to be distributed with a surface density $\sigma(x,y)$ g/cm^2 over the horizontal plane $z = 0$. To calculate the gravitational field of this coating at a point Q in $z = 0$, we choose a set of circular cylindrical coordinates (r,ϑ,z) whose axis is vertical and whose origin is placed at Q (Fig. 8-2). At a point P on the axis, the gravitational potential will be

$$U_P = -G \int_0^\infty \int_0^{2\pi} \frac{\sigma(r,\vartheta)}{\sqrt{r^2 + z^2}}\, r\, d\vartheta\, dr$$

and the gravitational field at P is $-\boldsymbol{\nabla} U_P$.

Here we must mark an important distinction between the two terms *gravitational field* and *gravity effect*. The earth's gravitational field has only

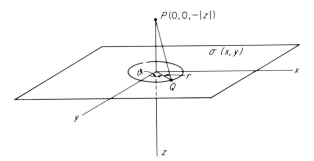

**Fig. 8-2   The "equiva-
lent stratum."**

one direction, for indeed it is by this field that our reference system is oriented. *Vertical* is by definition the direction of g, and horizontal is perpendicular to this direction. (Actually, the horizontal is established in practice from the local equipotential surface with the aid of bubble levels.) Gravitational fields due to anomalous bodies may act in various directions with reference to the vertical, depending upon where we stand in relation to the body. But these small, local disturbances can be measured only in combination with the gravitational field of the earth, which is vastly larger. The measuring instrument, which is leveled in the total gravitational field, can therefore respond only to the component of the disturbing field which is in the direction of total g, or in other words, to its vertical component. To describe this component, we use the term *gravity effect*, which is really the change in the earth's gravitational field strength caused by the local masses. We shall be reminded of this fact if we make a practice of using the symbol Δg to signify gravity effects, so that the symbol g may be appropriated to the earth's gravitational field strength. Implicit in our statement that Δg is measured in the direction of normal g is the assumption that the direction of g is not affected by the presence of the local masses, or in other words, that the direction of the vertical remains virtually undisturbed within the anomalous region. This will be true if $\Delta g \ll g$, a condition invariably fulfilled in practice.

Since the potential U is due to masses that are locally distributed on $z = 0$, we write for the gravity effect at P

$$\Delta g_P = \frac{\partial U_P}{\partial z} = Gz \int_0^\infty \int_0^{2\pi} \frac{\sigma(r,\vartheta)}{(r^2 + z^2)^{3/2}} \, r \, d\vartheta \, dr \qquad z < 0$$

the positive sign being used because Δg is to be measured in the direction of g. We wish to evaluate this integral in the limit as $z \to 0$. Normally this would lead to a singularity at Q, and so we divide the range of integration into two parts: one a small circle at Q of radius ϵ, and the other the remainder of the plane $z = 0$. If the radius ϵ is chosen small enough that throughout the circle $\sigma(r,\vartheta)$ does not change appreciably from its value at Q, the first integral becomes

$$T_1 = G\sigma(Q)z \int_0^\epsilon \int_0^{2\pi} \frac{r \, d\vartheta \, dr}{(r^2 + z^2)^{3/2}}$$

$$= 2\pi G\sigma(Q)\left(1 - \frac{z}{\sqrt{\epsilon^2 + z^2}}\right)$$

Although ϵ may be as small as we please, we suppose it to be finite, so that $\lim_{z \to 0} T_1 = 2\pi G\sigma(Q)$. In addition, the second integral is now nonsingular, and so $\lim_{z \to 0} T_2 = 0$. Therefore we have finally

$$\Delta g(Q) = 2\pi G\sigma(Q)$$

But the position of Q in $z = 0$ is arbitrary, so that we may write

$$\Delta g(x,y) = 2\pi G\sigma(x,y) \tag{8-6}$$

Let us suppose that the gravity effect $\Delta g(x,y)$ on $z = 0$ is produced by an unknown distribution of matter below this plane. Then whatever the array of masses may actually be, its effect at any point in $z \leq 0$ would be exactly the same if it were replaced by the surface distribution on $z = 0$ given by (8-6). This density coating is therefore called the *equivalent stratum* for the unknown distribution of matter in $z > 0$.

8-4   *The Continuation of Potential Fields*

Green's theorem states that if U and W are continuous functions within a volume V, with first and second derivatives that are both continuous and integrable, then

$$\int_V (U\nabla^2 W - W\nabla^2 U) \, d^3r_0 = \int_S \mathbf{n} \cdot (U\nabla W - W\nabla U) \, d^2r_0 \tag{8-7}$$

where the surface S encloses the volume V. The restrictions on U and W are satisfied if we let U be the gravitational potential due to the masses within V and let W be the function $1/|\mathbf{r} - \mathbf{r}_0| = 1/R$, where $\mathbf{r}$ is the position vector of a point P outside V and $\mathbf{r}_0$ is the position vector of a point Q within V.

At any point P outside V

$$U_P(\mathbf{r}) = -G \int_V \frac{\rho(\mathbf{r}_0) \, d^3r_0}{|\mathbf{r} - \mathbf{r}_0|}$$

and within V, from (8-5),

$$\nabla^2 U(\mathbf{r}_0) = 4\pi G\rho(\mathbf{r}_0)$$

Therefore, upon substituting for $\rho(\mathbf{r}_0)$, we find that

$$U(\mathbf{r}) = -\frac{1}{4\pi} \int_V \frac{\nabla^2 U(\mathbf{r}_0) \, d^3r_0}{|\mathbf{r} - \mathbf{r}_0|}$$

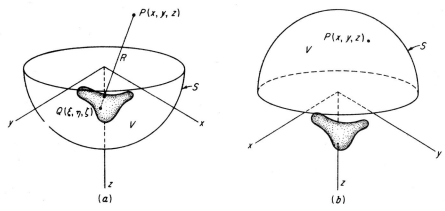

Fig. 8-3 The continuation theorem.

Since P lies outside V, $\nabla^2 W = 0$ everywhere, and the left-hand side of (8-7) becomes

$$-\int_V \frac{\nabla^2 U(\mathbf{r}_0)}{|\mathbf{r} - \mathbf{r}_0|}\, d^3r_0 = 4\pi U(\mathbf{r})$$

Thus

$$U(\mathbf{r}) = \frac{1}{4\pi} \int_S \left[U \frac{\partial}{\partial n}\left(\frac{1}{R}\right) - \frac{1}{R}\left(\frac{\partial U}{\partial n}\right) \right] d^2r_0 \qquad (8\text{-}8)$$

Now if we assume that all the masses lie within a finite region of the half-space $z > 0$, we may allow S to become a very large hemisphere in $z > 0$ closed by the plane $z = 0$ (Fig. 8-3). If the radius is made large enough, the integrand in (8-8) vanishes as R^{-3} everywhere on the curved surface of S, and the integral reduces to

$$U(\mathbf{r}) = \frac{1}{4\pi} \int_{-\infty}^{\infty} \int_{-\infty}^{\infty} \left[-U \frac{\partial}{\partial \zeta}\left(\frac{1}{R}\right) + \frac{1}{R}\left(\frac{\partial U}{\partial \zeta}\right) \right]_{\zeta=0} d\xi\, d\eta \qquad z \le 0 \quad (8\text{-}9)$$

where $R = \sqrt{(x - \xi)^2 + (y - \eta)^2 + (z - \zeta)^2}$. On the other hand, since there are no masses in $z < 0$, $\nabla^2 U = 0 = \nabla^2 W$ throughout this region; therefore if we close S above $z = 0$ rather than below, we find that (8-8) becomes

$$0 = \frac{1}{4\pi} \int_{-\infty}^{\infty} \int_{-\infty}^{\infty} \left[U \frac{\partial}{\partial \zeta}\left(\frac{1}{R}\right) + \frac{1}{R}\left(\frac{\partial U}{\partial \zeta}\right) \right]_{\zeta=0} d\xi\, d\eta \qquad (8\text{-}10)$$

Therefore if we add (8-10) and (8-9), we obtain[1]

$$U(\mathbf{r}) = \frac{1}{2\pi} \int_{-\infty}^{\infty} \int_{-\infty}^{\infty} \left(\frac{1}{R} \frac{\partial U}{\partial \zeta} \right)_{\zeta=0} d\xi\, d\eta$$

[1] For a more rigorous proof of this result, we refer to Kellogg (1), chap. 5.

Since U is the gravitational potential due to masses located within $z > 0$, we may put

$$\frac{\partial U}{\partial z} = \Delta g$$

and therefore

$$U(\mathbf{r}) = \frac{1}{2\pi} \int_{-\infty}^{\infty} \int_{-\infty}^{\infty} \frac{\Delta g(\xi, \eta)}{R} \, d\xi \, d\eta \qquad z \leq 0 \qquad (8\text{-}11)$$

By differentiating (8-11), we find that

$$\Delta g(\mathbf{r}) = \frac{\partial U(\mathbf{r})}{\partial z} = \frac{|z|}{2\pi} \int_{-\infty}^{\infty} \int_{-\infty}^{\infty} \frac{\Delta g(\xi, \eta)}{R^3} \, d\xi \, d\eta \qquad z \leq 0 \qquad (8\text{-}12)$$

showing that if Δg is known everywhere on $z = 0$, it is automatically determined everywhere in $z \leq 0$. The calculation of Δg at any level above the ground by means of (8-12) is known as *upward continuation*, and it consists of a straightforward numerical integration of the surface data.

A more practical although more suspect use of (8-12) consists in devising a distribution of Δg on a horizontal plane *below $z = 0$*, which, *assuming no masses intervene between the surface and this level*, will produce the observed gravity effect on $z = 0$. This is known as continuation downward, and it requires the inversion of (8-12) since it is now the left-hand side of the equation that is known, whereas the function under the integral sign is to be determined. The inversion may be accomplished through the use of the Fourier integral, as follows: Let us use subscripts 0 and z to denote values of Δg taken on $z = 0$ and at depth z, respectively. Then taking the Fourier transform of both sides of (8-12) with respect to x and y, we have

$$\int_{-\infty}^{\infty} \int_{-\infty}^{\infty} \Delta g_0(x, y) e^{i(px + qy)} \, dx \, dy = \frac{z}{2\pi} \int_{-\infty}^{\infty} \int_{-\infty}^{\infty} \Delta g_z(\xi, \eta) \, d\xi \, d\eta$$

$$\times \int_{-\infty}^{\infty} \int_{-\infty}^{\infty} [(x - \xi)^2 + (y - \eta)^2 + z^2]^{-\frac{3}{2}} e^{i(px + qy)} \, dx \, dy$$

Changing the variables x and y on the right-hand side into $x - \xi = r \cos \vartheta$ and $y - \eta = r \sin \vartheta$, and putting $p = u \cos \varphi, q = u \sin \varphi$ under the integral sign, the second integral on the right-hand side becomes

$$e^{i(p\xi + q\eta)} \int_0^{\infty} \int_0^{2\pi} (r^2 + z^2)^{-\frac{3}{2}} e^{iur \cos(\vartheta - \varphi)} r \, d\vartheta \, dr$$

$$= 2\pi e^{i(p\xi + q\eta)} \int_0^{\infty} J_0(ur)(r^2 + z^2)^{-\frac{3}{2}} r \, dr$$

$$= 2\pi e^{i(p\xi + q\eta)} e^{-uz}/z$$

Since $u = \sqrt{p^2 + q^2}$, we get, finally,

$$\int_{-\infty}^{\infty} \int_{-\infty}^{\infty} \Delta g_0(x, y) e^{i(px + qy)} \, dx \, dy = e^{-\sqrt{p^2 + q^2}\, z} \int_{-\infty}^{\infty} \int_{-\infty}^{\infty} \Delta g_z(\xi, \eta) e^{i(p\xi + q\eta)} \, d\xi \, d\eta$$

If we use the symbol $F_0(p,q)$ to denote the Fourier transform of $\Delta g_0(x,y)$ taken with respect to x and y, and $F_z(p,q)$ for the transform of $\Delta g_z(x,y)$, the above equation may be abbreviated into the symbolic form

$$F_0(p,q) = \exp\left(-\sqrt{p^2 + q^2}\, z\right) F_z(p,q)$$

Transposing the exponential term in this equation, we find therefore that

$$F_z(p,q) = \exp\left(\sqrt{p^2 + q^2}\, z\right) F_0(p,q) \qquad (8\text{-}13)$$

and thus

$$\Delta g_z(\xi,\eta) = \frac{1}{4\pi^2} \int_{-\infty}^{\infty} \int_{-\infty}^{\infty} F_0(p,q) \exp\left[\sqrt{p^2 + q^2}\, z - i(p\xi + q\eta)\right] dp\, dq$$

Since $F_0(p,q)$ is calculable from $\Delta g_0(x,y)$, this constitutes the formal solution of the downward continuation problem.

A serious difficulty with the convergence of this solution, however, is posed by the presence of the exponential factor in the integrand. *Unless the function $F_0(p,q)$ attenuates more rapidly than $exp\left(-\sqrt{p^2 + q^2}\, z\right)$, the integral $\iint |F_z(p,q)| \, dp \, dq$ does not exist,* and the downward continuation integral has no solution. It is therefore imperative that the surface gravity effect $\Delta g_0(x,y)$ should be a smooth function whose Fourier spectrum attenuates with the shorter "wavelengths" more rapidly than the exponential term rises. If it does not, it is because near-surface masses are making a contribution to the gravity effect which ought to have been excluded according to our basic assumption. Before continuation commences, therefore, it is usually necessary to devise a high-cut filter which suppresses the near-surface contributions to the residual gravity field. The "cutoff wavelength" of the filter is determined by the value of z.

Two methods of high-cut filtering are applied to gravity data in practice. In one, the residual gravity map is "digitalized" by replacing the contoured information with a set of point values taken at equal intervals in x and y. This automatically excludes from the Fourier transform of Δg all wavelengths shorter than twice the station interval. This cutoff position is controlled by adjusting the distance between the points, which makes this approach somewhat inflexible in view of the large amount of data preparation involved. The other method consists of the use of a mathematical filter. If we define

$$\Delta \bar{g}_0(x,y) = \frac{1}{4\pi\gamma} \int_{-\infty}^{\infty} \int_{-\infty}^{\infty} \Delta g_0(\xi,\eta) e^{-[(x-\xi)^2 + (y-\eta)^2]/4\gamma} \, d\xi \, d\eta \qquad (8\text{-}14)$$

and take the Fourier transform of both sides, we find that

$$\bar{F}_0(p,q) = e^{-\gamma(p^2 + q^2)} F_0(p,q)$$

Thus the integral transform (8-14) provides the necessary attenuation of $F_0(p,q)$ at the shorter wavelengths. If we substitute $\Delta \bar{g}_0(x,y)$ for $\Delta g_0(x,y)$

in (8-12) and then take Fourier transforms of both sides, we get

$$e^{-\gamma(p^2+q^2)}F_0(p,q) = \exp\left(-\sqrt{p^2+q^2}\,z\right)F_z(p,q)$$

of which the Fourier inversion is

$$\Delta g_z(\xi,\eta) = \frac{1}{4\pi^2}\int_{-\infty}^{\infty}\int_{-\infty}^{\infty} F_0(p,q)\exp\left[\sqrt{p^2+q^2}\,z - \gamma(p^2+q^2)\right.$$
$$\left. + i(p\xi+q\eta)\right]dp\,dq \quad (8\text{-}15)$$

This new integral converges properly over a limited range of z and can be evaluated. Thus a solution of the downward continuation problem exists, not for the residual gravity field itself, but for the quantity $\Delta\bar{g}_0(x,y)$ defined by (8-14), which is a smoothed-out version of $\Delta g_0(x,y)$. In this case the filtering is controlled by the value assigned to γ, which will ordinarily be set according to the value of z.

The application of smoothing techniques plays an essential role in downward continuation. Their use is fully justified by the observation that a solution cannot be found without them. Only the smoother parts of a gravity field can be continued downward to any considerable depth; the remainder must be left at the surface. We do not wish to imply by this statement that the smoother part of Δg_0 *must* originate at a greater depth, only that it *may* do so. The origin of any potential field is always ambiguous to this extent. The procedures for calculating the integrals will be discussed in detail in Sec. 9-10.

Since the magnetic scalar potential is also harmonic in free space, it can be continued either upward or downward according to the same theoretical principles that are applied to the gravitational field. Note first of all that if we subtract (8-9) from (8-10), we obtain the following formula for the upward continuation of the potential:

$$U(\mathbf{r}) = \frac{|z|}{2\pi}\int_{-\infty}^{\infty}\int_{-\infty}^{\infty}\frac{U(\xi,\eta)}{R^3}\,d\xi\,d\eta \quad z \leq 0$$

This integral is of little practical use in gravity interpretations, since Δg, not U, is the quantity that is related to the equivalent stratum. In magnetics, however, the *equivalent stratum* is connected with the scalar potential $A(\mathbf{r})$, and therefore both the potential A and the vertical component Z of the field intensity are used in continuation. The integral formulas are

$$A(x,y,z) = \frac{1}{2\pi}\int_{-\infty}^{\infty}\int_{-\infty}^{\infty}\frac{Z(\xi,\eta,0)}{R}\,d\xi\,d\eta \quad (8\text{-}16)$$

where $Z = -\dfrac{\partial A}{\partial z}$

or

$$A(x,y,z) = \frac{|z|}{2\pi}\int_{-\infty}^{\infty}\int_{-\infty}^{\infty}\frac{A(\xi,\eta,0)}{R^3}\,d\xi\,d\eta \quad (8\text{-}16a)$$

both of which are valid in $z \leq 0$. Neither the total field intensity H nor its horizontal component X can be used directly in upward or downward continuation. For further discussion of the continuation of magnetic fields, see Sec. 11-4.

8-5 Differentiation of the Potentials

By means of the integrals (8-11) or (8-16), derivatives of the potentials can be calculated at any point by numerical integration of Δg or of Z over the plane $z = 0$. In the half-space $z \leq 0$, differentiations may be carried out directly upon the integrals (8-11) or (8-16) themselves, but in the half-space $z > 0$, the integrals must first be inverted, then differentiated. To illustrate this point, we calculate the first vertical derivative of the gravity effect above and below the plane $z = 0$.

In $z \leq 0$ we get from (8-11)

$$- \frac{\partial \Delta g_z}{\partial z} = - \frac{\partial^2 U}{\partial z^2} = \frac{1}{2\pi} \int_{-\infty}^{\infty} \int_{-\infty}^{\infty} \frac{\Delta g_0}{R^3} \, dx \, dy - \frac{3|z|}{2\pi} \int_{-\infty}^{\infty} \int_{-\infty}^{\infty} \frac{\Delta g_0}{R^5} \, dx \, dy$$

No generality is lost by placing P directly above the origin of coordinates, since the origin may be taken anywhere. Then by changing to polar coordinates,

$$- \left(\frac{\partial \Delta g}{\partial z} \right)_P = \frac{1}{2\pi} \int_0^{\infty} \int_0^{2\pi} \frac{(r^2 - 2z^2)}{(r^2 + z^2)^{5/2}} \Delta g_0(r,\vartheta) r \, d\vartheta \, dr \qquad (8\text{-}17)$$

It is often useful to evaluate this expression in the limit as $z \to 0$. This raises a difficulty, because when $z = 0$, the integral becomes singular at $r = 0$. To circumvent this, we draw a small circle of radius ϵ around the origin and write

$$- \left(\frac{\partial \Delta g}{\partial z} \right)_P = \int_0^{\epsilon} + \int_{\epsilon}^{\infty} \frac{r^2 - 2z^2}{(r^2 + z^2)^{5/2}} \Delta g_0(r) r \, dr$$

where

$$\Delta g_0(r) = \frac{1}{2\pi} \int_0^{2\pi} \Delta g_0(r,\vartheta) \, d\vartheta$$

i.e., the average of Δg_0 around the circle of radius r. If ϵ is chosen small enough that Δg_0 does not sensibly change from its value at 0, then the first term on the right-hand side will be, approximately,

$$\Delta g_0(0) \int_0^{\epsilon} \frac{(r^2 - 2z^2)}{(r^2 + z^2)^{5/2}} r \, dr = \Delta g_0(0) \left[\frac{z^2}{(\epsilon^2 + z^2)^{3/2}} - \frac{1}{(\epsilon^2 + z^2)^{1/2}} \right]$$

and hence

$$\lim_{z \to 0} - \frac{\partial \Delta g}{\partial z} = - \frac{\Delta g_0(0)}{\epsilon} + \int_{\epsilon}^{\infty} \frac{\Delta g_0(r)}{r^2} \, dr \qquad (8\text{-}18)$$

In principle at least, integrals of the type (8-17) or (8-18), in which Δg_0 is not a mathematical function, can be evaluated approximately by making up templates or overlays having convenient divisions of r and ϑ and by estimating the average value of Δg_0 within each segment. In practice, the accuracy of these calculations is limited by the inability of the segment charts to include the contributions to the integrals that originate from beyond a certain distance from the origin.

However, there is another difficulty which is even more of a problem: The weighting function used in (8-18) attaches great importance to values clustered immediately around the reference point, virtually to the exclusion of the rest of the map. Not only does this overstrain the accuracy of contouring, but it makes the calculations highly susceptible to erratic, near-surface influences. To suppress this objectionable tendency and at the same time retain the more significant features of the derivative map, it is usually advisable to employ a suitable high-cut filter, as is done in downward continuation. The filter is devised with a view to suppressing the effects of near-surface heterogeneity while preserving as much as possible the significant geological effects.

To evaluate $-\dfrac{\partial \Delta g}{\partial z}$ in $z > 0$, we simply differentiate both sides of (8-13) with respect to z, getting

$$\frac{\partial}{\partial z} F_z(p,q) = \sqrt{p^2 + q^2}\, \exp\,(\sqrt{p^2 + q^2}\, z) F_0(p,q)$$

The Fourier inversion of the right-hand side of this equation is the mathematical solution of the problem, but the integral is divergent. To restore good behavior, it is necessary first to eliminate from F_0 the contributions from values of p and q that lie beyond a certain maximum, which will depend very largely upon the value taken for z. This, of course, brings us back once again to filtering. The filter will be designed on the same principle as the filter for downward continuation, since it has the same function to perform. Thus it turns out that in practice the mathematical expression for the derivative is the same no matter whether the plane $z = 0$ is approached from above or below. The derivation of this expression will be discussed in more detail in Sec. 9-9.

8-6   The Expansion of Gravitational Fields in Multipoles

The gravitational potential at any point outside matter distributed with a density $\rho(\mathbf{r}_0)$ throughout a volume V is

$$U(\mathbf{r}) = -G \int_V \frac{\rho(\mathbf{r}_0)\, d^3 r_0}{|\mathbf{r} - \mathbf{r}_0|}$$

Now
$$\frac{1}{|\mathbf{r} - \mathbf{r}_0|} = \frac{1}{\sqrt{r^2 + r_0^2 - 2 r r_0 \cos \gamma}}$$

$$= \frac{1}{r} \sum_{l=0}^{\infty} \left(\frac{r_0}{r}\right)^l P_l(\cos \gamma) \qquad r_0 \leq r$$

where P_l is the Legendre polynomial of order l. Therefore

$$U(\mathbf{r}) = -G \sum_{l=0}^{\infty} r^{-(l+1)} \int_V \rho(\mathbf{r}_0) r_0{}^l P_l(\cos \gamma) d^3 r_0 \qquad (8\text{-}19)$$

According to the addition theorem for Legendre functions,

$$P_l(\cos \gamma) = \frac{4\pi}{2l+1} \sum_{m=-l}^{m=l} y_l{}^{-m}(\vartheta_0, \varphi_0) y_l{}^m(\vartheta, \varphi)$$

where

$$y_l{}^m(\vartheta, \varphi) = \left[\frac{2l+1}{4\pi} \frac{(l-|m|)!}{(l+|m|)!}\right]^{1/2} e^{im\varphi} P_l{}^{|m|}(\cos \vartheta)$$

ϑ and φ being the spherical polar angles of the vector $\mathbf{r}$ (Fig. 8-4). The function $y_l{}^m$ is called the surface harmonic of order l, and $P_l{}^{|m|}$ is the associated Legendre function, which is defined as

$$P_l{}^{|m|}(z) = (1-z^2)^{|m|/2} \left(\frac{d}{dz}\right)^{|m|} P_l(z)$$

Substituting this expression into (8-19) gives

$$U(\mathbf{r}) = -4\pi G \sum_{l=0}^{\infty} (2l+1)^{-1} r^{-(l+1)} \sum_{m=-l}^{m=l} y_l{}^m(\vartheta, \varphi) \int_V \rho(\mathbf{r}_0) r_0{}^l y_l{}^{-m}(\vartheta_0, \varphi_0) \, d^3 r_0$$

$$= -\sum_{l=0}^{\infty} \sum_{m=-l}^{m=l} b_l{}^m y_l{}^m(\vartheta, \varphi) r^{-(l+1)} \qquad (8\text{-}20)$$

where

$$b_l{}^m = 4\pi G (2l+1)^{-1} \int_V \rho(\mathbf{r}_0) r_0{}^l y_l{}^{-m}(\vartheta_0, \varphi_0) \, d^3 r_0$$

These $b_l{}^m$ constitute, for a given l, the $2l+1$ possible "reduced multipole moments" of the body. They depend upon the shape of V, and since they

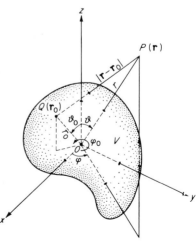

Fig. 8-4 Spherical coordinates used in the expansion of gravitational fields in multipoles.

can in principle be determined uniquely from the external potential field, they can be used as a means of making direct interpretations. Since there is an infinite number of these moments, however, they obviously cannot all be determined. In actual situations the series (8-20) is rapidly convergent, and the first three terms are usually sufficient to specify the potentials to within the accuracy of practical measurements.

It is convenient to convert the series (8-20) into one whose terms are all real, by writing

$$U(\mathbf{r}) = - \sum_{l=0}^{\infty} \sum_{m=0}^{l} \frac{B_l{}^m Y_l{}^m}{r^{l+1}} \tag{8-21}$$

where

$$B_l{}^m Y_l{}^m = b_l{}^{-m} y_l{}^{-m} + b_l{}^m y_l{}^m$$

Thus

$$Y_l{}^m(\vartheta,\varphi) = \cos m\varphi P_l{}^{|m|}(\cos \vartheta)$$

and

$$B_l{}^m = \frac{G(l-m)!}{(l+m)!} \int_V \rho(\mathbf{r}_0) r_0{}^l \cos m\varphi_0 P_l{}^{|m|}(\cos \vartheta_0)\, d^3r_0$$

If this series is truncated at the nth term, that is, if we write

$$- U(\mathbf{r}) = \sum_{l=0}^{n} \sum_{m=0}^{l} \frac{B_l{}^m Y_l{}^m}{r^{l+1}} + R_n$$

then we may calculate an upper bound for the remainder R_n as follows: Let a be the maximum radius of V and let ρ_m be an upper bound for the mean density within V. Then, since $|\cos m\varphi P_l{}^{|m|}(\cos \vartheta)| \leq 1$, it follows that

$$|B_l{}^m| \leq 4\pi G\rho_m \int_0^a r_0{}^{l+2}\, dr_0$$

$$= \frac{3GMa^l}{l+3}$$

if M is the total mass contained within V. Therefore

$$\left| \frac{B_l{}^m Y_l{}^m}{r^{(l+1)}} \right| \leq \frac{3GMa^l}{(l+3)r^{l+1}} \leq \frac{GM}{r}\left(\frac{a}{r}\right)^l$$

and consequently

$$R_n \leq \frac{GM}{r} \sum_{l=n+1}^{\infty} \left(\frac{a}{r}\right)^l$$

$$= \frac{GM}{r}\left(\frac{a}{r}\right)^{n+1}\left(1 - \frac{a}{r}\right)^{-1} \tag{8-22}$$

The expression (8-22) provides a means of measuring the convergence of the series (8-21) and the error involved in the neglect of terms beyond the nth.

8-7 On the Uses of Moments in Gravity Interpretations

In practice, it is usually permissible to truncate the series (8-21) at $l = 2$. Thus we set

$$- U(\mathbf{r}) = \sum_{l=0}^{2} \sum_{m=0}^{l} \frac{B_l{}^m Y_l{}^m}{r^{l+1}}$$

and derive $\Delta g(\mathbf{r})$ from this expression by differentiation. The terms in this series are further simplified by placing the origin 0 at the center of mass of the body, so that the coefficients in which $l = 1$ will vanish, as well as the coefficient B_2^1. Thus we are left with only the following terms:

$$- U(r,\vartheta,\varphi) = \frac{B_0^0}{r} + \frac{B_2^0(3 \cos^2 \vartheta - 1)}{2r^3} + \frac{3B_2^2 \cos 2\varphi \sin^2 \vartheta}{r^3}$$

which, if we transform into rectangular coordinates, become

$$- U(x,y,z) = \frac{B_0^0}{r} + \frac{B_2^0(3z'^2 - r^2)}{2r^5} + \frac{3B_2^2(x'^2 - y'^2)}{r^5} \tag{8-23}$$

We regard this expression as an exact representation of the anomalous gravitational potential of the mass M. The x', y', and z' axes in this expression are so chosen that they coincide with the principal axes of symmetry of V. In such a frame, the moments can be calculated easily. We shall call these the *body axes* of V.

The data to which the formula (8-23) is to be fitted are given, for practical purposes, on a horizontal plane. If one of the principal directions in the body happens to be vertical, the horizontal will be a plane of constant z. If, however, the body is tilted in some way, the body axes of V must be rotated (the moments themselves will remain invariant) in order to make the ground surface a plane of constant z. Two angles are needed to perform this rotation: (1) an azimuth ω to specify the direction of the major horizontal axis of the body with respect to the survey base line and (2) an angle of dip ψ (Fig. 8-5). The transformation which brings planes of constant z into a horizontal position and the x axis into line with the base of the survey is the following:

$$x' = x \cos \omega + y \sin \psi$$

$$y' = -x \sin \omega \cos \psi + y \cos \omega \cos \psi + z \sin \psi$$

$$z' = x \sin \omega \sin \psi - y \cos \omega \sin \psi + z \cos \psi$$

where the positive sense of both ω and ψ is anticlockwise, looking toward the origin in the negative direction of z and of x, respectively. If we substitute for x', y', and z' into (8-23) and differentiate with respect to z, we obtain

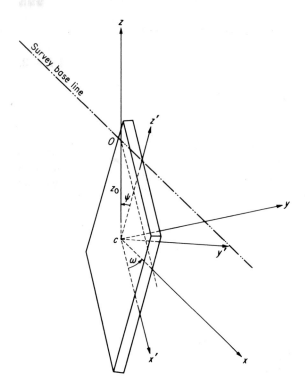

Fig. 8-5 Illustrating body axes x', y', z' and spatial axes x, y, z for an inclined, three-dimensional body.

the following expression for Δg on the plane $z = z_0$:

$$\Delta g(x,y) = \frac{B_0^0 z_0}{R^3}$$

$$+ 3B_2^0[(3\cos^2\psi - 1)z_0^3 + (5\sin^2\omega\sin^2\psi - 2\cos^2\psi - 1)x^2 z_0$$

$$+ (5\cos^2\omega\sin^2\psi - 2\cos^2\psi - 1)y^2 z_0 - 2\sin\omega\sin\psi\cos\psi(x^3 + xy^2 - 4xz_0^2)$$

$$+ 2\cos\omega\sin\psi\cos\psi(y^3 + x^2 y - 4yz_0^2) - 10\sin\omega\cos\psi\sin^2\psi(xyz_0)]/R^7$$

$$+ B_2^2[-3\sin^2\psi z_0^3 + (5\cos^2\omega - 5\sin^2\omega\cos^2\psi + 2\sin^2\psi)x^2 z_0$$

$$+ (5\sin^2\omega - 5\cos^2\omega\cos^2\psi + 2\sin^2\psi)y^2 z_0$$

$$+ 2\cos\omega\sin\psi\cos\psi(y^3 + x^2 y - 4yz_0^2)$$

$$- 2\sin\omega\sin\psi\cos\psi(x^3 + xy^2 - 4xz_0^2)$$

$$+ 10\sin\omega\cos\omega(1 + \cos^2\psi)xyz_0]/R^7 \qquad (8\text{-}24)$$

This is the formula that we are required to fit to the data. We should not be dismayed by its cumbersome length, since it leads to some remarkably simple analysis, as we shall see in Chap. 10.

The aim in gravity interpretation is to calculate values for B_0^0, B_2^0, and B_2^2 by fitting the formula (8-24) to a set of observational data. When these coefficients are known, they will yield the dimensions of whatever model we

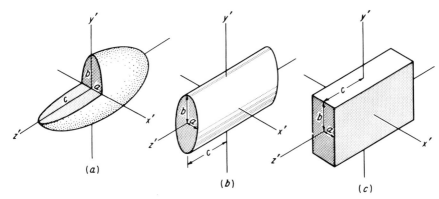

Fig. 8-6   Interpretational models: (*a*) **Triaxial ellipsoid;** (*b*) **elliptical cylinder;** (*c*) **rectangular block.**

may choose to represent the shape of the body. We must be careful, however, to limit ourselves to shapes which have only three parameters, if we are to obtain unambiguous solutions. The following are examples: (1) A triaxial ellipsoid, with semiaxes a, b, and c and a uniform density ρ. The x', y', and z' directions coincide with the three principal axes (Fig. 8-6a).

$$B_0^0 = \frac{4\pi G\rho abc}{3} \qquad B_2^0 = \frac{B_0^0(2c^2 - a^2 - b^2)}{20} \qquad B_2^2 = \frac{3B_0^0(a^2 - b^2)}{20}$$

(2) An elliptic cylinder of length $2c$, with semiaxes a and b and a uniform density ρ. The x' and y' directions coincide with the elliptic axes, and the z' direction coincides with the cylindrical axis (Fig. 8-6b).

$$B_0^0 = 2\pi G\rho abc \qquad B_2^0 = \frac{B_0^0(8c^2 - 3a^2 - 3b^2)}{24} \qquad B_2^2 = \frac{B_0^0(a^2 - b^2)}{32}$$

(3) A rectangular block with sides $2a$, $2b$, and $2c$ and having a uniform density ρ. The x', y', and z' axes lie in the three perpendicular directions (Fig. 8-6c).

$$B_0^0 = 8G\rho abc \qquad B_2^0 = \frac{B_0^0(2c^2 - a^2 - b^2)}{6} \qquad B_2^2 = \frac{B_0^0(a^2 - b^2)}{24}$$

Into any of these sets of formulas values of the moments may be inserted to yield the three dimensions a, b, and c. Note, however, that in order to do this, the value of ρ (the excess density, or density contrast) must be assumed. This is the ambiguity that still resides in the method of interpretation by moments.

8-8   Calculation of the Excess Mass

Gauss' theorem provides a very simple means of determining the excess mass responsible for a given anomaly in Δg when the observations are made on a horizontal plane. The theorem may be stated as follows: If $\mathbf{F}$ is a vector function which is analytic within and on a closed surface S containing a volume V, then

$$\int_V \nabla \cdot \mathbf{F} \, d^3 r_0 = \int_S \mathbf{F} \cdot \mathbf{n} \, d^2 r_0 \qquad (8\text{-}25)$$

where $\mathbf{n}$ is the unit outward normal vector on S. Let us put $\mathbf{F} = -\nabla U$, where U is the gravitational potential due to masses distributed with an excess density $\rho(\mathbf{r}_0)$ in V Then the left-hand side of (8-25) becomes

$$= -\int_V \nabla^2 U(\mathbf{r}_0) \, d^3 r_0 = -4\pi G \int_V \rho(\mathbf{r}_0) \, d^3 r_0 = -4\pi G M$$

where M is the total excess mass contained within the volume V.

Gauss' theorem is true regardless of whether or not V is partially empty. Let us therefore choose for the surface S a hemisphere of very large radius R in $z > 0$, closed by the plane $z = 0$, as shown in Fig. 8-3a. Then in the limit as $R \to \infty$, the right-hand side of (8-25) becomes

$$-\int_S \frac{\partial U}{\partial n} \, d^2 r_0 = -\int_{-\infty}^{\infty} \int_{-\infty}^{\infty} \left(\frac{\partial U}{\partial z}\right)_{z=0} dx \, dy - \lim_{R \to \infty} 2\pi R^2 \int_{\pi/2}^{\pi} \frac{\partial U}{\partial R} \sin \vartheta \, d\vartheta$$

The first term on the right-hand side is $-\int_{-\infty}^{\infty} \int_{-\infty}^{\infty} \Delta g(x,y) \, dx \, dy$, and the second may be evaluated as follows: If r_0 is the position of the center of mass of the anomalous material, which we assume to be distributed within a finite volume, then as R becomes large, $U(R) \to -GM/|\mathbf{R} - \mathbf{r}_0|$, which $\doteq -GM/R$ if $R \gg r_0$. The second term therefore tends, in the limit, to the value

$$-2\pi R^2 \frac{\partial}{\partial R} \frac{GM}{R} \int_{\pi/2}^{\pi} \sin \vartheta \, d\vartheta = -2\pi G M$$

Gathering terms from both sides of (8-25), we get

$$\int_{-\infty}^{\infty} \int_{-\infty}^{\infty} \Delta g(x,y) \, dx \, dy = 2\pi G M \qquad (8\text{-}26)$$

The excess mass M may therefore be found by integrating its gravity effect over the horizontal plane. There is no equivalent result in magnetic field theory.

8-9   Locating the Center of Mass

The position of the center of mass of M on the plane $z = 0$ may be found by applying the following theorem due to Kogbetliantz (2). Referring to Fig. 8-7, the gravity effect at $P(x,y,0)$ due to the mass element in V at $Q(\xi,\eta,\zeta)$ is

$$d \, \Delta g = G\rho(\mathbf{r}_0)\zeta[(x - \xi)^2 + (y - \eta)^2 + \zeta^2]^{-\frac{3}{2}} \, d^3 r_0$$

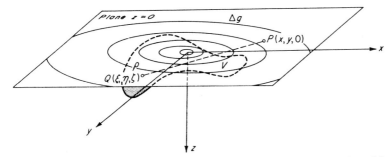

Fig. 8-7 **Locating the center of mass of a buried three-dimensional body.**

Now consider the integral

$$dN = \int_{-\infty}^{\infty} \int_{-\infty}^{\infty} [(x - \xi) + i(y - \eta)]\, d\,\Delta g(x,y)\, dx\, dy$$

If we let $x - \xi = r \cos \vartheta$, $y - \eta = r \sin \vartheta$, then

$$dN = G\rho(\mathbf{r}_0)\zeta\, d^3r_0 \int_0^{\infty} (r^2 + \zeta^2)^{-3/2} r^2\, dr \int_0^{2\pi} e^{i\vartheta}\, d\vartheta$$

which vanishes by symmetry. Therefore the real and imaginary parts of dN must vanish independently, giving

$$\int_{-\infty}^{\infty} \int_{-\infty}^{\infty} (x - \xi)\, d\,\Delta g(x,y)\, dx\, dy = 0 = \int_{-\infty}^{\infty} \int_{-\infty}^{\infty} (y - \eta)\, d\,\Delta g(x,y)\, dx\, dy$$

Let us consider the first of these two equations. It gives us

$$\int_{-\infty}^{\infty} \int_{-\infty}^{\infty} x\, d\,\Delta g(x,y)\, dx\, dy = \xi \int_{-\infty}^{\infty} \int_{-\infty}^{\infty} d\,\Delta g(x,y)\, dx\, dy$$
$$= 2\pi G\xi\, dm(\mathbf{r}_0)$$

according to (8-26). If we now integrate both sides of this equation over the volume V, we get

$$\int_{-\infty}^{\infty} \int_{-\infty}^{\infty} x\, \Delta g(x,y)\, dx\, dy = 2\pi G \int_V \xi\, dm = 2\pi G M \bar{x} \qquad (8\text{-}27a)$$

where $\bar{x}$ is the coordinate of the center of mass of M. Similarly,

$$\int_{-\infty}^{\infty} \int_{-\infty}^{\infty} y\, \Delta g(x,y)\, dx\, dy = 2\pi G M \bar{y} \qquad (8\text{-}27b)$$

Having calculated M from (8-26), $\bar{x}$ and $\bar{y}$ can now be determined.

8-10 *The Logarithmic Potential*

In some situations, the contour lines of Δg (or of ΔZ, or of ΔH) are strongly elliptical, indicating that the body is greatly elongated in relation to its transverse dimensions. In such cases, it is often more practical to treat the structure as though it were two-dimensional, rather than to try to force a fit to some volumetric distribution. Returning to the formula (8-1), if we now assume that ρ is independent of one of the coordinates (y, let us say), and if the body is infinitely elongated in this direction without changing its cross section anywhere, then

$$
U(x,z) = -G \int_S \rho(x_0,z_0) \, dx_0 \, dz_0 \int_{-\infty}^{\infty} [(x - x_0)^2 + (y - y_0)^2
$$
$$
+ (z - z_0)^2]^{-\frac{1}{2}} \, dy_0
$$
$$
= 2G \int_S \rho(x_0,z_0) \, \log_e R \, dx_0 \, dz_0 \tag{8-28}
$$

where $R = \sqrt{(x - x_0)^2 + (z - z_0)^2}$. The expression (8-28) is called the *logarithmic potential*, and the integration is to be carried out over the cross section S of the body.

The main purpose of introducing the logarithmic potential is to simplify the calculations by reducing the number of variables. When z is fixed, (8-28) is a function of only one variable; (8-1) is a function of two. The gravitational fields of two-dimensional bodies are fully determined by a single profile, whereas those of three-dimensional bodies must be presented as contoured maps. Since geophysical interpretation eventually consists in fitting theoretical models to observational data, it is obvious that the manipulations of the potential field functions can be carried out with far less labor on a single profile than on a contoured map. This undoubtedly explains the preference that has always been shown to two-dimensional models in the literature of applied geophysics. While this particular advantage has been somewhat depreciated recently by the increasing application of computing machines to geophysical problems, it has not yet been eliminated.

Two-dimensional bodies do not, of course, exist in nature. Every structure ends somewhere, and the important question concerning the use of these methods is how their accuracy is affected by finite strike lengths. This is not a question that is answerable categorically, since the objectives of geophysical work are not always the same. It can safely be said, however, that for general purposes the body must be highly elongated (roughly speaking, the length must be five or more times the greatest cross-sectional dimension) before two-dimensional methods can be safely applied. This means that the contour lines of Δg, ΔZ, or ΔH must be very strongly elliptical. Just how strongly elliptical they must be is a matter that will be discussed in greater detail in Chap. 10.

8-11   *The Expansion of Two-dimensional Fields in Multipoles*

The gravitational field due to two-dimensional bodies can be expanded into a series form analogous to (8-20). Referring to Fig. 8-1, we let

$$
U(\mathbf{r}) = 2G \int_S \rho(\mathbf{r}_0) \, \log_e |\mathbf{r} - \mathbf{r}_0| \, d^2r_0
$$

where $\mathbf{r}$ and $\mathbf{r}_0$ are now assumed both to lie in the same vertical plane. The gravity effect Δg associated with this potential is equal to the vertical derivative of (8-28) calculated at the surface of the ground. If the inclination of the polar axis of S ($\vartheta = 0$) to the ground surface is d, then

$$\Delta g(\mathbf{r}) = 2G \int_S \frac{\rho(\mathbf{r}_0)(r \sin \vartheta' - r_0 \sin \vartheta_0')}{r^2 + r_0^2 - 2rr_0 \cos (\vartheta' - \vartheta_0')} d^2 r_0$$

where $\vartheta' = \vartheta + d$ and $\vartheta_0' = \vartheta_0 + d$ are the vertical angles of r and r_0, respectively. If we expand the function

$$\frac{r \sin \vartheta' - r_0 \sin \vartheta_0'}{r^2 + r_0^2 - 2rr_0 \cos (\vartheta' - \vartheta_0')} = \frac{1}{r} \frac{\sin \vartheta' - (r_0/r) \sin \vartheta_0'}{1 + (r_0/r)^2 - 2(r_0/r) \cos (\vartheta' - \vartheta_0')}$$

by the binomial theorem into a series having the form

$$\frac{1}{r} \sum_{m=0}^{\infty} \left(\frac{r_0}{r}\right)^m T_m(\vartheta',\vartheta_0')$$

the coefficients T_m turn out as follows:

$$T_0 = \sin \vartheta' \qquad T_1 = \sin (2\vartheta' - \vartheta_0') \qquad T_2 = \sin (3\vartheta' - 2\vartheta_0')$$

and in general $T_m(\vartheta',\vartheta_0') = \sin [(m + 1)\vartheta' - m\vartheta_0']$. Thus we have

$$\Delta g(\mathbf{r}) = 2G \sum_{m=0}^{\infty} r^{-(m+1)} \int_S \rho(\mathbf{r}_0) r_0^m T_m(\vartheta',\vartheta_0') \, d^2 r_0$$

$$= -i \sum_{m=0}^{\infty} \frac{b_m e^{i[(m+1)\vartheta' - md]}}{r^{m+1}} \tag{8-29}$$

where $\qquad b_m = 2G \int_S \rho(\mathbf{r}_0) r_0^m e^{-im\vartheta_0} \, d^2 r_0$

provided it is understood that only the real part of the series (8-29) has significance.

This series is altogether analogous to the series (8-20) for the three-dimensional distribution. We have once again a term structure in which the coefficients are moments; only on this occasion they are moments of an area rather than of a volume. If this series is truncated at the nth term, the remainder has an upper bound as follows:

$$\Delta g(\mathbf{r}) = -i \sum_{m=0}^{n} b_m r^{-(m+1)} e^{i[(m+1)\vartheta' - md]} + R_n$$

$$|R_n| \le \frac{4Gm}{r} \left(\frac{a}{r}\right)^{n+1} \left(1 - \frac{a}{r}\right)^{-1}$$

where m is the total mass per unit length of the structure and a is its largest radius. Thus there is a means of measuring the convergence of (8-29).

This series plays the same role in interpretation in two dimensions as the series (8-20) does in three.

8-12   Calculation of Excess Mass and Moments of Two-dimensional Bodies

By methods exactly analogous to those used in Secs. 8-8 and 8-9, the mass per unit length and the location of the center-of-mass line of a two-dimensional body can be found from the profile of its gravity effect on a horizontal plane.   To apply Gauss' theorem in this case, we divide the body into vertical slices of unit length, the mass of one of which we shall call m.   The volume V to be used will be the D-shaped sector formed by the two semi-circles $x^2 + z^2 = R^2$ in $z > 0$, which are displaced by unit distance in the y direction (Fig. 8-8).   Since the contributions to the surface integral from the two flat sides of this sector will exactly cancel each other, the net contribution comes solely from the edges.   Thus

$$4\pi G m = \int_{-\infty}^{\infty} \left(\frac{\partial U}{\partial z}\right)_{z=0} dx + \lim_{R \to \infty} R \int_{0}^{\pi} \left(\frac{\partial U}{\partial R}\right) d\vartheta$$

and since $\partial U / \partial r \to 2Gm/R$ as $R \to \infty$, this reduces to

$$\int_{-\infty}^{\infty} \Delta g(x) \, dx = 2\pi G m \tag{8-30}$$

$\Delta g(x)$ is the profile of the gravity effect, measured in the direction perpendicular to the strike.

The center-of-mass line can be located also.   If in Fig. 8-9 we place both P and Q in the same principal section (i.e., by putting $y = \eta$) and regard the body as two-dimensional, then the gravity effect at P due to the line element at Q will be

$$d \, \Delta g(x) = \frac{2G\rho\zeta \, d\zeta}{(x - \xi)^2 + \zeta^2}$$

The integral

$$\int_{-\infty}^{\infty} (x - \xi) \, d \, \Delta g(x) \, dx$$

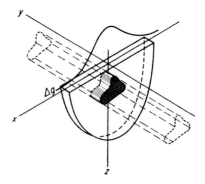

Fig. 8-8   Gauss' theorem in two dimensions.

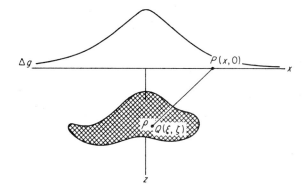

Fig. 8-9 Locating the center-of-mass line of a buried two-dimensional body.

vanishes by symmetry; so, if we integrate over S and apply (8-30), we get

$$\int_{-\infty}^{\infty} x\, \Delta g(x)\, dx = 2\pi G m \bar{x} \tag{8-31}$$

To determine the formula for the gravity effect of S, from which the values of the moments are to be found by fitting to observational data, we proceed exactly as we did in Sec. 8-7. Thus we begin by limiting the series (8-29) to terms including $m = 2$ (a step which is more difficult to justify a priori in two dimensions than it is in three) and by placing the origin of coordinates on the center-of-mass line which eliminates the term $m = 1$. If $\vartheta_0 = \pi/2$ is a plane of symmetry of S, then in the coefficients b_m the terms of odd symmetry in ϑ_0 will disappear, leaving

$$\Delta g(\mathbf{r}) = \sum_{m=0}^{2} b_m r^{-(m+1)} \sin\left[(m+1)\vartheta' - md\right]$$

where

$$b_m = 2G \int_S \rho(\mathbf{r}_0) r_0{}^m \cos m\vartheta_0\, d^2 r_0$$

Changing to rectangular coordinates, this becomes

$$\Delta g(x) = \frac{b_0 z_0}{r^2} + \frac{b_2[(3x^2 z_0 - z_0{}^3)\cos 2d - (x^3 - 3xz_0{}^2)\sin 2d]}{r^6} \tag{8-32}$$

where z_0 is the depth of the center-of-mass line.

Just as for three-dimensional bodies, the moments can be written down in terms of the linear dimensions of S. Thus for an elliptic cylinder having semiaxes a and b and a uniform density ρ,

$$b_0 = 2\pi G\rho ab \qquad b_2 = \tfrac{1}{4} b_0(a^2 - b^2)$$

For a prism having a rectangular cross section and a uniform density, with sides $2a$ and $2b$,

$$b_0 = 4G\rho ab \qquad b_2 = \tfrac{1}{3} b_0(a^2 - b^2)$$

To find a and b from the values of the moments, ρ must be assumed.   Thus some ambiguity remains, as always, in the relationship between the density and the size.

References

1. O. D. Kellogg, "Foundations of Potential Theory," Ungar, 1929.

2. E. G. Kogbetliantz, Quantitative Interpretation of Gravitational and Magnetic Anomalies, *Appl. Mathematics Quart.*, vol. 3, pp. 55–80, 1944.

chapter 9 The Reduction and Interpretation of Gravity Data

The small variations in g recorded by the gravity meter as it moves over the ground surface generally include several effects unrelated to geology. For instance, effects due to the oblateness of the earth, to changes in elevation, to isostatic compensation, and to various other causes, are irrelevant to geological interpretations. Our first concern, therefore, is to remove from the data, as carefully as possible, all these extraneous disturbances. Only after these preliminaries are accomplished can geophysical interpretation be seriously attempted. Because the magnitude of the corrections relative to the residues is large, however, the first steps in reducing the data are critical and must be done with the greatest care. A special characteristic of the gravity method, in fact, is the elaborate chain of reductions down which the data must pass before they attain a form suitable for interpretation. Chapter 9 will be very largely concerned with these processes.

9-1   The Latitude Effect

From world-wide geodetic measurements, the shape of the earth is known to be very nearly a spheroid. That spheroidal surface which gives the closest overall fit to mean sea level is called the *reference spheroid,* and its dimensions are by now accurately known. We also know from seismology that beneath the solid crust transverse changes in density appear to exist only in the upper part of the mantle, and while these are very broad in extent, they are low in amplitude. The densities of the materials composing the bulk of the mantle and the core vary essentially in the radial direction only, and these changes are due very largely to pressure. Only the solid crust, which constitutes less than 2 percent of the earth's total mass, appears to be significantly heterogeneous. Thus it would seem that a transversely-isotropic rotating spheroid should—in the first approximation at least—be a reasonably good mathematical model of the earth. According to (8-21), the formula for the gravitational field on the surface of such a body can be written in the form

$$g(\varphi) = \sum_{n=0}^{\infty} a_n \sin^2 n\varphi \qquad (9\text{-}1)$$

where φ is the geocentric latitude. The coefficients a_n are chosen by fitting world-wide pendulum measurements of gravity, which have been reduced to sea level, to the formula (9-1) by the method of least squares. It turns out (on the basis of data existing in 1930) that $a_0 = 978.0490$ gals; $a_1 = 0.0052884a_0$; $a_2 = -0.0000059a_0$; and a_3 and succeeding coefficients are insignificant. Thus the distribution of the main gravitational field on the reference spheroid is described by the formula [Heiskanen (1), Sec. 3-13]

$$g(\varphi) = 978.0490(1 + 0.0052884 \sin^2 \varphi - 0.0000059 \sin^2 2\varphi) \quad \text{gals} \quad (9\text{-}2)$$

to the accuracy of the pendulum measurements.

This formula may be used to correct all gravity data for the ellipticity of the earth. If tables of this function are not available, the north-south horizontal gradient of g may be computed from (9-2) as follows:

$$\frac{dg}{ds} = \frac{1}{R(\varphi)} \frac{dg}{d\varphi} \doteq \frac{1}{R_e} \frac{dg}{d\varphi}$$

$$= 13.07 \sin 2\varphi \quad \text{gu*/mile N–S}$$

where R_e is the equatorial radius. The correction at each gravity station is easily found by drawing a grid of east-west lines across the map, the spacing of which is equivalent to increments of 1 gu. In the middle latitudes, this

---

* 1 gravity unit (gu) $= 0.1$ mgal $= 10^{-4}$ gal. It is the unit most commonly used in geophysical surveys.

will be about 400 ft, and if higher accuracy is needed, the grid interval may be reduced.   Since the tendency is for g to increase from the equator to the poles, the latitude correction is added to the readings as one moves toward the equator.

9-2   Elevation Corrections: The Free-air Effect

Corrections based on formula (9-2) are strictly applicable only to observations taken on the reference spheroidal surface.   If an observation point lies above or below the reference spheroid, a correction must first be made for the vertical departure in order that formula (9-2) may apply.   If the radius of the reference spheroid is $R(\varphi)$ and the vertical departure $H \ll R$, we may write to a sufficient accuracy for the main gravitational field at that station

$$g(R + H) = g(R) + H \frac{\partial g}{\partial R}$$

The first term on the right-hand side is given by formula (9-2); thus the additional correction needed is $H \partial g/\partial R$, which must be subtracted.   To calculate this term, we make use of McCullagh's formula for the gravitational potential at any point outside of a spheroid of small eccentricity rotating with angular velocity Ω

$$-U(R) = \frac{GM}{R} + \frac{G}{2R^3} (C - A)(1 - 3 \cos^2 \varphi) + \frac{1}{2}\Omega^2 R^2 \cos^2 \varphi \quad (9\text{-}3)$$

where C and A are the axial and equatorial moments of inertia, respectively. We recognize in this formula the first three terms of the general expansion (8-21) in the center-of-mass coordinate system, plus the rotational term. Truncation at the third term is justified on the grounds that the series is known to be absolutely convergent, and we have already discovered that the main gravitational field is closely represented by a formula of this type if we neglect the coefficient a_2 in (9-2).   Retention of the a_2 term in McCullagh's formula would require us to extend the series (8-21) out to $l = 4$, and the additional set of coefficients would be moments of the fourth order whose values we are unable to determine by astronomical measurements.   Fortunately for the gravity method, the first two terms of the international formula provide enough accuracy for calculating the vertical gradient of g under almost all circumstances.

Differentiating (9-3) twice with respect to R, we get

$$\frac{\partial g}{\partial R} = + \frac{\partial^2 U}{\partial R^2} = - \frac{G}{R_e^5} [2M R_e^2 - 3(C - A)] - \frac{1}{2} \Omega^2$$
$$+ \left[\frac{9G(C - A)}{R_e^5} - \frac{1}{2} \Omega^2 \right] \cos 2\varphi$$

and substituting the astronomically determined values for C, A, M, and R_e gives, approximately,

$$\frac{\partial g}{\partial R} = -0.9416 - 0.0090 \cos 2\varphi \qquad \text{gu/ft}$$

Multiplying this expression by the departure H from the reference spheroid in feet gives the *free-air correction* which compensates for variations in the distance of the observation point from the center of attraction of the earth.[1] This correction has the same sign as H. The residue that remains after the free-air correction has been applied and after (9-2) has been subtracted from the measured value of g is variously called the "free-air gravity disturbance," the "free-air anomaly," "Faye's anomaly," or simply the "free-air gravity." We shall use the latter term.

9-3 Elevation Corrections: The Bouguer Correction

The free-air correction ignores the entire mass of material that lies between ground level and the reference spheroid. The gravity effect of this mass is included in the measurement of g, and its magnitude will of course depend upon the thickness H. Because the effect is positive, it opposes the free-air gradient and will therefore tend to diminish it. If due allowance is not made for this effect in reducing the values of g, they will bear a strong negative correlation with H. The profile of free-air gravity across the topographic section shown in Fig. 9-2 illustrates this point. The result of this negative correlation is an emphasis on topographic features at the expense of subsurface density changes. Since the latter are the objects of geophysical exploration studies, a further correction for the gravity effect of the terrain is essential.

To calculate the effect of terrain at the point A (Fig. 9-1), we must evaluate the integral

$$G \frac{\partial}{\partial z} \int_V \frac{\rho(\mathbf{r}_0) \ d^3 r_0}{|\mathbf{r} - \mathbf{r}_0|}$$

throughout the entire volume contained between the ground surface and the reference spheroid. Within limited areas, we generally ignore the earth's curvature and regard the reference spheroid as a flat surface. As a matter of additional convenience we replace the spheroid with local mean sea level by setting $H = h$, where h is the elevation of the ground surface above sea level,[2] and then we divide this volume into two parts by passing a horizontal surface through A. The contribution from the slab contained between the two horizontal surfaces is traditionally called the *Bouguer effect*, while the

---

[1] In mountainous regions, it may be necessary to add to this formula a third term $= +0.68 \times 10^{-7} H$ gu/ft, which takes into account the change in $\partial g / \partial R$ with H.

[2] A discussion of this approximation will be given in Sec. 9-11.

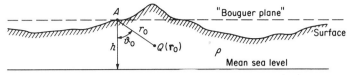

Fig. 9-1 The Bouguer and terrain correction.

remainder is called the *topographic effect;* and they are evaluated separately. If the density were constant throughout V, the Bouguer effect would be $= 2\pi G\rho h$. Since the density usually varies, we define a *Bouguer density* at A as follows:

$$\rho_B(A) = \frac{1}{2\pi h} \int_{slab} \frac{\rho(\mathbf{r}_0) \cos \vartheta_0}{r_0{}^2} d^3r_0 \tag{9-4}$$

and the Bouguer effect is then $2\pi G\rho_B h$.

As the position of A is arbitrary, the Bouguer density generally varies with the location. Since it is a weighted average, however, it must be smooth and can have no discontinuities. The correction at A that will be needed to compensate for the Bouguer effect is $0.1276\rho_B h$ gu when h is measured in feet, and it opposes the free-air correction. The two corrections are usually combined into a single elevation correction kh having the same sign as h. The elevation factor is

$$k = 0.9416 + 0.0090 \cos 2\varphi - 0.1276\rho_B - 0.68 \times 10^{-7} h \qquad \text{gu/ft} \tag{9-5}$$

On account of ρ_B, k will change from point to point. For practical reasons, however, it is the common practice to use a constant value within a limited area unless there are strong reasons to the contrary.

9-4 Topographic Corrections

To complete the calculation of the gravity effects of the terrain, we must determine the contributions to g from topography on the *Bouguer plane*. In Fig. 9-1 this residuum is represented by the material above the level surface through A, as well as by the material missing from below this surface. Both of these topographic features will reduce gravity at A: the hills because they exert an attraction having an upward component at A, and the valleys because they represent material having a downward gravitational force that has been removed. All topographic corrections, therefore, whether they compensate for hills or for valleys, will be positive.

The topographic correction is given by the integral

$$T(A) = G \int_v \frac{\rho(\mathbf{r}_0) \cos \vartheta_0}{r_0{}^2} d^3r_0 \tag{9-6}$$

in which the volume v is contained between the land surface and the Bouguer plane. To evaluate this integral, both the density distribution of the surface materials and the shape of the land surface must be known.   The first problem is solved by assuming ρ to be equal to the Bouguer density, and the second is handled by the use of either computing machines or templates. Methods programmed for digital computers depend essentially upon fitting mathematical surfaces to point values of the elevation [for details, see Kane (2)].   Alternatively, templates or overlay charts may be used to divide up the volume into a number of sectors approximating simple geometrical shapes, whose gravitational effect at the point of reference is known. Details of these procedures may be found in Dobrin (3), chap. 11.   In either case it is necessary to have reasonably accurate topographic maps of the area.

It is perhaps unnecessary to add that topographic corrections are very time-consuming.   In areas of low-to-moderate relief these corrections rarely exceed a very few gravity units; unless the heights (or depths) of the topographic features exceed one-twentieth of their distance from the reference point, they are usually omitted.

When both the latitude and the elevation corrections (including the topographic correction, when warranted) have been applied, the residual is called the *Bouguer gravity*.   This is the small component of g which is supposed to have its origins within the first few hundreds or thousands of feet of rock, and which therefore constitutes the raw material for geophysical and geological interpretation.

9-5   *Calculation of the Bouguer Density*

This is one of the more problematical steps in the sequence of data reduction. Direct sampling of the rock densities over large areas is obviously out of the question, and so ρ_B must be found from the gravity observations themselves.   There is no direct method of ascertaining the Bouguer effect by the analysis of gravity data alone, since the geological contributions are unknown; but it is usually possible to tell if a gross error has been committed in making the elevation reductions.   Figure 9-2 illustrates this point.   This drawing was made from a set of observations made in an area in which the bedrock consists of a fairly dense calcareous shale covered with glacial till, which for the most part is made up of a light sandy clay.   The mean elevation is approximately 600 ft above mean sea level.   In the diagram, a number of profiles of Bouguer gravity has been calculated by assuming different values for the Bouguer density.   Some of these profiles clearly follow the topography, while others mirror it; but there is one profile ($\rho_B = 1.9$ g/cm³) which shows no visible correlation with the topography at all.   Of all the Bouguer gravity profiles this is the one we should prefer to use for geophysical interpretation, because it contains no "anomalies" which are associated with the terrain and which might easily either be mistaken as evidence of subsurface density changes or else confuse existing geological

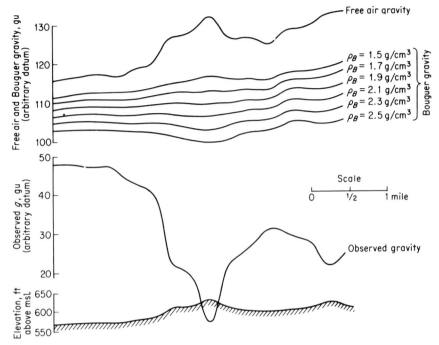

Fig. 9-2 **Profiles of free-air and Bouguer gravity.**

effects. This is the basis of the "density profiling" method suggested by Nettleton (4). If we are satisfied with the logic of this approach, we may at once state the following two corollaries:

1. This method samples only materials which lie between the lowest and highest topographic levels; consequently the reference surface for the Bouguer reduction is not sea level nor any other horizontal datum, but the smooth surface which passes through the lowest topographic points in the region. In corroboration of this statement, we remark that the Bouguer density arrived at in Fig. 9-2 by profiling is characteristic of the surface deposits, in spite of the fact that nearly 600 ft of bedrock lies between the gravity stations and sea level.

2. The rationale used in selecting ρ_B is to minimize the correlation (either positive or negative) between topographic *relief* and Bouguer gravity *anomalies*. The regional trends in either set of observations are discounted in the process. This is a reasonable point of view, provided that it is the short-range gravity patterns to which our attention is drawn, since the density profiling method produces the residual gravity map least contaminated by surface effects and the regional trends are usually to be removed in any case (Sec. 9-7). It seems to be the only system for making Bouguer reductions in areas in which the densities are not known beforehand, although it will fail if the land forms are closely related to density changes

below the base of the topography.   The only evidence for such occurrences will in general be unrealistic values for the Bouguer density.

Density profiles may be calculated at isolated features as we have described, and a mean value taken for the elevation factor within the area. This is a practice that is commonly followed, but it ignores variations in ρ_B. In areas where surface conditions vary, particularly if the gravity anomalies are likely to be less than 1 mg in amplitude, the changes in ρ_B very often cannot be ignored.   In such cases, k must be determined as a smooth function of x and y throughout the area.   This can best be done by finding the smooth function of x and y which minimizes the covariance of the Bouguer gravity anomalies with the topographic relief.   The essential preliminary to this calculation is first of all to remove the trends from the elevation and the Bouguer gravity field, so that we are left with residual values δh and $\delta(\Delta g)$, both of which are trend-free.   The elevation factor, according to the method of density profiling, is the function $k(x,y)$, which causes the quantity

$$\iint [\delta(\Delta g) + k(x,y)\ \delta h]^2\ dx\ dy$$

to have a minimum value throughout the area.   If ρ_B is constant, k will have the value

$$k = -\ \frac{\displaystyle\sum_{i=1}^{N} \delta(\Delta g)_i\ \delta h_i}{\displaystyle\sum_{i=1}^{N} (\delta h_i)^2} \tag{9-7}$$

which may not be the best local value to use at every point in the area but which will be the best constant value over the area as a whole.   If ρ_B varies, $k(x,y)$ will be a polynomial of a low order whose local value will be used for correcting the data at each station.   Such values should be nearly optimum everywhere.

9-6   Effects of Isostasy

As a matter of general observation, it is known that there are wide regional disturbances of Bouguer gravity over many parts of the earth which are often associated with uplifted or depressed regions of the earth's crust (e.g., mountain ranges, ocean troughs, rifts, plateaus, etc.) but which cannot be accounted for wholly by the topography itself.   This and other parallel lines of evidence have pointed the way toward the theory that the earth's crust is in an average sense supported in hydrostatic equilibrium between buoyant and gravitational forces.   According to the theory, anomalous loads such as mountain ranges are supported by the extra buoyant forces supplied by an excess of material of lighter density at the base of the crust that displaces the denser substratum, whereas unusual deficiencies in crustal matter such as the great ocean trenches are prevented from buckling upward by reductions in the normal buoyant forces caused by a deficiency of lighter material at the base. The excess of lighter material is supposed to be brought about by a local thickening of the crust, and the deficiency, by a thinning of the crust. These variations in crustal thickness ("roots" and "antiroots") give rise to broad anomalies in the Bouguer gravity at the surface. These almost invariably are negative over plateaus and positive over basins. If the gravity survey is very broad in extent and if its aim is to reconnoiter for regional features, it is sometimes necessary to separate out the isostatic effect where it may be significant. This is often particularly important along mountain fronts or near continental margins.

The method for making the isostatic correction follows closely that of making the terrain correction from a topographic map. The area is divided up into small segments, each of which may be considered as a prism having a simple cylindrical shape, whose gravity effect is therefore calculable. The length of the prism is determined by the supposition that the crust has little or virtually no rigidity and that each column therefore fully supports its own weight. Some assumption must be made about the average densities of crust and substratum, as well as about the mean crustal thickness. Since these are prior assumptions, it is often prudent to carry out the calculations for more than one set of values [see Heiskanen (1), chap. 6].

When a correction for isostasy is applied to the Bouguer gravity, the residue is usually called the *isostatic anomaly*. Except in reconnaissance surveys over very broad and anomalous regions, however, the isostatic effect, if present, is usually removed as a part of the regional trend.

9-7 The Regional-residual Separation

The picture that usually emerges after making the reductions leading to the preliminary Bouguer gravity map is one that shows the superposition of disturbances of noticeably different orders of size. The larger features generally show up as trends which continue smoothly over very considerable areas, and they are caused by the deeper heterogeneity of the earth's crust. Superimposed on these trends, but frequently camouflaged by them, lie the smaller, local disturbances, which are secondary in size but primary in importance. These are the *residual anomalies*, which may provide the direct evidence for reservoir-type structures or mineral ore bodies. To be interpreted, these anomalies must somehow be removed from the regional background. In some respects, the problem as usually conceived is analogous to filtering, but with the difference that it is the residue or "noise" (the part ordinarily removed by the filtering) that is required for subsequent analysis, whereas in communications theory this part of the signal is usually of little interest. Thus the normal methods of mathematical filtering (smoothing, averaging, etc.) often fail to provide the best solution because they ignore the structure of the residual field, usually regarding it as being essentially "random."   Apparently no sound methods of numerical filtering exist when the residual anomalies have a biased structure.   Thus the problem of regional-residual analysis is necessarily interpretative, the main emphasis being as a general rule on defining the forms and shapes of the residual effects.

One of the oldest and most traditional methods of making the separation is by visually smoothing the contour lines or a number of profiles, or preferably both at once.   The process works roughly by successive approximations in the following way: A suitable network of profiles is chosen and smoothed, and the new set of contour lines so derived is further smoothed.   The set of profiles which emerge from these contours is then examined with a view to further smoothing, and so on until an acceptable regional gravity map is obtained.   This method has the merit of being highly flexible, since it allows the interpreter to incorporate into the process his personal judgment or sense of "rightness" about the forms of the residual anomalies.   The use of this personal bias is subjective, but it is entirely reasonable.   If two numbers are sought when only their sum is known, the number of solutions is infinite; but if something can be assumed about one of them, then the range of admissible solutions is limited.   In this case, judgment that is based upon experience with anomalies over known structures, and also upon a direct knowledge of local geological conditions, has a legitimate bearing on the solution of the problem, even though it may profoundly affect the outcome of the interpretations.   Since this information is ordinarily too subtle to be abstracted into mathematical language, most of the processing work must be done manually.   As an example of an operation of this kind we have chosen a gravity survey in Garber County, Oklahoma.   The preliminary Bouguer gravity map is shown in Fig. 9-3, and the regional-residual separation in Fig. 9-4.   Profile A-A, shown in Fig. 9-5, illustrates the "bias" that was used in performing this separation.   Taken in this way, the residual anomaly is interpreted as being due to a fault, probably within the granitic basement.   Taken altogether without bias, it would have been difficult to arrive at a satisfactory interpretation of the very large residual anomaly, considering what is known of the geology of that region.

There are some circumstances, however, in which smoothing methods cannot be used.

1. When the ground is very hilly and the surface materials are inhomogeneous, the Bouguer density is likely to vary.   Under these conditions the residual anomalies in the preliminary Bouguer map are likely to be due to topography rather than to subsurface structures.

2. If the regional trend is very strong, residual anomalies are easily missed by visual methods.

3. When residual anomalies are very large, the regional trends are often difficult to discern.   This is particularly true in very large-scale surveys, although it is important only if the regional gravity map itself is required.

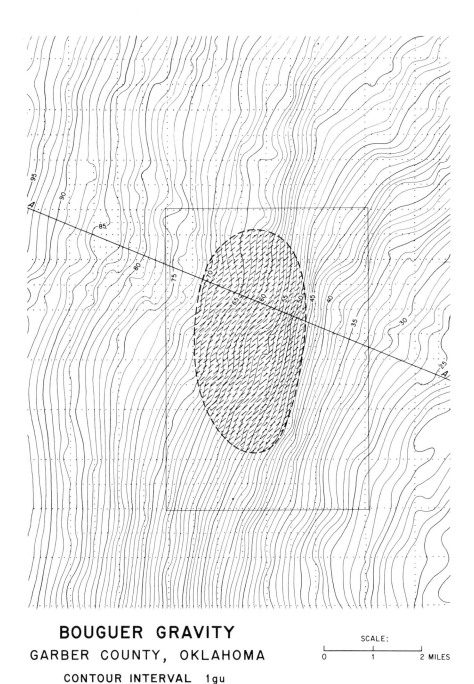

BOUGUER GRAVITY

GARBER COUNTY, OKLAHOMA

CONTOUR INTERVAL 1gu

SCALE:

0   1   2 MILES

Fig. 9-3 Bouguer gravity, Garber County, Oklahoma. The shaded zone represents the producing area. (*E. V. McCollum Co.*)

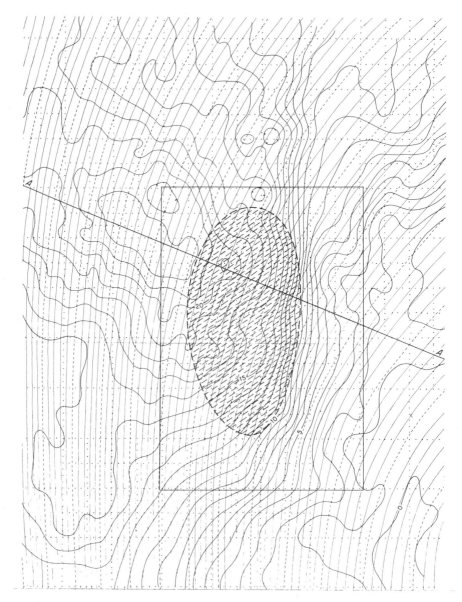

REGIONAL-RESIDUAL ANALYSIS
BOUGUER GRAVITY

GARBER COUNTY, OKLAHOMA
CONTOUR INTERVAL 1gu

SCALE:

0   1   2 MILES

Fig. 9-4   Regional-residual separation, Garber County, Oklahoma.

246

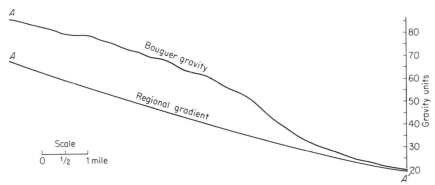

Fig. 9-5 The "bias" used in the regional-residual analysis shown in Fig. 9.4.

Under conditions such as these, and indeed in any situation in which for some reason judgment is likely to fail, the regional effect may be found by fitting polynomials to the data by the method of least squares. These do not contain any bias; hence the trends which they estimate are always averages. This means that the residual anomalies will be about equally positive and negative. One special advantage of polynomial fitting, apart from the time-saving feature, is that the elevation corrections may be programmed simultaneously with the regional-residual separation, thus eliminating one stage in contouring and drafting. For details, we refer to a paper by Grant (5).

A third approach to the problem which has been widely used for many years is to take a local average value of the Bouguer gravity field as an estimate of the regional effect. The "moving average" of a nonmathematical function is supposed to estimate its mean value, and if the residuals are unbiased, this should also estimate its trend. In practice, the moving average of Δg at a given point is taken from the average values of Bouguer gravity on circles of specified radii about that point. There is a good deal of arbitrariness in the selection of the sizes of the circles, profoundly affecting the outcome of the calculations. Generally speaking, the weighting coefficients attached to the different circles are so calculated that the regional effect could be estimated exactly if it coincided locally with a simple quadratic (or occasionally a cubic) surface. In this respect the averaging method differs little in principle from the polynomial method, the real difference being that the former reduces the regional effect within a small area to a very simple mathematical form, whereas the latter method approximates the regional trend throughout the entire area with a more complicated function. Both methods have the same defect in their mutual disregard of the structure of the residual field. A number of acceptable moving-average formulas (for "weighting" the circles) can be found, for example, in Kendall (6), chap. 29.

After the regional correction has been applied, we are left with a map of the residual gravity anomalies. This, presumably, is the "geological residue" of the total gravity picture. Accordingly, it constitutes the basic material for geophysical interpretation. In the solution of the inverse problem in potential theory, we shall identify the gravity effects (Δg values) of our theoretical models with the residual Bouguer gravity anomalies.

9-8   Derivatives

An additional device that sometimes proves to be effective in situations where for some reason the regional-residual separation is difficult to apply is the vertical derivative map. This makes direct use of the fact that nearby sources, even though they may be small, have a greater influence on gravity gradients than on gravity itself. Accordingly, there is good reason to expect that local features will show up more prominently on the map of one of the derivatives of Δg than on the map of Δg itself, thus pointing out the places where residual anomalies may be found. This device has been found to be particularly useful in resolving residual anomalies which overlap, especially since the traditional methods of separation will almost certainly fail in such cases. For further discussions of the application of derivative maps, we refer to Dobrin (3), chap. 12.

In practice, the derivatives are calculated from Bouguer gravity values by the numerical integration procedures discussed in Sec. 8-5. Several schemes have been proposed for programming digital machines or for constructing templates to perform computations of this kind, but since all these should in principle give comparable results, we do not wish to review them. The points of difference among them are mainly to be found in the system used for weighting the observations. An example of the kind of result that may be achieved with vertical derivatives is the vertical gravity gradient map of Garber County, Oklahoma (Fig. 9-3), which is shown in Fig. 9-6. We notice immediately the coincidence between the maximum horizontal gradient in the residual map (Fig. 9-4) and the strong positive anomaly in the vertical gradient field.

If the gravity stations are distributed on a regular grid, derivative calculations may be programmed for digital computers and carried out very readily [see, for example, Henderson (7)]. For irregular distributions of points, the calculations may be performed with the aid of templates, as suggested in Sec. 8-5. In referring back to this section, we are reminded that some filtering or smoothing of the data is necessary if the integrals are to converge. The problem of filter design will be taken up in the next section.

Derivative maps should not be construed as being equivalent to residual gravity maps. In spite of the proved resolving power of this method, the results are very difficult to interpret quantitatively. This is largely because

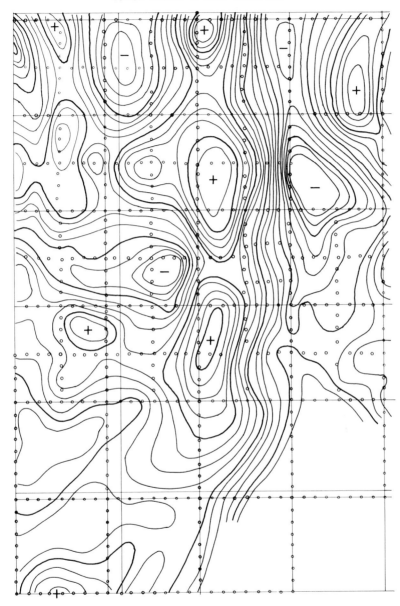

BOUGUER GRAVITY

VERTICAL GRADIENT

Fig. 9-6 Map of the vertical gradient of Δg, Garber County, Oklahoma.   (E. V. McCollum Co.)   The area covered by this map is indicated by the solid outline in Fig. 9-3 and Fig. 9-4. (1 Eötvös unit = 10^{-9} sec^{-2})

the instability of interpretations due to imperfections in the data rises with each differentiation.   Operating numerically on the Bouguer gravity data does not improve the amount of information available from these data.   In fact, one of the most familiar results of the calculus of finite differences is that the reliability of the calculated derivatives of numerically defined quantities deteriorates with each order of differentiation.   Derivative methods, properly viewed, are a complement to, rather than a substitute for, the regional-residual separation.   Their chief function is to indicate the presence of anomalies under difficult circumstances.   Once this is known quantitative geophysical interpretations should be based upon the residual gravity anomalies themselves.

9-9   *Downward Continuation and the Interpretation of Residual Anomalies*

The direct approach to the quantitative interpretation of residual Bouguer anomalies, when specific models are not used to represent the geological structures, is sometimes based upon the concept of the *equivalent stratum*. This is particularly relevant when an interface between two materials having different formation densities has been identified at some known or estimated depth by means other than gravity surveying.   The object of the interpretation is then to determine the relief on this surface compatible with the observed variations in Δg.

The usual method is to calculate the equivalent stratum at the given depth and then to replace it with the topographic surface $h(x,y)$, which is given by

$$\sigma(x,y) = \Delta\rho\, h(x,y)$$

where $\Delta\rho$ is the difference in the formation densities between the two media. This formula can be explained by referring to Fig. 9-7.   Suppose that $h(x,y)$ is the vertical departure of the interface at any point from its mean depth d. Then the gravity effect at $P(x,y)$ due to the undulation in this interface will be

$$\Delta g(x,y) = G\,\Delta\rho \left[\frac{\partial}{\partial z} \int_{-\infty}^{\infty} \int_{-\infty}^{\infty} d\xi\, d\eta \int_{d}^{d+h(\xi,\eta)} \frac{d\zeta}{R} \right]_{z=0}$$

$$\doteq G\,\Delta\rho\, d \int_{-\infty}^{\infty} \int_{-\infty}^{\infty} \frac{h(\xi,\eta)\, d\xi\, d\eta}{[(x-\xi)^2 + (y-\eta)^2 + d^2]^{3/2}} \quad (9\text{-}8)$$

provided that $|h| \ll d$.   This is identical with the expression for the upward continuation of the gravity effect at depth d whose values are given by

$$\Delta g_d(x,y) = 2\pi G\,\Delta\rho\, h(x,y) \quad (9\text{-}9)$$

Thus by continuing the residual Bouguer gravity field downward and finding Δg_d, an equivalent topographic surface can be constructed at that depth

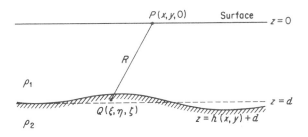

Fig. 9-7 The use of downward continuation when a single interface is involved.

which will account for the residual gravity anomalies at the ground surface. Whether this gives a correct picture of the shape of the discontinuity will depend, of course, on how much of the residual gravity field does actually originate from that interface and whether a part of the effect has been removed with the regional trend.   In one respect at least there is invariably a good deal of ambiguity in the interpretations, since it often happens that several interfaces actually contribute to the field.   The difficulties posed by such a contingency can be resolved only with the aid of additional subsurface data.

Another application of the continuation integral actually makes use of the fact that when the disturbing masses have been passed by, the propositions upon which this integral is based no longer hold true, and so the equivalent stratum is no longer a valid solution of the inverse potential problem.   The evidence for this is that the solution becomes unstable and begins to oscillate and diverge.   The onset of instability indicates that a solution is being sought at too great a depth.   This can be useful in gauging the maximum depth of the masses in the sense of an upper limit only.   The drawback is, however, that for sources having a fairly wide lateral distribution, instability does not become apparent until depths are reached which may be too great by a very large factor.   Downward continuation does not usually provide the most economical approach to the solution of maximum-depth problems.

Continuation methods also apply to the deep interpretation of gravity data by "stripping" [Hammer (8)].   This consists in calculating and removing from the data (usually according to the "equivalent-stratum" principle) the gravity effects of the uppermost layers whose boundaries have been defined by seismic methods or by drilling.   The purpose in doing so is to reveal the gravity effects of deeper masses not penetrated by the drill nor defined by seismic interpretations.

The theory of downward continuation was discussed in Sec. 8-4.   Ultimately it reduces to the relation (8-13) between the Fourier transforms of the measured gravity effect and its downward continuation.   We noted that the inversion of the Fourier integrals at the lower level can be accomplished only if $F_0(p,q)$ attenuates with rising values of the wave numbers p and q more rapidly than the exponential factor increases, so that the integral

$\iint |F_z(p,q)|\ dp\ dq$ exists. Two techniques used to ensure the fulfilment of this condition are (1) digitizing and (2) analytical smoothing. In the first, the residual gravity field is replaced by a set of point values at a regular network of stations, the mesh size of which is adjusted to the depth of continuation. In the second, a controllable mathematical filter is applied to the Δg values everywhere. We shall describe both methods, taking the latter one first.

Let us begin with the integral (8-15). If we expand it into its fully inverted form, putting $z = d$, we have

$$\Delta g_d(\xi,\eta) = \frac{1}{4\pi^2} \int_{-\infty}^{\infty} \int_{-\infty}^{\infty} \exp\left[\sqrt{p^2 + q^2}\, d - \gamma(p^2 + q^2)\right] \int_{-\infty}^{\infty} \int_{-\infty}^{\infty}$$
$$\Delta g_0(x,y) e^{i[p(x-\xi)+q(y-\eta)]}\ dx\ dy\ dp\ dq$$

Let us transform from rectangular to cylindrical polar coordinates by putting

$$x + iy = re^{i\vartheta} \qquad \xi + i\eta = r_0 e^{i\vartheta_0} \qquad p + iq = ue^{i\varphi}$$

and integrate with respect to φ. We obtain the result

$$\Delta g_d(r_0,\vartheta_0) = \frac{1}{2\pi} \int_0^{\infty} \int_0^{2\pi} \int_0^{\infty} \Delta g_0(r,\vartheta) e^{du-\gamma u^2} J_0(u\bar{R}) ur\ du\ d\vartheta\ dr$$

where $\qquad \bar{R} = \sqrt{r^2 + r_0^2 - 2rr_0 \cos(\vartheta - \vartheta_0)}$

We may now set $r_0 = 0$. This involves no loss of generality, since the origin may be placed anywhere. Then using (9-9) we obtain the formula of Bullard and Cooper (9)

$$2\pi G \Delta\rho h(0) = (2\pi)^{-1} \int_0^{\infty} \int_0^{2\pi} \int_0^{\infty} \Delta g_0(r,\vartheta) e^{du-\gamma u^2} J_0(ur) ur\ du\ d\vartheta\ dr$$
$$= \int_0^{\infty} \Delta g_0(r) K(r;\gamma) r\ dr \qquad (9\text{-}10)$$

where $\quad K(r;\gamma) = \int_0^{\infty} e^{du-\gamma u^2} J_0(ur) u\ du$

The weighting function $K(r;\gamma)$ converges over a limited range in d (which depends upon the value of γ), so that the integral can be computed. A table of the quantity $rK(r;\ \frac{1}{4})$ has been given by Bullard and Cooper.

To compute the integral (9-10), it is convenient to rewrite it in the following way:

$$2\pi G\ \Delta\rho h(0) = \int_0^{\infty} \Delta g_0(r) K(r;\gamma) r\ dr = \int_0^{\infty} e^{-w} \Delta g_0(2\sqrt{\gamma w}) K^*(w;\gamma)\ dw$$

where $r^2 = 4\gamma w$ and $K^*(w;\gamma) = 2\gamma e^w K(2\sqrt{\gamma w};\gamma)$. This allows us to make use of the Gauss-Laguerre quadrature formula [Kopal (10), chap. 6], which gives

$$2\pi G\ \Delta\rho h(0) \doteq \sum_{j=1}^{m} M_j(\gamma)\ \Delta g_0(r_j) \qquad (9\text{-}11)$$

where w_j's are the m roots of the Laguerre polynomial $L_m(w_j) = 0$ and where $r_j = 2\sqrt{\gamma w_j}$. The weighting coefficients $M_j(\gamma)$ are given by

$$M_j(\gamma) = 2\gamma H_j(m)e^{w_i}K(r_j;\gamma)$$

where

$$H_j(m) = \frac{(m!)^2}{w_j[L'_m(w_j)]^2}$$

The H_j's have been tabulated and are included in Kopal's book, and a table of values of M_j is given in an appendix. Thus the relief h on the interface directly beneath the origin, which is associated with the smoother parts of the gravity effect at the ground surface, is given as a weighted sum of ring averages of Δg. The choice of the number of rings is very largely at the discretion of the interpreter, but under ordinary circumstances it probably should be not less than six nor more than ten.

In addition to choosing the number of rings, the interpreter apparently also has somewhat of a free hand in choosing the value of the attenuation constant γ. It is quite obvious, however, that the value should be made as small as is practicable without destroying the convergence of (9-10). The smallest value that can reasonably be assigned will be such that the weighting distribution $K(r;\gamma)$ changes in sign within a distance which is not less than the average spacing between the gravity stations, for otherwise we would find ourselves reading far more information into the contours than is actually there. The change in K from positive to negative values occurs near the point $r(2z\sqrt{\gamma})^{-1} = 1.2$. Consequently, if the average distance between stations is s, it follows that γ should be $\geq s^2/6d^2$. The value may be taken at a convenient number not too far removed from this lower limit.

Sometimes an estimate of the maximum value of h (or of the closure of the relief on the interface) is of nearly as much value to the geologist as a contour map of $h(x,y)$. This can usually be calculated very roughly but with little effort. For if Δg_{max} represents the amplitude of the residual anomaly pattern on the surface of the ground, we may choose x and y axes which lie in the directions of maximum and minimum horizontal gradient in Δg (or as close as possible to these two directions if they are not perpendicular), and we may measure the distances x_m and y_m between the points at which Δg takes the value $\frac{1}{2}\Delta g_{max}$ in these two directions. The Fourier transform of Δg will probably have its maximum value somewhere near the wavelengths whose magnitudes are $2x_m$ in the direction of x and $2y_m$ in the direction of y. The wave numbers p and q to which these wavelengths correspond are π/x_m and π/y_m, respectively. If we then assume that the greater part of h corresponds to wave numbers of about these magnitudes, we may write, approximately,

$$2\pi G \Delta\rho\, h_{max} \approx \Delta g_{max} \exp\left(\pi d \sqrt{\frac{1}{x_m{}^2} + \frac{1}{y_m{}^2}}\right) \tag{9-12}$$

This estimate will be very crude but should be at least of the right order of magnitude.   The above formula also provides a means of roughly estimating the maximum gravity effect that can be expected from a topographic feature of a given size which lies at a given depth.

One of the most popular formulas for continuing the gravity field downward from a set of regularly spaced point values of Δg is that of Henderson (7), and at first glance it seems to be devised on an entirely different principle than the method of Fourier transforms.   It is based upon the thought that the field below the ground surface (but above the disturbing masses) should to a very large extent be predictable from values of the field at a succession of different levels *above* the ground.   Therefore, if the gravity effect is continued vertically upward to four consecutive levels, each one grid unit apart, the five values at each node may be extrapolated downward by the use of the Lagrangian interpolation formula

$$
\Delta g_z(0) = \frac{(z + a)(z + 2a)(z + 3a)(z + 4a)}{4!a^4} \Delta g_0(0)
$$
$$
- \frac{z(z + 2a)(z + 3a)(z + 4a)}{1.3!a^4} \Delta g_{-a}(0)
$$
$$
+ \cdots + \frac{z(z + a)(z + 2a)(z + 3a)}{4!a^4} \Delta g_{-4a}(0)
$$

where a is the size of the grid unit.   This leads to an expression for the downward continuation of Δg_0 at a depth of s units below the origin of coordinates of the following type:

$$
\Delta g_s(0) \doteq \sum_{j=1}^{m} A(r_j, s) \, \Delta g_0(r_j) \tag{9-13}
$$

where the $\Delta g_0(r_j)$'s are ring averages of Δg on circles of preselected radii r_j, and the A's are a set of numerical coefficients calculated from the upward continuation integral.   The values of r_j are chosen to satisfy certain performance criteria, and in practice the number of rings is limited to ten.   The calculation of (9-13), like (9-11), can very easily be programmed for automatic computation when the points are regularly spaced.

To compare the two methods—which must certainly have some features in common since they are both designed to do the same thing—we suppose that the gravity effect on $z = 0$ is a double cosine term, i.e., that

$$
\Delta g_0(x,y) = A \cos px \cos qy
$$

The downward continuation of this function to a depth of one grid unit a below the origin of coordinates is in fact

$$
\Delta g_a(0) = A e^{ua} \qquad \text{where} \qquad u = \sqrt{p^2 + q^2}
$$

Henderson's method gives

$$\Delta g_a(0) = A e^{ua}[1 - (1 - e^{-ua})^5]$$

while according to Bullard and Cooper we get

$$\Delta g_a(0) = A e^{ua - \gamma u^2}$$

Numerically, the two expressions are almost identical at all wavelengths down to $\lambda = 2a$ if $\gamma = 1/6a^2$. Thereafter, the Bullard and Cooper formula gives distinctly lower values. [It must be stressed, of course, that the above expressions represent the *theoretical* accuracy of the two methods. Their actual performance can be gauged only by applying the formulas (9-11) and (9-13), which depend to some extent on the selection of values for r_j.] From this we may infer that with smooth gravity fields the two different methods, although seemingly dissimilar, should give very much the same results; but if much near-surface disturbance is present, numerical smoothing is apt to be superior to upward continuation as a means of suppressing it.

9-10   *The Use of Density Logs in Continuation*

The major source of ambiguity in the determination of subsurface structures by the use of downward continuation is due to the existence of multiple interfaces. Indeed, the complex nature of the density logs obtained from boreholes makes it seem highly probable that in most cases residual anomalies are the superposition of effects originating from several interfaces at once. With some modifications and under certain conditions, the method of continuation can be extended to include general density-depth relationships such as those found by well-logging methods.

Referring to Fig. 9-8, we shall assume that from a minimum depth d downward, the entire column has participated uniformly in a structural movement whose magnitude at any point is $h(x,y)$. We shall also assume that below a certain depth D there are no further density contrasts of any significance. Let $\rho(z)$ be the density-depth relationship obtained from seismic observations or from borehole measurements. Within reasonable distances from the hole, we shall assume that this relationship does not change laterally except due to structural movements involving displace-

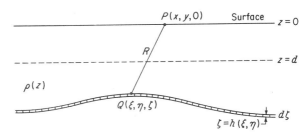

Fig. 9-8 The use of downward continuation when a continuous density log is available.

ments of the column as a whole. This may seem a rather unusual model from a geological viewpoint, but it aims at representing the other extreme from a single interface. In most cases we might expect one strong discontinuity to produce most of the gravity disturbance with additional contributions from several weaker interfaces.

The gravity effect due to the vertical variation $h(x,y)$ in the origin of $\rho(z)$ may be calculated as follows: Let the average formation density in $0 \leq z \leq d$ be ρ_0, and let us call

$$\Delta\rho(z) = \rho(z) - \rho_0$$

Now consider the change in gravity at the surface due to the relief h in the thin stratum at depth $\zeta \geq d$, whose thickness is $d\zeta$

$$d\,\Delta g(x,y) = G\,\Delta\rho(\zeta)d\zeta \int_{-\infty}^{\infty} \int_{-\infty}^{\infty} \left\{ \frac{\zeta - h(\xi,\eta)}{[(x-\xi)^2 + (y-\eta)^2 + (\zeta-h)^2]^{3/2}} - \frac{\zeta}{[(x-\xi)^2 + (y-\eta)^2 + \zeta^2]^{3/2}} \right\} d\xi\,d\eta$$

which, to a sufficient approximation, can be written as

$$d\,\Delta g(x,y) = -G\,\Delta\rho(\zeta)\,d\zeta \left(1 + \zeta\frac{\partial}{\partial\zeta}\right) \int_{-\infty}^{\infty} \int_{-\infty}^{\infty} \frac{h(\xi,\eta)\,d\xi\,d\eta}{[(x-\xi)^2 + (y-\eta)^2 + \zeta^2]^{3/2}}$$

provided that $|h| \ll d$. To find the total gravity effect at the surface, this expression is integrated with respect to ζ from d downward, i.e.,

$$\Delta g(x,y) = -G\int_d^{\infty} \Delta\rho(\zeta) \left(1 + \zeta\frac{\partial}{\partial\zeta}\right) \int_{-\infty}^{\infty} \int_{-\infty}^{\infty} \frac{h(\xi,\eta)\,d\xi\,d\eta}{[(x-\xi)^2 + (y-\eta)^2 + \zeta^2]^{3/2}}\,d\zeta$$

Once more, we are confronted with an integral equation which must be inverted. Following the traditional path and taking the Fourier transform of both sides with respect to x and y, we get

$$F(p,q) = (2\pi)^{-1}GH(p,q)X(p,q) \tag{9-14}$$

where H is the Fourier transform of h and where

$$X(p,q) = -\int_d^{\infty} \Delta\rho(\zeta) \left(1 + \zeta\frac{\partial}{\partial\zeta}\right) \frac{\exp\left(-\sqrt{p^2 + q^2}\,\zeta\right)}{\zeta}\,d\zeta$$

$$= \sqrt{p^2 + q^2} \int_d^{\infty} \Delta\rho(\zeta) \exp\left(-\sqrt{p^2 + q^2}\,\zeta\right)\,d\zeta$$

To find h, we have therefore to transpose (9-14) and perform the Fourier inversion.

As in downward continuation, the short-wavelength contribution to Δg, if any, must be taken out of the spectrum if we are to secure the convergence of the Fourier integral. In this section we shall consider only the use of numerical filtering. Thus we begin by applying the transformation (8-14) to the left-hand side of (9-14), transpose, and then transform to cylindrical polar coordinates. We then obtain

$$2\pi Gh(0) = \int_0^\infty \Delta g_0(r)K'(r;\gamma)r\,dr \qquad (9\text{-}15)$$

where

$$K'(r;\gamma) = \int_0^\infty e^{-\gamma u^2}[X(u)]^{-1}J_0(ur)u\,du$$

$$X(u) = u\int_d^\infty \Delta\rho(z)e^{-uz}\,dz$$

and where

$$\Delta g_0(r) = \frac{1}{2\pi}\int_0^{2\pi}\Delta g_0(r,\vartheta)\,d\vartheta \qquad \text{as before}$$

In practice, we do not as a rule make use of the minor details of the density-depth relation but replace $\rho(z)$ with a number of constant formation densities by dividing up the log into shorter sections. Thus if we put

$$\rho(z) = \rho_n \qquad z_{n-1} \leq z \leq z_n$$

where

$$z_n = d + \sum_{i=1}^n h_i \qquad \text{and} \qquad D = z_N$$

we find that in (9-15)

$$X(u) = \sum_{n=0}^N \Delta\rho_n e^{-uz_n} \qquad \Delta\rho_n = \rho_{n+1} - \rho_n$$

With the help of this formula it is then possible to compute the function $K'(r;\gamma)$. Since $\rho(z)$ is unique, however, the function cannot be tabulated. But for this difference, the remainder of the analysis proceeds exactly as it does in downward continuation, and using the Gauss-Laguerre quadrature formula, we get

$$2\pi Gh(0) \doteq \sum_{j=1}^m M_j'(\gamma)\,\Delta g_0(r_j) \qquad (9\text{-}16)$$

where

$$M_j'(\gamma) = 2\gamma H_j(m)e^{r_j^2/4\gamma}K'(r_j;\gamma)$$

and where the r_j's are the same as those used in (9-11).

The difference between the formulas (9-16) and (9-11) can probably best be illustrated by the use of an example. The density log shown in Fig. 9-9 is taken from a dry well and is made up partly from measurements on core samples and partly from values converted from a continuous velocity log according to Fig. 7-7. The dotted line shows the average formation densities used to approximate the density log. These average values were used to compute the function $X(u)$, which thereupon enables us to evaluate

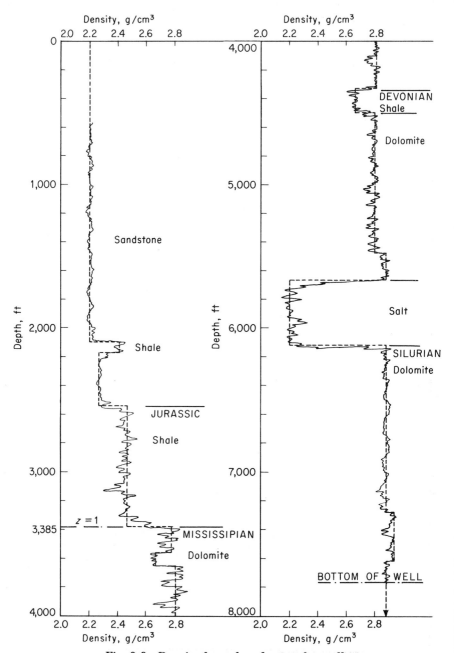

Fig. 9-9   Density log taken from a dry well.

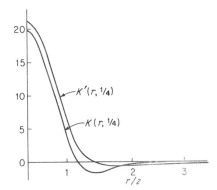

Fig. 9-10 **The weighting distributions used for downward continuation to a single interface** $[K(r)]$ **and to the 2,100-ft level of the section represented by Fig. 9-9** $[K'(r)]$.

the weighting function $K'(r;\gamma)$. In Fig. 9-10 this function (with $\gamma = \frac{1}{4}$) is plotted, together with the function $K(r;\frac{1}{4})$ for $z = 3,385$ ft (the major discontinuity appearing in the formation densities), for comparison. The value $\gamma = \frac{1}{4}$ was chosen because the best value for s was thought to be about 3/4 mile. This diagram shows very clearly the outward spreading of influence to greater distances from the reference point as a result of including the deeper horizons.

A Bouguer gravity residual map made within a few miles of this well is shown in Fig. 9-11, which includes also the station locations. If we interpret these data as being the gravity effects due to topography on the erosional unconformity at the base of the Jurassic, we may determine the amount of relief that is compatible with a density contrast of 0.35 g/cm³ by continuing the residual gravity downward according to (9-11) to the 3,385-ft depth. The structure lines on this surface inferred by the method of downward continuation are shown in Fig. 9-12.

If on the other hand we were to assume that all the materials below the Tertiary section (i.e., below 2,100 ft) had participated in the structural movement in an entirely conformable manner, we would use instead the formula (9-16) together with the density log. In this case the structure lines on any of the pre-Tertiary interfaces are as shown in Fig. 9-13. These two drawings represent two different extremes of direct interpretation. If the residual gravity map gives an essentially correct picture of the geological contribution from within the sediments, then the two interpretations should provide upper and lower limits on the amount of structure that is compatible with the gravity observations.

Incidentally, without adding to their difficulty, differential compaction can be allowed for in the calculations by introducing a suitable "form factor" in $X(u)$. For instance, the relation

$$h_\zeta(\xi,\eta) = e^{a(\zeta-d)}h_d(\xi,\eta)$$

would allow for an upward attenuation of structure, the constant a being so chosen that the value $\exp a \, (D - d)$ represents a reasonable ratio of deep

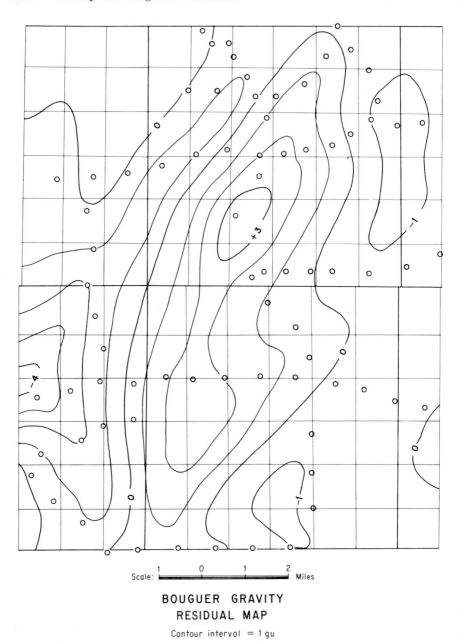

Scale: |— 1 — 0 — 1 — 2 —| Miles

BOUGUER GRAVITY
RESIDUAL MAP

Contour interval = .1 gu

Fig. 9-11   Residual Bouguer gravity map of an area close to the well indicated
in Fig. 9-9.

STRUCTURE ON TOP OF PALEOZOIC
Contour interval = 25 ft

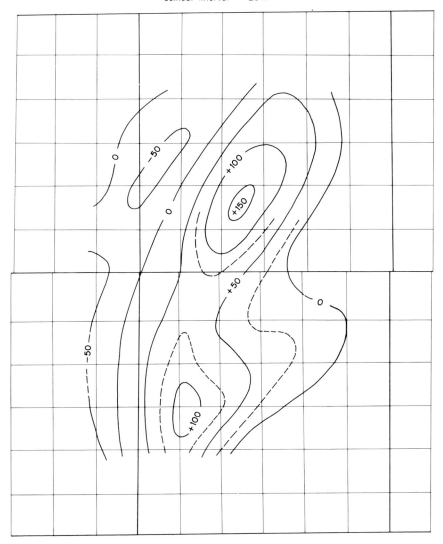

$z = 3,385\,\text{ft} \qquad \Delta\rho = +0.35\ \text{g/cm}^3$

Fig. 9-12 Calculated structure at 3,385-ft level from downward continuation to top of Paleozoic.

PRE—TERTIARY STRUCTURE

Contour interval = 25 ft

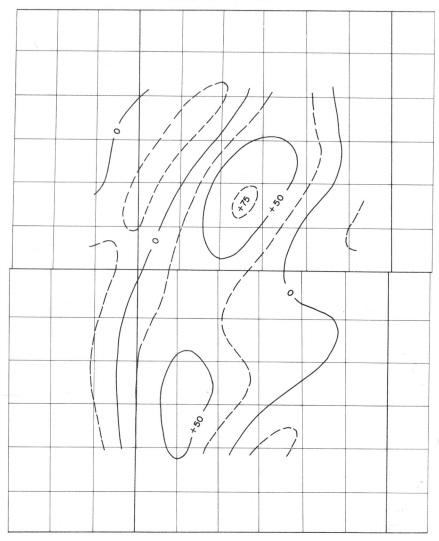

z_m = 2,100 ft

Fig. 9-13   Calculated structure of pre-Tertiary section from downward continu-ation using the density log shown in Fig. 9-9.

to shallow amplitude.　If the substitution is made into (9-15), we find that

$$X(u) = \frac{ue^{-ad}}{u - a} \sum_{n=0}^{N} \Delta\rho_n e^{-(u-a)z_n} \qquad \Delta\rho_n = \rho_{n+1} - \rho_n$$

We may proceed as before to compute $h_d(0)$ from (9-16).　From this value the structure at any lower level within the range $d \leq z \leq D$ may be determined.

9-11　Maximum-depth Rules

Various rules for estimating the maximum possible depth of a density interface from observed gravity anomalies have been derived from the continuation integral.　For instance, according to (9-12) we may write

$$d \leq \frac{x_m y_m}{\pi \sqrt{x_m^2 + y_m^2}} \ln \frac{2\pi G|\Delta\rho h|_{max}}{|\Delta g|_{max}}$$

or, what amounts to the same thing,

$$d \leq \frac{|\Delta g|_{max}}{|\Delta g'|_{max}} \ln \frac{2\pi G|\Delta\rho h|_{max}}{|\Delta g|_{max}}$$

where $\Delta g'$ is the horizontal gradient of Δg.　The inequality sign is used because the formula assumes that the entire gravity effect is concentrated at a single Fourier wavelength.　If upper bounds are set upon the density contrast and the structural relief on the interface, a limiting value can be calculated for d.　This formula was given by Bullard and Cooper (9).

By working directly from the continuation integral rather than from its Fourier transform, Smith (11) was able to show that if $\Delta\rho$ does not change in sign,

$$d \leq \frac{96\pi 5^{-5/2} G|\Delta\rho|_{max}}{|\Delta g''|_{max}}$$

where $\Delta g''$ is the second horizontal derivative of Δg.　Since ordinarily $|\Delta g''|$ will not be less than one-half of the second vertical derivative $|\partial^2 \Delta g/\partial z^2|$, we may also write

$$d \leq \frac{0.22|\Delta\rho|_{max}}{|\partial^2 \Delta g/\partial z^2|_{max}}$$

when measurements are in grams per cubic centimeter, gravity units, and feet.

These formulas and others like them should be used with restraint.　The values which they yield are seldom plausible values for d, because they correspond to extreme densities or structures.　The formulas apply only if the density contrast has the same sign for all the masses contributing to the gravity effect, and they do not as a rule work effectively for compact masses such as mineral ore bodies.   They are, however, useful for classifying anomalies thought to be due to flat-lying structures.

9-12   The Geoid

In Sec. 9-3 we replaced the reference spheroid with mean sea level for the purpose of making elevation reductions.   It is of some interest to us now to examine the nature of this approximation.   In the first place mean sea level, or in fact any horizontal found by using spirit levels, is gravitationally an equipotential surface.   But wherever the earth's crust is heterogeneous, the equipotential surfaces are deformed by the changes in density of the rocks (Fig. 9-14).   Thus the surveyor's "level" is in fact a highly complicated surface whose detailed features are largely determined by the geology of the earth's crust.   Because of this the reference spheroid, although smooth, is not a level surface.   It therefore cannot be used in surveying.

The height of one point above another, when measured by the use of levels, is the distance between the equipotential surfaces which pass through the two points, measured in the direction of g.   This is sometimes also called the difference in *orthometric* height.   If these differences are accumulated and referred back to mean sea level (MSL) at some convenient coastal station, the elevations of all inland points will be approximately determined with reference to the equipotential surface which coincides with MSL at that station.   We say "approximately" determined because the equipotential surfaces at the different inland elevations are not exactly parallel.   However, the error that may accrue as a result of making this approximation is unlikely to exceed 2 m anywhere in the world,[1] and so we may disregard it. For practical purposes, land elevations are based on the equipotential surface that coincides with MSL.   To describe this surface, the name *geoid* is used. In land-covered areas, the geoid is a surface that would be determined by the level of sea water in narrow canals cut through the continents.

It is clear that the details of the shape of the geoid are complicated and that the reference spheroid is at best a simple approximation to it.   The geoid will generally lie outside of the reference spheroid in land-covered areas and inside it over the oceans. H, the elevation above the reference

[1] The greatest discrepancy is likely to occur across the Tibetan plateau, where it is calculated to be less than 3 m.

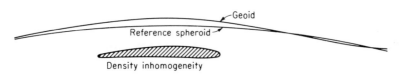

Fig. 9-14   The geoid.

spheroid, and h, the elevation based on MSL, are therefore not the same. The discrepancy is of very little practical importance in gravity surveying, but it is of great interest in physical geodesy. The chief reason for this interest is that land surveys are based on the geoid, but maps can be drawn only on a mathematical surface such as the spheroid. The difference between the two surfaces leads to the distortion of lengths and an uncertainty in the measurement of horizontal distances.

The departure of the geoid from the reference spheroid can be found approximately by a method due to Bruns [Helmert (12)]. We begin by putting

$$W = U_0$$

where W is the gravitational potential of the earth at the geoid and U_0 is the geopotential at the reference spheroid given by McCullagh's formula (9-3). Notice that this equation disregards the effects of masses which lie outside of the geoid. If we extrapolate U from the reference spheroid to the geoid we may write, approximately,

$$U = U_0 + g(\varphi)\,\Delta h$$

where Δh is the height of the geoid above the reference spheroid (Fig. 9-15) and $g(\varphi)$ is given by the international gravity formula (9-2). Thus

$$g(\varphi)\,\Delta h = U - W = T \tag{9-17}$$

in which T is called the *disturbing potential*. This is the irregular part of the earth's gravitational potential on the reference spheroid, which is due very largely to inhomogeneity of the earth's crust and upper mantle. The problem of determining Δh thus reduces to that of finding T from gravity measurements.

The first attempt to solve this problem was made by Stokes (13), who derived a method of calculating T when the source of this potential is assumed to lie *inside* the geoid, so that T obeys Laplace's equation everywhere outside. If the reference surface is assumed to be a sphere of radius a, the disturbing potential is given approximately by the surface integral

$$T(0) = \frac{1}{4\pi a} \iint_{R=a} \gamma(\vartheta,\varphi) F(\psi)\, dS \tag{9-18}$$

where γ is the free-air gravity anomaly reduced to the geoid, ψ is the angular distance from 0, and F is a weighting function. For details of the deriva-

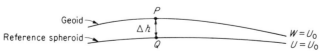

Fig. 9-15 **Bruns' method for determining the height of the geoid above the reference spheroid.**

tions and for the form of F, we refer to Jeffreys (14), chap. 4. If the reference surface is approximated by a *plane* (which may be used only locally where ψ does not exceed 1°), the solution may be found directly from (8-11). In polar coordinates

$$T(0) = \frac{1}{2\pi} \int_0^\infty \int_0^{2\pi} \gamma(r,\vartheta) \, d\vartheta \, dr \tag{9-19}$$

This is also the form to which (9-18) reduces when $a \to \infty$.

The implication of the assumption that all the masses contributing to T lie inside the geoid has been a vexed question in physical geodesy for more than a century. Various compensating adjustments to Stokes' solution have been tried, but the current trend seems to be leading toward the abandonment of the geoid in favor of the earth's physical surface as the place where the calculations should be made. Prominent among the contributions to the theory and practice of "three-dimensional" geodesy is the work of Molodensky (15), who has laid the foundations for this new approach. His solution is more complicated than that of Stokes and takes the form of a Fredholm integral equation which must be solved numerically. It is not possible to attempt a résumé of Molodensky's derivations here.

We shall conclude this very brief discussion of the geoid by remarking that Δh probably does not exceed 50 m anywhere in the world. Moreover, it changes very gradually in relation to the elevation h. Local geological bodies are not likely to raise or lower the geoid by more than a few centimeters, and it is only the very large, regional geological features which have any appreciable effect upon it. Thus the approximation used in Sec. 9-3 of putting $H = h$ is perfectly permissible in gravity prospecting.

References

1. W. A. Heiskanen and F. A. Vening-Meinesz, "The Earth and its Gravity Field," McGraw-Hill, New York, 1960.
2. M. F. Kane, A Comprehensive System of Terrain Corrections Using a Digital Computer, *Geophysics*, vol. 27, pp. 455–462, 1962.
3. M. B. Dobrin, "Introduction to Geophysical Prospecting," 2d ed., McGraw-Hill, New York, 1960.
4. L. L. Nettleton, "Geophysical Prospecting for Oil," McGraw-Hill, New York, 1940.
5. F. S. Grant, A Problem in the Analysis of Geophysical Data, *Geophysics*, vol. 22, pp. 309–344, 1957. See also W. C. Krumbein, Trend Surface Analysis of Contour-type Maps with Irregular Control-point Spacing, *Jour. Geophys. Research*, vol. 64, pp. 823–834, 1959.
6. M. G. Kendall, "Advanced Theory of Statistics," Griffin, London, 1946.
7. R. G. Henderson, A Comprehensive System of Automatic Computation in Magnetic and Gravity Interpretation, *Geophysics*, vol. 25, pp. 569–585, 1960.
8. S. Hammer, Deep Gravity Interpretation by Stripping, *Geophysics*, vol. 28, pp. 369–378, 1963.

9. E. C. Bullard and R. I. B. Cooper, Determination of the Masses Required to Produce a Given Gravitational Field, *Royal Soc. Proc.*, ser. *A*, vol. 194, pp. 332–347, 1948.

10. Z. Kopal, "Numerical Analysis," Chapman and Hall, London, 1960.

11. R. A. Smith, Some Depth Formulae for Local Gravity and Magnetic Anomalies, *Geophys. Prosp.*, vol. 7, pp. 55–63, 1959.

12. F. R. Helmert, Die mathematischen und physikalischen Theorieen der höheren Geodäsie, Teubner, Leipzig, vol. 1, 1880, vol. 2, 1884.

13. C. G. Stokes, On the Variation of Gravity and the Surface of the Earth, *Cambridge Philos. Soc. Trans.*, vol. 8, p. 672, 1849.

14. H. Jeffreys, "The Earth," Cambridge, London, 1952.

15. M. S. Molodensky, V. F. Eremeev, and M. I. Yurkina, "Methods for Study of the External Gravitational Field and Figure of the Earth," Moscow, 1960 (transl. by Israel Program for Scientific Translations, Jerusalem, 1962).

chapter 10 The Quantitative Interpretation of Gravity Anomalies

In this chapter we shall discuss a special type of gravity interpretation problem, viz., the quantitative interpretation of isolated, residual, Bouguer anomaly patterns. *Quantitative* interpretation means finding out the position, size, and shape of a gravitating mass through analysis of its potential field. It means, therefore, solving the inverse problem of potential field theory. And although much careful thought has been given to this question, it must not be forgotten that a *complete* solution of the inverse problem is impossible in both practice and theory. Therefore we may seek approximate solutions only, and we should be wasting our time in trying to refine these calculations beyond a very elementary stage, because the necessary amount of information is lacking in the potential field measurements no matter how carefully they are made. Add to this the further uncertainties engendered by the limited number of observations, and it is clear why nearly all methods of quantitative interpretation are based upon highly idealized representations of geological bodies, such as half-planes, steps, cylinders, and the like. The purpose of these simple models is to provide information *of a particular kind*, such as depth of burial, average dip, or depth extent.

Generally, the kind of information yielded by elementary models is all that the survey can be expected to provide. But sometimes (although all too infrequently) a gravity survey is carried out with such precision that more detailed information can be obtained by more elaborate methods. Ordinarily this implies at least 100 gravity stations within the anomaly pattern and a regional gradient that is neither too complex nor too steep. Under these conditions, however, no other potential field method is capable of yielding as much information about buried masses from surface measurements as the gravity method can. In mining exploration in particular, a comparatively modest outlay on geophysical work very often repays itself many times over by avoiding expensive and unnecessary drilling.

10-1 *Reduction of Residual Values to a Horizontal Plane*

Formulas for the direct interpretation of gravity anomalies are necessarily based upon the distribution of the potential on a flat plane. However, it should be remembered that the measurements themselves are taken on the ground surface, which may be uneven. This will distort the shape of the field to some extent, and in cases of severe topography the distortion may have a significant effect. This is entirely a geometrical problem, not to be confused with variations in the Bouger density which we assume to have already been taken into account. It can largely be removed by applying a first-order correction as follows:

$$(\Delta g)_{z=0} = (\Delta g)_{z=h} - h\left(\frac{\partial \Delta g}{\partial z}\right)_{z=0} \tag{10-1}$$

where h is the departure of the ground surface from the closest horizontal plane, which we define to be $z = 0$. To calculate the vertical gradient term, we should in theory use Δg values on the plane $z = 0$; but in practice, since we do not have them, we may use the measured Δg values on $z = h$ instead. The method of making this calculation has been discussed in Sec. 8-5.

10-2 *Calculation of the Excess Mass*

One of the simplest calculations that can be made from a residual gravity anomaly is the total mass excess (or defect) of the causative body. This is sometimes also of greater practical importance than any other calculation, particularly in problems of mining geophysics.

The basis of the calculation is the formula (8-26), which involves the integration of Δg over the (x,y) plane. It is of interest to point out that this formula contains no reference either to the shape or to the density of the anomalous body. However, two practical difficulties arise in connection with performing the calculations. One is due to the fact that data exist only within a limited area, although in theory the integrations are to be carried out over the entire horizontal plane. The other is due to the uncertainty in amplitude of Δg in cases where a regional correction has been made; for nearly always the tendency is to bias the residual anomaly downward, cutting off its tails and thus making a very substantial reduction in the value of the integral. These two effects can be compensated by applying a correction to the integral as follows: Suppose that the total available range of integration in x is from $-X$ to X, and in y, from $-Y$ to Y, when the origin is placed somewhere near the center of the anomaly. The values given to X and Y will represent either the outer limits of the data or the distances at which the residual anomaly pattern becomes obscured by neighboring influences. Then we write

$$2\pi GM = I + \mathcal{R}(X,Y)$$

where
$$I = \int_{-X}^{X} \int_{-Y}^{Y} \Delta g(x,y) \ dx \ dy$$

and where $\mathcal{R}(X,Y)$ is a remainder term.

The remainder $\mathcal{R}$ is calculated on the assumption that the entire mass M is concentrated at its mass center. This is a rather satisfactory approximation for a distribution of any shape, provided we are sufficiently far from it. Accordingly, if the center of mass of M is at $(\bar{x},\bar{y},\bar{z})$, we may write

$$\mathcal{R} = 2\pi GM - GM\bar{z} \int_{-X}^{X} \int_{-Y}^{Y} [(x - \bar{x})^2 + (y - \bar{y})^2 + (z - \bar{z})^2]^{-3/2} \ dx \ dy$$

which after some simple manipulation reduces to

$$\mathcal{R} = 2\pi GM - 4GM \tan^{-1} \frac{XY}{\bar{z}\sqrt{X^2 + Y^2}}$$

provided that $|\bar{x}| \ll X$ and $|\bar{y}| \ll Y$. Finally, if we put $R = \sqrt{X^2 + Y^2}$, we have

$$4GM = \left(\tan^{-1} \frac{XY}{\bar{z}R} \right)^{-1} I \tag{10-2}$$

To calculate the quantity in brackets, we need to know $\bar{z}$. However, the correction term is usually small enough that $\bar{z}$ need not be determined with great accuracy. The following simple rule is usually adequate: The depth to the center of a sphere is $\bar{z} = 0.65w_{1/2}$, where $w_{1/2}$ is the diameter of the gravity anomaly at an amplitude which is one-half of the maximum value. The depth to the center of an infinitely long circular cylinder is $\bar{z} = 0.5w_{1/2}$, where $w_{1/2}$ is the width of the gravity profile at one-half of the maximum amplitude. For a body which is too elongated to be approximated by a sphere and too short to be approximated by an infinitely long cylinder, $\bar{z}$

will lie somewhere between these two limits.   For practical purposes, we shall assume that $\bar{z}$ will have a value near $0.6w_{\frac{1}{2}}$, where $w_{\frac{1}{2}}$ is the "half-width" of the narrowest profile of Δg passing through the point of maximum anomaly.   The factor may be adjusted upward or downward within the limits prescribed for a sphere or for a cylinder, depending upon the ellipticity of the contours of Δg.

The integral in (10-2) is determined numerically.   If the points are distributed in a regular manner, Simpson's rule (or an equivalent quadrature formula) may be used.   If the points are not distributed at equal intervals, it will be necessary to use a template instead.   In exactly the same approximation as (10-2), we may write

$$GM = \left(1 - \frac{\bar{z}}{R}\right)^{-1} I$$

where $I = \dfrac{1}{2\pi} \displaystyle\int_0^R \int_0^{2\pi} \Delta g(r,\vartheta)r \, d\vartheta \, dr$, and R is the maximum radius within which Δg values are thought to be reliable.   To compute the integral I in this case, we rewrite it as follows:

$$I = \int_0^R \Delta g(r)r \, dr \doteq \sum_{j=1}^m W_j \, \Delta g(r_j)$$

where $\qquad \Delta g(r_j) = \dfrac{1}{2\pi} \displaystyle\int_0^{2\pi} \Delta g(r_j,\vartheta) \, d\vartheta$

and where the W_j's are a set of weighting coefficients.   The procedure is to make up a table of values of $\Delta g(r_j)$ by taking averages of Δg around circles of radius r_j.   To obtain maximum accuracy, the r_j's may be taken as the abscissas of a Gauss-type quadrature formula, whose weighting coefficients would then be identified with the W_j's. I is found by summing over the m rings, the choice of m being very largely a matter of discretion.

An actual example may be helpful at this stage.   Figure 10-1 is a map of the residual gravity effect obtained from a survey made near Noranda, Quebec (1).   A regional gradient of approximately 1 gu/100 ft running southwestward, plus some extraneous local disturbances, have been removed from the data.   The anomaly pattern is associated with a massive body of base metal sulfide (mainly pyrite) which has displaced volcanic rocks of middle Precambrian age.   The excess mass of the body is found by summing up Δg values along the picket lines according to Simpson's rule and then applying Simpson's rule for a second time to the row of sums.   This gives the value of I, which turns out to be 8.744×10^6 gu-ft^2.   Since $X = 1,200$ ft, $Y = 800$ ft, and $\bar{z} = 165$ ft ($w_{\frac{1}{2}} = 275$ ft), it also turns out that

$$\tan^{-1} \frac{XY}{\bar{z}R} = 1.33$$

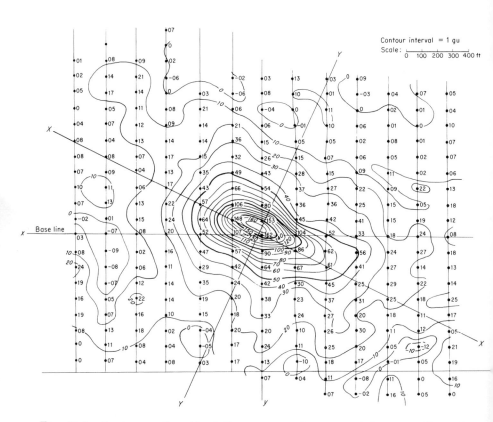

Fig. 10-1 Contours of residual Bouguer gravity on a claim near Noranda, Quebec.

According to formula (10-2) $4GM = 6.58 \times 10^6$ gu-ft^2, and therefore $M = 2.56 \times 10^6$ tons. Since this represents the excess mass, we must multiply by the ratio of the mineral density to the density contrast between the mineral and the host rock in order to calculate the total tonnage. In this case the average density of core samples of the mineral was 4.6 ± 0.5 g/cm^3, and for the host rock the density was about 2.7 g/cm^3. Thus the total tonnage of the sulfide body works out to be about $6.3 \pm 0.7 \times 10^6$ tons.

Exactly analogous methods may also be used to correct the estimate of the mass per unit length of two-dimensional bodies for missing flanks. If the coordinates of the center-of-mass line are $(\bar{x}, \bar{z})$ and if X is the greatest distance from 0 within which the Δg values are deemed to be reliable, we may write

$$4GM = \left(\tan^{-1} \frac{X}{\bar{z}}\right)^{-1} \int_{-X}^{X} \Delta g(x) \, dx$$

provided that $|\bar{x}| \ll X$. The point 0 should be placed near the point of maximum Δg. In this case $\bar{z} = 0.5 w_{1/2}$.

10-3 The Method of Characteristic Curves: The Ribbon Model

The next step beyond calculating excess mass is to make a direct interpretation of the residual anomaly in terms of a theoretical model. The problem is to select the dimensions and density contrast of the model so that its gravity effect agrees as closely as possible with the residual anomaly pattern. This is largely an exercise in curve or surface fitting, but it is not at all a simple one. It is made difficult by the fact that the potential fields even of the simplest models are described by formulas too unwieldy to be used in any direct fitting processes.

To give an example of the difficulty in the way of any direct approach to curve fitting, let us consider a model frequently used to represent dikes or veins. Ordinarily, a dike or a vein is thought to be a body that is rather narrow in proportion to its length, and the model often used to represent it is a slender ribbon, of which a diagram is shown in Fig. 10-2. Because gravity effects, more perhaps than other types of potential field observations, are sensitive to the strike lengths of such bodies, it is not permissible to assume that the ribbon is infinitely long. Accordingly we give it a strike

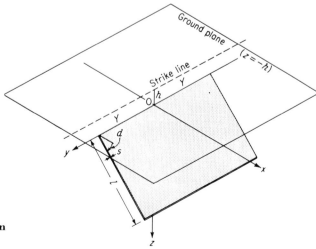

Fig. 10-2 The ribbon model.

length of $2Y$. Then, using the nomenclature of Fig. 10-2 and assuming a density contrast $\Delta\rho$, the formula for the gravity effect along a profile taken transversely across the center of the ribbon is

$$\Delta g(x) = 2G\,\Delta\rho\,s\left[\frac{1}{2}\sin d\left(\ln\frac{A-Y}{A+Y} - \ln\frac{B-Y}{B+Y}\right)\right.$$
$$+\cos d\left(\tan^{-1}\frac{Y(l+h\sin d - x\cos d)}{A(x\sin d + h\cos d)}\right.$$
$$\left.\left. - \tan^{-1}\frac{Y(h\sin d - x\cos d)}{B(x\sin d + h\cos d)}\right)\right]\quad(10\text{-}3)$$

where $\qquad A = \sqrt{(x - l\cos d)^2 + (h + l\sin d)^2 + Y^2}$
and $\qquad B = \sqrt{x^2 + h^2 + Y^2}$

Not until $Y > 10l$ (or $10h$, whichever is the greater) does this formula agree numerically with the formula for an infinitely long ribbon to within the accuracy of practical measurements.

One can imagine the difficulties involved in trying to fit a formula such as (10-3) to a set of observational data without having the least idea of what values to use for Y, l, h, or d. Until we have made estimates of these parameters, any such trial-and-error calculations cannot conceivably begin. A clue about how this problem might be approached is given in Fig. 10-3, which shows profiles of the gravity effect across an infinitely long ribbon in nine different configurations. Since lengths are arbitrary, we

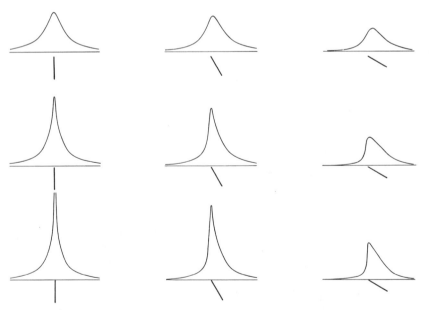

Fig. 10-3   Profiles of the gravity effect across ribbons of infinite length, drawn to scale.

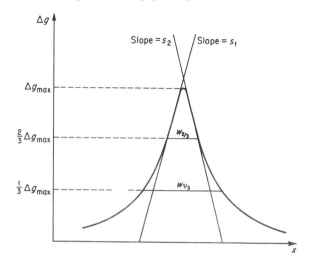

Fig. 10-4 "Characteristic estimators" for the ribbon model.

allow l to be our unit and scale all distances in terms of l. This leaves two parameters free, viz., the dip d and the ratio h/l. Each of these quantities is given three different values in the diagrams. It is obvious that changing the dip affects the symmetry of the profiles, while changing the depth of overburden affects their sharpness. If we could somehow define measures of these two properties, these measures might be used to indicate the attitude and depth of the ribbon.

Two measurements which can be used for this purpose and which are easy to make in practice are shown in Fig. 10-4. They are (1) the ratio of the maximum horizontal gradient on either limb of the profile, which is used for estimating the dip of the ribbon, and (2) the ratio of the width of the profile at two-thirds and one-third of the maximum value of Δg, which is used for estimating the depth ratio h/l. We shall call these two characteristic measures the *estimators* of the dip and the depth, and when discussing them individually, we shall refer to the slope ratio as the *skewness* and the width ratio as the *sharpness* of the gravity profile.

If we now fix the value of Y and draw lines of constant d and h/l on a plot of skewness versus sharpness, we get two families of intersecting curves, known as the *characteristic curves* for the ribbon model. Two actual examples of these curves are shown, for $Y = 0.5l$ and $Y = l$, in Figs. 10-5 and 10-6, respectively. For every gravity profile there is a corresponding point somewhere on each of these sheets. By interpolating between the nearest pairs of characteristic curves, we find values for d and h/l immediately. In practice, however, we should be reminded that skewness and sharpness can never be observed with perfect precision, but that each measurement has associated with it a small range of uncertainty. Thus a real profile will correspond to a "rectangle of uncertainty" on the characteristic curves, which implies certain limits on d and h/l rather than precise values. The

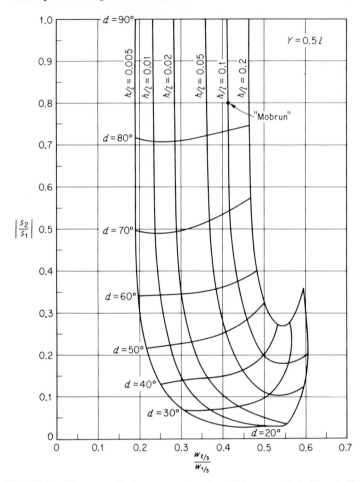

Fig. 10-5   Characteristic curves for the ribbon model, $Y = 0.5l$.

resulting uncertainties will be minimized if estimators can be found which cause the two families of curves to be as nearly as possible orthogonal to each other.

Once d and h/l are known, or their limits established, we shall require an additional set of curves to determine values for the ribbon width l and for the density-thickness product $\Delta\rho\, s$ in conventional units.   The necessary information is provided by a graph of the quantity $\Delta g_{max}/2G\,\Delta\rho\,s$ versus $w_{1/3}/l$, Δg_{max} being the amplitude of the residual anomaly and $w_{1/3}$ the width of the profile at one-third of the amplitude, with d and h/l as parameters.   Examples of such curves for $Y = 0.5l$ and $Y = l$ are shown in Figs. 10-7 and 10-8, respectively.   These are the "complementary sets" to the characteristic curves illustrated in Figs. 10-5 and 10-6.   Since the product $\Delta\rho\, s$ appears as a single factor in the formula (10-3), it is not possible to decompose it.   This product is called the *surface density* of the ribbon, and

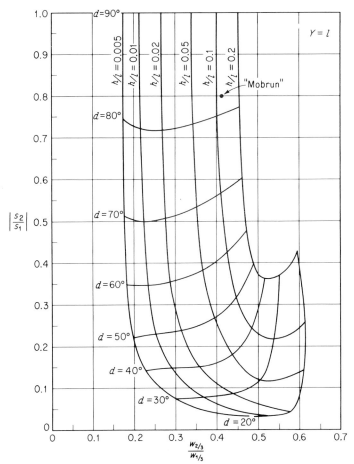

Fig. 10-6   Characteristic curves for the ribbon model, $Y = l$.

it is normally measured in grams per square centimeter.   To find s, we must assume $\Delta\rho$, or vice versa.   This is exactly where the density-size ambiguity, fundamental to all inverse-potential solutions, makes its appearance in the ribbon interpretation problem.

To find the right set of characteristic curves to use in a particular instance, we must know Y/l.   This parameter is estimated from Fig. 10-9.   The ordinate in this diagram is the ratio of $w_{\frac{1}{3}}$ to the half-width (i.e., the width at one-half the amplitude) of the anomaly pattern *in the strike direction*.   This ratio is a natural estimator for Y/l, and we shall refer to it as the *ellipticity* of the anomaly.   The abscissa is the sharpness of the profile perpendicular to the strike.   The shaded zones in Fig. 10-9 represent the total variation of ellipticity at a given sharpness for a full range of d and for $0.01 \leq h/l \leq 0.2$.

Let us now take up a practical illustration of the use of characteristic

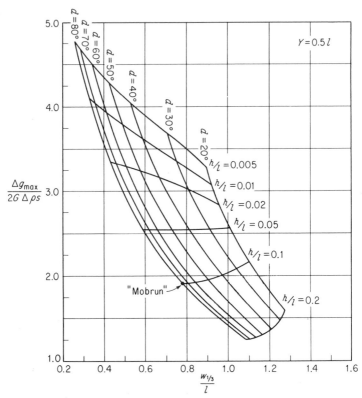

Fig. 10-7 Complementary set of curves to accompany Fig. 10-5.

curves. The interpretation of the anomaly pattern shown in Fig. 10-1 is obviously a three-dimensional problem. If we draw a profile through the center of the anomaly pattern perpendicular to its strike (Fig. 10-10), we find that

$$s_1 = 0.083 \text{ gu/ft} \qquad s_2 = -0.066 \text{ gu/ft}$$
$$w_{\frac{2}{3}} = 185 \text{ ft} \qquad w_{\frac{1}{3}} = 445 \text{ ft}$$

and

$$\Delta g_{\max} = 17.2 \text{ gu}$$

In addition to these measures, we find from the profile in the direction of the axis of the anomaly that $(w_{\frac{1}{2}})_y = 670$ ft. Thus the values of the various estimators are

$$\text{Skewness} = |s_2/s_1| = 0.80$$

$$\text{Sharpness} = w_{\frac{2}{3}}/w_{\frac{1}{3}} = 0.415$$

$$\text{Ellipticity} = w_{\frac{1}{3}}/(w_{\frac{1}{2}})_y = 0.66$$

According to Fig. 10-9, $Y/l = 0.6$. Thus we shall have to interpolate between the values of h, l, and d that are found from characteristic curves for $Y/l = 0.5$ and for $Y/l = 1$.

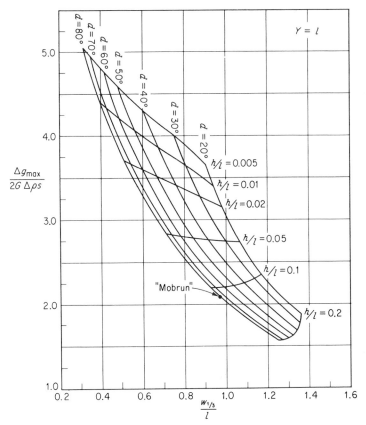

Fig. 10-8   Complementary set of curves to accompany Fig. 10-6.

The characteristic curves for $Y/l = 0.5$ (Fig. 10-5) show that $d = 83°N$ and $h/l = 0.10$. From the complementary set (Fig. 10-7) it follows that

$$\frac{\Delta g_{\max}}{2G\ \Delta\rho\ s} = 1.88 \quad \text{and} \quad \frac{w_{\frac{1}{3}}}{l} = 0.78$$

from which we deduce that $\Delta\rho\ s = 6.9 \times 10^3$ g/cm² and $l = 570$ ft. Thus it turns out that $h = 0.1l = 57$ ft. If we assume that $\Delta\rho = 1.9$ g/cm³, we find that $s = 118$ ft.

The characteristic curves for $Y/l = 1$ (Fig. 10-6) indicate that $d = 82°N$ and $h/l = 0.12$. From the complementary set (Fig. 10-8) we find that

$$\frac{\Delta g_{\max}}{2G\ \Delta\rho\ s} = 2.05 \quad \text{and} \quad \frac{w_{\frac{1}{3}}}{l} = 0.99$$

hence $\Delta\rho\ s = 6.3 \times 10^3$ g/cm², $l = 450$ ft, and $h = 52$ ft. Therefore, by interpolating between the two sets of results, we finally deduce that $d = 83°N$, $l = 550$ ft, $h = 56$ ft, $s = 116$ ft, and $2Y = 660$ ft.

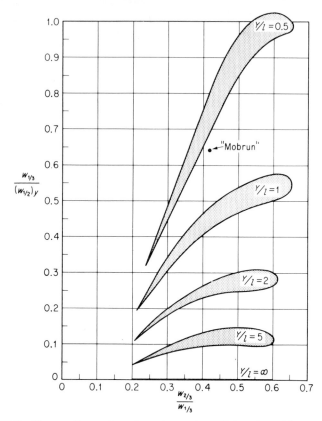

Fig. 10-9   Curves for estimating the ratio Y/l for the ribbon model.

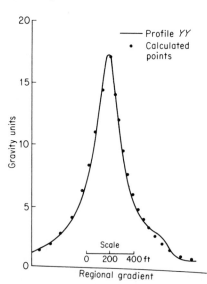

Fig. 10-10   Observed and theoretical profiles of Δg across the Mobrun body, Noranda, Quebec.

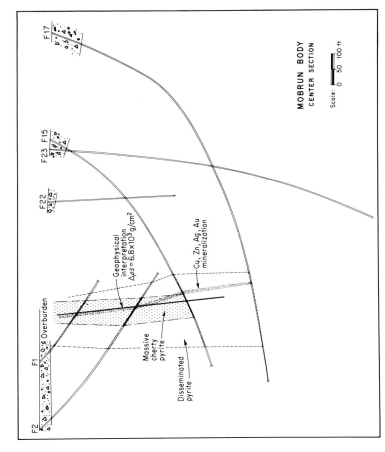

Fig. 10-11 Center section, Mobrun sulfide body. [After Seigel (1).]

281

Unless there is access to the subsurface, the only check we can make upon the solution is provided by the gravity data themselves. When we insert the above values into the formula (10-3) and calculate the theoretical profile, an almost perfect correspondence is obtained to the observations, as shown in Fig. 10-10. In this instance, however, extensive exploratory drilling has been carried out, and the results are shown in Fig. 10-11. The anomaly proved to be due to a massive deposit of pyrite and smaller quantities of chalcopyrite and other minerals. The drill intersected sulfides over a distance slightly greater than 100 ft, which according to the value found for $\Delta\rho$ s indicates an average density contrast very close to the value determined from core samples. We note also the other respects in which the geophysical interpretation is confirmed by drilling. The dip, depth extent, and overburden thickness all appear to have been predicted correctly. The excess mass of the body $2\,\Delta\rho s\,lY$ turns out to be 2.45×10^6 tons, which compares well with our earlier (and probably more accurate) calculation, which was based upon Gauss' theorem.

There are two points upon which we wish to remark. The first is that in spite of the fact that the mean thickness of this body is approximately twice its depth of burial, a very good fit to its gravity effect can be made by using a thin ribbon model. This would indicate that the gravity method is very insensitive to the thickness of bodies having this tabular shape and that there is no way to resolve this ambiguity except by the drill. The second point is that, as Fig. 10-9 clearly indicates, a ribbon model must have a strike length that is twenty or more times its width before the ends can be safely disregarded in the calculations. The implication of this fact is unmistakable: Interpretation techniques based upon two-dimensional models ought to be used with a great deal of restraint in gravity work and their limiting nature kept constantly in mind.

10-4   The Step Model

One problem in gravity analysis in which two-dimensional methods can generally be used with some safety is the interpretation of faults. The model customarily used to represent a fault for the purpose of direct interpretation is a paradigm of mathematical abstraction compared with the complex geo-

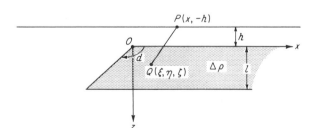

Fig. 10-12 The step model.

logical structure which it is supposed to represent. It is pictured in cross section in Fig. 10-12. The anomalous material has the form of a flat step which goes to infinity in the positive direction of x and is assumed to have a uniform density contrast throughout its entire bulk. Since most faults have a strike length that is many times their throw, the step is assumed to be two-dimensional. The model is therefore characterized by three parameters, viz., the density contrast $\Delta\rho$, the dip d, and the depth-to-vertical displacement ratio (since distances are arbitrary) h/l.

The formula for the profile of gravity effect across the step is

$$\Delta g(x) = 2G \,\Delta\rho \int_0^l \int_{\zeta \cot d}^\infty \frac{(h + \zeta)\,d\xi\,d\zeta}{(x - \xi)^2 + (h + \zeta)^2}$$

$$= \pi G \,\Delta\rho\, l + 2G \,\Delta\rho \int_0^l \tan^{-1}\left(\frac{x - \zeta \cot d}{h + \zeta}\right) d\zeta$$

Changing the variable of integration to $u = (x - \zeta \cot d)/(h + \zeta)$ and integrating by parts, we get (putting $l = 1$)

$$\Delta g(x) = 2G \,\Delta\rho \left\{\frac{\pi}{2} + (h + 1)\,\tan^{-1}\frac{x - \cot d}{h + 1} - h\,\tan^{-1}\frac{x}{h}\right.$$

$$+ (x \sin^2 d + h \sin d \cos d) \ln\left[\frac{(x - \cot d)^2 + (h + 1)^2}{x^2 + h^2}\right]^{\frac{1}{2}}$$

$$\left. - (x \sin d \cos d + h \cos^2 d)\left(\tan^{-1}\frac{x - \cot d}{h + 1} - \tan^{-1}\frac{x}{h}\right)\right\} \quad (10\text{-}4)$$

which we may write as $\Delta g(x) = 2G \,\Delta\rho\, f(x;h,d)$. The object is now to find two properties of the function $f(x;h,d)$ which may be used as estimators for

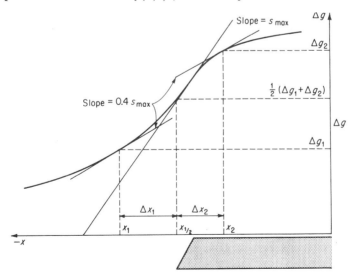

Fig. 10-13　Characteristic estimators for the step model.

h and *d*. For the step model this proves to be an especially difficult task, because there are so few attributes of gravity profiles across faults which can be measured reliably in practice. We must keep in mind too that in the great majority of cases only the central portion of the profile will be free from neighboring disturbances, and the measures used must therefore be found within the central region. And although there is a clear temptation to use the total amplitude of the anomaly to estimate the throw of the fault, it is seldom that the asymptotes can be found with such accuracy.

The estimators that we select for fault interpretation are illustrated in Fig. 10-13. First, we measure the maximum horizontal gradient s_{max}, which we can usually do with sufficient precision. We then locate the two points x_1 and x_2, at which the slope of the profile is equal to $0.4s_{max}$. We choose the ratio 0.4 because this usually places x_1 and x_2 close to the two "knees" of the profile, where the points of tangency can be most accurately determined. This provides us with a characteristic length $x_2 - x_1$ and a characteristic measure of the amplitude $\Delta g_2 - \Delta g_1$. Next we find the point $x_{\frac{1}{2}}$ at which the amplitude falls exactly half-way between Δg_1 and Δg_2, and form the two ratios

$$k_1 = \frac{x_2 - x_{\frac{1}{2}}}{x_{\frac{1}{2}} - x_1} \quad \text{and} \quad k_2 = \frac{\Delta g_2 - \Delta g_1}{(x_2 - x_{\frac{1}{2}})s_{max}}$$

It happens that k_1 is more sensitive to change in *h* than in *d*, while k_2 responds more strongly to *d* than to *h*. The complete set of characteristic curves calculated from (10-4) is shown in Fig. 10-14. Measurements of k_1 and k_2

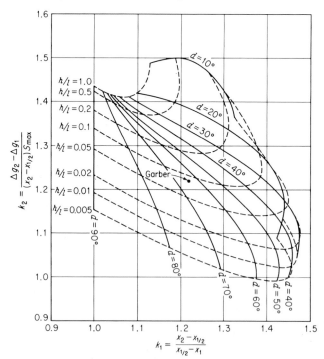

Fig. 10-14   Characteristic curves for the step model.

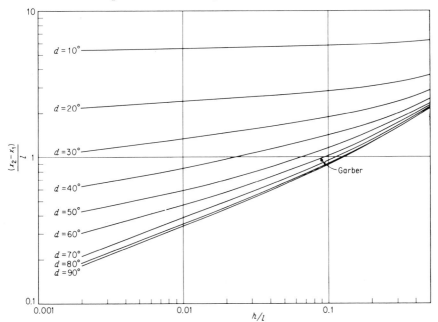

Fig. 10-15   Complementary curves for estimating *l*, step model.

on the residual gravity profile may be used with these curves to estimate both h/l and d.

To separate l from h, we turn to Fig. 10-15, which shows a plot of $(x_2 - x_1)/l$ against h/l, with d as a parameter. Since $x_2 - x_1$ can be measured and both h/l and d are now known, these curves will give us l and hence also h. If we wish also to find $\Delta\rho$, we may do so from Fig. 10-16, which shows $(\Delta g_2 - \Delta g_1)/2G\,\Delta\rho\,l$ versus h/l, with d as a parameter.

A very simple rule may be used in fault interpretations when the anomaly is weak or when the stations are too widely spaced to determine the finer details of the anomaly pattern. The rule is based upon the expression for the gravity profile across a very thin horizontal half-plane whose thickness is l and whose density contrast is $\Delta\rho$. If the half-plane lies at a depth h, then at a horizontal distance x from its edge

$$\Delta g(x) = 2G\,\Delta\rho\,l\left(\frac{\pi}{2} + \tan^{-1}\frac{x}{h}\right) \tag{10-5}$$

and from this formula it is a very easy matter to establish the following identities:

$$h = \frac{0.56(\Delta g_2 - \Delta g_1)}{s_{\max}}$$

$$x_2 - x_1 = 2.45h$$

and

$$\Delta\rho\,l = 14(\Delta g_2 - \Delta g_1)$$

when Δg is measured in gravity units, l in feet, and $\Delta\rho$ in grams per cubic centimeter. These formulas are based upon the assumption that $h \gg l$, but they may be expected to give reasonable solutions provided h does not become much less than l.

In Sec. 9-7 we illustrated the use of bias in the removal of the regional effect from a gravity survey made over the Garber oil field. According to the nature of the bias used, we concluded that the residual anomaly was probably caused by faulting. If we apply to the residual gravity profile shown in Fig. 10-17 the tests enumerated above, we obtain the following results: The maximum horizontal gradient is 8.0 gu/mile, the distance $x_2 - x_1$ is 11,000 ft (more or less), and the amplitude $\Delta g_2 - \Delta g_1$ is 11.7 gu. Our "rule-of-thumb" formula (10-5) gives a depth $h = 4,400$ ft, and a density contrast times throw = 165 g/cm^3 $\times$ ft. This would place the top of the fault in a stratigraphic region where no major density contrasts are likely to exist and in which there appears to be no evidence of faulting.

The picture that emerges from the analysis of characteristic curves is even less encouraging. In addition to the measures given above, we find that $k_1 = 1.22$ and $k_2 = 1.22$. This combination (marked in Fig. 10-14) indicates that $h/l = 0.09$ and $d = 65°$. Since it is usually unsatisfactory

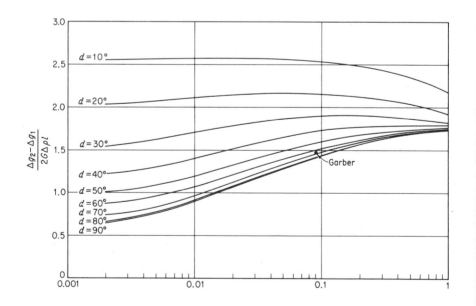

Fig. 10-16 **Complementary curves for estimating $\Delta\rho$, step model.**

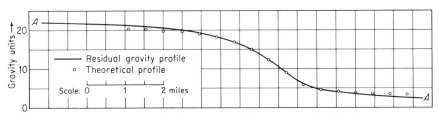

Fig. 10-17 Profile of residual Bouguer gravity across the Garber structure, Garber County, Oklahoma, and the theoretical profile determined from characteristic curves.

to rely upon a single set of measurements, we estimate upper and lower bounds on s_{max}. These turn out to be 0.00157 gu/ft and 0.00142 gu/ft, respectively, with corresponding limits on k_1 of 1.21 and 1.24 and on k_2 of 1.21 and 1.25. From these limits we infer that d lies between 50 and 70°, while h/l falls between 0.08 and 0.14. Then from Figs. 10-15 and 10-16 we discover that $h = 1,200$ ft while l lies between 8,000 and 13,000 ft. The corresponding limits on $\Delta\rho$ are $+0.022$ and $+0.016$ g/cm³.

These values cannot be reconciled with the known stratigraphy of this region, and in consequence we are forced to reexamine the "bias" used in making the regional-residual separation. If the residual anomaly is not due to a fault, it must have closure. This compels us to introduce a broad inflection into the regional gradient, which now assumes a shape somewhat characteristic of a very deeply buried contact, which certainly would now be located within the granitic basement. The new residual anomaly, whose amplitude is now reduced to less than 10 gu, can be accounted for by a known asymmetrical anticline at the post-Ordovician unconformity. Incidentally, the closure on this structure as given by Gish and Carr (2) is nearly 300 ft, which is quite compatible with an anomaly of this magnitude. We shall beg the question of whether a causal connection might exist between the anticline and the deep basement fault, for this involves geological considerations which lie outside our competence to discuss. The only point to be made here is that if the method of characteristic curves should lead to an unreasonable result, as it does in this case, it is not the curves that are at fault but the subjective decisions that were made in the regional-residual analysis or in the choice of a model. These decisions are generally rather easy to revise. Thus a certain "feedback" will inevitably exist between regional-residual analysis and interpretation, but without the help of characteristic curves or similar numerical aids it is difficult to control it in such a way that the interpretations converge as quickly as possible toward a reasonable answer.

10-5 The Numerical Calculation of Gravity Profiles

The end result of the method of characteristic curves is a set of values for the parameters of the model we have chosen to represent the anomalous body.   The next step is to verify that these values constitute a satisfactory *geophysical* solution, by comparing the gravity effect of the model with the residual anomaly.   To make these calculations, values for the parameters may be substituted into formulas such as (10-3) and (10-4) and the functions computed for a number of different values of x.   To compute such formulas a large number of times, however, the use of a computing machine becomes highly advantageous.   If a machine is not available, it is generally much easier to integrate numerically than to use formulas such as (10-3).

A simple graticule for performing the integrations graphically for *two-dimensional* bodies is constructed as follows: If in the expression (8-28) for the gravity potential of a two-dimensional body we set $x = z = 0$ and keep ρ constant, we shall have

$$\Delta g(0) = 2G\,\Delta\rho \iint_S \frac{\zeta\,d\xi\,d\zeta}{\xi^2 + \zeta^2} \tag{10-6}$$

Let us place the origin 0 at the point on the ground surface where we wish to make the calculation.   This involves no loss in generality, since the origin may be placed anywhere.   Changing now to polar coordinates by setting $\xi = r\sin\vartheta$ and $\zeta = r\cos\vartheta$, we get

$$\Delta g(0) = 2G\,\Delta\rho \iint_S \cos\vartheta\,d\vartheta\,dr$$

which we may write approximately as

$$\Delta g(0) \doteqdot 2G\,\Delta\rho \sum_S\sum \Delta(\sin\vartheta)\,\Delta r$$

Thus by drawing up a template having uniform increments in $\sin\vartheta$ and in r, the gravity effect at any point P is determined by counting up the number of segments $\Delta(\sin\vartheta)\,\Delta r$ into which S is divided when the origin of the template is placed at P and the line $\vartheta = 0$ is laid along the horizontal.

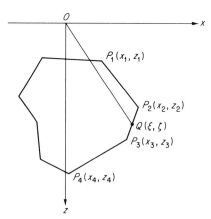

Fig.   10-18   Cross section of a two-dimensional body represented by means of a polygon.

A profile of the gravity effect is obtained by repeating the calculations at different positions of P.

The integrations can also be programmed for automatic machines. By using an idea proposed by Talwani, Worzel, and Landisman (3), the cross section of the two-dimensional body is replaced with an n-sided polygon as illustrated in Fig. 10-18. Assigning pairs of coordinates (x_i, z_i) to each of the n vertices, the formula for the kth side will be

$$\xi = a_k \zeta + b_k$$

where $\quad a_k = \dfrac{x_{k+1} - x_k}{z_{k+1} - z_k} \quad$ and $\quad b_k = \dfrac{x_k z_{k+1} - x_{k+1} z_k}{z_{k+1} - z_k}$

The gravity effect at 0 is given by (10-6), and if we integrate first with respect to ξ, this becomes

$$\Delta g(0) = 2G \, \Delta\rho \oint \tan^{-1} \frac{\xi}{\zeta} \, d\zeta$$

$$\doteq 2G \, \Delta\rho \sum_{k=1}^{n} \int_{z_k}^{z_{k+1}} \tan^{-1}\left(a_k + \frac{b_k}{\zeta} \right) d\zeta$$

$$= 2G \, \Delta\rho \sum_{k=1}^{n} \frac{b_k}{1 + a_k^2} \left[\frac{1}{2} \ln \left(\frac{x_{k+1}^2 + z_{k+1}^2}{x_k^2 + z_k^2} \right) \right.$$

$$\left. + a_k \left(\tan^{-1} \frac{x_{k+1}}{z_{k+1}} - \tan^{-1} \frac{x_k}{z_k} \right) \right] \quad (10\text{-}7)$$

Shifting 0 horizontally to a new location is equivalent to adding a fixed increment to each of the x_k's, while the z_k's will remain unchanged. It must be remembered also that $(x_{n+1}, z_{n+1}) \equiv (x_1, z_1)$. This formula leads to very rapid calculation of gravity profiles when an automatic machine is available.

Let us look at an illustration of the use of these methods. A detailed gravity survey was made on a valley glacier in British Columbia, one of the objects of which was to determine the profile of the glacier prior to surveying a line for a tunnel to pass underneath. This is a problem that is practically two-dimensional, and we shall treat it as such by dealing only with a single profile that traverses the glacier perpendicular to its flow axis. The reduction of the data was complicated by a time dependence in the positions and elevations of the gravity stations due to progressive ablation, and by exceptionally large contributions from the extremely irregular terrain. The latter problem was dealt with in a rather rough way by computing the gravity effects of the mountains as if they were made up of large tabular masses. This was thought to be accurate to about ± 2 mgals only, but in view of the large negative anomaly due to the ice this was considered satisfactory. The Bouguer gravity profile with terrain corrections added is shown in Fig. 10-19.

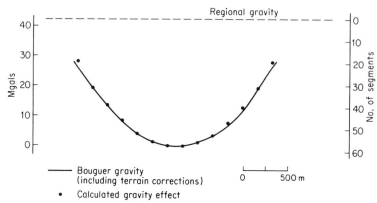

Fig. 10-19   Observed and calculated profiles of Bouguer gravity across the Salmon Glacier.

The first and most critical step in the "indirect" interpretation process is to find a suitable starting hypothesis.   Here we make direct use of the fact that the width of the ice can be measured, plus the well-founded conjecture that the cross section undoubtedly tapers with depth.   A rough idea of the maximum depth of the ice may be deduced from the formula for the maximum gravity effect measured on the base of an inverted, isosceles triangular prism.   If the height (or depth) of the triangle is Wh and the width of the base is $2W$, then at the center of the base

$$\Delta g_{\max} = 4G\, \Delta\rho\; W \left(\frac{h}{h^2+1}\right)\left(\frac{\pi}{2} + h \ln h\right) \tag{10-8}$$

If we draw a graph of the function $\Delta g_{\max}/4G\, \Delta\rho\; W$ (Fig. 10-20), we can use it to estimate h from measured values of $\Delta g_{\max}$ and W.   For we observe that $\Delta g_{\max} = -40$ mgals and $2W = 2.7$ km, and if we assume for $\Delta\rho$ a value -1.7 g/cm^3, we find that $\Delta g_{\max}/4G\, \Delta\rho\; W = 0.65$.   According to Fig. 10-20, this implies a value of 0.67 for h, giving a maximum depth of ice equal to 900 m.   This will not be accurate, but it will at least serve as a starting point; and we can now draw a cross section for the glacier with some assurance that its gravity effect will not widely disagree with the measured values.   It then remains to alter this shape by degrees to improve the correspondence between the observed and calculated profiles.   If we add to the thickness of the glacier beneath each point on the profile an amount given by $(\Delta g_{\mathrm{obs}} - \Delta g_{\mathrm{calc}})/2\pi G\, \Delta\rho$, the overall discrepancy between the observed and calculated gravity effects should be reduced.   A few repetitions of this process should quickly lead to a satisfactory correspondence.

In making these adjustments, we discover that the *scale* of the calculated Δg values is fixed by the value assumed for $\Delta\rho$, although the *shape* of

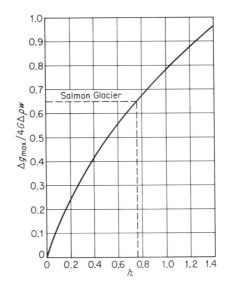

Fig. 10-20 The ratio $\Delta g_{max}/4G\,\Delta\rho W$ versus h on the base of an inverted, two-dimensional, triangular prism.

the gravity profile is not affected by this quantity. The best practice, therefore, is to direct our efforts toward reproducing the shape of the gravity profile, adjusting the density contrast at each stage of the calculations to make the amplitudes of the two curves coincide. When a match to the field curve has at last been found, $\Delta\rho$ is also determined. For instance, several adjustments with the aid of a template were needed to reach the final form of the cross section shown in Fig. 10-21. The shape of the computed gravity profile across this section is indicated by the heavy dots in Fig. 10-19, where the values given on the right-hand scale refer to the number of segments of the graticule which were covered by the glacier at each position. By adjusting this scale, we can make the amplitudes of the observed and calculated gravity profiles coincide, and when we do this, the profiles as a whole appear to match rather well. It happens that in this case the scale constant of the overlay was 32 m, and the maximum number of segments

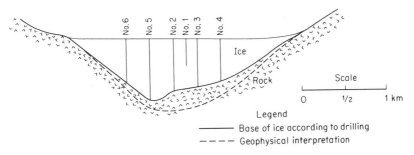

Fig. 10-21   Cross section of the Salmon Glacier.

counted was 56.8. Setting $2G \Delta\rho \times 3{,}200 \times 56.8 = -40 \times 10^{-3}$ gal gives $\Delta\rho = -1.65$ g/cm^3. This is a trifle smaller than the value we had originally assumed, indicating that the valley rocks are possibly somewhat lighter than we had supposed them to be. A number of holes later drilled through the ice along this profile (Fig. 10-21) have shown that the gravity interpretation has overestimated the depth of the valley by about 10 percent. In view of the wide uncertainties in the terrain corrections, this agreement is quite satisfactory.

A useful technique in which a computing machine is used to perform the trial-and-error adjustments in this particular type of problem has been developed by Bott (4). The supposed form of the valley floor is replaced by a series of horizontal steps, not necessarily all of the same width, so that the glacier may be thought of as consisting of a number of two-dimensional, narrow rectangular prisms standing on edge. The total gravity effect of all of these prisms is computed only at those points which lie at the centers of the top edges. The disagreement between these calculations and the observed effect is used to adjust the depth of each prism individually, ignoring the effects of its neighbors. The calculations are repeated with the new valley profile, and so on through several trials. The process appears to be highly convergent, and a small number of iterations will usually bring the computed gravity profile into close agreement with the observational data.

10-6 Numerical Calculation of Gravity Profiles Using End Corrections

As we have remarked, interpretational techniques based on the logarithmic potential have only a very limited application to actual problems. Corrections for finite strike length can be incorporated into the calculations by drawing the graticule to slightly different specifications. If the body extends only to a distance Y along its strike in either direction from the origin of coordinates, the gravity effect at this point will be, instead of (10-6),

$$\Delta g(0) = 2G \Delta\rho \, Y \iint_S \frac{\zeta \, d\xi \, d\zeta}{(\xi^2 + \zeta^2) \sqrt{\xi^2 + \zeta^2 + Y^2}}$$

which in polar coordinates is

$$\Delta g(0) = 2G \Delta\rho \, Y \iint_S \frac{\cos \vartheta \, d\vartheta \, dr}{\sqrt{r^2 + Y^2}} \tag{10-9}$$

We may write this integral approximately as

$$\Delta g(0) \doteq 2G \Delta\rho \, Y \sum_S \sum \Delta(\sin \vartheta) \, \Delta \, [\ln \, (r + \sqrt{r^2 + Y^2})]$$

and the method of calculation then consists in drawing up a template having equal increments in $\sin \vartheta$ and in $\ln (r + \sqrt{r^2 + Y^2})$ and counting up the segments intercepted by S as before. Notice, however, that the scale of this chart is determined by the value assigned to Y, and so the radii will depend upon the strike length. The calculations are to be compared only against the profile that traverses the body across its center at right angles to the strike direction.

The expression (10-9) cannot be integrated in closed form around a polygon. Profiles using end corrections must be programmed for automatic computation according to the formula

$$\Delta g(0) \doteq 2G \, \Delta\rho \sum_{k=1}^{n} \int_{z_k}^{z_{k+1}} \tan^{-1} \frac{Y(a_k\zeta + b_k)}{\sqrt{(1 + a_k^2)\zeta^2 + 2a_k b_k \zeta + b_k^2 + Y^2}} \, d\zeta$$

$$(10\text{-}10)$$

in which a_k and b_k have the same meanings as in Sec. 10-5. However, the integrals must now be evaluated numerically. Unfortunately, this places some restriction upon the economic advantages of automatic computations in gravity interpretations.

10-7   Some Simple Formulas

The method of characteristic curves, to be used successfully, depends upon knowing the shape of the anomaly rather well. Where the number of stations is so small that the anomaly shape is not well defined, the method is difficult to apply. Although it is still possible to make interpretations of a sort even under these conditions, the approach is necessarily based upon a few very simple shapes, such as spheres and cylinders, whose geological interpretation is apt to be a trifle obscure. Nevertheless, such methods play a definite role in applied geophysics because, quite apart from their application to poorly controlled data which are inadequate for a more detailed analysis, they are also used in the planning of gravity surveys and sometimes as a preliminary to the final interpretation. The following is a partial list of some of the simple rules for preliminary interpretations, which we give as illustrations only:

1. *Sphere* (compact mineralization, salt domes, etc.). Excess mass M, depth to center d.

For a profile through the center of the anomaly pattern

$$\Delta g(x) = GMd(x^2 + d^2)^{-3/2}$$

Consequently $d = 0.65w_{1/2}$, if $w_{1/2}$ is the "half-width" of the profile, and $M = 0.65\Delta g_{\max}(w_{1/2})^2$ ton, if $\Delta g_{\max}$ is measured in gravity units and $w_{1/2}$ in feet.

2. *Vertical ribbon* (thin dike, mineralized shear, etc.). Thickness s, width l, density contrast $\Delta\rho$, depth of burial h, strike length infinite.

$$\Delta g(x) = G\,\Delta\rho\,s\,\ln\left(\frac{x^2 + (h + l)^2}{x^2 + h^2}\right) \qquad s \ll l$$

Therefore,

$$\Delta g_{\max} = 2G\,\Delta\rho\,s\,\ln\left(1 + \frac{l}{h}\right)$$

$$w_{\frac{1}{2}} = 2h\,\sqrt{1 + \frac{l}{h}}$$

and

$$\int_{-\infty}^{\infty} \Delta g(x)\,dx = 2\pi G\,\Delta\rho\,sl = i_x$$

(For "flank correction," see Sec. 10-2.)   Putting these formulas together, we form the dimensionless quantity

$$f(a) = \frac{\Delta g_{\max}\,w_{\frac{1}{2}}}{i_x} = \frac{2\,\sqrt{1 + a}}{\pi a}\,\ln(1 + a)$$

where $a = l/h$.   The graph of $f(a)$ is shown in Fig. 10-22.   Since $f(a)$ can be measured from the profile, a can be determined from this graph.   Once a is known, h is found from the formula

$$h = \frac{w_{\frac{1}{2}}}{2\,\sqrt{1 + a}}$$

Knowing h, then $l = ah$ and $\Delta\rho\,s = 7.84\,i_x/l$, when Δg is measured in gravity units, distances are measured in feet, and $\Delta\rho$ is in g/cm^3.

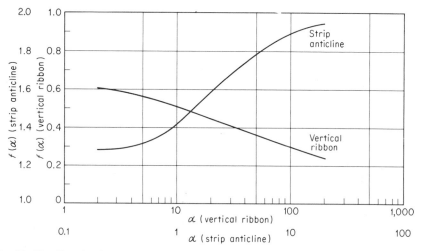

Fig. 10-22   Graphs for making rapid estimates of two-dimensional ribbons and anticlines.

3. *Strip anticline* (buried ridge, depression, etc.) Height t, basal width $2w$, density contrast $\Delta\rho$, depth of burial h, strike length infinite.

$$\Delta g(x) = 2G \, \Delta\rho \, t \left(\tan^{-1} \frac{x+w}{h} - \tan^{-1} \frac{x-w}{h} \right) \qquad \text{for} \qquad t \ll h$$

Thus
$$\Delta g_{max} = 4G \, \Delta\rho \, t \, \tan^{-1} \frac{w}{h}$$

$$w_{\frac{1}{2}} = 2 \sqrt{h^2 + w^2}$$

and
$$\int_{-\infty}^{\infty} \Delta g(x) \, dx = 2\pi G \, \Delta\rho \, t \, w = i_x$$

(For flank correction, see Sec. 10-2.) From these formulas, we find that

$$f(a) = \frac{\Delta g_{max} \, w_{\frac{1}{2}}}{i_x} = \frac{4 \sqrt{1 + a^2}}{\pi a} \tan^{-1} a$$

where $a = w/h$. The graph of $f(a)$ is shown in Fig. 10-22. Finding a value for a from this graph gives

$$h = \frac{w_{\frac{1}{2}}}{2 \sqrt{1 + a^2}}$$

Finally, $w = ah$ and $\Delta\rho \, t = 7.84 i_x/w$, when Δg is measured in gravity units, distances are measured in feet, and $\Delta\rho$ is in g/cm^3.

4. *Thin circular disk* (reef, boss, bedrock depression, etc.) Thickness t, radius a, density contrast $\Delta\rho$, depth of burial h. The formula for the profile of the gravity effect involves elliptic integrals and cannot be expressed in a simple, closed form. For $t \ll h$, however,

$$\Delta g_{max} = 2\pi G \, \Delta\rho \, t \left[1 - \frac{h}{\sqrt{a^2 + h^2}} \right]$$

and
$$\int_{-\infty}^{\infty} \Delta g(x,0) \, dx = 4\pi G \, \Delta\rho \, t(\sqrt{a^2 + h^2} - h) = i_x$$

The "half-width" of the profile is not a simple function of a and h, but to a sufficient accuracy we can write

$$w_{\frac{1}{2}} \doteq 1.24h(1 + 0.86a) \qquad 0.5 < a < 2.5$$

where $a = a/h$. This expression is of course an approximation, but it is adequate within the range specified. For values of a outside of this range, values of $w_{\frac{1}{2}}$ will have to be calculated numerically. From these formulas, we may set

$$f(a) = \frac{\Delta g_{max} \, w_{\frac{1}{2}}}{i_x} = \frac{0.62(1 + 0.86a)}{\sqrt{1 + a^2}} \qquad 0.5 < a < 2.5$$

and plot $f(a)$ versus a. From this graph we can find a, and hence also h. Finally, we shall have $a = ah$ and $\Delta\rho \, t = 7.84 \, \Delta g_{max}/(1 - 1/\sqrt{1 + a^2})$, when Δg is measured in gravity units, distances are measured in feet, and $\Delta \rho$ is in g/cm^3.

10-8   Quantitative Interpretation in Three Dimensions

The methods of characteristic curves and graticules are both limited to what is essentially a "two-dimensional" interpretation of residual gravity anomalies.   That is, these methods are able to deal with only a single profile of the residual gravity field.   Thus they make use of only a small amount of the information actually present in a gravity survey, and the great bulk of it is largely unused.   The aim of three-dimensional methods in gravity interpretation on the other hand is to fit a theoretical potential field to the entire network of points that determines the anomaly, and although this requires a little extra labor, the results are often more gratifying.

The three-dimensional approach is simple and direct—in principle. The aim is to fit the formula (8-24) to a network of Δg values and by so doing to calculate values for the three moments B_0^0, B_2^0, and B_2^2.   If we can assume a value for the density contrast $\Delta \rho$, the dimensions of any of the theoretical models described in Sec. 8-7 can be ascertained from these moments.   In spite of the unwieldy appearance of (8-24), the analysis presents no very grave problems.

First, we choose a frame of coordinates in which the x axis is in the direction $\omega = 0$.   This direction will be along the major axis of the anomaly pattern, which ought to be evident from the shapes of the contours.   Then according to (8-24) we find that at the point $Q(0,0,z)$

$$\Delta g(0,0) = \frac{B_0^0}{z_0{}^2} + \frac{3(2B_2^0 - \Gamma_2 \sin^2 \psi)}{z_0{}^4}$$

where

$$\Gamma_2 = 3B_2^0 + B_2^2$$

and

$$\left(\frac{\partial \, \Delta g}{\partial y}\right)_{\substack{x=0 \\ y=0}} = - \frac{4\Gamma_2 \sin 2\psi}{z_0{}^5}$$

In addition to these two quantities, we have

$$i_x = \int_{-\infty}^{\infty} \Delta g(x,0) \, dx = \frac{2B_0^0}{z_0} + \frac{4\Gamma_2 \cos 2\psi}{3z_0{}^3}$$

and

$$i_y = \int_{-\infty}^{\infty} \Delta g(0,y) \, dy = \frac{2B_0^0}{z_0} + \frac{4(3B_2^0 - B_2^2 - \Gamma_2 \sin^2 \psi)}{3z_0{}^3}$$

By a little rearranging, these formulas give

$$3 \tan 2\psi = - \frac{4z_0{}^2 \left(\frac{\partial \, \Delta g}{\partial y}\right)_{\substack{x=0 \\ y=0}}}{4z_0 \, \Delta g(0,0) + i_x - 3i_y}$$

$$16\Gamma_2 = 3z_0{}^3[4z_0 \, \Delta g(0,0) + i_x - 3i_y] \sec 2\psi \qquad (10\text{-}11)$$

and

$$8B_2^0 = z_0{}^3 \left[i_x + i_y - \frac{4GM}{z_0} + \frac{z_0{}^2}{2} \left(\frac{\partial \, \Delta g}{\partial y}\right)_{\substack{x=0 \\ y=0}} \tan \psi \right]$$

where z_0 is the root of

$$\Delta g(0,0)z_0{}^2 - \tfrac{3}{4}(i_x + i_y)z_0 + 2GM = 0 \qquad (10\text{-}12)$$

Since $B_0^0 = GM$, these formulas constitute a complete solution of the inverse potential problem in three dimensions.

Before we can apply (10-11) and (10-12), however, we must first apply corrections to i_x and i_y for their missing tails. These corrections are

$$i_x \doteq \bar{i}_x + \frac{GMz_0}{X^2}$$

where $\qquad \bar{i}_x = \int_{-X}^{X} \Delta g(x,0)\, dx \qquad \text{for } X \gg z_0$

and a similar expression for i_y. These change (10-12) to

$$\left[\Delta g(0,0) - \frac{3GM(X^2 + Y^2)}{4X^2Y^2}\right] z_0{}^2 - \tfrac{3}{4}(\bar{i}_x + \bar{i}_y)z_0 + 2GM = 0 \quad (10\text{-}13)$$

which will give two roots for z_0. Often one of these can be eliminated by inspection. After z_0 is found i_x and i_y are calculated and substituted into the formulas (10-11). The derivative $(\partial\, \Delta g/\partial y)_{\substack{x=0 \\ y=0}}$ may be found from a finite-difference formula at the point Q.

The formulas (10-13) and (10-11) constitute the solution of the three-dimensional fitting problem, but in order to apply them, we must find the point $Q(0,0,z_0)$. This is the point which lies directly above the center of mass, the coordinates of which are found from (8-27). Again, however, we must be careful to make corrections for missing tails in evaluating these integrals. The correction, calculated on the usual assumption that M is concentrated at its mass center, gives in rectangular coordinates

$$4GM\bar{x} \doteq \left(\tan^{-1}\frac{XY}{\bar{z}R} - \frac{\bar{z}Y}{RX}\right)^{-1} \int_{-X}^{X}\int_{-Y}^{Y} \Delta g(x,y)x\, dx\, dy \quad (10\text{-}14)$$

and a similar expression for $\bar{y}$, or in circular coordinates

$$2\pi GM\,\frac{\bar{x}}{\bar{y}} \doteq \left(1 - \frac{3}{2}\frac{\bar{z}}{R} + \frac{5}{4}\frac{\bar{z}^3}{R} \cdots\right)^{-1} \int_0^R \int_0^{2\pi} \Delta g(r,\vartheta)r^2 \frac{\cos}{\sin}\vartheta\, dr\, d\vartheta$$

In the latter case, the integrals may be computed by making up tables of the average values of $\Delta g(r,\vartheta)\,\frac{\cos}{\sin}\,\vartheta$ taken around circles of radius r and integrating numerically by applying a suitable quadrature formula.

10-9 An Example

Three-dimensional interpretation can be attempted only with data of high quality, but when it is applicable, it can be very effective. An example which will illustrate this is the anomaly shown in Fig. 10-1. The various steps in the interpretation are given below.

1. *Calculation of GM.* This has already been done in Sec. 10-2. The result obtained was $GM = 1.645 \times 10^6$ gu-ft².

2. *Location of* $Q(0,0,z_0)$. By applying Simpson's rule to observations along the picket lines we find that

$$\int_{-X}^{X} \int_{-Y}^{Y} \Delta g(x,y)x \, dx \, dy = 155.3 \times 10^6 \text{ gu-ft}^3$$

and
$$\int_{-X}^{X} \int_{-Y}^{Y} \Delta g(x,y)y \, dx \, dy = -38.9 \times 10^6 \text{ gu-ft}^3$$

Since $X = 1,200$, $Y = 800$, and $\bar{z} = 165$ ft,

$$\tan^{-1} \frac{XY}{\bar{z}R} - \frac{\bar{z}Y}{RX} = 1.326 - 0.076 = 1.250$$

and
$$\tan^{-1} \frac{XY}{\bar{z}R} - \frac{\bar{z}X}{RY} = 1.326 - 0.172 = 1.154$$

Therefore according to (10-14) $\bar{x} = +19$ ft and $\bar{y} = -5$ ft from the origin.

3. *Calculation of* z_0. New x and y axes are now taken through the point Q, the direction of Ox following the long axis of the anomaly pattern. By drawing profiles in these two directions through the point Q, we find that

$$\bar{i}_x = 13,892 \text{ gu-ft} \qquad \text{and} \qquad \bar{i}_y = 6,832 \text{ gu-ft}$$

when $X = 1,200$ ft and $Y = 800$ ft. Since $\Delta g(0,0) = 16.8$ gu and $GM = 1.645 \times 10^6$ gu-ft², it follows from (10-13) that $z_0 = 285$ ft or $z_0 = 826$ ft. According to our earlier estimate, the smaller root is obviously the correct one to use.

4. *Calculation of* ψ. Now that we have z_0, we can apply corrections to $\bar{i}_x$ and $\bar{i}_y$ for missing tails. This gives

$$i_x = 14,218 \text{ gu-ft} \qquad \text{and} \qquad i_y = 7,564 \text{ gu-ft}$$

By using a Lagrange 5-point formula at Q, we find also that

$$(\partial \, \Delta g/\partial y)_{\substack{x=0 \\ y=0}} = -0.0327 \text{ gu/ft}$$

Thus
$$4z_0 \, \Delta g(0,0) + i_x - 3i_y = 10,678 \text{ gu-ft}$$

and $\tan 2\psi = 0.331$. Consequently $\psi = 9°$, corresponding to a dip of 81°N.

5. *Calculation of* B_2^0 *and* B_2^2. We may now substitute for z_0 and ψ directly into the formulas (10-11) to calculate the two quadrupole moments. Thus we get

$$B_2^0 = -4.39 \times 10^9 \text{ gu-ft}^4 \qquad \text{and} \qquad B_2^2 = 6.20 \times 10^{10} \text{ gu-ft}^4$$

The disparity between these two values suggests that the body is narrow in proportion to its length.

6. *Calculation of structure.* To calculate the physical dimensions of the body from the values of its moments, we must introduce a specific shape to represent it. If we assume that the body is an ellipsoid having semiaxes a, b, and c, we find, according to Sec. 8-7, that

$$a^2 - b^2 = \frac{20B_2^2}{3B_0^0} = 25.1 \times 10^4 \text{ ft}^2$$

and

$$2c^2 - a^2 - b^2 = \frac{20B_2^0}{B_0^0} = -5.33 \times 10^4 \text{ ft}^2$$

remembering that $B_0^0 = GM = 1.645 \times 10^6$ gu-ft². If we set $\Delta\rho = 1.9$ g/cm³ (since some assumption about the density must be made at this stage), we get for the third equation

$$abc = \frac{3B_0^0}{4\pi G \Delta\rho} = 10.17 \times 10^6 \text{ ft}^3$$

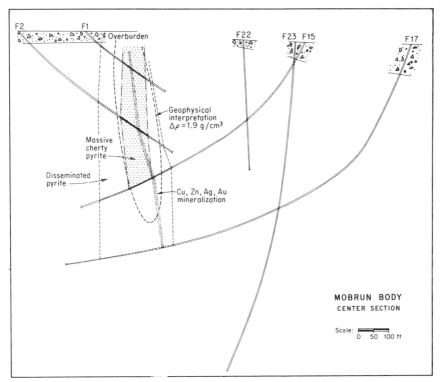

Fig. 10-23   Center section of the Mobrun sulfide body [after Seigel (1)] with geophysical interpretation.

Solving for a, b, and c, we get

$$a = 500 \text{ ft} \qquad b = 62 \text{ ft} \qquad c = 320 \text{ ft}$$

Tentatively, therefore, we interpret the body to be about 1,000 ft in length, slightly more than 100 ft in maximum width, and having a maximum depth of just over 600 ft. It strikes 26°S from the survey base line and dips 81°N. The slight protrusion of the upper edge above the ground surface indicates that the value we have assumed for $\Delta\rho$ may have been a bit too small and that the body lies beneath a very shallow cover. These findings are illustrated in cross section in Fig. 10-23, together with coring sections

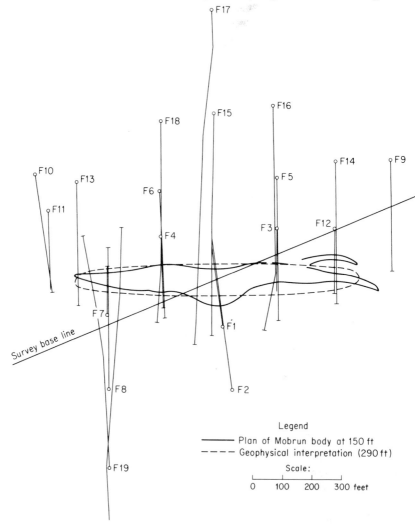

Legend

——— Plan of Mobrun body at 150 ft
– – – Geophysical interpretation (290 ft)

Scale:

0   100   200   300 feet

Fig. 10-24   Plan of Mobrun sulfide at 150-ft depth.   [After Seigel (1).]

taken with the diamond drill, and in plan in Fig. 10-24, together with an outline of the sulfide body at the 150-ft depth. Had a slightly larger value been used for $\Delta\rho$, the corroboration of the geophysical interpretation by drilling would have been almost perfect. There is no evidence in the gravity map of dense material below a depth of 600 ft, and this was confirmed by drilling.

10-10 The Adjustment of Solutions

To verify solutions obtained by three-dimensional methods, it is no longer sufficient to compare profiles, since the values for a, b, c, ψ, and z_0 are calculated from the entire anomaly pattern. Any adjustments to be made in these values should reduce the discrepancies, not on a single line of points, but over a two-dimensional network. By the use of a method proposed by Gassmann (5), the adjustments can be determined by setting up a number of normal equations as follows:

$$\Delta g_{obs}(x_i,y_i) = \Delta g_{calc}(x_i,y_i) + \Delta a \left(\frac{\partial \Delta g}{\partial a}\right)_{x_i,y_i} + \Delta b \left(\frac{\partial \Delta g}{\partial b}\right)_{x_i,y_i} + \cdots \quad (10\text{-}15)$$

where the coefficients $\partial \Delta g/\partial a$, etc., as well as Δg_{calc}, are found from (8-27). Thus if there are N points, there will be N linear equations[1] in the five unknowns Δa, Δb, Δc, $\Delta \psi$, Δz_0. Solving the equations by the method of least squares gives a set of "improvements" to be added to the parameters a, b, c, etc. These improvements ought to be such that they reduce the differences between the observed and calculated values of Δg at the N stations. Thus when the calculations are repeated, the next set of "improvements" should be smaller than the first. By continuing the iterative process until it ceases to converge, a "best" fit to the observational data will eventually be found. As an illustration, the parameter values found by successive iterations using 49 gravity stations from Fig. 10-1 are

$$\rho = 1.90 \text{ g/cm}^3 \qquad z_0 = 290 \text{ ft}$$

a, ft	b, ft	c, ft	ψ, deg
500 (trial)	62 (trial)	320 (trial)	9 (trial)
438	69	291	9
425	72	286	10
425	73	286	10

Thus the final interpretation is only slightly changed from the initial solution shown in Fig. 10-23.

---

[1] It is advisable to "weight" the equations somewhat in favor of larger values of Δg, so that the number of equations actually used in the calculations is $> N$.

Much the same techniques can also be applied to the adjustment of solutions found by the method of characteristic curves.

10-11   The Calculation of the Gravity Effect of Three-dimensional Bodies

Numerical methods are also available for calculating the gravity effects of three-dimensional bodies having an arbitrary form. In the majority of these methods, the body is divided into horizontal laminae of a uniform thickness, and the effect of each layer is calculated in turn. Thus at the point 0 (Fig. 10-25) we may write

$$\Delta g(0) = G\,\Delta\rho \int z\,dz \iint\limits_{S(z)} (r^2 + z^2)^{-3\!/\!2} r\,d\vartheta\,dr$$

$$\doteq G\,\Delta\rho \sum_{j=1}^{m} W_j z_j \iint\limits_{Sj} (r^2 + z_j{}^2)^{-3\!/\!2} r\,d\vartheta\,dr$$

where the W_j's are the weighting coefficients appropriate to the quadrature formula used for the integration in z. (Note that they have the dimension of length.) In most cases uniform intervals in z are used, since this lends itself conveniently to calculations over interfaces represented by depth contours, and the W_j's will then become the Cotesian numbers or the weights used in Simpson's rule. The integrals appearing in this expression are evaluated over the m horizontal surfaces $z_j = $ const or, equivalently, around the m contours which represent the body. If the integrations are to be done by hand, special charts prepared by Goguel (6) will be helpful in reducing

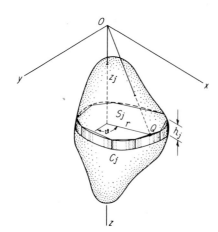

Fig. 10-25   Approximation of a three-dimensional body by polygonal laminae.

the total amount of labor. If machines are available for carrying out the computations, a method proposed by Talwani and Ewing (7) may be used, in which the m laminae are replaced with polygons. For if the contour C_j is replaced with an n-sided polygon whose vertices are (x_k, y_k), we may write

$$W_{jz_j} \iint_{S_j} \frac{d\xi \, d\eta}{(\xi^2 + \eta^2 + z_j^2)^{3/2}} = W_{jz_j} \oint_{C_j} \frac{\xi \, d\eta}{(\eta^2 + z_j^2) \sqrt{\xi^2 + \eta^2 + z_j^2}}$$

$$\doteq W_{jz_j} \sum_{k=1}^{nj} \int_{y_k}^{y_{k+1}} \frac{(a_k\eta + b_k) \, d\eta}{(\eta^2 + z_j^2) \sqrt{(1 + a_k^2)\eta^2 + 2a_k b_k \eta + b_k^2 + z_j^2}}$$

$$= W_j \sum_{k=1}^{nj} \left\{ \tan^{-1} \frac{z_j(b_k y_{k+1} - a_k z_j^2)}{x_{k+1}[(1 + a_k^2)z_j^2 + b_k^2] - (a_k^2 z_j^2 + b_k^2) \sqrt{x_{k+1}^2 + y_{k+1}^2 + z_j^2}} \right.$$

$$\left. - \tan^{-1} \frac{z_j(b_k y_k - a_k z_j^2)}{x_k[(1 + a_k^2)z_j^2 + b_k^2] - (a_k^2 z_j^2 + b_k^2) \sqrt{x_k^2 + y_k^2 + z_j^2}} \right\} \qquad (10\text{-}16)$$

where

$$a_k = \frac{x_{k+1} - x_k}{y_{k+1} - y_k} \qquad b_k = \frac{x_k y_{k+1} - x_{k+1} y_k}{y_{k+1} - y_k}$$

This expression is easily programmed. The method of calculation consists in replacing the m depth contours of V with polygons and assigning pairs of coordinates (x_k, y_k) to each of their vertices. Shifting the origin to a new location merely amounts to adding fixed increments to x_k and y_k.

While these methods can in principle be applied to volumes having any shape whatever, they are most easily applied to domelike masses having flat bottoms and convex sides.

10-12   On the Use of Moments in Two Dimensions

The direct interpretation of gravity anomalies by the use of moments is also applicable to two-dimensional bodies. Although this approach is less compact than the method of characteristic curves, it is also more general, since it estimates the general features of the body without making any prior assumptions about its shape. The basis of the method is to estimate numerical values for the two moments b_0 and b_2 of the cross section of the body by fitting the formula (8-32) to the gravity profile.

In order to extract b_0 and b_2 from (8-32), we note the following identities at the point $Q(0, z_0)$:

$$\Delta g(0) = \frac{b_0}{z_0} - \frac{b_2 \cos 2d}{z_0^3}$$

and

$$\left(\frac{\partial \, \Delta g}{\partial x} \right)_{x=0} = \frac{3b_2 \sin 2d}{z_0^4}$$

Recalling also that $b_0 = 2Gm = \pi^{-1} \int_{-\infty}^{\infty} \Delta g(x)\, dx$ we can, by a little rearrangement, show that

$$\tan 2d = \frac{-z_0^2 (\partial \Delta g / \partial x)_{x=0}}{3[z_0\, \Delta g(0) - 2Gm]} \tag{10-17}$$

and
$$b_2 = z_0^2 [z_0\, \Delta g(0) - 2Gm]\, \sec 2d$$

To find the point $Q(0, z_0)$ we make use of the integrals (8-30) and (8-31). However, a "missing flank" correction usually needs to be made. If the limits of the data are at a distance X on either side of the approximate center of the profile and if the mass center lies at $(\bar{x}, \bar{z})$, then as we have seen in Sec. 10-2,

$$2b_0 = 4Gm \doteq \left(\tan^{-1} \frac{X}{\bar{z}} \right)^{-1} \int_{-X}^{X} \Delta g(x)\, dx$$

provided that $X \gg |\bar{z}|$. To exactly the same approximation in $\bar{z}/X$, we get from (8-31)

$$4Gm\bar{x} \doteq \left(\tan^{-1} \frac{X}{\bar{z}} - \frac{\bar{z}}{X} \right)^{-1} \int_{-X}^{X} \Delta g(x)x\, dx$$

To apply these corrections, we estimate $\bar{z}$ from the "half-width" of the profile, i.e., $\bar{z} = 0.5 w_{\frac{1}{2}}$.

Returning now to (10-17), all that remains to be found is z_0, the depth to the center of the body. According to the formula (8-32), we observe that

$$\Delta g(x) - \Delta g(-x) = -2b_2 \sin 2d(x^3 - 3xz_0^2)(x^2 + z_0^2)^{-3}$$

This function has a zero at $x = \sqrt{3}\, z_0$. Thus by "folding" the profile on itself at $Q(0, z_0)$ and taking differences, z_0 may be determined. This is provided, of course, that $d \neq 90°$, for otherwise the profile would be symmetrical.

Having now found b_0 and b_2, the linear dimensions a and b may be determined according to the formulas given in Sec. 8-12 for any of a number of different two-dimensional models.

10-13   Remarks

The methods we have discussed for the quantitative interpretation of gravity anomalies fall generally into three groups: (1) Downward continuation, for the calculation of topography on a single density interface or on a number of conformable interfaces. (2) Characteristic curves, for fitting specific models to the observational data, with provision for making adjustments to the solutions by numerical procedures. (3) Direct methods of interpretation of isolated anomalies, by the calculation of quadrupole moments. This list is not by any means exhaustive, for it fails to include a variety of proposals that have appeared in the literature from time to time for solving special problems such as the use of special density functions, etc. The range of these problems is very wide. However, the principles used in attacking the inverse potential problem under most of the circumstances that are met with in practice are believed to be contained within the methods we have discussed. To add anything further would take us too deeply into the area of special applications.

References

1. H. O. Seigel, Discovery of Mobrun Copper Ltd. Sulphide Deposit, Noranda Mining District, Quebec, pp. 237–245 in "Methods and Case Histories in Mining Geophysics," Sixth Commonwealth Mining and Metall. Cong. Proc., Mercury, Montreal, 1957.

2. W. G. Gish and R. M. Carr, Garber Field, Garber County, Oklahoma, pp. 176–191 in "Structure of Typical American Oil Fields," Am. Assoc. Petroleum Geologists, Tulsa, 1929.

3. M. Talwani, J. L. Worzel, and M. Landisman, Rapid Computations for Two-dimensional Bodies with Application to the Mendocino Submarine Fracture Zone, *Jour. Geophys. Research*, vol. 64, pp. 49–59, 1959.

4. M. H. P. Bott, The Use of Rapid Digital Computing Methods for Direct Gravity Interpretation of Sedimentary Basins, *Geophys. Jour.*, vol. 3, pp. 63–67, 1960.

5. F. Gassmann, Auswertung geophysikalischer Sondierungen mit Hilfe von Potentialfeldern, *Schweizerische Mineralog. u. Petrog. Mitt.*, vol. 28, pp. 335–352, 1948.

6. J. Goguel, Calcul de l'attraction d'un polygone horizontal de densité uniforme, *Geophys. Prosp.*, vol. 9, pp. 116–127, 1961.

7. M. Talwani and M. Ewing, Rapid Computation of Gravitational Attraction of Three-dimensional Bodies of Arbitrary Shape, *Geophysics*, vol. 25, pp. 203–225, 1960.

chapter 11   The Quantitative
Interpretation of Magnetic Anomalies

The immediate purpose of magnetic surveys is to detect rocks or minerals possessing unusual magnetic properties, which reveal themselves by causing disturbances or *anomalies* in the intensity of the earth's magnetic field. Maps of magnetic anomalies are often used in a qualitative way to aid in making regional geological interpretations. Very often the large amount of complex detail which they contain is a bar to the use of quantitative methods of evaluation. But sometimes individual magnetic anomalies are found which stand out so clearly that they can easily be separated from neighboring effects, and which are so simple in appearance that they seem to be due to a single, magnetized body. It is in circumstances such as these that quantitative methods of interpretation can be used most effectively to deduce the shape and size of the magnetic formation. Chapter 11 will be very largely taken up with this question. To give a geological identity to these volumes requires some understanding of the roles played by different factors in determining magnetic properties, and we shall consider some of these factors in Chap. 12.

11-1   Corrections Applied to Magnetic Data

Because disturbances in the geomagnetic field due to the presence of mag-netized rocks are often fairly strong, many magnetic field maps, especially aeromagnetic maps which span large areas, appear to be dominated by residual anomalies.   On a local scale the horizontal gradient of H_0 (which ranges between 0 and about 10 gammas/km from the magnetic equator to the pole) is often scarcely discernible.   Large-scale regional anomalies are seldom found either, probably because strong differences in bulk magnetic properties are unlikely to be sustained throughout very large volumes of rock; moreover, the higher temperatures that prevail at greater depths within the earth's crust would forbid such large contrasts from surviving for very long in any case.   Consequently, the more prominent features on mag-netic maps tend to be rather local in extent, originating from within the top few kilometers of crustal rock, where the contrasts in magnetic properties are probably greatest.   They also tend to overlap on their flanks, making the map as a whole exceedingly complex in its details.   The disentanglement of these features is of the first importance in making magnetic interpretations, and it requires a good deal of careful study.   Analyses based upon intuitive ideas of smoothness, which are commonly employed in gravity interpreta-tions, often fail in magnetics for several reasons.   As a general rule, the work of separating anomalies, to be done effectively, requires some famil-iarity with the shapes of the magnetic fields of theoretical models.

Apart from the vexatious matter of separating anomalies, the reductions that must be applied to magnetic data are rudimentary.   The vertical gradient of $\mathbf{H}_0$ is only about -0.03 gamma/m at the magnetic poles and about half of this amount at the magnetic equator, so that the effect of elevation changes is generally negligible.   Sometimes in very mountainous regions, a correction is made for changes in $\mathbf{H}_0$, for which a sufficiently accurate expression is $-0.047\ H_0$ gamma/m, where H_0 is the local value of the geomagnetic-field intensity in oersteds.   The correction is positive north of the magnetic equator and negative south of it.   It can be performed with sufficient accuracy from barometric heights.

Terrain effects, due to the magnetization of surrounding outcropping rocks, may cause a significant change in the field intensity, especially in regions of igneous outcrop.   But information on the magnetic properties of these rocks is almost never available in the detail that would be required to calculate this effect.   While it is perfectly possible in principle to draw a set of templates for making such corrections, their actual value—except in a very few special situations—is somewhat doubtful.

11-2   Reduction to a Horizontal Plane

While the corrections of magnetic data for changes in $\mathbf{H}_0$ due to elevation are negligible, local magnetic fields can be appreciably distorted by irregularities in the shape of the land surface. This effect is generally more apparent in magnetics than it is in gravity, since a dipolar field changes more rapidly with position than a polar field. Consequently, the need for making a suitable correction as a preliminary to detailed interpretation is correspondingly more pressing in magnetics. Since the problem arises only in connection with magnetic surveys made on the ground, in which it is generally the vertical component of the total-field intensity that is measured, we shall consider only correction of this component.

Let us use the vector **H** to denote the total magnetic field intensity and let X and Z be, respectively, the horizontal and vertical components of **H**. To reduce measurements of Z from the surface $z = h(x,y)$ to the horizontal plane $z = 0$, we write, approximately,

$$Z(x,y,0) \doteq Z(x,y,h) - h\left(\frac{\partial Z}{\partial z}\right)_{z=h}$$

Generally, the derivative must be calculated from the map of $Z(x,y,h)$. Since $\nabla^2 Z = 0$ for $z \leq h$, it follows directly from the results obtained in Sec. 8-5 that a formula exactly analogous to (8-18) may be used to calculate $\partial Z/\partial z$. The numerical processes used to calculate this formula when the data are given at a regular network of stations and also when they are given at irregularly distributed points have already been discussed in Chap. 9.

11-3 Magnetic Anomalies

The scalar magnetic potential at an external point due to a volume V of rock which is magnetized with a dipole moment per unit volume **M** is

$$A(\mathbf{r}) = - \int_V \mathbf{M} \cdot \nabla \frac{1}{|\mathbf{r} - \mathbf{r}_0|} \, d^3 r_0$$

If **M** is constant and if its direction is denoted by the symbol κ, so that $\mathbf{M} \cdot \nabla = M \, \partial/\partial \kappa$, we may write

$$A(\mathbf{r}) = - M \frac{\partial}{\partial \kappa} \int_V \frac{d^3 r_0}{|\mathbf{r} - \mathbf{r}_0|} \tag{11-1}$$

The directional derivative $\partial/\partial \kappa$ is in the direction of **M**, and in a coordinate system in which the positive direction of Oz is vertically downward, it has the components

$$\frac{\partial}{\partial \kappa} = \cos \tau \sin \iota \frac{\partial}{\partial x} + \cos \tau \cos \iota \frac{\partial}{\partial y} + \sin \tau \frac{\partial}{\partial z} \tag{11-2}$$

Here τ is the dip of the magnetization **M** measured clockwise from the horizontal when the direction of **M** is to the right (Fig. 11-1). ι is the angle measured counterclockwise from the y axis to the projection of **M** onto the

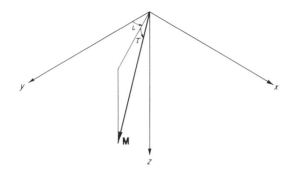

Fig. 11-1   Definitions of ι and τ.

horizontal plane, looking downward.   If the body is inductively magnetized in the earth's magnetic field and has a negligible permanent magnetic moment in any other direction, τ will be equal to i, the magnetic inclination, and ι will be equal to λ, the magnetic declination of Oy.   If the body has no permanent magnetic moment, $M = kH_0$ where k is the volume magnetic susceptibility.[1]

In the region outside V, the intensity of the magnetic field is given by the negative gradient of $A(\mathbf{r})$.   Since this is an *anomalous* magnetic field which is superimposed upon the main geomagnetic field, we write

$$\mathbf{\Delta H(r)} = -\nabla A(\mathbf{r})$$

Thus the total-field intensity in the vicinity of the magnetized rocks will be

$$\mathbf{H} = \mathbf{H_0} + \mathbf{\Delta H}$$

plus any disturbances due to other nearby bodies.   *Total-field* magnetometers measure the magnitude of $\mathbf{H}$ without sensing its direction, and unless $|\Delta H| \ll H_0$, this direction changes continuously within the vicinity of the magnetized body.   Knowing neither the magnitude nor the direction of $\mathbf{\Delta H}$, it is impossible to determine the direction of $\mathbf{H}$ and therefore the direction of the measurement; and lacking this information, it is not practically possible to compare anomalies with the fields of theoretical models.   As a result, interpretations of total-field measurements are not really feasible unless $|\Delta H| \ll H_0$, in which case $\mathbf{H}$ and $\mathbf{H_0}$ are presumed to lie very nearly in the same direction.   In practice this is not an unduly severe limitation on airborne surveys, where anomalies greater than about $0.2\ H_0$ are uncommon, but it can create a real difficulty in the interpretation of measurements made on the ground over shallow, magnetized bodies.

When $|\Delta H| \ll H_0$, the component of $\mathbf{\Delta H}$ measured by the total-field magnetometer is that which lies in the direction of the main geomagnetic

---

[1] This argument disregards the effect of demagnetization, which we shall discuss in Sec. 11-5 and which may be significant if $k > 0.01$ emu.

field.   If this component is ΔT, it is clear that

$$\Delta T(\mathbf{r}) = -\frac{\partial}{\partial a} A(\mathbf{r})$$

where a signifies the direction of $\mathbf{H}_0$.   In other words,

$$\frac{\partial}{\partial a} = \cos i \sin \lambda \frac{\partial}{\partial x} + \cos i \cos \lambda \frac{\partial}{\partial y} + \sin i \frac{\partial}{\partial z} \tag{11-3}$$

where i is the magnetic inclination and λ is the declination of the y axis measured counterclockwise to magnetic north.   The range of i over the entire world is from -90 to $+90°$; and of λ, from 0 to 360°.   If the body is magnetized by induction uniformly throughout its volume and has no permanent magnetic moment and negligible demagnetization, then

$$\Delta T(\mathbf{r}) = kH_0 \frac{\partial^2}{\partial a^2} \int_V \frac{d^3r_0}{|\mathbf{r} - \mathbf{r}_0|} \tag{11-4}$$

If, on the other hand, the permanent magnetic moment $\mathbf{M}$ dominates the induced magnetization, or if k becomes large,

$$\Delta T(\mathbf{r}) = M \frac{\partial^2}{\partial a \, \partial \kappa} \int_V \frac{d^3r_0}{|\mathbf{r} - \mathbf{r}_0|} \tag{11-5}$$

If the body is uniformly magnetized by induction, has negligible permanent magnetic moment, and is two-dimensional, then

$$\Delta T(\mathbf{r}) = -2kH_0 \frac{\partial^2}{\partial a^2} \int_S \ln |\mathbf{r} - \mathbf{r}_0| \, d^2r_0 \tag{11-6}$$

provided that k is not large.

It is possible, of course, that the permanent and the induced magnetic moments may be comparable in magnitude but different in direction.   In that event the *total-field anomaly* ΔT will be given by (11-5), where κ is now the direction, and M is the magnitude, of the vector $\mathbf{M} + k\mathbf{H}_0$.   However, the difficulties that stand in the way of making quantitative interpretations based upon (11-5) when the direction of κ is unknown are prohibitive, and the problem is rarely attempted.

Ground magnetic surveys in which only the vertical (or the horizontal) component of $\mathbf{H}$ is measured present a much easier problem of interpretation. In this case we may write

$$\Delta Z(\mathbf{r}) = -\frac{\partial}{\partial z} A(\mathbf{r})$$

and

$$Z = Z_0 + \Delta Z$$

ΔZ being the vertical component of $\Delta \mathbf{H}$, and Z_0 of $\mathbf{H}_0$.   In this case there is no ambiguity concerning the direction of measurement, and so for an inductively magnetized body having no permanent moment and a moderate susceptibility

$$\Delta Z(\mathbf{r}) = kH_0 \frac{\partial^2}{\partial z\, \partial a} \int_V \frac{d^3r_0}{|\mathbf{r} - \mathbf{r}_0|} \tag{11-7}$$

while for a body whose permanent moment is important or whose susceptibility is large,

$$\Delta Z(\mathbf{r}) = M \frac{\partial^2}{\partial z\, \partial \kappa} \int_V \frac{d^3r_0}{|\mathbf{r} - \mathbf{r}_0|} \tag{11-8}$$

For two-dimensional bodies, the formulas for the *vertical-field anomaly* are

$$\Delta Z(\mathbf{r}) = -2kH_0 \frac{\partial^2}{\partial z\, \partial a} \int_S \ln |\mathbf{r} - \mathbf{r}_0|\, d^2r_0 \tag{11-9}$$

for bodies having no permanent moment and moderate susceptibility, and

$$\Delta Z(\mathbf{r}) = -2M \frac{\partial^2}{\partial z\, \partial \kappa} \int_S \ln |\mathbf{r} - \mathbf{r}_0|\, d^2r_0 \tag{11-10}$$

for bodies whose permanent moment or susceptibility is large.

11-4 The Continuation of Magnetic Fields

There is a place for upward and for downward continuation in the interpretation of magnetic observations. However, the uses of continuation maps deviate in some particulars from their uses in gravity interpretations.

1. Upward continuation is sometimes used in order to simplify the appearance of magnetic maps by suppressing local features. The proliferation of local magnetic anomalies often obscures the regional picture with an overabundance of detail. Upward continuation will smooth out these disturbances without impairing the main regional features. Since $\nabla^2 A = 0$ in free space, it follows from (8-16) that

$$\Delta Z(x,y,z) = \frac{|z|}{2\pi} \int_{-\infty}^{\infty} \int_{-\infty}^{\infty} \frac{\Delta Z(\xi,\eta,0)}{R^3}\, d\xi\, d\eta \qquad z < 0 \tag{11-11}$$

where $R = \sqrt{(x - \xi)^2 + (y - \eta)^2 + z^2}$. Only the vertical magnetic field is used in continuation (see Sec. 8-4).

2. Downward continuation is used for a purpose exactly opposite to upward continuation, namely, to increase the resolution of weak anomalies. Since (11-11) is exactly similar in form to (8-12), the theory for the downward continuation of ΔZ is the same as for Δg, and all tables and charts derived for the gravity problem are applicable to the continuation of vertical magnetic fields.

It has been proposed by Peters (1) that if we can assume that the magnetic material is vertically polarized, downward continuation can also be

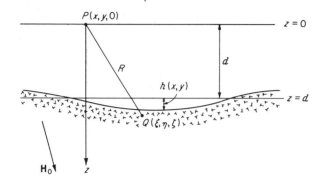

Fig. 11-2 Downward continuation of magnetic fields (Peters' method).

used to determine the shapes of interfaces between uniformly but differently magnetized media. Before we elaborate further upon this proposal, we should remark that in the light of our knowledge of rock magnetism the assumption of perfect uniformity for magnetic properties throughout large areas of the ground is highly tenuous, and the use of downward continuation for the purpose of drawing structural maps should be used with strong reservations.

The basis of Peters' suggestion is as follows: In Fig. 11-2, let us suppose that the crystalline rocks are uniformly magnetized and that the depth to the upper surface of these rocks is given by a mean value d plus a smaller term $h(x,y)$, which represents the topographic relief on the igneous surface. Let k be the mean volume magnetic susceptibility contrast between the crystalline and the noncrystalline rocks, and let $\mathbf{H}_0$ be the geomagnetic field intensity. The magnetic anomaly that will be produced at $z = 0$ because of changes in the depth of the magnetized rocks will be given by the potential

$$A_0(x,y) = kH_0\left[\frac{\partial}{\partial a}\int_{-\infty}^{\infty}\int_{-\infty}^{\infty}d\xi\,d\eta\right.$$

$$\left.\int_{d}^{d+h(\xi,\eta)}\frac{d\zeta}{\sqrt{(x-\xi)^2+(y-\eta)^2+(z-\zeta)^2}}\right]_{z=0}$$

$$\cong kH_0\left[\frac{\partial}{\partial a}\int_{-\infty}^{\infty}\int_{-\infty}^{\infty}\frac{h(\xi,\eta)\,d\xi\,d\eta}{\sqrt{(x-\xi)^2+(y-\eta)^2+(z-d)^2}}\right]_{z=0}$$

$$\text{if } |h| \ll d$$

The vertical-field anomaly will therefore be

$$\Delta Z_0(x,y) = -\left[\frac{\partial}{\partial z}A(x,y,z)\right]_{z=0}$$

$$= kH_0\left[\frac{\partial}{\partial a}(z-d)\right.$$

$$\left.\int_{-\infty}^{\infty}\int_{-\infty}^{\infty}\frac{h(\xi,\eta)\,d\xi\,d\eta}{[(x-\xi)^2+(y-\eta)^2+(z-d)^2]^{3/2}}\right]_{z=0} \quad (11\text{-}12)$$

Now the vertical-field anomaly on $z = 0$ corresponding to a magnetic potential $A_d(x,y)$ at $z = d$ is, according to Sec. 8-4,

$$\Delta Z_0(x,y) = \frac{1}{2\pi} \left[\frac{\partial}{\partial z} (z - d) \int_{-\infty}^{\infty} \int_{-\infty}^{\infty} \frac{A_d(\xi,\eta) \, d\xi \, d\eta}{[(x - \xi)^2 + (y - \eta)^2 + (z - d)^2]^{3/2}} \right]_{z=0}$$
(11-13)

Equation (11-13) can be equated to (11-12) only if $\partial/\partial a = \partial/\partial z$. Thus for vertical magnetization we may set

$$A_d(x,y) = 2\pi k H_0 h(x,y)$$
(11-14)

at $z = d$. This is the *equivalent stratum* for magnetic fields.

To calculate $h(x,y)$ at $z = d$, it is necessary to determine the anomalous magnetic potential $A_d(x,y)$ at that level from measurements taken at $z = 0$. If we take the Fourier transform with respect to x and y of both sides of (11-13), we have

$$\begin{aligned}
F_0(p,q) &= \int_{-\infty}^{\infty} \int_{-\infty}^{\infty} \Delta Z_0(x,y) e^{i(px+qy)} \, dx \, dy \\
&= \frac{1}{2\pi} \left[\frac{\partial}{\partial z} \exp \left[(z - d) \sqrt{p^2 + q^2} \right] \int_{-\infty}^{\infty} \int_{-\infty}^{\infty} A_d(\xi,\eta) e^{i(p\xi+q\eta)} \, d\xi \, d\eta \right]_{z=0} \\
&= \sqrt{p^2 + q^2} \exp \left(-d\sqrt{p^2 + q^2} \right) B_d(p,q)
\end{aligned}$$

where $F_0(p,q)$ is the Fourier transform of ΔZ on $z = 0$ and $B_d(p,q)$ is the Fourier transform of A on $z = d$. Transposing this equation gives

$$B_d(p,q) = (p^2 + q^2)^{-1/2} \exp \left(d\sqrt{p^2 + q^2} \right) F_0(p,q)$$
(11-15)

and taking the Fourier inversion gives $A_d(x,y)$.

The Fourier inversion of (11-15) presents the same difficulties with convergence as the analogous problem in the gravity method, only in a more intensified form. Because of their generally "noisy" character (which is due to their higher sensitivity to near-surface features), magnetic data *must* be smoothed before any downward continuation can be done. Thus values of ΔZ may be taken at a regular grid of $m \times n$ points on $z = 0$, and a formula analogous to (9-13) may be used to carry the field downward. Alternatively, we may use the analytical smoothing method of Bullard and Cooper, and by following the procedure laid down in Sec. 9-9, we obtain

$$2\pi k H_0 h(0) = \int_0^{\infty} \Delta Z_0(r) N(r;\gamma) r \, dr$$

where
$$\Delta Z_0(r) = (2\pi)^{-1} \int_0^{2\pi} \Delta Z_0(r,\vartheta) \, d\vartheta$$

and where
$$N(r;\gamma) = \int_0^{\infty} e^{du - \gamma u^2} J_0(ur) \, du$$

A quadrature formula similar to (9-11) may be constructed from this expression, i.e.,

$$h(0) \doteq \sum_{j=1}^{m} W_j^* \Delta Z_0(r_j)$$

for which the weighting coefficients W_j^* can be calculated in the same way as those used in the gravity problem.

Because of its inherent limitation to vertical-field anomalies and to vertical polarizations, Peters' method applies in principle only to the interpretation of ground magnetic surveys made in the higher magnetic latitudes. Since most magnetic surveys are made from aircraft using a total-field instrument, however, there is a much greater practical need for continuation methods which can be used with total-field measurements. To continue total magnetic fields either upward or downward, it is necessary first to extract the vertical component ΔZ from ΔT and then to calculate the continuation of ΔZ. Both of these operations can be combined into a single calculation.

The problem of calculating the vertical field component ΔZ from a map of ΔT was first solved by Hughes and Pondrom (2). The details of the method will depend upon whether we wish to find ΔZ itself at a lower level or whether we wish to find A for substitution into (11-14). To find ΔZ, we begin by writing the integral

$$A(x,y,z) = \frac{1}{2\pi} \int_{-\infty}^{\infty} \int_{-\infty}^{\infty} \frac{\Delta Z_d(\xi,\eta)}{R} \, d\xi \, d\eta \qquad (11\text{-}16)$$

whereas to find A, we start with

$$A(x,y,z) = \frac{d-z}{2\pi} \int_{-\infty}^{\infty} \int_{-\infty}^{\infty} \frac{A_d(\xi,\eta)}{R^3} \, d\xi \, d\eta \qquad (11\text{-}17)$$

where $R = \sqrt{(x-\xi)^2 + (y-\eta)^2 + (z-d)^2}$ and where $z < d$

To find ΔZ on $z = d$ from measurements of ΔT on $z = 0$, we put

$$\Delta T_0(x,y) = \left(-\cos i \frac{\partial A}{\partial x} + \sin i \frac{\partial A}{\partial z} \right)_{z=0}$$

when Ox is directed toward magnetic north. Substituting (11-16) for A, we get

$$\Delta T_0(x,y) = \frac{1}{2\pi} \int_{-\infty}^{\infty} \int_{-\infty}^{\infty} \frac{(x-\xi) \cos i - d \sin i}{R^3} \Delta Z_d(\xi,\eta) \, d\xi \, d\eta$$

and taking Fourier transforms of both sides gives

$$W_0(p,q) = \left(\frac{p \cos i}{\sqrt{p^2+q^2}} - \sin i \right) \exp\left(-d\sqrt{p^2+q^2} \right) F_d(p,q)$$

where $W_0(p,q)$ is the Fourier transform of $\Delta T_0(x,y)$. Now if we invert this equation and apply some smoothing, we obtain

$$F_d(p,q) = \left(\frac{p \cos i}{\sqrt{p^2 + q^2}} - \sin i \right)^{-1} [\exp -\gamma(p^2 + q^2) + d\sqrt{p^2 + q^2}] W_0(p,q)$$

and if we change to polar coordinates and integrate we get, finally,

$$\Delta Z_d(0) = \frac{1}{2\pi} \int_0^\infty \int_0^{2\pi} \Delta T_0(r,\vartheta) S(r,\vartheta;\gamma) r \, d\vartheta \, dr$$

where
$$S(r,\vartheta;\gamma) = \int_0^\infty e^{-\gamma u^2 + du} I(ur,\vartheta) u \, du$$

and where
$$I(ur,\vartheta) = \frac{1}{2\pi} \int_0^{2\pi} \frac{e^{iur\cos(\vartheta-\varphi)}}{\cos i \cos \varphi - \sin i} d\varphi$$

Unless $d = 0$, the integrals cannot be evaluated analytically. The solution given by Hughes and Pondrom for $d = 0$ is

$$\Delta Z_0(0) = \frac{\Delta T_0(0)}{\sin i} + \frac{\cos i}{2\pi} \int_0^\infty \int_0^{2\pi} \Delta T_0(r,\vartheta) \left[\frac{\cos \vartheta - \cos i}{(1 - \cos i \cos \vartheta)^2} \right] \frac{1}{r} d\vartheta \, dr$$

The above paragraph gives some indication of the difficulties involved in trying to convert a map of ΔT anomalies into a map of ΔZ anomalies. To go from ΔT to A from (11-17) is even more difficult, although not altogether impossible. However, it would be appropriate at this point to remind ourselves that the inference of structural relief from magnetic anomalies is based upon the premise that the magnetic susceptibility of the crystalline rocks is uniform and that remanent magnetization can be disregarded, and it is exceedingly difficult to visualize this situation ever arising in practice. To overcome this difficulty, Peters suggested using only those anomalies which *might* be due to buried topography, the criterion for selection being that they are apt to be very much smaller in amplitude than the effects due to changes in the magnetic properties of the crystalline rocks. Thus it is usually necessary to make an interpretative study of aeromagnetic maps with a view to selecting low-amplitude anomalies before applying Peters' method.

Baranov (3) has gone even farther and has suggested using the Poisson relation (8-4) to calculate what he has termed a "pseudo-gravity" map from aeromagnetic data. The basis of the idea is that by defining a "pseudo-density" contrast $\Delta\rho = kH_0/G$, we can write, according to (8-11),

$$A_0(x,y) = \left(\frac{\partial U}{\partial a} \right)_{z=0} = \frac{1}{2\pi} \int_{-\infty}^\infty \int_{-\infty}^\infty \Delta g_d(\xi,\eta) \left(\frac{\partial}{\partial a} \frac{1}{R} \right)_{z=0} d\xi \, d\eta$$

Thus $\Delta T_0(x,y) = \left(-\frac{\partial A}{\partial a} \right)_{z=0} = -\frac{1}{2\pi} \int_{-\infty}^\infty \int_{-\infty}^\infty \Delta g_d(\xi,\eta) \left(\frac{\partial^2}{\partial a^2} \frac{1}{R} \right)_{z=0} d\xi \, d\eta$

The method of Fourier transforms, applied in the usual way, will give the "pseudo-gravity" field Δg_d at depth d, which corresponds to known values of ΔT at $z = 0$ if no magnetic materials are located between the two levels. The purpose in making this calculation is to position the peaks of the anomaly patterns more directly above the bodies that cause them. One might expect that this will simplify the appearance of aeromagnetic maps somewhat, but the advantage of this as far as interpretation is concerned is doubtful. It is very difficult in any case to carry out the numerical processes in the lower magnetic latitudes, and the method fails (as all other continuation methods do) if the magnetization does not lie very closely in the direction of $\mathbf{H}_0$.

11-5 Demagnetization

Before we proceed with a study of the magnetic fields of theoretical models, we should consider the problem of demagnetization. This has for a long time been a vexed question in the literature of geophysics, and it relates chiefly to the fact that, on account of the magnetization of permeable bodies, the magnetic field inside the body is not the same as the ambient field. Thus it is not very fair to equate the induced magnetization $\mathbf{M}$ to the product of volume susceptibility and external magnetic field intensity.

Suppose that a body whose volume magnetic susceptibility is k is placed in a magnetic field whose intensity is $\mathbf{H}_0$. Inside the body, there will be a magnetic field $\mathbf{H}_2$, which we can represent by a scalar potential A_2. Now according to (8-5a),

$$\nabla^2 A_2 = -\nabla \cdot \mathbf{H}_2 = 4\pi \nabla \cdot \mathbf{M}$$

i.e.,

$$\nabla \cdot (\mathbf{H}_2 + 4\pi \mathbf{M}) = 0$$

But $\mathbf{M} = k\mathbf{H}_2$, and accordingly we may write

$$\frac{1}{\mu_0} \nabla \cdot (\mu \mathbf{H}_2) = 0 \tag{11-18}$$

where $\mu/\mu_0 = 1 + 4\pi k$ is called the *magnetic permeability*.[1]

Outside the body, in addition to $\mathbf{H}_0$, there will be a secondary magnetic field due to $\mathbf{M}$, which we may call $\mathbf{H}^{(S)}$ and which may be represented by a scalar potential $A^{(S)}$. Thus the total potential outside the body is

$$A_1 = A_0 + A^{(S)}$$

and the external magnetic field intensity will be

$$\mathbf{H}_1 = -\nabla A_1$$

---

[1] The term μ_0 appears in this definition in order to keep the permeability (as defined) a dimensionless number in whatever system of units is employed. μ_0 is generally referred to as the *magnetic permeability of free space*.

On both sides of the boundary, therefore, $\nabla \times \mathbf{H} = 0$. Now, according to Stokes' theorem, we may write

$$\oint_C \mathbf{H} \cdot d\mathbf{l} = \int_S \mathbf{n} \cdot (\nabla \times \mathbf{H})\, dS = 0$$

when the contour C encloses the area S. Thus if we choose for C the thin loop shown in Fig. 11-3 that includes an arbitrary segment of the boundary, we can, by making the loop as thin as we please, ignore the contribution to the line integral from the end portions bc and da and write

$$\int_a^b (H_1)_{\parallel}\, dl + \int_c^d (H_2)_{\parallel}\, dl = 0$$

Since this is true over any length of path, it must be true that

$$(H_1)_{\parallel} = (H_2)_{\parallel} \tag{11-19}$$

or in other words, that the tangential component of $\mathbf{H}$ is continuous at the boundary.

Outside the boundary, moreover,

$$\nabla^2 A_1 = -\nabla \cdot \mathbf{H}_1 = 0$$

while (11-18) holds inside. The two equations are equivalent to the single statement that

$$\nabla \cdot (\mu\mathbf{H}) = 0$$

everywhere. According to Gauss' theorem,

$$\int_S \mathbf{n} \cdot (\mu\mathbf{H})\, dS = \int_V \nabla \cdot (\mu\mathbf{H})\, dV = 0$$

where the surface S encloses the volume V. If V is the thin wafer shown in Fig. 11-4 which contains a patch of the boundary, we may disregard the contribution to the surface integral from the curved edge by making the wafer as thin as we please. Therefore in the limit the contributions to the

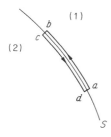

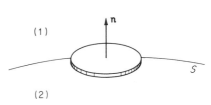

Fig. 11-3 Continuity of magnetic field intensity across an interface.

Fig. 11-4 Continuity of the flux of magnetic intensity across an interface.

surface integral from the flat faces must balance, and this can happen for an arbitrary S only if

$$\mu_0(H_1)_\perp = \mu(H_2)_\perp \qquad (11\text{-}20)$$

or in other words, if the product of the permeability and the normal component of **H** is continuous at the boundary.

The boundary-value problem consists in finding expressions for A_2 and $A^{(S)}$ which satisfy the appropriate field equations as well as the boundary conditions (11-19) and (11-20). If the body is a homogeneous sphere and if **H**$_0$ is uniform, the solution can be found in many textbooks on electricity and magnetism (see, for example, Stratton (4), chap. 4). It turns out that

$$\mathbf{H}_2 = \mathbf{H}_0 - \frac{4\pi}{3}\,\mathbf{M}$$

$$= \frac{\mathbf{H}_0}{1 + (4\pi/3)k}$$

so that the *apparent magnetic susceptibility* of the sphere is $k_a = k/(1 + 4\pi k/3)$. This is the value which, multiplied by the *external* field intensity H_0, gives the magnetization M. The additional term $4\pi k/3$ in the denominator of this expression represents the effect of demagnetization—in other words, of the reduction in the magnetic field intensity inside the sphere due to magnetization—and the coefficient $4\pi/3$ is accordingly called the *demagnetization factor* for the sphere. This number varies with different shapes, ranging from zero for needlelike bodies magnetized along their axes to 4π for flat plates magnetized transversely. Thus demagnetization has the effect of reducing the magnetic moment of the body in the ratio $1 + Nk$, where $0 \le N \le 4\pi$. Since the initial susceptibilities (i.e., values appropriate to small ambient fields) of rock-forming minerals seldom exceed 0.01 emu, however, the effect is usually not noticeable. Only in magnetite bodies, where k may be as large as 0.5 emu, might demagnetization become significant.

A point worth making in this connection is that for thin, sheetlike bodies, the demagnetization factor is greater for transverse than for parallel magnetization. Accordingly, the apparent magnetic susceptibility is always less in the direction transverse to the plate than parallel to it. Thus sheetlike bodies have a natural magnetic aeolotropy, the tendency of which is to deflect the magnetization **M** out of the direction of **H**$_0$ and into the plane of the body. The effect of this aeolotropy begins to appear if k exceeds a value of about 0.02 emu, in which case the direction of **M** is controlled by the dip of the body as well as by the direction of **H**$_0$. When this occurs, a special correction for the demagnetizing effect may be necessary.

According to boundary-value theory, the magnetic field inside a susceptible body can be uniform only if the body is bounded by a surface of the second degree, such as an ellipsoid. For various reasons, it is more convenient to use shapes such as sheets or prisms as interpretational models, and so it is simply assumed that such bodies are uniformly magnetized. The assumption will obviously fail near edges and corners, but it should be satisfactory throughout the body as a whole. Unfortunately the approximations introduced thereby are difficult to test rigorously, but the volumes affected are generally so small for bodies having simple shapes that the effects of nonuniform magnetization near edges and corners are probably negligible.

11-6 Characteristic Curves: The Step Model

We now turn to the "inverse" potential problem in magnetics, namely, the interpretation of magnetic anomalies. In the last analysis, all quantitative magnetic interpretations are "indirect," in the sense that they depend upon the use of models. Because the magnetic moment of a body cannot be found independently of its shape, no direct approach to interpretation is open in magnetics. We are therefore forced to depend very largely upon trial-and-error "curve-matching" methods. The practical problems inherent in this approach are enhanced by the fact that the direction of magnetization adds two new variables in addition to the dimensions of the model. Thus the number of combinations of parameters possible for each model is so large that a comprehensive portfolio of theoretical profiles becomes a practical impossibility. A more compact way in which to present much the same information is by making up sets of characteristic curves for each interpretational model. To illustrate the application of the method of characteristic curves to magnetic interpretations, we shall consider the step model and the ribbon model, and in addition two other models, viz., the tabular body and the vertical prism.

The chief interest in fault interpretation probably arises in connection with the analysis of aeromagnetic maps. Accordingly, we shall consider the total-field anomaly ΔT, measured in the direction of the geomagnetic field, over the two-dimensional step (Fig. 10-12). The formula is derived from (11-6), but first we should draw attention to a means of combining the inclination i and the strike declination λ into a single variable. We note that if the body strikes in the y direction, then $\partial/\partial y = 0$, and therefore, since $\nabla^2 = 0$, $-\partial^2/\partial x^2 = \partial^2/\partial z^2$ in free space. It therefore follows that

$$\frac{\partial^2}{\partial a^2} = (\cos^2 i \sin^2 \lambda - \sin^2 i) \frac{\partial^2}{\partial x^2} + 2 \sin i \cos i \sin \lambda \frac{\partial^2}{\partial z \, \partial x}$$

and if we define a new angle β by the relation

$$\tan \beta = \frac{\tan i}{\sin \lambda}$$

then
$$\frac{\partial^2}{\partial a^2} = (1 - \cos^2 i \cos^2 \lambda) \left(\cos 2\beta \frac{\partial^2}{\partial x^2} + \sin 2\beta \frac{\partial^2}{\partial z \, \partial x} \right)$$

Accordingly, from (11-6),

$$\frac{\Delta T}{2kH_0(1 - \cos^2 i \cos^2 \lambda)} = \sin d \cos (d - 2\beta) \ln \left[\frac{(x - l \cot d)^2 + (h + l)^2}{x^2 + h^2} \right]^{1/2}$$

$$+ \sin d \sin (d - 2\beta) \left(\tan^{-1} \frac{x - l \cot d}{h + l} - \tan^{-1} \frac{x}{h} \right) \quad (11\text{-}21)$$

where k is the contrast in the volume magnetic susceptibilities of the two contiguous media.

The difficulty with this formula for the construction of characteristic curves is that it produces so many different forms. Three examples from the same magnetic latitude are shown in Fig. 11-5a, b, and c. These profiles are calculated for the same fault in three different declinations. It is clearly not possible to define a single set of properties that will encompass this extreme variety of shapes, and we must divide the profiles into two recognizably distinct types. In the first, the magnetization is in a direction more normal than parallel to the fault plane $(2\beta - \pi/4 \leq d \leq 2\beta + \pi/4)$ when the first term of (11-21) dominates the second and the profile is rather symmetrical, as shown in Fig. 11-5c. The second type is the converse, in which the magnetization lies in a direction that is more parallel than normal to the fault plane $(2\beta - 3\pi/4 \leq d \leq 2\beta - \pi/4$, in the northern hemisphere) and which leads to the rather antisymmetrical profile of Fig. 11-5a. These will merge, of course, into a profile such as the one shown in Fig. 11-5b, which is predominantly neither of the one type nor the other.

We shall consider, by way of example, only profiles of the first type, which we shall call *transversely-magnetized* step profiles. Estimators which can be used for their interpretation are indicated in Fig. 11-6. Dip is indicated by, among other things, the ratio of the maximum slope of either

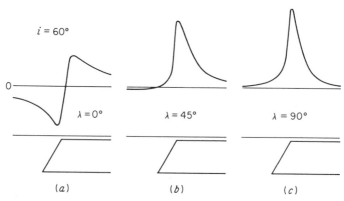

Fig. 11-5 **Profiles of the anomalous total magnetic intensity across a step, at magnetic latitude 60°.**

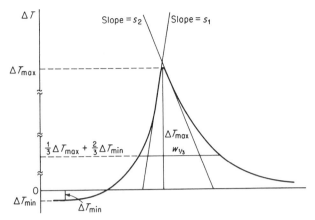

Fig. 11-6 Dip and depth estimators for the transversely-magnetized step model.

limb. For depth we may use the ratio of the maximum positive to the maximum negative amplitude. The shape of the profile is determined by the three parameters β, d, and h/l, of which β is known in advance. Since it is possible to deal with only two parameters at a time, separate sheets of curves will have to be assembled for different values of β. Where β does not fall at a tabulated value, we would normally interpolate between the results derived from the two sets of curves whose β values are closest to the observed value.

The characteristic curves for the total-field anomaly over a transversely-magnetized step when $\beta = 80°$ are shown in Fig. 11-7. The ratio $|\Delta T_{max}/\Delta T_{min}|$ is shown plotted against the ratio $|s_2/s_1|$, where s_1 and s_2 are the maximum horizontal gradients of the anomaly profile, s_1 being the greater. The major problem in using these curves will lie in finding the zero level of the anomaly, since faults tend to produce profiles without well-defined minima and which may sustain negative values over very great distances (note Fig. 11-5 in this respect). To choose the zero level correctly requires some experience in the recognition of fault anomaly patterns.

The fault interpretation is completed by establishing susceptibility contrast and throw. Throw is found from the set of complementary curves shown in Fig. 11-8, which gives the ratio $w_{⅓}/l$ ($w_{⅓}$ is the width of the profile where it acquires one-third of its total peak-to-trough amplitude above the minimum value) as a function of h/l, with d as a parameter. Since $w_{⅓}$ can be measured and h/l and d are known, l can be found and therefore also h. A second set of curves, giving the ratio $(\Delta T_{max} - \Delta T_{min})/2kH_0(1 - \cos^2 \lambda \cos^2 i)$ as a function of h/l, may be used to find the difference in the apparent volume susceptibilities of the two formations, if this information is desired.

For profiles of the second type, which we may term *dip-magnetized* step profiles, finding a pair of suitable estimators which can be used in practice is a great deal more problematical. The problem is usually made worse by the tendency of nearby disturbances to mask the flanks of the anomaly, thereby making it very difficult to estimate the background level. A generally satisfactory method of interpreting this type of anomaly does not yet appear to have been found.

A simpler although less accurate rule for the interpretation of magnetized steps also exists. It is based upon the formula for the total magnetic intensity across a very thin horizontal step buried at a depth h. If l is the thickness of the step and M its magnetic moment per unit volume, the

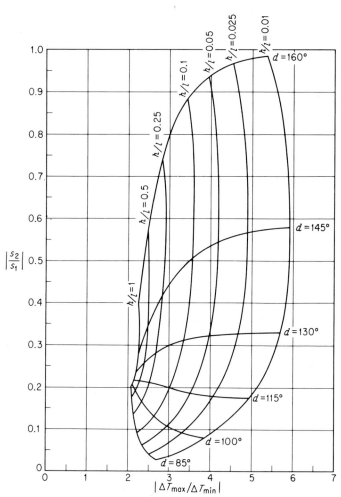

Fig. 11-7   **Total-field characteristic curves for the transversely-magnetized step model, $\beta = 80°$.**

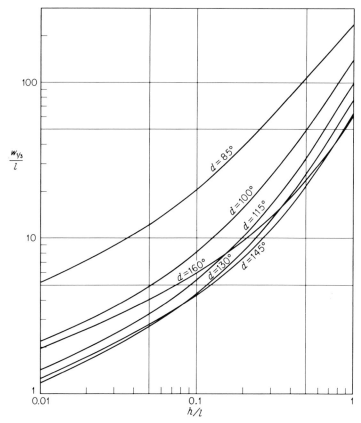

Fig. 11-8 Complementary curves for estimating the depth to a transversely-magnetized step from its total-field characteristics, $\beta = 80°$.

formula for the total-field anomaly across the step is

$$\Delta T(x) = 2Ml(1 - \cos^2 \lambda \cos^2 i) \frac{x \cos 2\beta + h \sin 2\beta}{x^2 + h^2} \qquad l \ll h$$

From this formula it is easy to verify that if $w_{\frac{1}{2}}$ is the width of the anomaly between the two points at which ΔT lies half-way between its maximum and minimum values [i.e., where $\Delta T = \frac{1}{2}(\Delta T_{\max} + \Delta T_{\min})$] and if L is the horizontal distance between the points of maximum and minimum value of ΔT, then

$$w_{\frac{1}{2}} = \frac{2h}{\sin 2\beta} \qquad \text{and} \qquad L = \frac{2h}{\cos 2\beta}$$

and also $\Delta T_{\max} - \Delta T_{\min} = 2M(1 - \cos^2 \lambda \cos^2 i)l/h$. From these simple formulas, it is possible to estimate the depth h and the product Ml. For the transversely-magnetized step we would use $w_{\frac{1}{2}}$, since L would normally not be measurable.   For the dip-magnetized step we would use L.   This simple method is generally satisfactory provided that h is not $< l$.

11-7   The Ribbon Model

The ribbon model is very often used in magnetic interpretations because intrusive igneous rocks in the forms of dikes or veins very frequently carry magnetic minerals.   We shall consider a uniformly magnetized ribbon having a small thickness s, a width l, a total strike length $2Y$, and a dip d, striking in the direction Oy (Fig. 11-9).   The depth of burial is h, and d is measured clockwise from the horizontal when the x component of $\mathbf{M}$ is directed to the right.   As we remarked at the end of Sec. 11-5, when a body becomes very thin in relation to its other dimensions, we can no longer neglect demagnetization; so that even if the ribbon has no permanent moment, $\mathbf{M}$ will not necessarily lie in the direction of $\mathbf{H}_0$.   Thus we should derive the formulas for its magnetic field components from (11-5) or (11-8) rather than from (11-4) or (11-7).

According to (11-8), the formula for the vertical-field anomaly over the ribbon is

$$\Delta Z(x,y) = Ms \sqrt{1 - \cos^2 \iota \cos^2 \tau}\, [f(x, y + Y) - f(x, y - Y)] \quad (11\text{-}22)$$

where

$$
\begin{aligned}
&f(x, y + Y) \\
&= \frac{y + Y}{\sqrt{C^2 + (y + Y)^2}} \left[\frac{(x - l \cos d) \sin Q - (h + l \sin d) \cos Q}{C^2} \right] \\
&\quad - \frac{y + Y}{\sqrt{D^2 + (y + Y)^2}} \left(\frac{x \sin Q - h \cos Q}{D^2} \right) + \left[\frac{(A + l) \cos d}{\sqrt{C^2 + (y + Y)^2}} \right. \\
&\quad \left. - \frac{A \cos d}{\sqrt{D^2 + (y + Y)^2}} \right] \left[\frac{KB - (y + Y) \sin Q}{B^2 + (y + Y)^2} \right] \\
&\quad - K \sin d \left[\frac{1}{\sqrt{C^2 + (y + Y)^2}} - \frac{1}{\sqrt{D^2 + (y + Y)^2}} \right]
\end{aligned}
$$

with

$$A = h \sin d - x \cos d \qquad B = x \sin d + h \cos d$$

$$C^2 = (x - l \cos d)^2 + (h + l \sin d)^2 \qquad D^2 = x^2 + h^2$$

$$Q = d - \gamma \qquad \text{and} \qquad K = \cos \gamma \cot \iota$$

where

$$\gamma = \tan^{-1} \frac{\tan \tau}{\sin \iota}$$

There are six parameters in this formula, and after scaling linear distances in units of l, five of them are left "free."   These are the two ratios

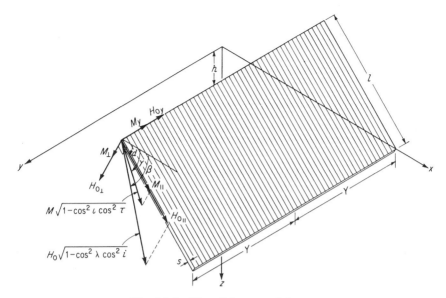

Fig. 11-9   The ribbon model.

h/l and Y/l and the three angles γ, ι, and d.   Fortunately, the effect of the strike length on the shape of the profile is small except when $Y < l$; otherwise the problem would be intractable.   As it is, the number of cases required to cover a suitable range of values in the five parameters prohibits any thought of making up a complete working set of theoretical profiles. For the ribbon model, the method of characteristic curves is a practical imperative.

Before we proceed to develop characteristic curves for the ribbon model, let us consider a special case.   When $Y \to \infty$ (11-22) reduces to the formula for a very long ribbon

$$\Delta Z(x) = 2Ms \sqrt{1 - \cos^2 \iota \cos^2 \tau}\, f(x;h,l,d) \tag{11-23}$$

where $$f(x;h,l,d) = \frac{(x - l \cos d) \sin Q - (h + l \sin d) \cos Q}{(x - l \cos d)^2 + (h + l \sin d)^2}$$

$$- \frac{x \sin Q - h \cos Q}{x^2 + h^2}$$

Indeed, this expression could have been derived more simply from (11-10). If we now let l become very much larger than h, (11-23) simplifies even further to the formula for the vertical magnetic field intensity over a half-plane

$$\Delta Z(x) = 2Ms \sqrt{1 - \cos^2 \iota \cos^2 \tau}\, \frac{h \cos Q - x \sin Q}{x^2 + h^2} \tag{11-23a}$$

This simple formula lends itself readily to elementary rules. For instance,

$$\frac{w_{\frac{1}{2}}}{h} = 2 \sec Q \quad \text{and} \quad \frac{L}{h} = 2 \csc Q$$

where $w_{\frac{1}{2}}$ is the *half-width* of the anomaly profile (i.e., the distance between points where $\Delta Z = \frac{1}{2} \Delta Z_{max} + \frac{1}{2} \Delta Z_{min}$) and L is the horizontal distance between the extremes of ΔZ. To find h, therefore, we have only to determine the value of Q. This can be done independently from the relations

$$\frac{w_{\frac{1}{2}}}{L} = \tan Q \quad \text{and} \quad \left| \frac{\Delta Z_{max}}{\Delta Z_{min}} \right| = \frac{1 + \cos Q}{1 - \cos Q}$$

Thus h is easily calculated.

Having found Q, the dip d is also determined if we know the direction of **M**. Alternatively, if d is known, we can use Q to determine the direction γ. Let us suppose that the sheet is inductively magnetized in the earth's field and that it has no permanent magnetic moment. As usual, we let i be the inclination of $\mathbf{H}_0$ and λ the declination of Oy. If we then resolve $\mathbf{H}_0$ into three components, the first of which $H_{0\perp}$ is transverse to the sheet in the sense of d, the second H_{0y} is in the direction of the strike, and the third $H_{0\parallel}$ is in the plane of the sheet (see Fig. 11-9), we have

$$H_{0\parallel} = H_0(\sin i \sin d + \sin \lambda \cos i \cos d)$$
$$H_{0y} = H_0 \cos \lambda \cos i$$

and
$$H_{0\perp} = H_0(\sin i \cos d - \sin \lambda \cos i \sin d)$$

Now if k is the volume magnetic susceptibility of the sheet, the three components of **M** will be

$$M_x = M_\parallel \cos d - M_\perp \sin d \qquad M_y \qquad \text{and} \qquad M_z = M_\parallel \sin d + M_\perp \cos d$$

where $\quad M_\parallel = kH_{0\parallel} \qquad M_y = kH_{0y} \qquad \text{and} \qquad M_\perp = \dfrac{kH_{0\perp}}{1 + 4\pi k}$

But according to our definition of ι and τ

$$M_x = kH_0 \sin \iota \cos \tau \qquad \text{and} \qquad M_z = kH_0 \sin \tau$$

so that if $f = 1/(1 + 4\pi k)$, we have

$$\sin \iota \cos \tau = \sin \lambda \cos i (\cos^2 d + f \sin^2 d) + \sin i \sin d \cos d(1 - f)$$
$$\sin \tau = \sin \lambda \cos i \sin d \cos d (1 - f) + \sin i (f \cos^2 d + \sin^2 d)$$

and
$$\tan (d - \gamma) = f \tan (d - \beta) \tag{11-24}$$

To find d we have to substitute the appropriate values for k and β into (11-24).

From the formula (11-23a) we further note that

$$\Delta Z_{max} - \Delta Z_{min} = \frac{2H_0 ks \sqrt{1 - \cos^2 \iota \cos^2 \tau}}{h}$$

from which it follows that

$$ks = \frac{(\Delta Z_{max} - \Delta Z_{min})w_{\frac{1}{2}} \cos Q}{4H_0 \sqrt{1 - \cos^2 \iota \cos^2 \tau}} \qquad (11\text{-}25)$$

It is easily shown from the expressions for $\sin \iota \cos \tau$ and for $\sin \tau$ that

$$(1 - \cos^2 \iota \cos^2 \tau)^{\frac{1}{2}}$$
$$= (1 - \cos^2 \lambda \cos^2 i)^{\frac{1}{2}}[1 - (1 - f^2) \sin^2 (d - \beta)]^{\frac{1}{2}} \qquad (11\text{-}26)$$

Thus (11-24) and (11-26) are a pair of nonlinear simultaneous equations for d and k, and they must be solved by successive approximations. One method is to begin by solving (11-24) for d with $k = 0$ and then to use this result in (11-26) to find a value for k. At this stage a value must be assigned to the thickness s, and the value of k so obtained may then be used to correct d, and so on until the solution converges. A simpler approach to this problem has been proposed by Gay (10), but it requires knowing the value of k.

The above procedures are applicable only if the anomaly pattern is sharply peaked and very elongated. Nonetheless, the interpretation of strike- and depth-limited bodies presents such awkward problems in practice that the simpler formulas are often used for general guidance even when they are not strictly valid.

Now let us turn to the finite ribbon. In order to illustrate the effects of changing the dip, depth, and direction of magnetization, we show in Fig. 11-10 a series of transverse profiles across a two-dimensional ribbon at different values of those parameters. The effects of shortening the strike length are relatively minor until Y becomes $<l$. It is clear from this diagram (1) that the *symmetry* of the profile is determined very largely by the difference between d and γ, rather than by either one of these two angles individually, and (2) that the *sharpness* of the profile is determined by the depth to the top of the ribbon. Quantitative measures of skewness and sharpness are shown in Fig. 11-11. These measures are: (1) the ratio $|s_2/s_1|$ of the maximum horizontal gradients of the principal profile and (2) the ratio $w_s/w_{\frac{1}{2}}$ of the horizontal distance between the points of inflection to the width of the principal profile where it reaches an amplitude that lies one-third of the way up from the minimum to the maximum value. The *principal profile* is the one which passes through the peak of the anomaly in a direction perpendicular to the strike of the ribbon and is easily located in practice. Oddly enough, the measure of sharpness used in gravity interpretations, $w_{\frac{2}{3}}/w_{\frac{1}{3}}$, appears to be ineffective in magnetics.

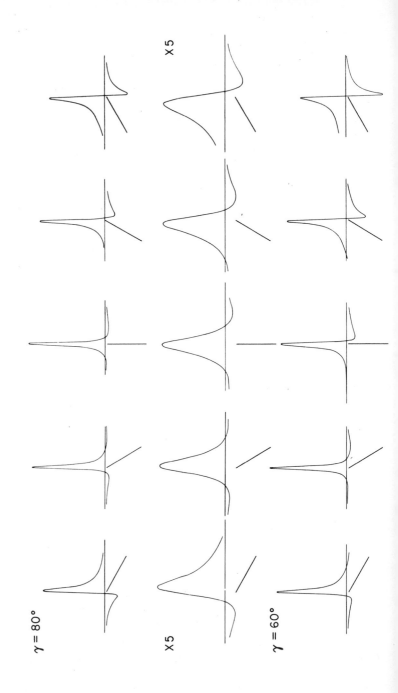

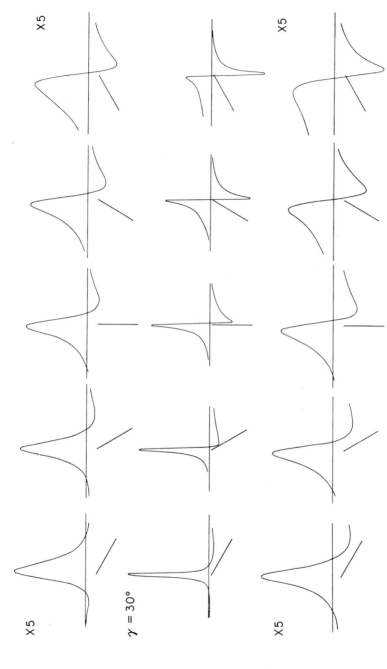

Fig. 11-10 Profiles of the anomalous vertical magnetic intensity across two-dimensional ribbons of various dips and depths, and for various directions of magnetization.

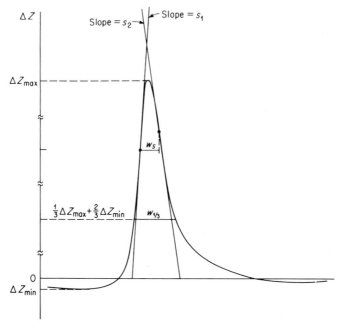

Fig. 11-11   Dip and depth estimators for the ribbon model.

We have still to determine Y/l.   The natural estimator for this parameter is the ratio $w_{\frac{1}{2}}/(w_{\frac{1}{2}})_y$, $w_{\frac{1}{2}}$ being the "half-width" of the principal profile and $(w_{\frac{1}{2}})_y$ the width of the anomaly pattern at the half-amplitude level measured in the strike direction.   We shall sometimes refer to this ratio as the *ellipticity* of the anomaly.

Having decided upon the estimators to use, it is clear that the number of parameters needed for the ribbon model creates a major problem in the organization of characteristic curves.   It is solved in the following way: (1) Separate portfolios are made up for different values of γ, since this is usually the parameter whose value is most readily assumed.   (2) Within each portfolio, a number of diagrams similar to Fig. 11-12 are provided which show ellipticity versus the ratio $\bar{s}w_s/\mathrm{AMPL}$, $\bar{s}$ being the mean of $|s_1|$ and $|s_2|$ and AMPL denoting the amplitude of the anomaly, i.e., $\Delta Z_{\max} - \Delta Z_{\min}$.   The shaded zones of Fig. 11-12 represent loci of constant Y/l for $20° \le d \le 90°$ $(0° \le Q \le 70°)$ and   for $0.05 \le h/l \le 0.5$ when $\gamma = 90°$ and $K = 0.2$. This type of diagram may be used directly to estimate Y/l.   A few of them may be needed to cover a suitable range of K values, making interpolations necessary.   (3) The characteristic curves within each portfolio are plots of $|s_2/s_1|$ versus $w_s/w_{\frac{1}{3}}$ for constant values of Y/l and K.   Because the effect of Y is small unless $Y/l < 1$, in practice all curves of constant K are combined onto a single sheet in order to reduce the volume of data.   A sample set is shown in Fig. 11-13. (4) The complementary curves give the ratio $(\Delta Z_{max} - \Delta Z_{min})h/Ms \sqrt{1 - \cos^2 \iota \cos^2 \tau}$ versus $w_{\frac{1}{3}}/h$, from which the depth h, the depth extent l, and the surface magnetization Ms may be found. Figure 11-14 shows the complementary sheet to Fig. 11-13.

The recommended procedure in using these curves is as follows: (1) First select the value to use for γ. If this is not known and it is thought that the body may be inductively magnetized, use the value of β and correct later for demagnetization. Pick out the two portfolios having the two nearest γ values. (2) In each of the two portfolios, go to the appropriate diagram(s) for Y/l. (3) Select from each portfolio the two sets of characteristic curves whose K values lie nearest to the value of $\cos \beta \cot \lambda$. From each of these sets, estimate d and h/l. Thus two values each for d and h/l will be derived from each portfolio. (4) Then, from the complementary curves, find the four corresponding values for each of h, l, d, and Ms. Interpolate first with respect to K, then with respect to γ. This leaves us with a final set of values for these four parameters and also for Y.

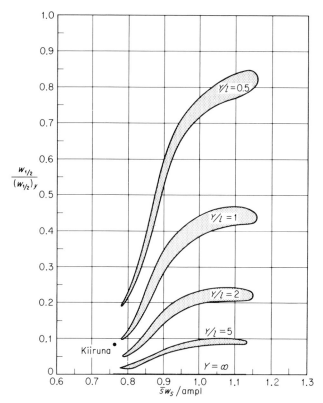

Fig. 11-12 Diagram for estimating the ratio Y/l of a ribbon from its vertical magnetic field anomaly. The shaded zones represent the loci of constant Y/l.

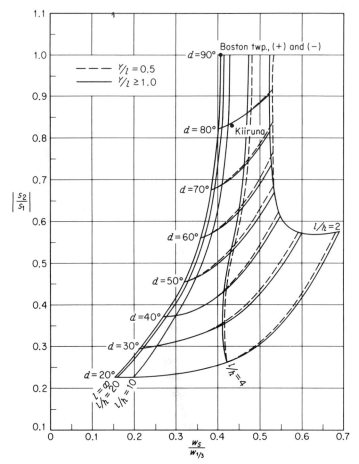

Fig. 11-13   Vertical-field characteristic curves for the ribbon model; $\gamma = 90°$, $K = 0.2$.

This is as far as we can go without knowing the thickness s. s cannot be found explicitly by using the ribbon model, and we shall have to consider other means of determining it.   Before we do this, however, let us illustrate by two examples what has been achieved up to this point.

The first example is the famous magnetic anomaly at Kiirunavaara in northern Sweden originally described by von Carlheim-Gyllenskjöld (5), the principal profile of which is shown in Fig. 11-15.   It is due to a vein of approximately 20 percent magnetite, about 85 m thick and dipping 55°N. The magnetic inclination at Kiirunavaara is 76° and the declination of the axis of the anomaly is 18°, so that $\beta = 85°$ in this case.   The total amplitude of the anomaly is 0.718 oersted, or 71,800 gammas.   The value of $K = \cos\beta$

cot λ is 0.22. The other measurements taken from this profile are

$$s_1 = 800 \text{ gammas/m} \qquad s_2 = -660 \text{ gammas/m}$$
$$w_s = 75 \text{ m} \qquad w_{\frac{1}{2}} = 125 \text{ m} \qquad w_{\frac{1}{3}} = 174 \text{ m}$$

and from adjacent magnetic profiles it would appear that $(w_{\frac{1}{2}})_y$ may be about 1,500 m. To find w_s, we use a Lagrange 5-point finite-difference formula and operate on the station values themselves. w_s is taken to be the distance between the two zeros of the finite-difference value for the second horizontal derivative of ΔZ.

From these measurements, we obtain the following values for the characteristics: ellipticity $= w_{\frac{1}{2}}/(w_{\frac{1}{2}})_y = 0.083$; $\bar{s}w_s/\text{AMPL} = 0.763$; $|s_2/s_1| = 0.83$; and $w_s/w_{\frac{1}{3}} = 0.43$. To solve for the parameters, we start

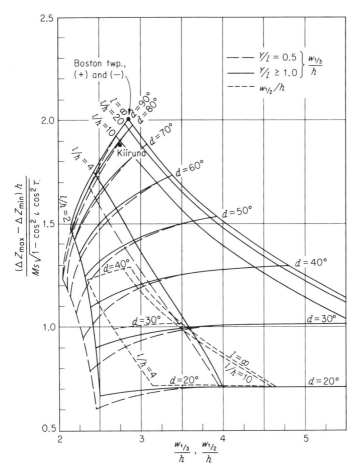

Fig. 11-14 Complementary curves for estimating the depth and surface magnetization of a ribbon from its vertical-field characteristics; $\gamma = 90°$, $K = 0.2$.

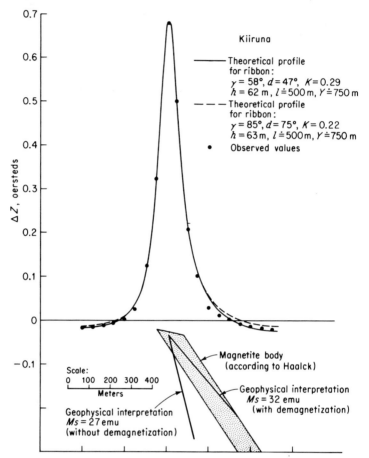

Fig. 11-15 Principal profile of the magnetic anomaly at Kiirunavaara, Sweden. [After Haalck (6).]

out with the two portfolios $\gamma = 90°$ and $\gamma = 80°$. We shall give illustrations only from the first.

For $\gamma = 90°$, $K = 0.2$, we find from Fig. 11-12 that $Y/l = 1.5$. The same ratio is found when $\gamma = 80°$. Moving now to Fig. 11-13, we find that $d = 79°$ and $h/l = 0.12$ when $\gamma = 90°$ and $K = 0.2$, whereas the curves for $\gamma = 80°$ give the values $d = 69°$ and $h/l = 0.15$. Applying these two sets of values to the complementary curves, we find two values each for d, h, l, Y, and Ms. By interpolating between them, we get as our first solution a body which dips at 74°N, lies at a depth of 60 m, and has a strike length in excess of 1,000 m and a depth extent (although this is a difficult parameter to estimate from magnetics) of about 400 m. The surface magnetization is 27 emu when s is in meters, and if this is regarded as an induced magnetization, the apparent surface susceptibility is 46 emu. Putting $s = 85$ m, the apparent value of k is 0.54 emu. As it happens, this lies well within the range of sample values measured by von Carlheim-Gyllenskjöld, and it compares well with the value (0.57 emu) calculated by Haalck (6). When we compare the theoretical profile for a ribbon having these values with the observational data, we find that on the whole the fit is rather good (Fig. 11-15). It is clear that the true dip of the Kiiruna iron formation is disguised by demagnetization or by remanent magnetization, or perhaps by both.

To correct for demagnetization, we have to solve (11-24) putting $Q = -11°$, which is the value indicated by the characteristic curves. Before we do this, however, we shall make one rather minor change in the formula by replacing k with $k(l - s)/(l + s)$. This corresponds to a demagnetizing factor $4\pi l/(l + s)$ across the ribbon and $4\pi s/(l + s)$ parallel to the ribbon, which are what we obtain by solving the boundary-value problem for a thin elliptical cylinder in a uniform field (for details of this

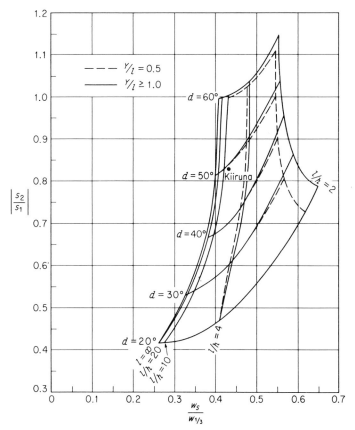

Fig. 11-16 Vertical-field characteristic curves for the ribbon model; $\gamma = 60°$, $K = 0.2$.

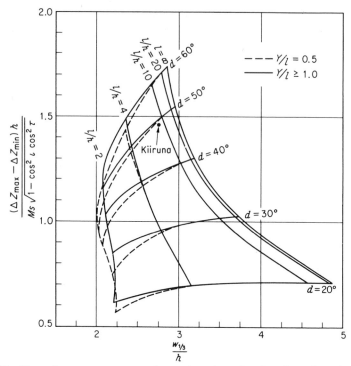

Fig. 11-17   Complementary curves for estimating the depth and surface magnetization of a ribbon from its vertical field characteristics; $\gamma = 60°$, $K = 0.2$.

solution, see Sec. 14-12).   If we now put $\beta = 85°, k = 0.54$ emu, $Q = -11°$, and use the values of l and s found previously, we find that $\gamma = 58°$.   Therefore we turn to the characteristic curves for $\gamma = 60°$ (Figs. 11-16 and 11-17) and for $\gamma = 50°$ and repeat the interpretation.   It turns out that $d = 47°$, $h = 63$ m, $l \doteq 400$ m, and $Ms = 32$ emu ($k = 0.57$ emu).   The theoretical profile for this ribbon (Fig. 11-15) again agrees very well with the field observations.   Although the correction for demagnetization has improved the dip estimate, in neither case is the interpreted value correct.   In spite of this, good agreement with the field observations can be obtained.

The chief point to be noted in this example is the very large part that a strong magnetization plays in the determination of dip.   As a matter of fact, when the susceptibility of a formation is as large as it appears to be at Kiirunavaara, it is very difficult indeed to make any strong prediction at all about the dip; for under these conditions the magnetic moment is deflected almost into the plane of the body, and as Fig. 11-10 suggests, the dip can be determined only in relation to the direction of the magnetic moment.   For this reason, dip is generally an ambiguous parameter in the interpretation of strong magnetic anomalies.

An average volume susceptibility of 0.57 emu sustained throughout a width of 85 m suggests that the Kiiruna iron formation may be coarse-grained and banded in structure, since this far exceeds the values predicted by any of the formulas of Sec. 12-5 for disseminated magnetite.   If this is the case, then the true susceptibility of the ore may well be greater than the calculated average value, making the demagnetization effect even greater than it at first appears.   It is, of course, also possible that the magnetization may be partly remanent.   If so, its direction apparently does not differ markedly from that of the combined external and demagnetizing field. Since it is clearly impossible to distinguish in any way between anomalies due to the two different types of magnetization, we use the term *equivalent volume magnetic susceptibility* in reference to their combined effect.

In the second example two opposing magnetic moments occur in the same formation.   The profile shown in Fig. 11-18 is the principal profile of

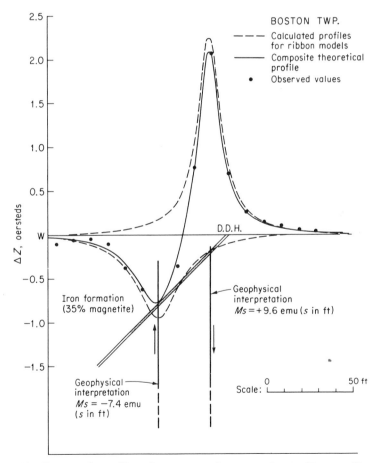

Fig. 11-18 Principal profile of a magnetic anomaly in Boston Township, Ontario, Canada. [After Ratcliffe (7).]

a magnetic anomaly discovered in northern Ontario, Canada, originally from an aeromagnetic survey [Ratcliffe (7)]. Its total amplitude is more than four times the intensity of the earth's magnetic field, which must rank it as one of the most intense anomalies ever measured with a magnetometer. It is due to a body that grades up to 30 percent by volume of magnetite and which comes to within a few feet of the ground surface. The magnetic inclination in this area is 78°, and the body strikes almost in the magnetic meridian, so that β is almost 90° and $K = \cos\beta \cot\lambda = 0.20$.

The most striking feature about this magnetic profile, apart from its great intensity, is the very deep minimum on its west flank. The maximum positive effect is $+2.17$ oersteds, or 217,000 gammas, while the maximum negative effect is $-75,000$ gammas. The ratio is 2.9; this suggests, according to rule of thumb, that the value of $Q = \gamma - d$ should be about $+60°$. For such a large anomaly, a difference of this amount between the direction of magnetization and the angle of dip would be almost unthinkable. Another contradiction appears when we measure the characteristics of the profile. We find that the maximum horizontal gradients are about equal in magnitude ($\pm 16,000$ gammas/ft) and the stations are closely enough spaced (the interval is 10 ft) that there cannot be very much dispute about this fact. This would indicate that d and γ are very nearly equal, as we should expect them to be. The other characteristic measures of the anomaly are

$$w_s = 13.4 \text{ ft} \qquad w_{1/2} = 20 \text{ ft} \qquad w_{1/3} = 35 \text{ ft} \quad \text{and} \quad (w_{1/2})_y = 90 \text{ ft}$$

Figures 11-12 and 11-13 show quite clearly that these values cannot be reconciled with the ribbon model. Decidedly, there is something queer about the shape of this profile.

It was suggested by Ratcliffe that the negative part of the anomaly might be caused by inverse remanent magnetization, and subsequent drilling confirmed that the iron formation, which consists of coarsely banded magnetite, is inversely remanently magnetized at its western edge. A strong normal remanent magnetization in the direction of the earth's field was also discovered in the eastern part of the formation. Figure 11-18 shows how the assumption that there is a positive *and* a negative source dispels the mystery surrounding the shape of the profile. Because the two anomalies are so intense, we shall assume that they are both symmetrical. A further reason for making this assumption is that dips in this area are known to be vertical or nearly vertical. Thus by starting at both ends of the profile and working toward the middle, we can decompose it uniquely into symmetrical positive and negative components. To each component we can fit a vertical ribbon model at $\gamma = 90°$, as indicated by Fig. 11-13. Both ribbons appear to be infinite in depth extent. The eastern ribbon appears to lie at a depth of only 8.3 ft and has a magnetization of $9.6/s$ emu when s is measured in feet. The western ribbon lies at a depth of 15 ft and has a magnetization of $-7.4/s$ emu, the negative sign indicating a polarity opposite to $\mathbf{H}_0$. There will be no correction for demagnetization, since $\gamma = \beta = d$. The composite profile of these two ribbon models matches the observational data very well. An overburden thickness of 8 ft above the eastern edge of the formation was confirmed by the drill.

The most remarkable feature of this interpretation is that it has been possible to achieve such satisfactory correspondence to the field data by using the ribbon model. This would suggest that the normally magnetized portion, seen from a distance of only 8 ft, appears to be rather thin. Since the effect of a finite thickness on the shape of the profile is likely to become apparent if s exceeds $2h$, we infer that the width of the magnetic zone is probably less than 16 ft. This implies a *minimum* volume magnetization of 0.6 emu, or a minimum volume susceptibility of about 1 emu. Such values can be achieved in massive crystalline magnetite, but they are impossible if the grains are disseminated. Since it is more likely that such coherence might be preserved in several narrow zones than in a single wide one, it is not surprising to discover such an intense magnetization in a banded iron formation. Nor, in fact, is the inverse magnetization such a mystery, when we consider the processes by which magnetite acquires a permanent moment (Sec. 12-5). One possible explanation is that the western part of the formation for some reason cooled more slowly than the rest and reached its Curie temperature when the eastern half was already highly magnetized. Thus the ambient field which caused the initial alignment of the magnetic domains would have had a polarity opposite to that of the geomagnetic field.

The situation described above is not uncommon. Even over thick, massive iron formations, the shapes of magnetic anomalies frequently seem to be controlled by relatively narrow zones of abnormally intense magnetization. Doubtless, this does explain why the ribbon model is successful in magnetic interpretations even for bodies that come almost to the ground surface. It also indicates that the true thickness of a magnetic formation is apt to be difficult to assess from geophysics. Nonetheless, it is occasionally necessary to try. It is essential, for instance, to know the equivalent susceptibility in order to correct for demagnetization. Moreover, the surface magnetization of a dikelike body is not in general a very revealing property. Volume magnetic moments or susceptibilities would be more meaningful as an index of composition.

11-8 The Tabular Model

If we wish to estimate thickness, clearly we must abandon the thin ribbon model. Let us suppose therefore that the ribbon thickness s becomes finite. Since this leaves us with one too many parameters, we shall also suppose that either the bottom or the ends of the body lie at an infinitely great distance. If indeed we are close enough to sense the finite thickness, then

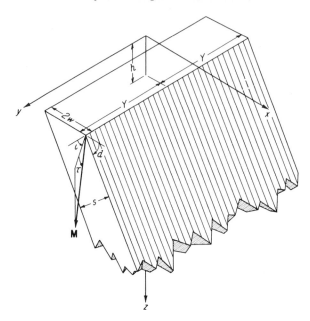

Fig. 11-19 The tabular model.

we might hope that the principal profile of the anomaly will be little affected by one or the other of the two larger dimensions. If both flanks of the profile become negative, clearly the depth extent is important; but if at least one of the flanks remains positive, the strike extent is more likely to play a part.

Let us consider the tabular body of finite length and infinite depth extent magnetized in the direction κ (Fig. 11-19). An approximate formula for the vertical component of magnetic intensity across it is

$$\Delta Z(x,y) = M \sqrt{1 - \cos^2 \iota \cos^2 \tau} \sin d \, [f(x + w, y + Y)$$
$$- f(x + w, y - Y) - f(x - w, y + Y) + f(x - w, y - Y)] \quad (11\text{-}27)$$

where
$$f(x + w, y + Y) = \cos Q \left\{ \tan^{-1} \left[\frac{y + Y}{B(x + w)} \right] \right.$$

$$- \tan^{-1} \left[\frac{(y + Y) A (x + w)}{B(x + w) \sqrt{D^2(x + w) + (y + Y)^2}} \right] \right\}$$

$$+ \sin Q \left\{ \tfrac{1}{2} \ln \left[\sqrt{D^2(x + w) + (y + Y)^2} - (y + Y) \right] \right.$$

$$\left. - \tfrac{1}{2} \ln \left[\sqrt{D^2(x + w) + (y + Y)^2} + (y + Y) \right] \right\}$$

with

$$A(x + w) = h \sin d - (x + w) \cos d$$
$$B(x + w) \doteq (x + w) \sin d + h \cos d$$
$$D^2(x + w) = (x + w)^2 + h^2$$
$$w = \tfrac{1}{2} s \csc d \qquad Q = \gamma - d$$

In the derivation of this formula terms of $O(Y/h)^{-2}$ and smaller have been disregarded, so that we may expect it to be accurate only if $Y \geq 3h$. Since it is to be used chiefly to estimate s rather than the other parameters, however, it is probable that this approximation can be tolerated. It should be possible to construct characteristic curves from this formula in much the same way as we did for the ribbon model. In this case we set $s = 1$ and scale all distances in units of thickness, since s is now the characteristic linear dimension of the model.

In the limiting case when the body becomes two-dimensional $(Y \gg s, h)$, formula (11-27) reduces to

$$\Delta Z(x) = M \sqrt{1 - \cos^2 \iota \cos^2 \tau} \, \sin d \left[\sin Q \ln \sqrt{\frac{(x - w)^2 + h^2}{(x + w)^2 + h^2}} \right.$$
$$\left. + \cos Q \tan^{-1} \frac{2wh}{x^2 - w^2 + h^2} \right] \quad (11\text{-}27a)$$

This formula is the sum of two terms, the first of which is antisymmetrical about $x = 0$, and the second, symmetrical. This remarkable property has been exploited in a variety of ways to make quantitative estimates of s, h, and d possible from field observations [Cook (8), Hutchinson (9), Gay (10), and probably many others]. The infinite, tabular model has been the object of much attention by mining geophysicists over many years because it imitates geological forms which have a very frequent occurrence in many areas.

11-9 *The Numerical Calculation of Magnetic-anomaly Profiles*

As a final stage in the fitting of theoretical models to magnetic field observations, minor adjustments may be made to the outlines of bodies in order to bring the calculated profiles into closer correspondence to the observational data. Like the corresponding problem in gravity interpretations, the integrations may be performed numerically with the aid of templates. Thus for the total-intensity anomaly ΔT across a two-dimensional body, for example, we may write

$$\Delta T(0) = 2M \left[(\cos^2 i \sin^2 \lambda - \sin^2 i) \frac{\partial^2}{\partial x^2} \right.$$
$$\left. - 2 \sin i \cos i \sin \lambda \frac{\partial^2}{\partial z \, \partial x} \right] \iint_S \ln \left[(x - \xi)^2 + (z - \zeta)^2 \right]^{1/2} d\xi \, d\zeta$$
$$= 2M(1 - \cos^2 \lambda \cos^2 i) \left(\cos 2\beta \iint_S \frac{\cos 2\vartheta}{r} \, d\vartheta \, dr \right.$$
$$\left. + \sin 2\beta \iint_S \frac{\sin 2\vartheta}{r} \, d\vartheta \, dr \right)$$

where $\beta = \tan^{-1}(\tan i/\sin \lambda)$. The integrals may be evaluated approximately by replacing them with sums; thus

$$\frac{\Delta T(0)}{2M(1 - \cos^2 \lambda \cos^2 i)} \doteq \cos 2\beta \sum_S \Delta(\sin 2\vartheta) \, \Delta(\ln r)$$

$$- \sin 2\beta \sum_S \Delta(\cos 2\vartheta) \, \Delta(\ln r) \quad (11\text{-}28)$$

The application of this formula requires the use of an overlay diagram with radii drawn at equal intervals of $\sin 2\vartheta$ and circles at equal increments of $\ln r$. The chart is set with its center at the point where the calculation is to be made and with the line $\vartheta = 0$ in the horizontal. The first term of (11-28) is calculated by counting up the number of segments which fall within the cross section S of the body. The chart is then rotated through $45°$, and the second term is found. To complete the calculations, of course, β must be known, but its value is assumed only in the final stages of the work, when it is relatively easy to make adjustments for remanent magnetic fields.

As in making gravity interpretations, the usual procedure is first of all to find a form for S whose magnetic field will match the shape of the anomaly profile. After each set of calculations the magnetic moment should be adjusted so that the amplitude of the theoretical curve coincides with that of the anomaly profile. When the shape has been thus determined, the final value of M will be the mean magnetic moment of the body. If it is entirely an induced magnetic moment, dividing by H_0 will give the apparent susceptibility contrast.

For calculating vertical-field anomaly profiles, we may write

$$\frac{-\Delta Z(0)}{M \sqrt{1 - \cos^2 \lambda \cos^2 i}} \doteq \sin \beta \sum_S \Delta(\sin 2\vartheta) \, \Delta(\ln r)$$

$$+ \cos \beta \sum_S \Delta(\cos 2\vartheta) \, \Delta(\ln r) \quad (11\text{-}29)$$

The chart for calculating this quantity is precisely the same as the one used for ΔT. The same is also true of the horizontal-field anomaly ΔX.

End corrections for finite strike length can be included in these computations by drawing the charts to slightly different specifications. The vertical-field anomaly at any point in the central traverse line will be given by

$$\frac{\Delta Z(0)}{M \sqrt{1 - \cos^2 \lambda \cos^2 i}} \doteq \cos \beta \sum_S \Delta(\cos 2\vartheta) \, \Delta[F(r;Y)]$$

$$+ \sin \beta \sum_S \Delta(\sin 2\vartheta) \, \Delta[F(r;Y)] - \sin \beta \sum_S \Delta(\vartheta) \, \Delta[G(r;Y)]$$

where
$$G(r;Y) = \frac{Y}{\sqrt{r^2 + Y^2}}$$

and
$$F(r;Y) = G(r;Y) - 2 \ln \frac{(\sqrt{r^2 + Y^2} - Y)}{r}$$

Corresponding expressions exist also for ΔT and ΔX. Two different charts must be employed in this calculation, one of which is used a second time with a 45° rotation. The limitations of indirect methods of interpretation begin to make themselves felt when one realizes that each new adjustment to the shape of the body involves two (or three) new calculations at every point on the profile.

This large demand on labor can be lessened by the use of digital computing machines. Like the corresponding problem in gravity interpretation, however, the calculations can be programmed easily only for two-dimensional bodies. As before, we approximate the outline of the cross section S with a polygon, to the vertices of which we assign pairs of coordinates (x_i, y_i). Then the kth side of the polygon will be represented by the formula $x = a_k z + b_k$ (Fig. 11-20), where

$$a_k = \frac{x_{k+1} - x_k}{z_{k+1} - z_k} \quad \text{and} \quad b_k = \frac{x_k z_{k+1} - x_{k+1} z_k}{z_{k+1} - z_k}$$

The vertical-field anomaly at the point (0,0) will be given by

$$\frac{\Delta Z(0)}{2M \sqrt{1 - \cos^2 \lambda \cos^2 i}} = \sin \beta \iint_S \frac{\zeta^2 - \xi^2}{(\xi^2 + \zeta^2)^2} \, d\xi \, d\zeta$$

$$+ \cos \beta \iint_S \frac{2\xi\zeta}{(\xi^2 + \zeta^2)^2} \, d\xi \, d\zeta$$

$$= \oint \frac{\xi \sin \beta - \zeta \cos \beta}{\xi^2 + \zeta^2} \, d\zeta$$

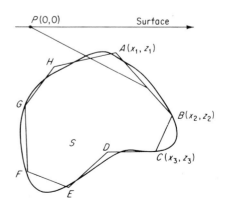

Fig. 11-20 The cross section of a two-dimensional body represented by a polygon.

where the line integral is taken around the polygon $ABCDE$. . . . Substituting into this integral $\xi = a_k \zeta + b_k$ we have

$$
\frac{\Delta Z(0)}{2M \sqrt{1 - \cos^2 \lambda \cos^2 i}} = \sum_{k=1}^{n} \frac{1}{1 + a_k^2} \left\{ (a_k \sin \beta + \cos \beta) \right.
$$
$$
\times \ln \sqrt{\frac{(1 + a_k^2)z_{k+1}^2 + 2a_k b_k z_{k+1} + b_k^2}{(1 + a_k^2)z_k^2 + 2a_k b_k z_k + b_k^2}} - (a_k \cos \beta - \sin \beta)
$$
$$
\left. \times \left[\tan^{-1}\left(\frac{(1 + a_k^2)z_{k+1}}{b_k} + a_k \right) - \tan^{-1}\left(\frac{(1 + a_k^2)z_k}{b_k} + a_k \right) \right] \right\} \quad (11\text{-}30)
$$

The calculations are performed very quickly once coordinates are assigned to the vertices. Displacing the origin to a new position on the surface of the ground is equivalent to adding a constant increment to each x_k.

11-10 Depth Estimates from Aeromagnetic Surveys

In exploring new areas for petroleum or minerals, particularly unmapped sedimentary basins, the airborne magnetometer is often used as a device for making preliminary estimates of the thickness of the sedimentary section. The premise is that sedimentary rocks are nonmagnetic, so that any magnetic anomalies must originate from within the igneous crystalline complex. Calculation of the depth to the magnetic material therefore yields an upper limit to the total thickness of the sedimentary strata. Since in this application only the depth of the source is required and the details of its shape are of little direct interest, the use of elementary models such as poles and dipoles is rather common.

One of the chief difficulties with aeromagnetic interpretations is that the instrument is placed as a rule so far above the magnetic body that the body no longer appears to be two-dimensional no matter how elongated it may be. Therefore two-dimensional models are of little value in aeromagnetic interpretations, and neither is the majority of characteristic curves used for interpreting ground surveys. For this and other reasons, the models that have achieved widespread use in aeromagnetics are different from those most often used for interpreting ground surveys.

A model introduced some years ago by Vacquier, Steenland, Henderson, and Zeitz (11), which has since then achieved very wide popularity, is the bottomless, vertical-sided prism of rectangular cross section. The thought that led to the use of this model is that the magnetic effects of a number of closely spaced sources will blend together to such an extent when viewed from a large distance that they will appear as a single anomaly pattern. Thus it would be difficult to distinguish between a region of banded igneous intrusions and a single block of uniformly magnetized material from magnetic observations made at a distance of several thousands of feet. The prism model serves only to outline very crudely the volumes of rock within which the magnetic minerals are found in greatest concentration and is in no way concerned with studying the geological occurrence of these minerals. The success of this model in predicting the depth of nonmagnetic overburden is due directly to the fact that at sufficiently great distances, potential fields are little affected by the details of the shapes of their sources.

The shape of the magnetic anomaly over a vertical prism is determined by five quantities, viz., the lengths of the two sides a and b, the depth h below the aircraft, the magnetic azimuth λ of side b, and the magnetic inclination i. By choosing a as the unit of measure, we shall have four parameters: h/a, b/a, λ, and i, of which the last two are either known or assumed beforehand. The object is to make use of certain characteristic features of the shape of the anomaly pattern to eliminate a and b in order to find h.

Four independent parameters lead to a very large number of individual cases. There is an obvious connection between b/a and λ which reduces this number by making it necessary to consider only values of λ between 0 and 90°. Even so, the number of cases for all values of i is staggering. Vacquier et al. calculated patterns of the total-field anomaly and also of the second vertical derivative of the total-field anomaly ($\partial^2 \Delta T/\partial z^2$) for a number

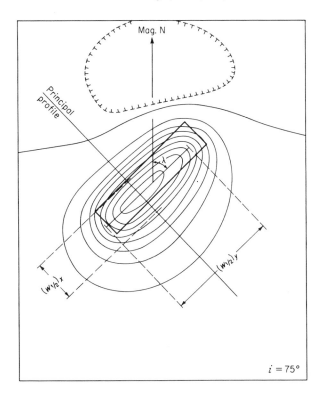

Fig. 11-21 Theoretical total-field anomaly over a vertical prism, showing the estimator for b/a. [After Vacquier, Steenland, Henderson, and Zeitz (11).]

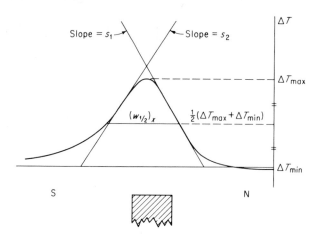

Fig. 11-22 Principal profile of the anomaly pattern shown in Fig. 11-21, showing the quantities used to estimate h/a when $i \geq 60°$.

of individual cases at magnetic inclinations ranging from 0 to $+90°$, and although these diagrams fill a small volume, they are by no means a complete working set. The use of these charts consists in comparing directly certain characteristic distances on the aeromagnetic anomaly pattern with equivalent lengths on the appropriate theoretical diagram. Since the latter is scaled in units of h, the ratio of these lengths will give h directly.

Much of this information can be distilled into sets of characteristic curves which effect a great reduction in the volume of numerical data. A good estimator for the ratio b/a is the ratio of the two widths of the anomaly pattern parallel to and transverse to its "strike" direction, at an amplitude which is half-way between the maximum and minimum values. We shall denote these widths with the symbols $(w_{\frac{1}{2}})_y$ and $(w_{\frac{1}{2}})_x$, respectively. The estimator for h/a in the higher latitudes ($|i| \geq 60°$) is the ratio of the mean maximum horizontal gradient transverse to the strike direction, multiplied by $(w_{\frac{1}{2}})_x$ and divided by the total amplitude ($\Delta T_{max} - \Delta T_{min}$). By using the mean of the maximum slopes on the two sides of the anomaly peak (both taken with a positive sign), we cancel out any linear regional gradients in the magnetic field and we also reduce the effect of dip, which is not allowed for in the model. The two estimators are illustrated in Figs. 11-21 and 11-22.

The set of characteristic curves for $i = 75°$, $\lambda = 45°$ is shown in Fig. 11-23, and the curves for determining h are shown in Fig. 11-24. Figure 11-23 is used to find b/a and h/a simultaneously, so that the effect of the shape of the prism (b/a) can be eliminated by choosing the appropriate curve of Fig. 11-24. To accommodate a full range of values of λ and i, about 30 sets of these graphs are needed. These will include over 1,200 different combinations of the four parameters i, λ, h/a, and b/a.

As a practical example of the use of these curves, we show in Fig. 11-25 a portion of an aeromagnetic survey made in southwestern Ontario, where the Precambrian rocks are covered with approximately 3,000 ft of Paleozoic sediments. The strike declination of the anomaly pattern is 48°, and the inclination of the geomagnetic field is 75°. The point which corresponds to this anomaly is marked in Fig. 11-23 and indicates a value of 3.5 for the ratio b/a and 1.25 for h/a. According to Fig. 11-24, the value of $(w_{1/2})_x/h$ is 2.52, and since $(w_{1/2})_x = 2.40$ miles, it follows that $h = 5,030$ ft. The height of the aircraft above the ground was 1,800 ft, and so we conclude that the depth of the magnetic rocks is about 3,200 ft. Since these rocks are assumed to lie within the Precambrian complex, we infer on the basis of the prism model that the *maximum* thickness of the sediment at this point is just over 3,000 ft. (The smaller anomaly on the southwest flank indicates a much

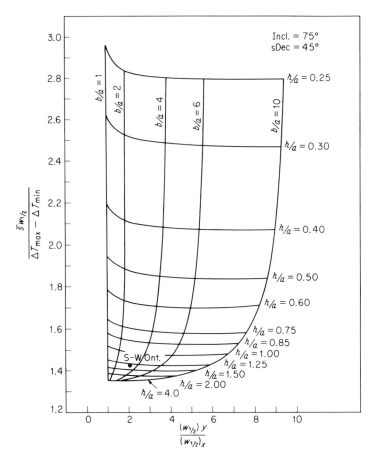

Fig. 11-23 Total-field characteristic curves for the vertical prism model; $i = 75°$, $\lambda = 45°$. **[Courtesy of** *Computer Applications and Systems Engineering* **(*C.A.S.E.*), Toronto, Canada.]**

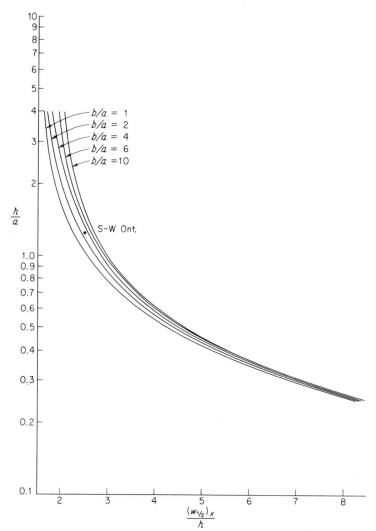

Fig. 11-24 Complementary curves for determining the depth of a prism from its total-field characteristics; $i = 75°$, $\lambda = 45°$. **(Courtesy of** *C.A.S.E.*, **Toronto, Canada.)**

smaller depth, which suggests that the prism is not a suitable model for this feature.)

In one sense, the bottomless prism model represents an extreme situation geologically, in which the magnetized rocks extend to a very great depth without any diminution in their magnetic properties. This would seem to suggest rocks having a plutonic origin. The other extreme would correspond to magnetized rocks in formations which have a limited depth extent, such as volcanic flows and sills, and for this the interpretational model is a thin, rectangular, horizontal plate. There is no point in trying to use a depth-limited prism, because aeromagnetic anomaly patterns are rarely so well determined that we could contemplate introducing a fifth parameter into the characteristic curves. As long as the thickness of the magnetized formation is not much greater than its vertical distance from the aircraft, the plate model will be adequate. It is assumed to be magnetized in the direction of the geomagnetic field with a "surface magnetiza-

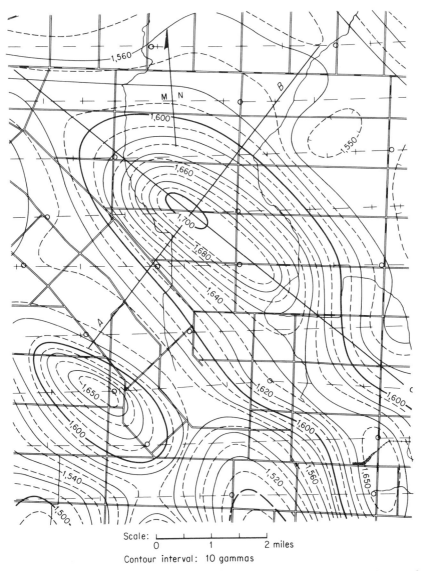

Scale: 0 — 1 — 2 miles

Contour interval: 10 gammas

Fig. 11-25 Portion of an aeromagnetic survey flown at an altitude of 1,800 ft over a part of southwestern Ontario, Canada. (Courtesy of *Hunting Survey Corporation*, Toronto, Canada.)

tion," which is the product of the average magnetic moment and the thickness.

The estimators for the horizontal plate model are the same as the ones we used for the prism. The characteristic curves are not shown because they resemble those of the prism, except that they are displaced upward. Thus, as we should expect, greater values for h will be deduced from the plate model than from the prism, for the same anomaly pattern. Depth-limited sources can sometimes be distinguished from those having a large depth extent by the greater sharpness of the anomaly peak. In doubtful or uncertain cases, the two models can be used to estimate upper and lower bounds for h.

As a second type of model for aeromagnetic interpretations, we shall consider a dipping sheet of finite strike extent. This is the same model that we have previously used to represent dikes or veins in the interpretation of ground surveys. We now introduce it in order to represent mineralized shear zones, fractures, or intrusive belts that are distinctly linear in general outline. Such features are very common in Precambrian rocks and most are much too narrow in proportion to their length to be represented by prisms. Thus we consider the sheet to be infinitely thin and to have a surface magnetization only. The two extremes of this model are a bottomless sheet $l \to \infty$ which corresponds to a zone of deep fracturing and a finite line of dipoles $l \to 0$ which represents a magnetic stringer formation.

We shall neglect demagnetization and assume that the plate is magnetized in the direction of $\mathbf{H}_0$. As we found out earlier, demagnetization is likely to have a pronounced effect upon the calculation of the dip of the model but only a minor effect upon the interpretation of depth. Dip is not a quantity likely to be easily interpretable from aeromagnetic anomalies in any case, and as far as depth studies are concerned, it holds little interest. Thus the formula used for the total-field anomaly is

$$\Delta T(x,y) = Ms(1 - \cos^2 \lambda \cos^2 i)[f(x, y + Y) - f(x, y - Y)] \quad (11\text{-}31)$$

where $f(x, y + Y)$

$$= \frac{y + Y}{\sqrt{C^2 + (y + Y)^2}} \left[\frac{(x - l \cos d) \cos Q - (h + l \sin d) \sin Q}{C^2} \right]$$

$$- \frac{y + Y}{\sqrt{D^2 + (y + Y)^2}} \left[\frac{x \cos Q - h \sin Q}{D^2} \right] - \frac{1}{B^2 + (y + Y)^2}$$

$$\times \left[\frac{A + l}{\sqrt{C^2 + (y + Y)^2}} - \frac{A}{\sqrt{D^2 + (y + Y)^2}} \right] [(\cos Q \cos d - \cos^2 \beta$$

$$+ K^2)(y + Y) + 2K \sin (d - \beta)B] - 2K \cos (d - \beta) \left[\frac{1}{\sqrt{C^2 + (y + Y)^2}} \right.$$

$$\left. - \frac{1}{\sqrt{D^2 + (y + Y)^2}} \right]$$

where $\qquad Q = 2\beta - d \qquad \tan \beta = \dfrac{\tan i}{\sin \lambda} \qquad K = \cos \beta \cot \lambda$

$$A = h \sin d - x \cos d \qquad B = x \sin d + h \cos d$$

$$C^2 = (x - l \cos d)^2 + (h + l \sin d)^2 \qquad D^2 = x^2 + h^2$$

This formula contains six parameters, h, l, Y, d, i, and λ. Two of these, i and λ, are assumed to be known, and a third, l, will be used as a unit of length. Thus three of the parameters are "free," viz., h/l, Y/l, and d. The strike length has little effect upon the principal profile characteristics as long as $Y \geq 2l$, so that characteristic curves may be drawn for various combinations of i and λ and for $Y/l = 0.5$, 1, and ∞.

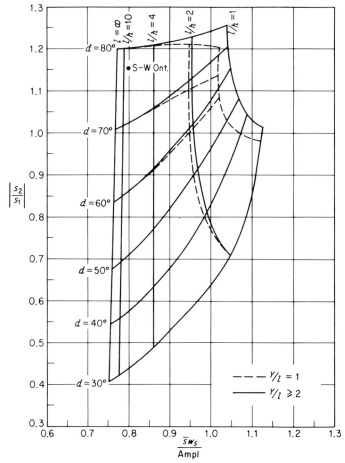

Fig. 11-26 Total-field characteristic curves for the ribbon model; $\gamma = 80°$, $K = 0.2$.

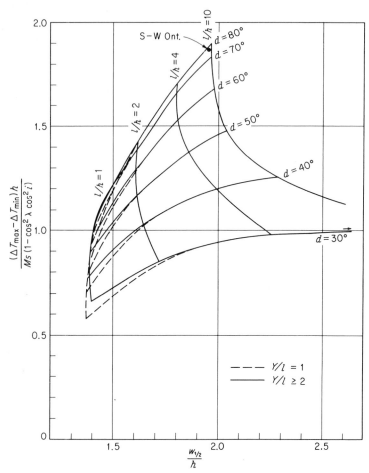

Fig. 11-27   **Complementary curves for estimating the depth and surface magnetization of a ribbon from its total-field characteristics; $\gamma = 80°$, $K = 0.2$.**

The characteristic estimator for the depth of the ribbon is the ratio $w_s \bar{s}/\text{AMPL}$, in which w_s is the horizontal distance between the inflection points of the principal profile, $\bar{s}$ is the mean of the two maximum horizontal gradients, and AMPL signifies the total amplitude of the anomaly $\Delta T_{\max} - \Delta T_{\min}$. Once again, the mean maximum horizontal gradient is used in order to minimize the effect of dip and to cancel out linear regional gradients. The characteristic estimator for the dip is, as usual, the ratio of the two maximum horizontal gradients (the positive one in the numerator) taken with a positive sign. A set of these curves is shown in Fig. 11-26. The complementary set is shown in Fig. 11-27.

We now wish to illustrate the importance of choosing carefully the model to use in an interpretation. According to Fig. 11-26, the ribbon which best fits the anomaly pattern shown in Fig. 11-25 (whose K value = 0.18) has a dip of 77° and a ratio $l/h = 10$. From a separate diagram, which is not shown, we also learn that $Y \doteq 3h$. Thus from Fig. 11-27 it follows that $w_{\frac{1}{2}}/h = 1.95$ and $(\Delta T_{max} - \Delta T_{min})h/Ms(1 - \cos^2 \lambda \cos^2 i) = 1.87$. Since $w_{\frac{1}{2}} = 2.40$ miles and $\Delta T_{max} - \Delta T_{min} = 166$ gammas, it turns out that $h = 6,500$ ft and $Ms = 11$ emu, with s in feet. It thus appears that the ribbon model implies a maximum sedimentary thickness of 4,700 ft, whereas the prism model indicates a maximum thickness of only 3,200 ft. But when we calculate and plot the principal profiles of these two best-fitting models and compare them with the observational values (Fig. 11-28), any ambiguity between them at once disappears. It is perfectly clear from these results that the magnetized zone has a thickness which can be discerned by a magnetometer operating at a distance of 5,000 ft or more. A little previous familiarity with the shapes of profiles across tabular bodies and ribbons would have indicated this fact and guided us to the better model at the outset.

A third model often used in aeromagnetic interpretations is the tabular body of large strike length discussed in Sec. 11-8. There are probably others also. We shall not extend our illustrations beyond the two models we have already discussed, but it is noteworthy that the results obtained from the southwestern Ontario anomaly with the tabular model match those obtained with the vertical prism very closely. This establishes beyond

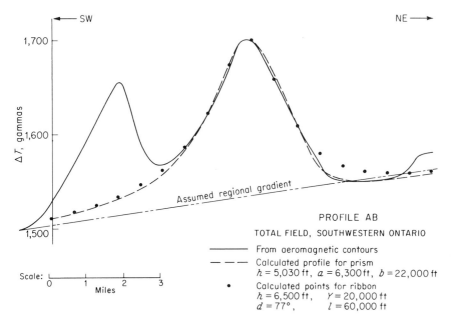

Fig. 11-28 Principal profile of the total-field anomaly pattern shown in Fig. 11-25.

much doubt that the "body" is a zone of appreciable thickness and supports the shallower depth as the more correct one.

Finally, we comment that the calculation of depth by aeromagnetic surveys depends very heavily at present upon the use of anomaly patterns and models. To obtain satisfactory results, both must be chosen carefully. Furthermore, the analyst must be prepared to go to great pains in separating features which overlap. Attention is being given to the question of using "roughness" in a quantitative manner by studying the spectral density functions of aeromagnetic maps, but only time will tell how effective this approach is likely to be. In the meanwhile, the method of characteristic curves seems to fulfill the purpose of making the maximum number and variety of models available for use with the minimum amount of numerical calculation.

References

1. L. J. Peters, The Direct Approach to Magnetic Interpretation and Its Practical Application, *Geophysics*, vol. 14, pp. 290–320, 1949.

2. D. S. Hughes and W. L. Pondrom, Computation of Vertical Magnetic Anomalies from Total Magnetic Field Measurements, *Am. Geophys. Union Trans.*, vol. 28, pp. 193–197, 1947.

3. V. Baranov, A New Method for Interpretation of Aeromagnetic Maps: Pseudo-gravimetric Anomalies, *Geophysics*, vol. 22, pp. 359–383, 1957.

4. J. Stratton, "Electromagnetic Theory," McGraw-Hill, New York, 1941.

5. von Carlheim-Gyllenskjöld, "Magnetic Survey of Kiirunavaara," Stockholm, 1910.

6. H. Haalck, Zür Frage der Erklärung der Kursker magnetischen und gravimetrischen Anomalie, *Gerlands Beitr. Zür Geophysik*, vol. 22, pp. 385–399, 1929.

7. J. H. Ratcliffe, The Boston Township Iron Range, pp. 195–210 in "Methods and Case Histories in Mining Geophysics," Sixth Commonwealth Mining and Metallurg. Cong., Mercury, Montreal, 1957.

8. K. L. Cook, Quantitative Interpretation of Magnetic Anomalies over Veins, *Geophysics*, vol. 15, pp. 667–686, 1950.

9. R. D. Hutchinson, Magnetic Analysis by Logarithmic Curves, *Geophysics*, vol. 23, pp. 749–769, 1958.

10. S. P. Gay, Standard Curves for Interpretation of Magnetic Anomalies over Long, Tabular Bodies, *Geophysics*, vol. 28, pp. 161–200, 1963.

11. V. Vacquier, N. C. Steenland, R. G. Henderson, and I. Zeitz, "Interpretation of Aeromagnetic Maps," Geol. Soc. America Memoir 47, 1951.

chapter 12　Rock Magnetism

In earlier chapters we discussed the theory of the magnetic method of geophysical exploration. By applying interpretation procedures such as those described in Chap. 11 to data obtained from magnetic surveys, we draw detailed conclusions about the magnetization of rocks within the ground, and assume that this property is ostensibly related to the lithology of the subsurface. Unfortunately, the connection between a rock's geological properties and its magnetization is rarely simple, and it is, at this stage of our knowledge, impossible to lay down any strict rules upon which to pattern a relationship. Even so, it is worth investigating the very considerable fund of information available on this subject, so that any geological inferences to be drawn from magnetic interpretations may at least be well-informed. In the past fifteen years a considerable effort has been directed toward research in rock magnetism (1) (2), mostly with a view to helping decide upon such large-scale geophysical and geological questions as whether or not continental drift and polar wandering have taken place. It is the purpose of this chapter to select from this research the points of greatest significance to magnetic interpretations.

12-1   *Ferromagnetism*

If a rock has an appreciable magnetization, it is because it is ferromagnetic. All materials are diamagnetic, i.e., when placed in a magnetic field, they will acquire a small moment which is in a direction opposing the field.   The cause of this effect is the Larmor precession of electron orbits (1), but the phenomenon is an exceedingly weak one and is totally negligible in geological surroundings.   Very many materials are also paramagnetic, i.e., when placed in a magnetic field they acquire a magnetization which is proportional to, and in the same direction as, the external field.   This is usually much stronger than the diamagnetic effect, but even so it is of such small magnitude that it may virtually always be disregarded.

Paramagnetism has its origin in the intrinsic magnetic moments of individual ions.   Normally these have a random orientation within the material, because of thermal agitation of the lattice structure.   If, however, the bonding between the atoms of a crystal is of such a nature that it affects the orientation of the orbital structures, an alignment of the magnetic moments may take place.   It will be opposed by the thermal agitation, of course, but below a certain temperature, called the Curie point, a natural alignment, and therefore a spontaneous macroscopic magnetization, will take place.   The phenomenon is not uncommon in crystalline materials, but it is most easily, and was earliest, recognized in iron.   It is thus known as *ferromagnetism.*

Although the underlying principle is simple enough to understand, ferromagnetic phenomena can be extraordinarily complicated in detail.   To start with, we must distinguish three ways in which alignment may take place.   If the magnetic moments of all the ions in the lattice tend to point in the same direction, the material is said to be truly ferromagnetic.   If, however, some of the moments tend to lie in a direction quite different from that of the others (usually in exactly the opposite direction), the material is said to be *ferrimagnetic.*   In the special case in which the ions of the lattice are divided into two exactly equivalent groups, or sublattices, which are magnetized oppositely, the material is said to be *antiferromagnetic.*   Accord-

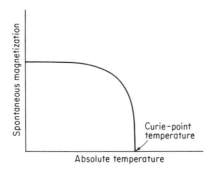

Fig. 12-1   **The spontaneous magnetization of a ferromagnetic crystal lattice below its Curie point.**

ing to simple theory, then, the macroscopic moment of an antiferromagnetic material must be zero.

Examples of each of these three types of ferromagnetism can be found among materials familiar in geology. Metallic iron, nickel, and cobalt are true ferromagnetic substances. Magnetite (Fe_3O_4) is the classic example of a ferrimagnetic material. Other ferrimagnetic substances that have a common occurrence in geological environments are pyrrhotite (Fe_7S_8), ulvöspinel (Fe_2TiO_4), and a form of hematite known as maghemite (γFe_2O_3). Of the antiferromagnetic materials hematite (αFe_2O_3) is probably the best known, but another example from among the common minerals is ilmenite ($FeTiO_3$).

Since in a ferromagnetic material the magnetic moments of all atoms are coupled together, we might expect crystals of these substances to exhibit large spontaneous magnetic moments at all times. As is so often the case in physics, however, this is not what is observed. Only when placed in a moderately strong magnetic field does a ferromagnetic crystal exhibit its maximum moment. The reason for this is that the crystal subdivides itself into numerous regions, known as *domains*, whose magnetic moments are oppositely, or at least differently, directed. This allows alignment of the electron-spin vectors of neighboring atoms (known as exchange coupling) to take place except in the boundaries between domains. At the same time it prevents the formation of strong magnetic fields in and around the crystal, which would lead to large amounts of stored magnetostatic energy. The crystal thus compromises, in fixing its internal structure, between a minimal exchange coupling energy (uniform alignment everywhere) and a minimal magnetostatic energy (random alignment). In this way, the total free energy of the crystal is held to a minimum value.

The domains which form in a ferromagnetic crystal are rather variable in shape and size. Some schematic examples are shown in Fig. 12-2. In

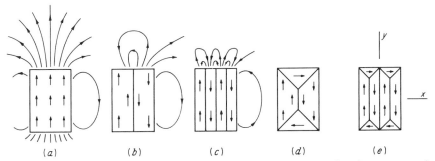

Fig. 12-2 **Schematic diagrams showing how the division of a ferromagnetic crystal into domains can reduce its magnetostatic energy.** The series (*a*) to (*d*) shows a successive reduction in energy; (*e*) shows a domain pattern similar in effect to (*d*), but which would be more likely to occur if the *y* direction of magnetization were preferred to the *x* direction.

large, nearly perfect crystals, the domains may have dimensions as large as several microns. However, in the small, imperfect grains of the ferrimagnetic substances found in rocks they are usually much smaller. Grains larger than about 0.1 μ are likely to contain at least two domains. The crystallographic directions in which the domains can be magnetized depend very much on the crystal structure and on the nature of the exchange coupling. In general there are several directions in which the domain magnetization may lie without causing a large increase in the internal energy. All crystals are aeolotropic, however, and in a crystal of low symmetry a single most favored direction (+ and −) will generally exist. According to the picture just described, a material made up of a large number of isolated ferromagnetic grains distributed throughout an inert matrix behaves in the following way: If the substance is isolated from external magnetic fields, the domains in each crystalline grain will arrange themselves in such a way that their internal field will vanish to a high degree. Thus the net moment of the material will be almost zero. If a magnetic field is then applied, a net magnetization will appear. Those domains whose magnetizations lie in the direction of the field will enlarge themselves, and the others will diminish by a movement of the domain boundary surfaces. The net magnetization which appears in each grain will then be just sufficient to produce an internal field equal and opposite to the external field, since this maintains the free energy at a constant, minimal amount.

To gain a rough idea of what the *net* or bulk moment induced in the material might be, we can think of the grains as spheroids and use the results discussed in Sec. 11-5. If each spheroid is uniformly magnetized in its axial direction with a magnetization M_g, a uniform magnetic field of intensity $-NM_g$ will be created inside it. The demagnetizing factor N depends on the ellipticity of the spheroid and varies from 4π for a thin circular disk, through $4\pi/3$ for a sphere, to zero for a very long needle. Thus if the external field H_0 is to be canceled in the interior of a grain, the magnetization will be H_0/N. The net magnetization of the material can then be found if the volume fraction of magnetic material is known and if a mean value can be assumed for N (averaged on a volume basis). If these parameters are p and $\bar{N}$ respectively, the net magnetization M of the bulk material will be

$$M = \frac{pH_0}{\bar{N}} \tag{12-1}$$

and thus the material will exhibit a bulk susceptibility k, which is given by

$$k = \frac{p}{\bar{N}} \tag{12-2}$$

The above picture is, of course, much oversimplified but is at least qualitatively correct. To explain the actual magnetic behavior of a real

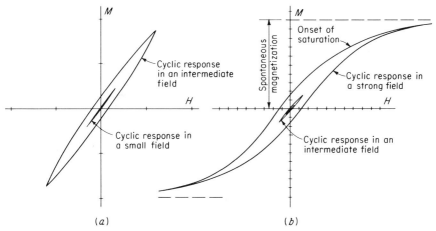

Fig. 12-3 Magnetizations induced by a cyclic magnetic field. (a) Small and intermediate peak field strengths; (b) strong field. Note the increasing hysteresis with increasing peak field and the saturation effect in a strong field.

ferromagnetic material, however, we must also recognize the presence of minute defects in the crystal lattice, such as dislocations, lattice vacancies, impurities, etc. These tend to restrict the movement of domain boundaries, so that the pattern of domains, and thus the grain's net magnetic moment, cannot change with complete freedom. Thus, if an aggregate of ferromagnetic grains is placed in a magnetic field, its induced magnetization will fall at least slightly below the value predicted by (12-1); and when the field is removed, the magnetization will not completely disappear. In a cyclical field of moderate intensity a plot of the magnetization versus the external field strength will be an ellipse such as that shown in Fig. 12-3a. This is the basis of the familiar hysteresis effect.

If the strength of the external field is increased, the induced magnetization will increase also. At low-to-moderate field intensities, the relationship is roughly one of simple proportionality. However, the induced magnetization clearly cannot exceed the spontaneous magnetization, and as the external field is increased, a limit is eventually reached beyond which a further increase in magnetization is not possible, and the material is said to be saturated. The approach to saturation seldom occurs suddenly but rather takes place gradually over a finite interval. The reason for this is that at higher field intensities it is possible to cause a rotation of those domain magnetizations which are not parallel with the field out of their most favored directions.

The residual magnetization which will remain in a material after the completion of a magnetizing cycle (or half-cycle) behaves somewhat differently as the maximum external magnetic-field strength is increased. At first it is negligible, increasing approximately as the square of the maximum

field strength.   It then slowly reaches a saturation value as the maximum
field intensity approaches a level sufficient to cause the maximum possible
movement of the domain walls.

12-2   *The Influence of Grain Size*

In the foregoing discussion we visualized each ferromagnetic grain as containing a very large number of domains.   Such will be the case if the grains
are comparatively large, perhaps greater than 10 μ in average diameter in
magnetite, for example.   When the grains are smaller than this, the magnetic properties of the material are somewhat altered.

As grain size decreases, it seems that domain boundaries find themselves
more deeply trapped.   Thus, the magnetization induced by a given external
field in a material containing fine-grained ferromagnetic constituents is
rather less than would be induced in a coarse-grained material having the
equivalent concentration.   Moreover, when the field is removed, a stronger
residual magnetization remains.

The hysteresis effect continues to increase with decreasing grain size
until the material contains only monodomain grains.   Obviously domain
wall movement then ceases to be a factor in determining magnetic properties, and magnetization occurs only by rotation or inversion of the spontaneous moment of those grains which are not magnetized in the direction
of the field.   Since a rather high energy barrier exists between the different
directions in which the magnetization is stable, it is understandable that a
strong field will be required to produce changes, and that once changed, the
magnetization does not easily return to its original state.   However, it is
important to bear in mind that monodomain grains are extremely small and
therefore susceptible to more violent thermal agitation than large grains.
In fact, it is the interplay between thermal agitation, grain size, and aeolotropy which determines the magnetic properties of aggregates of monodomain
grains.

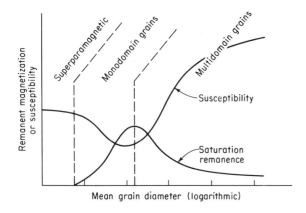

**Fig. 12-4 Variation of
susceptibility and satu-
ration remanence with
grain size (schematic only).**

Néel (3) has provided a theory which successfully accounts for the main properties of monodomain grains. According to the theory and to experimental observations, their susceptibility increases with decreasing grain size while hysteresis effects diminish, until a critical point is reached (whose value is highly dependent on temperature) below which thermal agitation maintains the magnetization in a completely free state. In this condition the material exhibits no hysteresis and behaves exactly like a paramagnetic substance, except that it is *much* more sensitive to external magnetic fields. A material in such a state is said to be superparamagnetic. Some sandstones that contain extremely fine-grained maghemite exhibit this property. Such grains are submicroscopic in size, being only barely visible under the electron microscope.

12-3 Magnetic Minerals

The minerals which have magnetic properties that are important in geology may for the most part be put into two geochemical groups, viz., the iron-titanium-oxygen group and the iron-sulfur group. The geochemical and magnetic properties of the iron-titanium-oxide system have been carefully and intensively studied, but they are not by any means fully known. The heavy lines on Fig. 12-5 indicate the two-component systems which have been investigated geochemically (4). The solubility gaps which occur at room temperature along with the approximate temperatures at which they close are also shown. There are very substantial regions in this diagram where stable solid solutions are not possible. Thus if time and temperature permit, the oxide materials commonly found in igneous rocks are likely to separate into two-component intergrowths. The basic magnetic properties of the pure minerals of this group have also been comparatively well investigated as have also the properties of many metastable solid solutions of intermediate composition (5).

Magnetite (Fe_3O_4) is ferrimagnetic. It crystallizes in the cubic system with an inverse spinel structure, and its formula can be written as

$$(Fe^{3+})_8[(Fe^{++})_8(Fe^{3+})_8](O^{--})_{32}$$

It has a Curie temperature of 580°C and a saturation magnetization (per gram) of 93 emu. The easy direction of magnetization is the 111 axis. At low temperature it has fairly complete solubility with ulvöspinel but only limited solubility with ilmenite or hematite.

Ulvöspinel (Fe_2TiO_4) is also ferrimagnetic, having the same structure as magnetite, in which $Fe^{++} + Ti^{4+}$ replaces $2Fe^{3+}$. However, it is paramagnetic at room temperature, indicating that the Curie temperature of this structural group falls rapidly as the composition moves from pure magnetite toward ulvöspinel.

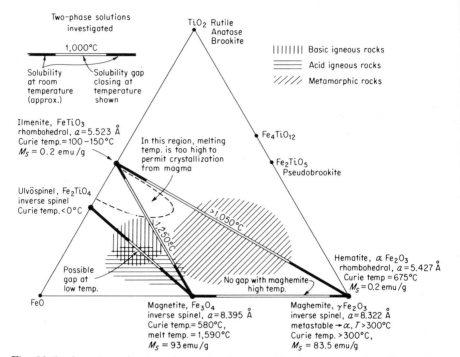

Fig. 12-5   A review of mineralogical and magnetic data concerning the system FeO, TiO₂, Fe₂O₃.

Ilmenite (FeTiO₃) has a rhombohedral structure. It is antiferromagnetic with a Néel temperature[1] of about 100 to 150°C, although it usually shows a small unbalance in the theoretically perfect cancellation, which gives it a saturation magnetization of about 0.2 emu.

There are two forms of Fe_2O_3: *hematite* (αFe_2O_3), which has a rhombohedral structure and is fundamentally antiferromagnetic, and *maghemite* (γFe_2O_3), which has an inverse spinel structure like magnetite but has one-ninth of the iron sites vacant. Hematite, like ilmenite, has a slight ferrimagnetization of about 0.5 emu and a Néel temperature of 675°C. Maghemite is ferrimagnetic and has a saturation magnetization (per gram) of 83 emu and an unmeasured Curie temperature. This is because pure maghemite is unstable at temperatures exceeding about 300°C when it converts spontaneously to hematite. It appears, however, that impurities can stabilize it at much higher temperatures, and its Curie point is then well above 500°C.

The other minerals of this group are comparatively rare, and less is known of their properties.

[1] The temperature above which thermal agitation prevents ordering in an antiferromagnetic substance is usually known as the Néel temperature rather than the Curie temperature.

Much less is known geochemically about the iron sulfur group. The compositional sequence is usually written FeS_{1+x}. When $x = 1$, we have the mineral *pyrite*, which has a cubic structure and is paramagnetic. $x = 0$ corresponds to the mineral *troilite*, which is antiferromagnetic and has a Curie temperature of about 600°C. Troilite has a hexagonal structure. The range $0 < x < 1.0$ includes the *pyrrhotites*, natural pyrrhotites occurring mostly from the middle to the lower end of this range. Pure pyrrhotite is often stated to have the composition Fe_7S_8 but also is sometimes identified with FeS. Between $x = 0.1$ and 0.94 the mineral is ferrimagnetic, having a saturation magnetization of about 60 emu (per gram) and a Curie temperature of 300 to 325°C. The range of solid solubility seems to be complete between $-0.16 < x < 1.0$, but the mineralogy has been very little investigated. The properties of pyrrhotites in sulfide ores are completely unstudied.

12-4   The Geochemistry of Magnetic Minerals

In order to discuss the origin of the magnetic minerals in rocks, we may consider the three fundamental geological processes: magmatic or volcanic crystallization, sedimentation, and metamorphism. We begin first with the crystallization process, since this involves no reference to either original minerals or structure.

The ultimate form of any igneous crystalline rock depends on three factors: the original composition of the melt, the rate of cooling during crystallization, and the occurrence of macroscopic changes in melt composition due to the addition or subtraction of material, such as the loss of volatile liquids that sometimes occurs while the melt is relatively permeable. For a melt having a given chemical constitution, there will be a certain mineral compositional structure that is in equilibrium with the residual liquid at each temperature. This structure is determined by the thermodynamic properties of the minerals themselves and changes continuously with the decreasing temperature. However, compositional changes in the already solidified mineral components take place extremely slowly, and equilibrium is approximated only in the liquid component and on mineral surfaces exposed to it. Growing crystals therefore become zoned, and previously solidified crystals may dissolve until complete solidification is eventually attained. Meanwhile, some of the crystals which had formed at higher temperatures become unstable at lower temperatures and may endeavor to purify themselves by slowly exsolving certain substances. Complex intergrowths of allied minerals may therefore develop, and small grains of secondary minerals may be precipitated within the larger host crystals. Magnetic minerals may be found in either form or as original unexsolved primary crystals. Generally speaking, the faster the rate of cooling, the fewer the number of secondary crystals relative to primary, since the pri-

Table 12-1   Influence of Cooling Rate on Exsolution Structures*

Rock type	Number of samples	Percent showing only a spinel or an orthorhombic Fe, Ti oxide lattice	Percent showing exsolution intergrowth between Fe, Ti oxides
Basalt	17	65	23
Dolerite	36	22	78
Gabbro	29	26	74

*After G. D. Nicholls (4). Data from W. H. Newhouse, Opaque Oxides and Sulphides in Common Igneous Rocks, *Geol. Soc. America Bull.*, vol. 47, pp. 1–57, 1936.

mary minerals have had no time in which to exsolve their unstable constituents. Table 12-1 illustrates this.

As we have previously stated, the bulk of the magnetic material in igneous rocks is in the form of iron-titanium-oxygen compounds or of iron-sulfur compounds. It is noteworthy that oxides and sulfides of other elements, although often of economic importance, are rather rare. The chemical explanation for the tendency of these two mineral groups to form from complex silicate melts does not seem to be known at present, and information relating the composition of the oxide or of the sulfide mineral fractionates to the petrology of the whole rock is very sparse indeed. The few published analyses suggest that the composition of the oxide minerals tends to lie in a fairly restricted area on the Fe-Ti-O diagram, the oxides in mafic rocks containing rather more titanium than those in silicic rocks. Figure 12-5 shows the regions. Perhaps the only other generalization that can be made is that mafic rocks tend to carry larger quantities of oxides and sulfides than silicic rocks.

Metamorphism can cause considerable change in the oxide constituents of an igneous rock. Magnetite may be oxidized toward hematite or to maghemite, the latter process occurring at lower temperatures than the former. Often the oxidation is incomplete, and only parts of the magnetite crystals (along surfaces such as cracks) are affected. Balsley and Buddington (8) have studied the composition of the oxides of some Appalachian metamorphic rocks which fall into quite a different region of Fig. 12-5 than most igneous rocks. It is, however, possible that many of these are highly metamorphosed sediments in which hematite has been reduced toward magnetite by regional metamorphism, rather than altered igneous rocks in which magnetite has been oxidized toward hematite by hydrothermal solutions. Stress often accompanies the higher temperatures of metamorphism, promoting the regrowth of many crystals. Curie temperatures may even be exceeded in very high-grade metamorphism.

The magnetic minerals of sedimentary rocks are largely the same as those of igneous rocks. The iron oxides occur either as detrital particles or else they are created from iron which has been precipitated into the sediment mainly as ferric hydroxide. Magnetite, ilmenite, and hematite are resistant to low-temperature weathering and are likely to occur in rather large particles in detrital sediments such as sandstones. They may also occur in glacial clays as minute grains with dimensions of the order of a micron or less. Ferric hydroxide tends to precipitate in both marine sandstones and clay sediments which have formed near shorelines, since it is insoluble in sea water. It is rarely found in limestones or in deep-water sediments. Ferric hydroxide ages to goethite or (much more rarely) to lepidocrocite, which may in the process of lithification dehydrate to hematite or maghemite, respectively. The grain sizes within these materials are extremely fine, being of the order of 0.1 μ or less. Metamorphism tends to promote the growth of these crystals, and if the environment is a reducing one, it will change the bulk composition of the oxides toward the FeO side of the composition scale.

Iron sulfide is also created during the lithification of sediments, particularly the argillaceous muds. Usually pyrite is formed in the process, but occasionally other sulfides are found. Metamorphism may mobilize this material and redeposit it elsewhere in the adjacent rocks.

12-5   *The Magnetization of Rocks*

The ferromagnetic minerals in a rock can become magnetized in a wide variety of ways. There is, of course, the induced magnetization that is determined by the earth's field and by the present state of the magnetic grains, but this is very often accompanied by a permanent magnetization which has been acquired continuously or intermittently during the rock's history. This permanent magnetization is given the name *natural remanent magnetization* (NRM), and we shall discuss some of the processes by which it may have been acquired.

Induced magnetization.   We have indicated in Sec. 12-1 that a rock which contains ferromagnetic minerals is susceptible, taking on a magnetization proportional to, and in the direction of, the ambient magnetic field. While this statement is generally true, certain exceptions must also be recognized. It is sometimes found, for example, that the magnetization does not bear a simple proportionality to the field intensity even for fields as weak as the earth's. In the majority of cases in which it does not, the susceptibility increases somewhat with increasing field strength. Moreover, aeolotropy is not uncommon, so that induced magnetic moments are sometimes deflected out of the direction of the ambient field. Many metamorphic rocks, in particular, exhibit a platy, crystalline grain structure which gives the average demagnetizing factor $\bar{N}$ in Eq. (12-1) different values in different directions. As a general rule, the bulk susceptibility will be smaller in the direction normal to the schistosity or gneissosity than in a

Table 12-2   **Susceptibility of Common Rocks***

Rock type	Number of samples	Percent having volume susceptibility, emu			
		$k < 10^{-4}$	$10^{-4} < k < 10^{-3}$	$10^{-3} < k < 4 \times 10^{-3}$	$k > 4 \times 10^{-3}$
Basic effusive	97	5	29	47	19
Basic plutonic	53	24	27	28	21
Granites and allied rocks	74	60	23	16	1
Gneisses, schists, slates	45	71	22	7	0
Sedimentary rocks	48	73	19	4	4

* From L. B. Slichter (6).

direction parallel to it.   Variations of several percent are not uncommon, and extreme cases arise in finely banded magnetite or pyrrhotite.   There the demagnetization factor is for the most part 4π in the direction transverse to the structure and zero in a direction parallel to it.   Thus the transverse susceptibility will be just $p/4\pi$, while the parallel susceptibility will be limited only by the defect structures in the ferromagnetic grains and by the appropriate demagnetizing factor of the outer envelope of the mineralized region.

Several attempts have been made to formulate statistical "laws" relating the bulk susceptibility of rocks to petrological parameters.   A useful guide, taken from "Handbook of Physical Constants" (6), is given in Table 12-2.   It clearly demonstrates the tendency of magnetite to concentrate in mafic rocks, which therefore exhibit a higher susceptibility than the more silicic types.

Mooney and Bleifuss (7) have measured the susceptibilities of a suite of Precambrian rocks from Minnesota, and the averages they obtained for different rock types are shown in Table 12-3.

Table 12-3

Rock type	No. of samples	Mean k, emu
Basalt	37	2.95×10^{-3}
Diabase	19	2.59×10^{-3}
Rhyolite	5	1.12×10^{-3}
Gabbro	37	0.99×10^{-3}
Granite	31	0.47×10^{-3}
Other acid intrusives	17	0.35×10^{-3}
Ely greenstone	15	0.09×10^{-3}
Slates	26	0.05×10^{-3}

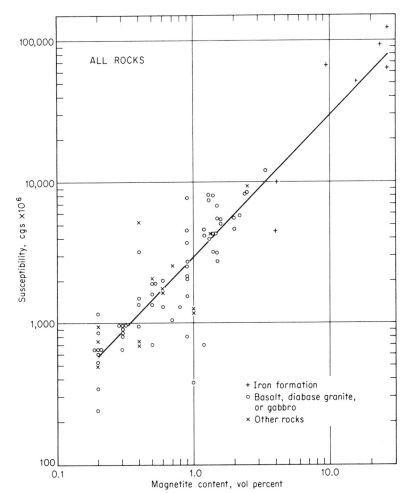

Fig. 12-6 Data from which the empirical formula for susceptibility $k = 2.89 \times 10^{-3} V^{1.01}$ **was derived.** [**Mooney and Bleifuss** (7).]

They also have derived a formula relating the bulk susceptibility to the volume of magnetite found by crushing, magnetic separation, and chemical analysis for iron. It is

$$k = 2.89 \times 10^{-3} V^{1.01} \qquad (12\text{-}3)$$

where V is the volume percentage of magnetite, and the data on which it is based are shown in Fig. 12-6.

Balsley and Buddington (8) have related the susceptibility of a suite of Adirondack rocks to the fractional volume of all the minerals visually identified as "magnetite," which would generally include any Fe-Ti oxide minerals of spinel structure. Their empirical formula for the bulk susceptibility is

$$k = 2.6 \times 10^{-3} V^{1.33} \qquad (12\text{-}4)$$

where V is the volume percentage of "magnetite." The data upon which this formula is based are shown in Fig. 12-7. It is interesting to note that throughout the range of V for which these two formulas are valid $(0.1 < V < 10)$ they give values which do not differ greatly from those predicted by Eq. (12-2). [When $V = 1$ percent, for instance, Eq. (12-4) gives the same value for k as Eq. (12-2) when the average demagnetizing factor is 3.8, a figure very close to the theoretical value for spherical grains.] The nonlinearity of (12-4) can probably be explained as being due to a diminishing grain size as V becomes much less than 1 percent and a general

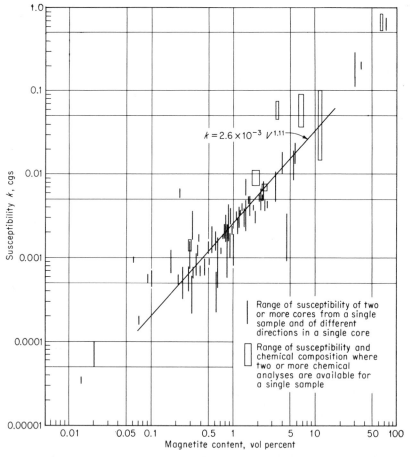

Fig. 12-7 **Data from which the empirical susceptibility formula $k = 2.6 \times 10^{-3} V^{1.33}$ was derived.** [Balsley and Buddington (8).]

decrease in the effective demagnetizing factor when the concentration is relatively large.

A similar formula has been determined by Bath (9) from analyses of magnetite-bearing iron ores. It is

$$k = 1.16 \times 10^{-3} V^{1.39} \tag{12-5}$$

where V is the volume percentage of "magnetite" determined by magnetic separation. The similarity in form between this expression and (12-4) is rather remarkable, and the difference in the proportionality constants can probably be explained by the tendency for the separation process to over-estimate the ferrimagnetic mineral content and for a visual or microscopic examination to underestimate it. Moreover, (12-5) ignores any contribution to the bulk susceptibility by hematite and other less susceptible oxide minerals which are very likely to be present.

Isothermal remanent magnetization (IRM). This is just the residual magnetization which is left after an external field is applied and removed from a magnetic material. As was explained in Sec. 12-1, it is of negligible intensity when the field is weak. Thus the earth's field does not produce significant IRM in rocks.

However, lightning strokes are accompanied by intense magnetic fields, and they can easily produce strong IRM. And although it may be thought that lightning is unlikely to have struck an outcrop, it is important to remember the very long exposure of many outcrops to the elements. Naturally the magnetic effects brought about by a lightning stroke are very local and also very irregular. It is unlikely that a single stroke will affect an area larger than a few hundreds of square feet.

Viscous remanent magnetization (VRM). The IRM caused by the application of a weak magnetic field during a short period of time is usually negligible. If, however, the substance remains exposed to the ambient field for a very long time, gradually more and more of the induced magnetization becomes irreversible. The increase in remanence is generally a logarithmic function of time and is known as a *viscous magnetization*. Rocks vary considerably in their magnetic "viscosity." The most sensitive are those which are on the verge of superparamagnetism. In some cases their remanence can be greatly changed in a few hours by a field that is no stronger than the earth's. Coarser-grained minerals also may exhibit this effect, especially when geological times are to be considered. The explanation of it is to be found in the thermal agitation which is always present in the crystal lattices. Because of its statistical nature it can, given enough time, move a domain wall over almost any obstacle. Solid diffusion of defect structures may also be a factor.

Thermoremanent magnetization (TRM). This is the most important mechanism to account for the permanent magnetization of igneous rocks. If a magnetic material is cooled from its Curie point even in

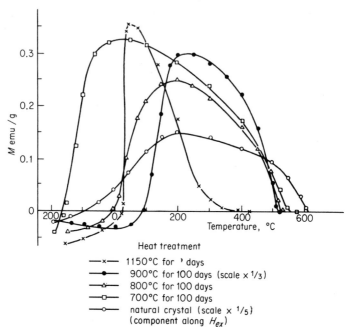

Fig. 12-8 Thermoremanence versus temperature for ilmenite-hematite crystals from Allard Lake, Quebec. [Carmichael, (12).] Compare these with the ideal simple curve of Fig. 12-1.

a weak magnetic field, it will acquire a relatively strong and very stable remanence. The properties of this phenomenon are rather remarkable and were first described by Thellier (10). In fields up to a few oersteds the magnetization is proportional to the magnetizing field strength. Moreover, partial TRMs are additive. By partial TRM we mean a magnetization produced by cooling a material from its Curie temperature to room temperature, while the magnetizing field has been applied only over a limited temperature interval. Thus it is found that the addition of two partial TRMs, the first having been acquired between the Curie point and temperature T and the second between T and room temperature, yields a sum that is equal to the total TRM produced in the specimen by cooling it directly from the Curie point down to room temperature in the same magnetic field. This is one of the very few cases in ferromagnetism where different processes add in a linear fashion, and it has considerable importance in paleo- and archeomagnetism. However, TRMs formed within a given temperature interval will not be the same for all temperatures. Materials made up of multidomain grains acquire most of their TRM within a few tens of degrees of the Curie point and change little thereafter. On the other hand substances containing monodomain grains tend to acquire their TRM over a comparatively broad temperature range.

One extraordinary property of TRM is its ability to become, in certain special cases, an inverted magnetization, i.e., one which is directed oppositely to the ambient magnetizing field. The circumstances and explanations of this phenomenon are rather complicated and have been outlined by Néel (11). The essential feature is an interaction between two ferromagnetic components which must be intimately associated either in a ferrimagnetic crystal lattice or in crystal intergrowths. The substance may then acquire a normal TRM near the Curie point, but as the temperature falls, it can change into a reversed magnetization. Figure 12-8 shows some measurements made by Carmichael (12) on a suite of ilmenite minerals which bear this out.

Chemical remanent magnetization (CRM). A magnetizing process somewhat akin to TRM is chemical remanent magnetization. It has not been as carefully studied as TRM, but the intensity of the effect seems to be proportional to the magnetizing field strength, and it appears to be a rather stable form of magnetization. It takes place whenever ferromagnetic grains grow or are transformed from one form to another at a temperature below their Curie point. It is probably the most important mechanism leading to permanent magnetization in many sedimentary and metamorphic rocks. Even in igneous rocks the iron oxide minerals may undergo a transition from one form to another during a slow exsolution or unmixing process and may then acquire a CRM.

Detrital remanent magnetization (DRM). This process can take place during the sedimentation of fine-grained, almost colloidal particles. Magnetite is well preserved during the weathering of rocks and may be among the detritus which slowly settles out of suspension. If the particle is a monodomain grain, it will have a rather large magnetic moment, which will tend in some degree to become aligned with the earth's field. Moreover, since the direction of the moment in the particle is likely to be fixed by its shape, the settling process may tend to deflect the direction of magnetization either toward or away from the horizontal. Thus, even if the grains are not closely aligned with the earth's field, the sediment will often acquire a net magnetization. Clearly the effect will be strongest when the particle size is small, so that the time of settling is long. Varved clays provide the best example of a DRM. Investigations of these clays reveal that the magnetic particles are extremely small and very numerous (of the order of $10^4/cm^3$). The NRM of the clay is then about 10^{-4} or 10^{-5} emu.

12-6 Paleomagnetism

It seems likely that the terrestrial magnetic field is generated by the self-exciting dynamo action of thermal convection currents within the earth's core. If this is so, it is also probable that the field is somewhat unstable and has changed considerably in amplitude—and even in polarity—in the past.   On the other hand, it seems from theoretical considerations that the field, if it existed at all, has always been essentially dipolar, the dipole axis coinciding fairly closely with the axis of rotation.   There is no definite evidence at present that the geomagnetic field has ever been absent for any considerable period, or that it has had a vastly greater magnitude than it has today.

Paleomagnetic analysis tends also to support theories of polar wandering and continental drift.   The NRM of most rocks lies in a direction quite different from that of the present earth's field, and when the data are carefully selected, a fairly systematic pattern evolves for the history of the geomagnetic field on each continent.   However it is not possible to reconcile these histories and also maintain the hypothesis of both a dipole field and fixed continents.   At present it seems easier to accept continental drift than to explain the results on the basis of a nondipolar geomagnetic field.

Paleomagnetic data also support the theoretical arguments that the earth's field has reversed its direction periodically during the past.   Analysis of certain lava sequences, baked contact zones, and other specially selected geological situations has indicated fairly frequent reversals, at least during certain geological periods.   The latest was probably about 1 million years ago.

12-7   The Demagnetization of Rocks

If a material is given a magnetization which is greater than that predicted by (12-1), the excess will gradually disappear.   The process is related to VRM, and it is often given the name viscous demagnetization.   Another slightly more general term for this phenomenon is relaxation.   Viscous demagnetization as the result of thermal agitation has been studied in the laboratory and has been found to be of limited importance at ordinary temperatures in multidomain grains of magnetite (13).   However, it seems quite possible that pressure changes and low-temperature solid diffusion and recrystallization may contribute to the amount of demagnetization which takes place during geological time intervals.

Demagnetization should be distinguished from remagnetization.   There is nothing to prevent a rock from acquiring several different types of magnetization during successive stages of its history.   New magnetizations need not necessarily wipe out old ones.   Thus a rock's NRM may be a combination of several different effects.

In studying the NRM of rocks, a numerical constant, known as Koenigsberger's Q ratio, is often quoted.   This is the ratio of a material's NRM to the magnetization induced by the earth's field at the sample location.   Because of relaxation processes of all types, older rocks are usually found to have smaller Q's than younger ones.   The range of values is very large, however, occasionally falling outside the limits $0.1 < Q < 100$.

12-8   *Examples of Rock Magnetization*

Very few examples exist in the literature in which the natural magnetization of a rock formation (both induced and remanent components) has been studied in detail.   Paleomagnetic research is concerned primarily with the direction of the NRM, and consequently most publications give only these data.   It is thus not possible for us to make general statements about in situ magnetizations.   Instead, we shall review some of the few cases in which fairly complete data have been obtained.

Strangway (14) has measured the magnetic properties of a suite of Precambrian diabase dikes.   These dikes are very numerous in the Canadian Shield and are easily recognizable.   Usually they have a thickness of several tens to a few hundreds of feet, dip almost vertically, and form distinct boundaries with their hosts.   They are often continuous over large distances and have a very uniform composition.

Figure 12-9 shows Strangway's measurements on two suites of samples from the "Abana" dike.   One set of samples was taken from surface outcrop, the other in a mine drift.   In both sets the directions of NRM are scattered, but they are by no means completely random.   The surface samples show a greater dispersion, suggesting that surface weathering has had an effect.

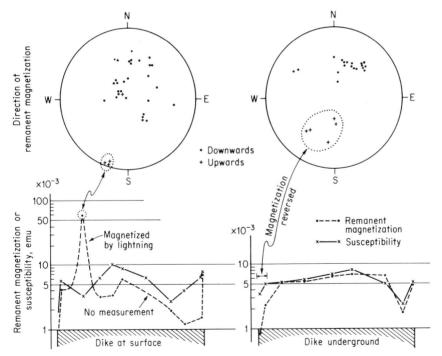

Fig. 12-9   NRM and susceptibility of the Abana diabase dike.   [After Strangway (14).]

Their mean direction is also rather more nearly parallel with the present earth's field.   In both sets, however, some anomalous magnetizations have been found.   Several of the underground samples exhibit reversed magnetizations, almost certainly due to physical or chemical processes and not to a reversal of the earth's magnetic field.   Among the surface samples one group which had an exceedingly strong remanence as well as an anomalous direction was found.   Its NRM was interpreted as caused by lightning.

The Q ratio of these rocks is of the order of unity.   Both the susceptibility and the NRM vary with position in the dike by a factor of about two. It is possible that there is a zone near one edge where the magnetite content is rather lower than it is elsewhere.   From examination of thin sections it was found that the diabase contains roughly 5 percent of Fe-Ti oxides, mostly in the form of rather large grains which have separated into intergrown lathes of magnetite and ilmenite.

Hood (15) has published a series of measurements on the Sudbury norite series.   The norite is a mafic igneous rock, with which the Sudbury copper-nickel ores are closely associated.   It outcrops in an approximately elliptical ring 37 miles long and 17 miles across.   The interior of the ring is an asymmetric synclinal basin of sedimentary and volcanic rocks which the norite and its associated formations apparently underlie conformably.   The whole structure appears to rest unconformably on older granites and volcanic rocks.   The norite itself is the lower member of the so-called Sudbury irruptive.   Above the norite is a silicic and for the most part nonmagnetic member, known as micropegmatite, and between the two is a complicated but fairly mafic section, known as the transition zone.

The results of sampling on one traverse across the south side of the irruptive are shown in Fig. 12-10.   The directions of the NRM are very closely grouped but somewhat displaced from the direction of the present earth's field.   The intensities of NRM and the susceptibilities are rather irregular, but in the norite the Q ratio remains everywhere much greater than unity.   Within the transition zone the NRM and susceptibility both decrease substantially, but it is the NRM which diminishes the most.   It seems likely that there is less magnetite in the rocks of the transition zone than in the norite and that it occurs in a form which is less able to acquire or hold a remanence.

Hawes (16) has made a study of the magnetization of the Spavinaw granite in Oklahoma.   This is an intrusive body which has an areal extent of several square miles but possesses only four actual outcrops.   It produces a substantial aeromagnetic anomaly and seems to be unusually magnetic for a granite.   Although the aeromagnetic anomaly does not give any indication that the formation is irregularly magnetized, a very detailed magnetometer survey on the outcrops revealed a highly irregular anomaly pattern there, indicating that the magnetization is not uniform over distances of more than about 50 ft.   Measurement of the NRMs of a large number of samples confirmed the expected irregularity and strength of the remanence. Hawes' results are summarized in Fig. 12-11. The NRM seems to be totally random and exceedingly strong. In fact, Q ratios greater than 100 are found. It is also noteworthy that the NRM intensities are far more scattered than the susceptibility values.

The most likely explanation of these results is magnetization of the outcrops by lightning. Since the four outcropping regions apparently stand out as small hills above the surrounding terrain, it does not seem

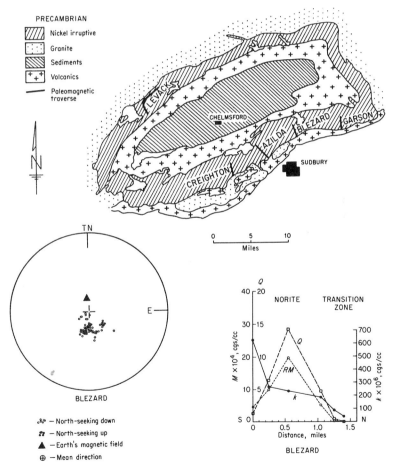

Fig. 12-10 NRM and susceptibility of the Sudbury irruptive on a traverse at Blezard. [After Hood (15).]

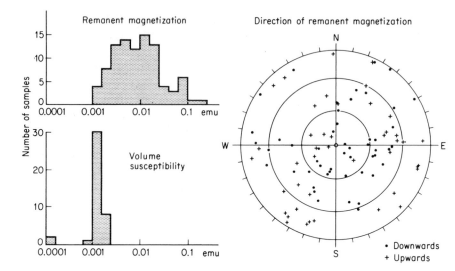

Fig. 12-11   NRM and susceptibility of the Spavinaw granite.   [After Hawes (16).]

improbable that they may have been struck several times during the past few hundreds or thousands of years.

Books (17) has analyzed a number of aeromagnetic anomalies in Montana and has measured the magnetization of specimens of the rocks which presumably caused them.   Two of his examples are shown in Figs. 12-12 and 12-13.   Both anomalies occur over buttes which are capped by igneous rocks, the first by a mafic lava and the second by a layered laccolith of so-called syenite-shonkinite.

The volcanic rock samples were found to have a very strong NRM, the directions of which were somewhat scattered but distinctly grouped about an azimuth of 149° and an inclination of 35° upward.   The mean intensity of their magnetization was 9.91×10^{-3} emu, and their mean Q ratio was of the order of 5.   To see if this would account for the observed aeromagnetic anomaly, a theoretical profile was calculated by the method of Henderson and Zeitz (18).   The induced magnetization was neglected in the calculation because it was relatively small.   The fit with the observed profile is sufficiently close that there seems to be no doubt that the remanence observed in the specimens is chiefly responsible for the aeromagnetic anomaly.   It would seem, however, that the mean intensity of the NRM throughout the formation as a whole is somewhat larger than that estimated from the sample measurements.

The second example reveals a different situation.   The NRM has a direction similar to that of the present earth's field and a mean intensity of 4.6×10^{-4} emu.   The induced magnetization has almost the same intensity, 5.1×10^{-4} emu.   A theoretical profile calculated on the basis of the sum of

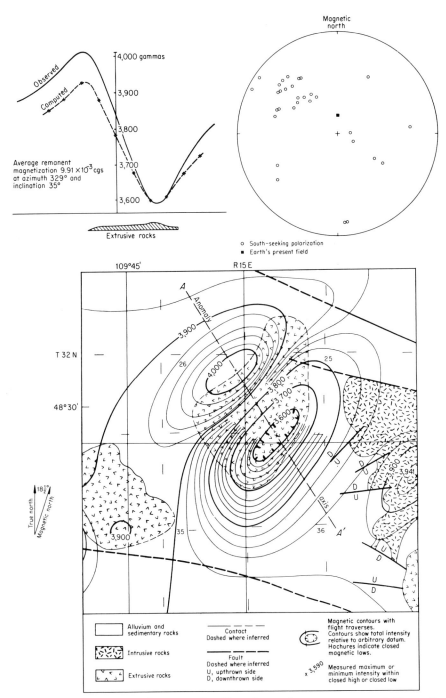

Fig. 12-12 **Aeromagnetic anomaly, directions of remanence, and theoretical anomaly profile over Boxelder Butte.** [After Books (17).]

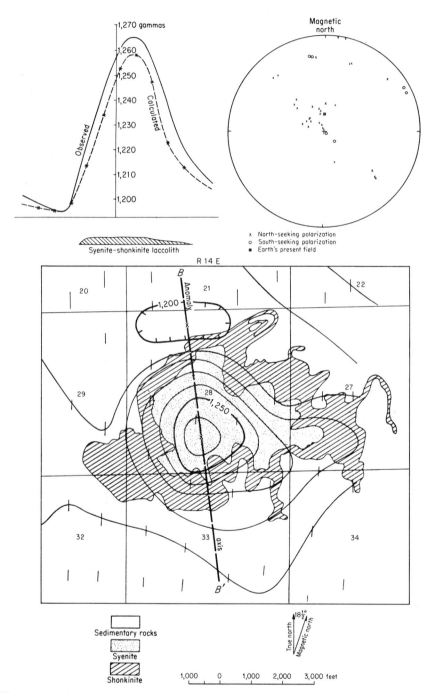

Fig. 12-13   Aeromagnetic anomaly, directions of NRM, and theoretical anomaly profile over Squaw Butte.   [After Books (17).]

these two magnetizations fits the observed anomaly rather well, although again it would seem that the formation magnetization is somewhat larger than the estimate derived from the sample measurements.

As a final example, it is worth noting some of the evidence which demonstrates that soils are often appreciably magnetic. Le Borgne (19) has shown that in many cases the uppermost soil layer has a much higher volume susceptibility than all the rest, often of the order of 5×10^{-4} emu. This he feels is due to the syngenetic formation of very fine-grained maghemite. Cook and Carts (20) have sampled several hundred soils in the United States and Panama, and they also find that moderately high soil susceptibilities are common. In agreement with Le Borgne they conclude that there is little or no relation between the susceptibility of a soil and that of its parent rock unless the soil contains large quantities of parent-rock fragments. A summary of their susceptibility measurements is shown in Fig. 12-14.

In Table 12-4 we show Cook and Carts' tabulation of the results of ten magnetic analyses by which they have attempted to discover the origin of

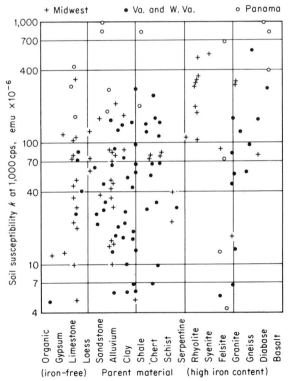

Fig. 12-14 **Correlation of soil susceptibility with parent material.** [**Cook and Carts (20).**]

Table 12-4

Area of origin	Great soil group	*k*, emu $\times 10^{-6}$ Original	*k*, emu $\times 10^{-6}$ Residual	Magnetic fraction, ppm	Composition of the most magnetic fraction			
C. Texas	Reddish chestnut	25	20	33	.35 magnetite	.6 quartz		
S. Texas	Rendzina	10	7	17	.45 magnetite	.55 quartz		
Texas coast	Sand	5	2	18	.5 magnetite	.5 quartz		
Texas coast	Semi-bog	5	5	?	.25 magnetite .5 quartz	.1 limonite .15 gypsum		
E. Texas	Yellow podzolic	10	3	?	.5 magnetite .09 ilmenite	.4 quartz .01 gypsum		
E. Texas	Yellow podzolic	15	2	215	.75 magnetite	.25 quartz		
C. Texas	Rendzina	350	330	1050	.1 magnetite	.9 quartz		
W. Texas	Reddish brown	180	35	75	.8 magnetite .05 quartz and gypsum	.15 ilmenite (?)		
N. New Mex.	Sand	80	5	700	.95 magnetite	.05 quartz		
C. Colo.	Brown	1080	200	2400	.7 magnetite	.3 quartz		

soil susceptibilities.   The majority of the large magnetic particles (i.e., those visible under the microscope) were extracted from each soil specimen and identified.   Nearly all particles turned out to be magnetite.   The susceptibility of the residuum was then measured to see what fraction of the original susceptibility had been contributed by the extract.   In four of the ten cases it was found to have produced less than half, and thus the susceptibility of these soils is probably due to ultra-fine-grained maghemite, as suggested by Le Borgne.   In the other six cases the magnetite content was more important.

Cook and Carts made no direct measurements of the NRM of the soils they sampled, but their work did indicate that it may be appreciable. Le Borgne has found the soils containing ultra-fine-grained magnetic material to be very susceptible to VRM, indicating that the particles are monodomain grains on the verge of superparamagnetism.   Such material, unless disturbed regularly by operations such as cultivation, will almost certainly exhibit an appreciable magnetic remanence.

References

1. T. Nagata, "Rock Magnetism," rev. ed., Maruzen, Tokyo, 1961.

2. E. Irving, "Paleomagnetism and Its Applications to Geological and Geophysical Problems," Wiley, New York, 1964.

3. L. Néel, Théorie du trainage magnétique des ferromagnétiques en grains fins avec applications aux terres cuites, *Annales Géophysique*, vol. 5, pp. 99–136, 1949.

4. G. D. Nicholls, The Mineralogy of Rock Magnetism, *Advances in Physics*, vol. 4, pp. 113–190, 1955.

5. S. Akimoto, T. Katsura, Magneto-chemical study of the generalized titano-magnetite in volcanic rocks *Jour. Geomagnetism and Geoelectricity*, Kyoto, vol. 10, pp. 69–90, 1959.

6. L. B. Slichter, Magnetic Properties of Rocks, pp. 293–297 in F. Birch (ed.), "Handbook of Physical Constants," Geol. Soc. America Spec. Paper 36, 1942.

7. H. M. Mooney and R. Bleifuss, Magnetic Susceptibility Measurements in Minnesota, Part II, Analysis of Field Results, *Geophysics*, vol. 18, pp. 383–393, 1953.

8. J. R. Balsley and A. F. Buddington, Iron-Titanium Oxide Minerals, Rocks, and Aeromagnetic Anomalies of the Adirondack Area, New York, *Econ. Geology*, vol. 53, pp. 777–805, 1958.

9. G. D. Bath, Magnetic Anomalies and Magnetizations of the Biwabik Iron Formation, Mesabi Area, Minnesota, *Geophysics*, vol. 27, pp. 627–650, 1962.

10. E. Thellier, Thèse, "Sur l'aimantation des terres cuites et ses applications géophysiques," Faculté de Science, Paris, *Annales l'Inst. Physique Globe*, vol. 16, p. 157, 1938.

11. L. Néel, L'inversion de l'aimantation permanente des roches, *Annales Géophysique*, vol. 7, pp. 90–102, 1951.

12. C. W. Carmichael, The Magnetic Properties of Ilmentite-Hematite Crystals, *Royal Soc. London Proc.*, ser. A, vol. 263, pp. 508–530, 1961.

13. Y. Shimizu, Magnetic Viscosity of Magnetite, *Jour. Geomagnetism and Geoelectricity*, Kyoto, vol. 11, pp. 125–138, 1960.

14. D. W. Strangway, Magnetic Properties of Diabase Dikes, *Jour. Geophys. Research*, vol. 66, pp. 3021–3032, 1961. "Magnetic Properties of Some Canadian Diabase Dike Swarms," Ph.D. thesis, University of Toronto, 1960.

15. P. J. Hood, Paleomagnetic Study of the Sudbury Basin, *Jour. Geophys. Research*, vol. 66, pp. 1235–1241, 1961.

16. J. Hawes, A Magnetic Study of the Spavinaw Granite Area, Oklahoma, *Geophysics*, vol. 17, pp. 27–55, 1952.

17. K. G. Books, Remanent Magnetism as a Contributor to Some Aeromagnetic Anomalies, *Geophysics*, vol. 27, pp. 359–375, 1962.

18. R. Henderson and I. Zeitz, Graphical Calculation of Total-intensity Anomalies of Three-dimensional Bodies, *Geophysics*, vol. 22, pp. 887–904, 1957.

19. E. Le Borgne, Susceptibilité magnétique anormale du sol superficiel, *Annales Géophysique*, vol. 11, pp. 399–419, 1955.

20. J. C. Cook and S. L. Carts, Magnetic Effect and Properties of Typical Topsoils, *Jour. Geophys. Research*, vol. 67, pp. 815–828, 1962.

part III ELECTRICAL CONDUCTION AND ELECTROMAGNETIC INDUCTION METHODS

chapter 13 Introduction to the Electrical Methods

The most highly variable of all the physical properties of minerals and rocks is their ability to conduct electricity. The electrical conductivities in naturally occurring minerals run through a range of magnitudes whose extreme values differ by a factor approaching 10^{20}. The most highly conductive minerals are the native metals, silver and copper; and below them the semi-metallic ore minerals, brine-filled sandstones, glacial clays, soils, shales, and limestones follow in roughly that order. At the lower end of the scale are the ionic crystals such as quartz, whose conductivities are so small that they can scarcely be measured. For most of the rocks, however, chemical composition is of such small importance in relation to other factors such as porosity and fracturing that it is impracticable to try to place the different species on the conductivity scale. On the other hand, since the factors which generally do determine the average conductivity of a rock formation very often are preserved throughout the bulk of the formation, it is natural that geophysical methods should be used to exploit these effects as an aid in mapping subsurface geological contacts. What marks the difference between the electrical methods and other geophysical methods is the peculiar difficulty in relating these measurements to recognizable lithological characteristics.

Not only is the simple process of ohmic conduction observed in earth materials, but more complicated electrical effects which encompass a very wide range of electrochemical phenomena are also observed. For example, electrical potentials may develop at interfaces between minerals which have different chemical potentials, or they may occur where minerals lie in contact with an electrolyte. They can also be caused by gradients in the concentration of certain solutes in interstitial waters, and they can even be produced by the motion of fluids in permeable materials. Detection of these steady-state effects forms the basis of the *spontaneous polarization* method. In addition, electrical charges sometimes accumulate on the interfaces between certain materials as a result of the flow of electrical current from an external source. The method of *induced polarization* has been devised to take advantage of this phenomenon in searching for metallic minerals and ground water.

13-1   The D-C Conduction (Resistivity) Method

Of all the electrical methods, the one which makes use of the conduction of direct current is probably the simplest to manipulate and certainly the simplest to understand. Essentially, it consists in the use of two pieces of apparatus: one, a source of electric current, and the other, a device for measuring potential differences. Each is connected to a pair of electrodes. When the current electrodes are inserted into the ground, a stationary current field is established; and because of the ohmic potential drop an electrical potential field is created also. This field can be distorted by subsurface zones of anomalous conductivity, and the object is to search for such "anomalies" in the electrical field with the pair of potential probes. It is always assumed that the current flow in the potential measuring circuit is negligible in comparison with the current flow in the ground, so that the potential electrodes themselves will have no disturbing effect upon the electrical field.

The deployment of the current and potential electrodes will depend entirely upon the type of problem confronting us. Broadly speaking, there are two major categories, which we distinguish as "vertical" and "horizontal" exploration problems. The first applies to situations in which the rocks tend to lie in flat and fairly uniform beds, such as are found in relatively undisturbed sedimentary basins. Here the object is to "sound" the depths of the strata by expanding the electrode configuration about a fixed location and observing the changes in the potentials. The second is connected with prospecting for mineral ore bodies, where the effort is directed toward finding anomalies in the electrical field. Within each of these

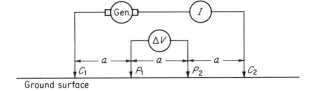

Fig. 13-1 A "Wenner" electrode array for resistivity surveying.

two categories there is a great variety of special procedures, only a very few of which we shall attempt to describe. Details of a number of others may be found in texts such as Heiland (1), Jakosky (2), and Sorokin (3).[1] For the purpose of this book, a single example from each category will suffice.

One of the best-known configurations for "depth sounding" is the Wenner array, in which the electrodes are equidistantly spaced along a line, the current electrodes on the outside and the potential probes in between (Fig. 13-1). Measurements of potential difference and current flow are made as the whole array is expanded symmetrically about a fixed center point. As the electrode spacing grows larger, the deeper strata will have an increasing effect upon the potentials, and much can be learned about the depths to these strata by observing the details of these changes.

In horizontal exploration, on the other hand, it is a more common practice to fix the current electrodes a great distance apart, astride the area to be surveyed. If the ground were of uniform conductivity, a reasonably constant electric-field intensity would be expected throughout the mid-region between them. Accordingly, the actual field pattern is mapped with the potential electrodes, used either singly or as a pair, so that contours of either the potential or the potential gradient may be traced. Since the direction of current flow is an important factor, it is desirable to give some consideration to the regional geological strike in placing the current electrodes to best advantage.

A technical problem which frequently arises in field practice is the occurrence of a spontaneous potential difference between the potential electrodes even in the absence of a supply of current. This is due to the formation of electrochemical emfs either on the surfaces of the electrodes or on interfaces within the ground. It is easily overcome in practice by taking a measurement with the current flowing in first one direction and then the other, or by using alternating currents of very low frequencies and measuring only the a-c component of the potential difference. If an alternating current is used, the frequency must be low enough that any potentials induced in the ground by the magnetic field of the current system will be negligible, in order that d-c theory may be used to interpret the results. This is easily achieved in practice, although if the frequency is not extremely low (<1 cps),

[1] For an extensive bibliography of early papers on electrical methods see J. N. Hummel, Ünterlagen der geoelektrischen Aufschietzungsmethoden, *Beitr. zür angew.* Geophysik, vol. 5, pp. 32–132, 1935–1936.

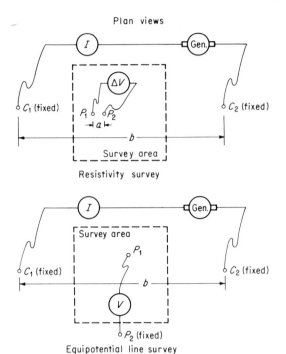

Plan views

Resistivity survey

Equipotential line survey

Fig. 13-2 Two variants of a common method of horizontal exploration. The current electrode separation *b* is often much more than a mile, while the distance through which the potential electrodes are moved between measurements is only of the order of 100 ft.

the possibility of small effects from induced polarizations cannot be entirely overlooked.

In the interpretation of d-c potential field measurements, the notion of *apparent resistivity* is frequently used. If the ground were uniform electrically, a single measurement of the potential difference and current flow would determine its resistivity. The formula is invariably a simple one having the form $\rho = f(\Delta \mathcal{V}/I, a)$, where ρ is the resistivity of ground, $\Delta \mathcal{V}$ is the difference in potential between the two potential probes, I is the current delivered by the source, and a is a characteristic linear dimension of the array. (In the Wenner array, for example, $\rho = 2\pi a \, \Delta \mathcal{V}/I$, where a is the interelectrode spacing.) If the ground is inhomogeneous, the same formula is used to define an *apparent resistivity*—the resistivity which the ground would have if it were homogeneous. In the presence of electrical inhomogeneities this quantity will change with the size or position of the array, and the interpretation is an endeavor to explain these variations.

The chief drawback of the d-c conduction method is its high sensitivity. If the conductivity changes by a factor of two, this can have a distinct effect upon the current field. But as we have noted earlier, this is a very small variation when we consider the enormous range of conductivities occurring in nature. In itself this is not a disadvantage, but it is detrimental to the ability of the method to discriminate between good and moderate conductors. If a zone is, let us say, thirty times more conductive than its host rock, it becomes practically indistinguishable from a perfect conductor. In the same way, if it is more resistive than its host rock, a similar ratio in the resistivities makes it appear like a perfect insulator. Consequently, the wide range of natural conductivities cannot be differentiated, and the "noise level" caused by minor geological variations tends to be uncomfortably high.

To try to overcome this difficulty, the electromagnetic-induction or the a-c conduction methods are often employed. This is especially true in prospecting for semimetallic ore minerals, which may be several orders of magnitude more conductive than their host rocks and where smaller variations of conductivity are of little economic significance. These methods are discussed in some detail in Chaps. 15 to 18, and at present it suffices to say that when an alternating current is used, the interaction between the current system and its magnetic field may be all-important, whereas in the d-c method it is nonexistent.

It is appropriate at this point to mention again the spontaneous and induced polarization methods. Spontaneous polarization is a general term applied to the steady-state generation of electrochemical potentials within the ground. These polarizations are observable only because they create a steady flow of conduction current. The same basic theory that is appropriate to d-c conduction methods is therefore applicable here. It is in the location and the physical nature of the current source that the two methods differ.

In the induced polarization method it is artificial potentials produced by applied current fields which are of interest. To observe them, either transient or alternating currents must be used. Fortunately, electromagnetic interactions are not of great importance in this method because the polarization effects have sufficiently large time constants. Consequently, d-c theory may again be used, although with some important modifications.

Quantitative interpretation techniques for the induced polarization method have only recently been attempted. A great deal of research has been carried out in order to elucidate the nature and behavior of the polarizations, and fortunately a good deal of this work has been published. We shall not attempt to review it in this text. Instead we shall refer the reader to the monograph edited by Wait (4) as well as to the papers by Vacquier et al. (5), Frische and von Buttlar (6), Seigel (7), and Marshall and Madden (8).

For the spontaneous polarization method there is still a considerable amount of doubt about the source mechanism and its actual location in the ground, although recent papers by Sato and Sato and Mooney (9) shed some new light on this subject. It is not yet possible, however, to construct an entirely satisfactory model of this phenomenon and devise quantitative interpretation techniques therefrom. The geophysical interpretation of spontaneous potentials still remains very largely an empirical process.

13-2 Ohm's Law

The physical principle underlying the method of d-c conduction (and for that matter almost all the electrical methods) is embodied in Ohm's law. In its original form, the statement of this law applied only to circuits. It stated that if a direct current is allowed to flow through a passive circuit element, the ratio of the potential drop across the element to the current flowing through it is a constant which is characteristic of that element. This quantity is defined to be its electrical resistance, which is thus given by the ratio

$$\text{Resistance } R \equiv \frac{-\Delta \mathcal{U}}{I}$$

The concept of an electrical resistance obviously needs to be broadened before it can be applied to current fields in voluminous media. This generalization is almost certainly well known to the reader.

As an obvious corollary to Ohm's law, if we consider a uniform electrical current flowing through a homogeneous cylinder in the direction of its axis, the resistance will be proportional to the length L and inversely proportional to the cross-sectional area A, i.e.,

$$R = \frac{\rho L}{A} \tag{13-1}$$

The constant of proportionality ρ is numerically equal to the resistance between opposite faces of a cube of unit dimensions cut from the material. It is called the *resistivity*, or sometimes the specific resistance of the material. In the mks system, the unit is the ohm-meter (ML^3/TQ^2).

The extension of this argument to small volumes follows directly. Instead of the total current flowing through a finite volume, however, we must consider the current density field **J**; and instead of the potential drop across the specimen, we must use the electric potential gradient **E**. Attention is then focused on a small rectangular parallelepiped having linear dimensions Δx, Δy, Δz which is situated at a point P and oriented so that Oz is in the direction of the current density vector **J** at P. If there are no discontinuities in the current field at P, the resistance across the parallelepiped can be written as

$$R = \frac{E \cdot \Delta z}{J \, \Delta x \, \Delta y}$$

But according to (13-1), $R = \rho \Delta z / \Delta x \, \Delta y$, and therefore

$$\rho = \frac{\mathbf{E}}{\mathbf{J}}$$

Since the argument continues to hold even as ΔV becomes infinitesimally small, it is meaningful to speak of the resistivity at the point P. Resistivity is therefore a physical property like density. If the material is isotropic, we have no reason to doubt that **E** and **J** will be in the same direction. The differential form of Ohm's law is therefore

$$\rho \mathbf{J} = \mathbf{E} \tag{13-2}$$

It is sometimes more convenient to write (13-2) in the form

$$\mathbf{J} = \sigma \mathbf{E} \qquad \text{where} \qquad \sigma \equiv \frac{1}{\rho}$$

The reciprocal resistivity σ is called the *conductivity*, and in the mks system the unit is mho per meter. From the point of view of the physical nature of electrical conduction, the conductivity is possibly the more fundamental quantity, since it is directly related to the density and mobility of charge carriers. However, this connection has no significance in macroscopic electrical field theory, and we shall use whichever quantity better suits our purpose at the moment.

The most significant feature of Ohm's law is that it is linear. This makes electromagnetic field theory a tractable subject. However, we must remember that it is an empirical law and that it should not be accepted unconditionally. It is well known, for instance, that many materials similar to those found in the earth do *not* obey Ohm's law at high current densities. On the other hand, the current densities used in most of the geophysical methods are so small (rarely exceeding 1 amp/m² except in the immediate vicinity of the current electrodes) that linearity can probably be assumed. Nonetheless, it should be stressed that the linear relationship is rarely ever checked, and therefore unsuspected side effects which have no explanation are always possible on the linear hypothesis.

Equation (13-2) applies to an isotropic medium. However there is no difficulty in reformulating it for aeolotropic media, in which the resistivity depends upon direction and in which **J** and **E** are no longer necessarily parallel. If we speak of the conductivity rather than the resistivity, we may write in the indicial notation used previously in Chap. 2

$$J_i = \sigma_{ik} E_k \tag{13-3}$$

where the conductivity σ_{ik} now appears not as a scalar, but as a second-rank tensor. Just which of the elements of this tensor will be the most significant will depend of course on the character of the aeolotropy. When $\sigma_{11} = \sigma_{22} = \sigma_{33} = \sigma$ and all other components vanish, we return to the isotropic case. Equation (13-3) represents Ohm's law for aeolotropic media in its differential form.

To gather some idea of how serious the effect of aeolotropy is likely to be in practice, we might consider a few limiting cases. Most rocks and

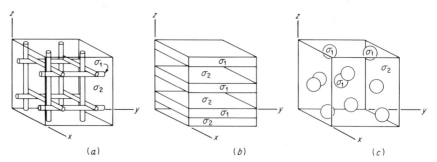

Fig. 13-3   Some extreme structures of a two-component material.

minerals are themselves extremely poor conductors. They pass electrical current almost entirely by electrolytic conduction through interstitial water rather than by electron transport through the rock itself. Ground waters, especially if they contain dissolved salts, have enormously greater conductivities than the rock materials which they permeate. On the other hand semimetallic minerals, if they are present, can be better conductors than electrolytes by several orders of magnitude. The current will simply take the path of least resistance, wherever it may lead.

Let us consider a material consisting of two components having conductivities σ_1 and σ_2, which are present in volume fractions p and $1 - p$, respectively. As one example we suppose that one of the components (conductivity σ_1) is arranged in the form of slender rods lying in equal numbers in each of the three coordinate directions (Fig. 13-3). Then if p is not too large, the conductivity in each direction will be, approximately,

$$\sigma_x = \sigma_y = \sigma_z = \frac{p\sigma_1}{3} + (1 - p)\sigma_2$$

If on the other hand the same substance is arranged in the form of plates lying perpendicular to Oz, then

$$\sigma_x = \sigma_y = p\sigma_1 + (1 - p)\sigma_2$$

and
$$\sigma_z = \frac{\sigma_1\sigma_2}{p\sigma_2 + (1 - p)\sigma_1}$$

It can further be shown (10) that if the material of conductivity σ_1 is a fairly small fraction of the whole and is present as a random scattering of small spheres throughout a host medium whose conductivity is σ_2, then

$$\sigma_x = \sigma_y = \sigma_z \doteq \frac{2\sigma_2 + \sigma_1 + 2p(\sigma_1 - \sigma_2)}{2\sigma_2 + \sigma_1 - p(\sigma_1 - \sigma_2)} \sigma_2$$

To illustrate these formulas, let us suppose that $\sigma_1 = 10\sigma_2$ and $p = 0.2$. For the rod structure we then find that $\sigma_x = \sigma_y = \sigma_z = 1.47\sigma_2$. For the plate structure, $\sigma_x = \sigma_y = 2.8\sigma_2$, while $\sigma_z = 1.22\sigma_2$. For the distribution of spheres $\sigma_x = \sigma_y = \sigma_z = 1.53\sigma_2$. If the contrast is larger, say $\sigma_1 = 100\sigma_2$, the rod structure gives $\sigma_x = \sigma_y = \sigma_z = 7.47\sigma_2$, while for the plates we obtain $\sigma_x = \sigma_y = 20.8\sigma_2$ and $\sigma_z = 1.25\sigma_2$. For the spheres the conductivity rises only a very little, $\sigma_x = \sigma_y = \sigma_z = 1.72\sigma_2$.

Not surprisingly we find that a platy structure imparts a strong aeolotropy to a conducting medium, whereas even in isotropic media the overall conductivity is strongly dependent on the geometry of the component materials. This warning must be taken seriously because many rocks and mineral deposits are well known to have a banded or a platy structure. In sediments small-scale bedding frequently restricts the interconnection of interstitial water across the bedding planes, while in sulfides a banded or sheared formation is rather common. Unfortunately, aeolotropy makes mathematical analysis so difficult that in all but a very few cases it is impracticable to try to include it in the theory. The best we can usually do is to try wherever possible to estimate the error in interpretation that may be introduced by this effect.

13-3 *Conductivity of Minerals and Rocks*

With the exception of the semimetallic minerals, almost all rock-forming minerals are compounds which have mixed ionic and covalent bonding. Consequently, they are fundamentally electrical insulators and in pure crystalline form their conductivity, if it is measurable at all, lies somewhere in the range 10^{-12} to 10^{-17} mho/m. Impurities and defects, which are invariably present in natural crystals, may be able to increase the conductivity by a few orders of magnitude, due, for instance, to the presence of electron donors or acceptors among the impurities. However, it turns out that the conductivities actually found by measurement on rocks in situ lie for the most part up in the range 10^{-1} to 10^{-8} mho/m—larger by several orders of magnitude than the values found in the laboratory. This unexpected result is due to the fact that even the tightest igneous and metamorphic rocks contain sufficient amounts of moisture in minute cracks and along grain boundaries to conduct electricity by electrolytic transport. Corroborative evidence for this is obtained by carefully drying specimens under heat and vacuum and observing the decrease in conductivity that accompanies dessication. As a result of this effect, the conductivity exhibited by a rock has really very little to do with its mineral composition. It depends rather upon its permeability and porosity and upon the conductivity of the fluid which it contains as a consequence of its present environment and past history.

A rather large number of conductivity measurements has been performed on samples of various rock types, and innumerable measurements have been made in situ in oil wells. Although it is rather old (1942), the

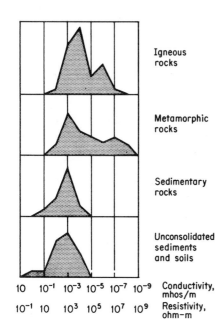

Igneous
rocks

Metamorphic
rocks

Sedimentary
rocks

Unconsolidated
sediments
and soils

| 10 | 10^{-1} | 10^{-3} | 10^{-5} | 10^{-7} | 10^{-9} | Conductivity, mhos/m |
| 10^{-1} | 10 | 10^3 | 10^5 | 10^7 | 10^9 | Resistivity, ohm-m |

Fig. 13-4 Histograms of resistivity measurements listed in "Handbook of Physical Constants" (11).

best easily available tables of measurements are given by Slichter and Telkes in the "Handbook of Physical Constants" (11). As a summary of these tables, histograms of the conductivities of igneous, metamorphic, and sedimentary rocks and of unconsolidated sediment and soils are shown in Fig. 13-4. These diagrams have no very simple statistical meaning, since the observations do not form any sort of representative or random sample. There is also little guarantee that the natural moisture was retained or restored to each sample before measurement, so that some measurements may show anomalously low conductivity values. Yet in spite of this, the general observations that we mentioned earlier are confirmed. The sedimentary rocks and unconsolidated materials are distinctly more conductive than most of the igneous rocks on account of their higher porosities and moisture content. Metamorphic rocks are more variable because they have a very wide range of porosities, extending from values similar to those of sediments to virtually zero among the crystalline materials that have been reformed under very high containing pressures. The soils appear to have conductivities similar to those of sedimentary rocks, in spite of their greater porosities. The explanation of this phenomenon is very likely that these materials are saturated by fresher, purer water than the older and more deeply buried sediments. This may also account for the fact that logging measurements in oil wells generally reveal resistivities in the range 1 to 1,000 ohm-m, which is decidedly lower than the great majority of the sample measurements. The general increase of temperature with depth may also be a factor since the conductivity of electrolytes is very temperature sensitive.

Because of the very widespread use of electrical logging methods in oil wells, a considerable amount of research has been done toward establishing a relation between the resistivity and the porosity, saturation, and other observable attributes of certain types of sedimentary rock. Those that have been the most extensively studied are the sandstones and carbonate rocks, which commonly form petroleum reservoirs. For clean rocks (i.e., those without clay in their pores) it has been found that a good estimate of the resistivity in situ is given by the following relationship:

$$\text{Resistivity } \rho = IF\rho_w$$

where F = the formation factor
I = the resistivity index
ρ_w = the resistivity of the water in the formation

In terms of porosity and saturation, the formation factor and the resistivity index are found to have the following forms:

$$F = aP^{-m} \quad \text{and} \quad I = S^{-n}$$

where P = the porosity of the rock formation
S = fractional amount of saturation of pore spaces
a, m, n = empirical constants

For clean sandstones, values between 0.6 and 1.0 for a, of 2.0 for m, and of 2.0 for n are generally accepted; and for carbonate rocks $a = 1.0$, $m = 1.8 - 2.6$, and $n = 1.5 - 2.2$, depending upon the type. Those which have large pores and cavities generally exhibit the larger values of m, while the larger values of n seem to be an attribute of the more dense and compact varieties.

The presence of clay in the pores of a rock has a considerable effect upon its conductivity. The clay minerals and other hydrous substances such as serpentine are generally found to be rather good conductors. In sandstones a small quantity of clay tends to make the rock much more conductive at low water saturations, and often in such circumstances a value of about 1.5 rather than 2 is indicated for the exponent m in the resistivity formula. Although a dry clay mineral is not itself an unusually good conductor, a small amount of excess water can make it so through the action of ion exchange.

Shales are generally worse conductors than the more porous rocks, and they are also often noticeably aeolotropic. Minute mica grains and other platy minerals tend to lie in parallel bedding planes, giving higher conductivities along these planes than across them. Ratios of parallel to transverse conductivity as high as four are not uncommon.

The determining factor in the electrical properties of all these materials is the conductivity of the electrolyte which permeates the rock. In dilute solutions this can be closely correlated with the total ion content, although the relationship between conductivity and composition is more complicated for higher solute concentrations. The stronger natural electrolytes, however, are brine. Figure 13-5 shows the relationship between the total ion content T and the conductivity of a number of natural surface waters. A formula $\sigma_{25°C} = T/100$ fits the data rather well and may be used to predict conductivities of other waters. The total ion content or ionic sum (measured in equivalents per million, epm) is just half the sum of the gram equivalent weights of the various ions present in 10^6 cm³ of water. The following are the most common:

$$(Na)_{23}{}^+, (Ca)_{40}{}^{++}, (Mg)_{24.3}{}^{++}, (Cl)_{35}{}^-, (HCO_3)_{61}{}^-, (CO_3)_{60}{}^{--}, (SO_4)_{96}{}^{--}$$

Another formula (12), of a type more commonly used by petroleum engineers, relates complex solutions to the equivalent concentrations of sodium chloride, giving for the net equivalent salt concentration

$$C_e = 0.95(Ca) + 2.00(Mg) + 1.00(Na) + 1.26(CO_3) + 0.27(HCO_3)$$
$$+ 1.00(Cl) + 0.50(SO_4)$$

Here all the ion concentrations are given in weight of ion per unit weight of

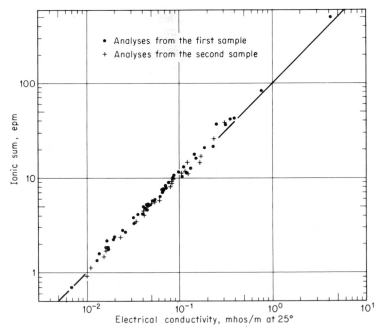

Fig. 13-5 **Conductivity and concentration of the individual ground-water analyses listed in two different publications.** [After Logan (12).]

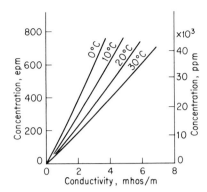

Fig. 13-6 Conductivity of saline water.

solvent, e.g., in parts per million (ppm). The conductivity of the solution is then obtained from Fig. 13.6, in which the conductivity of saline water is given as a function of concentration and temperature.

As a *very* rough guide, ionic sums in the range 1 to 30 epm are characteristic of lakes and stream waters, while values from 10 to 300 epm are more characteristic of soil solutions. Beyond this concentration lie the saline waters found at depth. For comparison, the concentration of ordinary sea water is about 600 epm, i.e., 0.6 N.

So far the metallic minerals have been disregarded, and we shall now discuss their properties. In nature, of course, they are rather rare; but since they very often are the object of electrical surveys, their relative importance is high. Much less is known about the processes of conduction in these materials than in porous rocks. In fact almost all our knowledge is entirely empirical, being the result of comparisons between mineralogical examinations and conductivity measurements. A rather large number of measurements on individual specimens is given in the "Handbook of Physical Constants," but the most comprehensive single study of this problem has been given by Parasnis (13). He has measured the conductivities of specimens of minerals of many different types, and although most of these were from ores found in Sweden, making the sampling more representative of Precambrian mineralization than of newer deposits, his results give some useful information about the conductivity of many of the more common metallic minerals. Table 13-1 reviews these results and adds some other measurements.

Pyrite, the most ubiquitous of the metallic sulfides, has the most variable conductivity. Except for remarking that it is nearly always a better conductor than porous but unmineralized rocks, there are few general statements that can safely be made about it. Even in its usual forms, the mineral conductivity of pyrite ranges over several orders of magnitude centered on or about 10^2 mhos/m.

Pyrrhotite seems to be an extremely good conductor both in pure mineral form and as an ore. Moreover, its conductivity seems to be comparatively

Table 13-1

	Conductivity σ, mhos per meter					
	Pure crystalline form		Natural mineralization			
			Approx. 100%		Approx. 50%	
Mineral	In direction giving maximum	In direction giving minimum	Mean	s	Mean	s
Pyrite FeS_2			10^2	1.2	(% mineralization uncertain)	
Pyrrhotite $FeS(Fe_7S_8)$			3×10^4	0.3	6×10^3	
Chalcopyrite $CuFeS_2$			2×10^3	0.4		
Arsenopyrite $FeAsS$			3×10^3	0.3		
Galena PbS	5×10^4		10		5	
Magnetite Fe_3O_4	10^4		3	1	(% mineralization uncertain)	
Graphite C	10^6	10^2	5×10^3		100 1.0 on shales containing only a few % C	
Hematite Fe_2O_3	insulator		0.1	0.8		

Mean = geometric mean = $\exp_{10} \dfrac{1}{n} \sum\limits_{n} \log_{10} \sigma_i$.

s = standard deviation of logarithm of conductivities (rough estimate).

constant, the variations covering only about one order of magnitude in the vicinity of 10^4 mhos/m. This is confirmed by experience in the field, which indicates that bodies which contain any substantial amount of pyrrhotite are nearly always extremely good conductors.

Chalcopyrite is a mineral that seems to be comparable to pyrrhotite, but it has a mean conductivity of only 2×10^3 mhos/m. The scatter of values is intermediate between pyrite and pyrrhotite, covering about two decades. The same remarks apply also to *arsenopyrite*.

Galena in small, pure specimens is apparently an excellent conductor, having a conductivity of 10^4 mhos/m. However, its cubic habit seems to make it a poor conductor in polycrystalline specimens even in fairly concentrated ores, since the grains are apparently unlikely to be interconnected. It appears that galena mineralization by itself is a poor conductor, but when mixed with other metallic minerals which tend to connect the grains, it may possibly conduct quite well.

Magnetite seems to be somewhat similar to galena in that its crystal conductivity is high, but in polycrystalline or even massive form it is not a good conductor. Again this is probably due to the tendency to form in discrete, euhedral crystals.

Graphite in pure crystalline form is an excellent conductor, having a conductivity of 10^6 mhos/m in the basal plane and 10^2 mhos/m across it. Polycrystalline graphite appears to be intermediate between these two values at 5×10^4 mhos/m. Its most remarkable feature is the amazing way in which it remains connected even when present in amounts of only a few percent. While 2 percent pyrite will have virtually no effect upon the conductivity of a rock, 2 percent graphite will give it a very noticeable conductivity increase.

Hematite is not normally a conductor. However, it appears that the presence of impurities can induce an appreciable conductivity at least in the specular form of this mineral—up to the range 1 to 10^2 mhos/m. Because of the large size of some hematite bodies this may have important consequences in electrical surveys.

Zincblende is also an insulator. This is certainly true of pure ZnS and probably also applies to the varieties containing iron (sphalerite). It is possible (although this has not actually been demonstrated) that impurities such as an excess of iron or other metals might make it a conductor.

Other minerals which have been found to be appreciably conductive but which have not been so well studied are *bornite* ($CuFeS_4$), *chalcocite* (Cu_2S), and *covellite* (CuS), which seem to have conductivities comparable to that of chalcopyrite, *ilmenite* ($FeTiO_3$), *molybdenite* (MoS_2), and the manganese minerals *hollandite* and *pyrolusite*, which seem to have conductivities that are at least as great as 1 mho/m. In addition, it is hardly necessary to mention that the native metals are excellent conductors, the conductivities approaching those of their commercial forms. Metallic silver has a conductivity of about 6×10^7 mhos/m; copper, 5.8×10^7 mhos/m; and iron, approximately 5×10^6 mhos/m.

The third class of conductors we shall call the *structural* conductors. These have little to do with rock type, porosity, or metallic mineral content but occur as a result of certain geological structures superimposed on the rocks. Chiefly, these structures are of the linear type, such as faults, shear zones, or contact fracture zones, where fracturing has increased the space available to ground water. Features such as these are of considerable importance to electrical surveys because they may be very well connected over great areas—a property which often compensates for their moderate bulk conductivity. In addition, they are often found to be the locus of alteration and weathering which has produced clay and other hydrous minerals. These minerals in turn can increase the effective conductivity of the water in the zones.

Such structures are often difficult to observe, except perhaps in mines.

Because they are essentially zones of weakness, they weather easily and thus are seldom visible in outcrops. Being narrow, they are often indicated in diamond drilling by nothing more than a few inches of lost core. However, the electrical methods, particularly the airborne electromagnetic methods, have demonstrated the great size and widespread occurrence of such features in a great many areas.

13-4   *Interpretation Models for the Electrical Methods*

Models used in the interpretation of electrical surveys are exceedingly simple. In virtually all cases the ground is supposed to consist of a small number of distinct regions, each of which is characterized by a uniform and isotropic electrical conductivity. The different regions are so well bonded to one another that currents may flow across the interfaces without encountering potential barriers. This very simple physical picture of what is undoubtedly a much more complex situation is a natural result of the compromise that must be made with geology in order to interpret the geophysical data quantitatively. In a very few instances the conductivity may be assumed to have a simple type of aeolotropy, when this is thought to serve a useful purpose, but for the most part it is not feasible to add additional conductivity parameters into the interpretations.

The forms assumed for the various regions into which the ground is divided will depend, of course, on the type of problem to be solved. For vertical exploration, a sequence of uniform layers overlying a uniform substratum is probably the most important and widely used model. This has been generalized to include conductivities which vary in an arbitrary manner with depth, and it has also been extended in scope to include transverse aeolotropy within the layers. As a variant on this model and in order to gather some idea of how critically the potentials depend upon the flatness of the layers, the case of two homogeneous media in contact along a dipping interface has also been worked out in some detail. With models such as these we try to evaluate the effects of occurrences such as a stratified sequence of sedimentary rocks, one or several layers of soil overlying bedrock, etc. The resemblance to geology is crude, partly because the number of layers which can be handled mathematically is small and partly because the natural layering is seldom as perfect as we suppose it to be. Surprisingly enough, however, the numbers that come out of the interpretations very often do give useful information about the subsurface geology, although admittedly it is difficult to identify the formations from their conductivity values.

In horizontal exploration the situation is somewhat more difficult. The shapes of the models used generally have to be described by level surfaces in systems of coordinates in which Laplace's equation is separable, and this severely limits their number. To give two examples, a sphere may be used if the conductor is thought to be roughly equidimensional, and a horizontal elliptic cylinder, if its strike is obviously large compared with its width. Neither one of these shapes disturbs the configuration of electric or magnetic fields in the same way as bodies of a less symmetrical outline might do, and so they do not give a very representative idea of the effect of a typical ore body. They are useful, however, for estimating the limits of detectability of compact or of sheetlike conductors by electrical or electromagnetic methods. This will become clearer when we examine these models in greater detail.

References

1. C. A. Heiland, "Geophysical Exploration," Prentice-Hall, Englewood Cliffs, N.J., 1940.

2. J. J. Jakosky, "Exploration Geophysics," Trija, Los Angeles, 1950.

3. L. W. Sorokin (transl. to German), "Lehrbuch der geophysikalischen Methoden zur Erkundung von Erdölvörkommen," Springer-Verlag, Berlin, about 1959.

4. J. R. Wait (ed.), "Overvoltage Research and Geophysical Applications," Pergamon, New York, 1959.

5. V. M. Vacquier, C. R. Kintzinger, and M. Lavergne, Prospecting for Ground Water by Induced Electrical Polarization, *Geophysics*, vol. 22, pp. 660–687, 1957.

6. R. H. Frische and H. von Buttlar, A Theoretical Study of Induced Electrical Polarization, Geophysics, vol. 22, pp. 688–706, 1957 (and discussion, vol. 23, pp. 144–153, 1958).

7. H. O. Seigel, Mathematical Formulation and Type Curves for Induced Polarization, *Geophysics*, vol. 24, pp. 547–565, 1959.

8. D. J. Marshall and T. R. Madden, Induced Polarization, a Study of its Causes, *Geophysics*, vol. 24, pp. 790–816, 1959.

9. M. Sato, Oxidation of Sulfide Ore Bodies, Parts I and II, *Econ. Geology*, vol. 55, pp. 928–961 and pp. 1202–1231, 1960.

   M. Sato and H. M. Mooney, The Electrochemical Mechanism of Sulfide Self Potentials, *Geophysics*, vol. 25, pp. 226–249, 1960.

10. J. C. Maxwell, "A Treatise on Electricity and Magnetism," 3d ed. (2 vol.), Clarendon, Oxford, 1892.

11. L. B. Slichter and M. Telkes, Electrical Properties of Rocks and Minerals, pp. 299–319 in F. Birch (ed.), "Handbook of Physical Constants," Geol. Soc. America Spec. Paper 36, 1942.

12. J. Logan, Estimation of Electrical Conductivity from Chemical Analysis of Natural Waters, *Jour. Geophys. Research*, vol. 66, pp. 2479–2483, 1961.

13. D. S. Parasnis, The Electrical Properties of some Sulphide and Oxide Minerals and their Ores, *Geophys. Prosp.*, vol. 4, pp. 249–278, 1956.

chapter 14

Quantitative Interpretation for the Direct-current Conduction Method

The *direct-current-conduction* method of geophysical exploration encompasses a varied assortment of systems and techniques adapted to a number of specific purposes. We have alluded to some of these in Sec. 13-1, and we shall consider others in due course, but it will not be possible within the limits of this text to write a comprehensive account of the numerous ramifications followed in practice.

The great diversity found among direct-current techniques is essentially a result of the different ways of deploying the current and potential electrodes. Such freedoms do not exist in, let us say, the gravity or magnetic methods, and although they confer upon the direct-current-conduction method an almost unrivaled flexibility, they also deprive it of any distinctly recognizable character. Indeed, it might be said that the only basic factor common among the several systems in use is a theoretical framework founded upon electrical potential theory.

Sometimes the interpretation process does not endeavor to go beyond identifying anomalies. Occasionally it aims farther than this and attempts to calculate the volumes and shapes of disturbing bodies. In any case a sound interpretation, however seriously it may be commended as a form of art, rests solidly upon experience and knowledge of the behavior of harmonic functions. The aim in this chapter is to trace a number of analytical developments upon which certain methods of interpretation in common use are based. Since the number of these solutions is small when compared with the variety of situations that occur in nature, we also propose to lay the groundwork for numerical methods which may be used to augment the few soluble cases. Only occasionally shall we pause along the way to demonstrate specific applications. We hope that, by careful reasoning, the application of these results to particular electrode configurations will seem, if not altogether self-evident, at least not entirely obscure.

14-1   General Equation for the Electrical Potential

Let us first consider the continuous flow of electrical currents in voluminous media. If the quantity of current flowing across a surface element of area δS which is perpendicular to the direction of current flow is $J \, \delta S$ amp, then the vector whose magnitude is J amp/m² and whose direction is everywhere in the direction of the current flow (provided that the flow is continuous) may be called the *current density*. The relation between the current density and the electric field intensity is given by Ohm's law

$$\mathbf{J} = \sigma \mathbf{E}$$

where the electric field intensity $\mathbf{E}$ is measured in volts per meter. In aeolotropic media the conductivity σ is a second-rank tensor, but in isotropic media it is a scalar. Since stationary electric fields are conservative, $\mathbf{E}$ may be related to a scalar potential function $\mathcal{V}$ as follows:

$$\mathbf{E} = -\nabla \mathcal{V}$$

where $\mathcal{V}$ is measured in volts. Combining this with Ohm's law gives

$$\mathbf{J} = \sigma \nabla \mathcal{V} \qquad (14\text{-}1)$$

To obtain an equation for the potential, we may apply the principle of the conservation of charge. This principle states that under ordinary conditions electrical charge may neither be created nor destroyed in macroscopic amounts, and it is based entirely upon experimental evidence. This means that across any surface S which encloses a volume within which there are no sources or sinks of electrical current

$$\int_S \mathbf{J} \cdot \mathbf{n} \, dS = 0$$

and hence, according to Gauss' theorem,

$$\nabla \cdot \mathbf{J} = 0$$

It follows at once from (14-1) that

$$\nabla \cdot (\sigma \nabla \mathcal{U}) = 0$$
and hence $$\nabla \sigma \cdot \nabla \mathcal{U} + \sigma \nabla^2 \mathcal{U} = 0 \qquad (14\text{-}2)$$

If σ remains constant throughout the region, then

$$\nabla^2 \mathcal{U} = 0 \qquad (14\text{-}3)$$

and the potential is in this case a harmonic function.

14-2   *Boundary Conditions for Direct Currents*

Two conditions which must be satisfied at any boundary separating different conducting media are (1) $\mathcal{U}$ must be continuous across the boundary and (2) the normal component of $\mathbf{J}$ must also be continuous.   The first condition states that the amount of work required to deposit a given charge on one side of the boundary should not differ from that required to deposit the same charge on the other side; this is merely a particularization of the principle of conservation of energy.   The second condition requires that the rate of accumulation of charge at the boundary, as elsewhere, should be zero (unless, of course, the boundary happens to contain sources of current); this formally restates the law of conservation of charge.   Both of these statements can also be deduced from purely formal reasoning.

1. First, let us consider the continuity of potentials.   Since $\mathbf{E} = -\nabla \mathcal{U}$ and $\nabla \times \nabla \equiv 0$, it follows that $\nabla \times \mathbf{E} = 0$.   Consider current circulating at the boundary in a thin loop as shown in Fig. 14-1.   According to Stokes' theorem,

$$\int_S \mathbf{n} \cdot \nabla \times \mathbf{E} \, dS = \oint \mathbf{E} \cdot d\mathbf{l} = 0$$

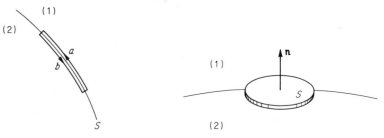

Fig. 14-1   **Fig. 14-2**

Therefore, in the limit as the width of the loop vanishes

$$\int_a \mathbf{E} \cdot d\mathbf{l} = - \int_b \mathbf{E} \cdot d\mathbf{l}$$

and hence $\mathcal{V}^{(1)} = \mathcal{V}^{(2)}$. Thus $\mathcal{V}$ is continuous across the boundary.

2. Now let us consider the continuity of normal current flows. Where no sources or sinks of current exist

$$\int_S \mathbf{J} \cdot \mathbf{n}\, dS = 0$$

across any closed surface S; and if we let S be the surface of a thin disk containing a small portion of an electrical interface, as shown in Fig. 14-2, then in the limit as the thickness of the disk vanishes, we must have

$$\mathbf{J} \cdot \mathbf{n}^{(1)} + \mathbf{J} \cdot \mathbf{n}^{(2)} = 0$$

Thus $\mathbf{J} \cdot \mathbf{n} = \sigma(\partial \mathcal{V}/\partial n)$ is continuous at the boundary.

14-3　The Uniqueness Theorem

The uniqueness theorem states that if a solution of Laplace's equation can be found which satisfies a given set of boundary conditions, then it is the correct and only possible solution appropriate to those conditions. Another way of stating the theorem is that wherever boundary conditions apply to Laplace's equation, the solution, if it exists at all, is unique. The proof is as follows: Let $\mathcal{V}_1$ and $\mathcal{V}_2$ be two different functions, both of which satisfy $\nabla^2 \mathcal{V} = 0$ within a given region V. In addition, let both of these functions satisfy the boundary conditions on the surface S which encloses V. Now let us consider the function $W = \mathcal{V}_1 - \mathcal{V}_2$. We know that $\nabla^2 W = 0$ within V and that $W = 0$ on S; therefore, according to Gauss' theorem,

$$\int_V \boldsymbol{\nabla} \cdot (W \boldsymbol{\nabla} W)\, dV = \int_S W \boldsymbol{\nabla} W \cdot \mathbf{n}\, dS = 0$$

However, the left-hand side of this equation can be written

$$\int_V \boldsymbol{\nabla} \cdot (W \boldsymbol{\nabla} W)\, dV = \int_V (W \nabla^2 W + \boldsymbol{\nabla} W \cdot \boldsymbol{\nabla} W)\, dV$$

and since $\nabla^2 W = 0$, it follows that $\boldsymbol{\nabla} W = 0$ as well. Therefore, W is a constant, but since it vanishes on S, this constant must be zero. If $W = 0$, then $\mathcal{V}_1 = \mathcal{V}_2$.

This theorem plays an important role in physical problems where Laplace's equation applies, since it allows us to employ any means to discover a solution, while assuring us that the solution, if found, is quite independent of the means we have used to find it.

14-4   Potentials about a Point Electrode at the Surface of a Uniform Ground

Let us suppose that an electrode whose effective dimensions are zero supplies current at the rate of I amp to a uniform ground at a point P on the surface. To avoid violation of the conservation principle, we must also assume that current is withdrawn from the ground at the same rate elsewhere, but for the present we may assume that the sink is located far enough away that it does not disturb the field near the source.

If the conductivity of the ground is uniform, then $\nabla^2 \mathcal{V} = 0$ everywhere. Furthermore, since all conditions are symmetrical with respect to the point P, we may surmise that $\mathcal{V}$ will be a function of $R = \sqrt{r^2 + z^2}$, the distance from P. Since $\nabla^2(1/R) = 0$, let us assume a solution of the form $\mathcal{V}(r,z) = C/R$. If with this simple function we are able to satisfy the boundary conditions, the uniqueness theorem guarantees that this is the correct and only solution of the problem.

If the air above the ground has no conductivity (and this is a good approximation), then the condition to be satisfied at the surface is $\partial \mathcal{V}/\partial z = 0$ when $z = 0$. This obviously is true for $\mathcal{V} = C/R$. To find C, we consider a small hemisphere with center at P and bounded by the surfaces $R = a$ and the plane $z = 0$. If the resistivity of the ground is ρ ohm-m, then the total current that flows outward across a unit area of this hemispherical surface is given by the current density function $J = \dfrac{1}{\rho}\left(\dfrac{\partial \mathcal{V}}{\partial R}\right)_{R=a} = \dfrac{C}{\rho a^2}$. The total current that flows from P, therefore, is $= 2\pi a^2 J$, and since this must be equal to I, we find that $C = I\rho/2\pi$. Hence

$$\mathcal{V}(r,z) = \frac{I\rho}{2\pi R} \tag{14-4}$$

To obtain the expression for the potential due to a pair of current electrodes, we need only superimpose two of these solutions. Because Laplace's equation is linear, the sum of two solutions must also be a solution. Thus the potential at a point P, distant R_1 from the source and R_2 from the sink, will be

$$\mathcal{V}(P) = \frac{I\rho}{2\pi}\left(\frac{1}{R_1} - \frac{1}{R_2}\right)$$

This leads immediately to the formula for the ground resistivity appropriate to any electrode configuration. For the Wenner system described in

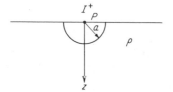

Fig. 14-3   A point source of current at the surface of a uniform ground.

Sec. 13-1, for instance, the difference in potential observed between the potential probes will be

$$\Delta\mho = \frac{I\rho}{2\pi}\left[\left(\frac{1}{a} - \frac{1}{2a}\right) - \left(\frac{1}{2a} - \frac{1}{a}\right)\right] = \frac{I\rho}{2\pi a}$$

Therefore the resistivity of the ground is given by

$$\rho = 2\pi a \frac{\Delta\mho}{I} \tag{14-5}$$

Similarly, the electric field **E** within the region between two remote current electrodes is

$$\mathbf{E} = -\nabla\mho = -\frac{I\rho}{2\pi}\nabla\left(\frac{1}{R_1} - \frac{1}{R_2}\right)$$

$$= -\frac{I\rho}{2\pi}\nabla\left\{\left[\left(\frac{b}{2} + x\right)^2 + y^2\right]^{-\frac{1}{2}} - \left[\left(\frac{b}{2} - x\right)^2 + y^2\right]^{-\frac{1}{2}}\right\}$$

in terms of rectangular coordinates taken about the center point between the electrodes with Ox directed toward the sink. If x and y are both small compared with the separation b of the current electrodes, this expression reduces to

$$\mathbf{E} \doteq -\frac{I\rho}{2\pi b}\nabla\left(1 - \frac{2x}{b}\right) = \frac{I\rho}{\pi b^2}\mathbf{i}_x$$

Thus to a first approximation in its binomial expansion, **E** is constant. To this approximation, the ground resistivity can be calculated from a measurement of **E**, since

$$\rho = \pi b^2 \frac{E_x}{I} \tag{14-6}$$

14-5 Potentials about a Point Electrode at the Surface of a Layered Ground

Now let us introduce some heterogeneity into the ground by bringing in a number of horizontal layers, each having different but homogeneous and isotropic electrical properties. Because of the presence of these layers, the potential will no longer have spherical symmetry about P, and we must now look for solutions of $\nabla^2\mho = 0$ which can be made to satisfy the continuity conditions at a number of horizontal boundaries. Such solutions are found by separating Laplace's equation in the cylindrical coordinate system whose origin is at P and whose z axis is vertical and positive in the downward sense. Since symmetry with respect to the coordinate ϑ still exists, we may write

$$\nabla^2\mho = \frac{\partial^2\mho}{\partial r^2} + \frac{1}{r}\frac{\partial\mho}{\partial r} + \frac{\partial^2\mho}{\partial z^2} = 0$$

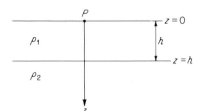

The complementary solution of this equation is formed from the characteristic functions of the separated variables.   By choosing only those which are well behaved when $r \to 0$, we obtain for the complementary function

$$\mathcal{V}(r,z) = \int_0^\infty [A(\lambda)e^{-\lambda z} + B(\lambda)e^{\lambda z}]J_0(\lambda r)\ d\lambda$$

to which we must add the particular solution $\mathcal{V}(r,z) = I\rho_1/2\pi R$ which applies within the close vicinity of P.   If no boundaries exist other than the surface $z = 0$, then $A(\lambda) = B(\lambda) = 0$.   If there are boundaries, then the coefficients $A(\lambda)$ and $B(\lambda)$ will be determined from the boundary conditions.

To illustrate, let us assume a single homogeneous layer whose resistivity is ρ_1 and whose thickness is h lying on top of a uniform half-space whose resistivity is ρ_2 (Fig. 14-4).   For the potential in the upper medium we may write

$$\mathcal{V}_1(r,z) = \frac{I\rho_1}{2\pi R} + \int_0^\infty [A(\lambda)e^{-\lambda z} + B(\lambda)e^{\lambda z}]\,J_0(\lambda r)\ d\lambda \qquad 0 \le z \le h$$

In the substratum there is no external source of current, and therefore the particular integral is not required.   Moreover, we must reject terms involving $e^{\lambda z}$, since the potential must remain finite when $z \to \infty$.   Consequently, we may write for the potential in $z \ge h$

$$\mathcal{V}_2(r,z) = \int_0^\infty C(\lambda)e^{-\lambda z}J_0(\lambda r)\ d\lambda \qquad z \ge h$$

The boundary conditions which these solutions must satisfy are the following:

$$\frac{\partial \mathcal{V}_1}{\partial z} = 0 \qquad z = 0$$

$$\mathcal{V}_1 = \mathcal{V}_2 \qquad z = h$$

$$\frac{1}{\rho_1}\frac{\partial \mathcal{V}_1}{\partial z} = \frac{1}{\rho_2}\frac{\partial \mathcal{V}_2}{\partial z} \qquad z = h$$

Thus three equations are available to determine the three unknown functions A, B, and C.   To apply these conditions, we make use of the Lipschitz integral identity [Watson (1), chap. 13]

$$\frac{1}{R} = \int_0^\infty e^{-\lambda z}J_0(\lambda r)\ d\lambda$$

Then, on substituting for $\mathcal{U}_1$ and $\mathcal{U}_2$ into these expressions, we arrive at the following linear equations:

$$A - B = 0$$

$$(1 + A)e^{-\lambda h} + Be^{\lambda h} = Ce^{-\lambda h}$$

and
$$-\frac{1}{\rho_1}[(1 + A)e^{-\lambda h} - Be^{\lambda h}] = -\frac{1}{\rho_2}Ce^{-\lambda h}$$

Solving, we get

$$A = B = \frac{I\rho_1 k}{2\pi(e^{2\lambda h} - k)} \qquad k = \frac{\rho_2 - \rho_1}{\rho_2 + \rho_1}$$

and therefore, on $z = 0$,

$$\mathcal{U}_1(r) = \frac{I\rho_1}{2\pi r}G(r;k) \tag{14-7}$$

where
$$G(r;k) = 1 + 2kr\int_0^\infty (e^{2\lambda h} - k)^{-1}J_0(\lambda r)\,d\lambda$$

This solution was derived originally by Stefanescu (2). Normally we take the layer thickness as the unit of length by setting $h = 1$, and the function $G(r;k)$ may then be evaluated numerically for given values of k. It is clear that the analysis can be extended to any number of layers, and in fact extensive tables for evaluating potentials over one, two, and three layers which rest conformably upon a uniform substratum are available.

14-6 Depth Determinations

With the aid of tables or graphs of the function $G(r;k)$ developed in the previous section, or of the equivalent functions for two or more layers, it is possible to interpret the results of expansions made with an electrode array such as the Wenner system (Fig. 13-1) in terms of the depths to the various boundaries. What we term an *expansion* consists of a series of resistivity determinations made over a wide range of electrode separations, in which the electrodes are kept in a straight line and the center point of the array is fixed. This technique is commonly used for finding the thicknesses of contrasting horizontal layers such as a water-saturated soil overlying a relatively impervious bedrock.

For the one-layer model the potential on the ground surface at a point distant r_1 from the current source and r_2 from the sink, when distances are measured in units of the layer thickness, is

$$\mathcal{U} = \frac{I\rho_1}{2\pi}\left(\frac{G(r_1;k)}{r_1} - \frac{G(r_2;k)}{r_2}\right)$$

Thus for the Wenner array, the *apparent resistivity* of the model, which is defined by the formula (14-5), has the theoretical value

$$\rho_a = 2\pi a\frac{\Delta\mathcal{U}}{I} = \rho_1[2G(a;k) - G(2a;k)] \tag{14-8}$$

If the ground consists of a single, uniform layer resting conformably upon a homogeneous substratum, we may endeavor to find the layer thickness by fitting the formula (14-8) to observed values of the apparent ground resistivity measured at various electrode spacings. The fitting process is greatly facilitated by preparing in advance a set of curves, on logarithmic paper, of the ratio ρ_a/ρ_1 plotted against the dimensionless interelectrode distance a/h for different values of k. The complete set for the single layer is shown in Fig. 14-5. The curve fitting is then carried out simply by plotting the apparent ground resistivity, which is calculated from the formula (14-5), against the interelectrode distance on a sheet of similar paper and matching these values to one of the theoretical curves by displacing the field profile in the direction of ρ_a/ρ_1 or of a/h. Such displacements, when applied to logarithmic quantities, merely fix the scale ratios. Then, when a suitable match has been found, h is given by the interelectrode distance which

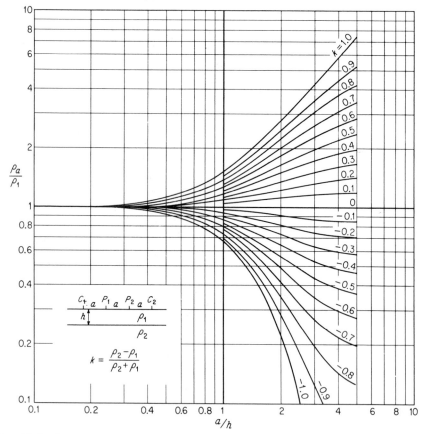

Fig. 14-5 Apparent resistivity curves for the single-layer model using the Wenner method. [After Mooney and Wetzel (3).]

falls at $a/h = 1$; and ρ_1, by the limiting value of apparent ground resistivity as $a \rightarrow 0$. The resistivity of the substratum is found from the value of k which is appropriate to the matching theoretical curve. This is entirely similar to the technique described in Chap. 4 for the interpretation of seismic surface-wave dispersion curves.

If the apparent ground resistivity values cannot be matched with any of the curves shown in Fig. 14-5, the indications are that the one-layer model is inadequate. Generally speaking, if the ground is known to be stratified, the number of layers that is sensed by the resistivity method is indicated by the number of inflections in the expansion profile. Additional sets of curves will be needed for these multilayer models. In principle, the method that we have just described can be extended to as many layers as we please, but when we stop to consider that each new layer introduces two new parameters into the curve-fitting process, it is evident that the number very soon reaches a practical limit. If n curves are needed to deal adequately with the one-layer model, n^3 will be required for two, n^5 for three, and so on. At about three layers a limit is reached beyond which the number of cases becomes altogether unwieldy, and the curve-matching method breaks down. Beyond three layers, the only recourse is to program a digital computer to calculate the apparent resistivities of specific ground models and then to proceed indirectly. A full set of theoretical curves for the Wenner electrode array for one, two, or three horizontal layers has been published by Mooney and Wetzel[1] (3). These are in a form which permits their routine use for depth determinations in all instances where the ground is known or thought to be horizontally stratified.

To give an example of the actual use of these curves, we show in Fig. 14-6 data from a Wenner expansion that was taken in the Fond du Lac area of Wisconsin (4). Superimposed on these data is the family of type curves from Mooney and Wetzel's compilation that seemed to provide the closest overall resemblance to the field curve. Interpolations from these curves give two boundaries, one at a depth of 70 ft and the other at 12 ft. The indicated resistivity of the outcropping material is about 70 ohm-m; of the middle layer, 7,000 ohm-m; and of the substratum, 0.7 ohm-m. This result is difficult to interpret in simple geological terms. Nearby water wells indicate a thickness of soil and drift of about 30 ft, and below this is a rather dense dolomite interbedded with sands and shale to a depth of about 300 ft. Beneath the dolomite are sandstones which appear to be quite porous. The conductivities interpreted from the two-layer model are not unreasonable for these formations, but the depths are at variance with what is known. Either there is strong aeolotropy present (Sec. 14-7), or else the saturation of some of the sandy beds in the dolomite with ground water (which is highly alkalic in this region) has masked the deeper layering.

---

[1] In these curves all distances have been scaled to $h = 6$ rather than $h = 1$, probably to make the best possible use of a standard graph paper.

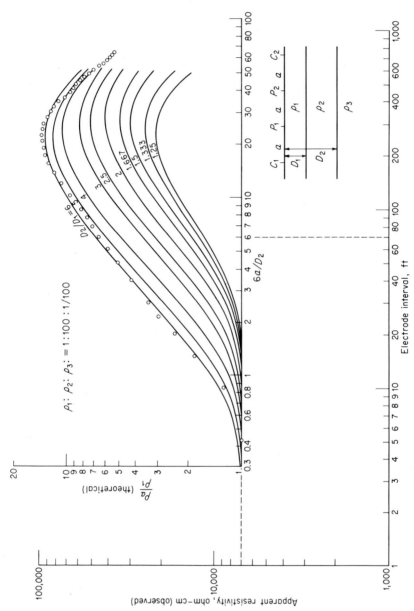

Fig. 14-6 The interpretation of a profile of apparent resistivity in the vicinity of Fond-du-Lac, Wisconsin.

Spicer (4), by employing geological considerations, has achieved a more satisfactory interpretation. It would thus appear that depth interpretations by resistivity sounding may not be fully determinate in practice from the electrical measurements alone.

Although it appears that some rather general stratigraphic information can be obtained by resistivity mapping under favorable conditions, it must be admitted that the resolving power of the method is not very high. This is particularly true of deeper boundaries, where substantial variations must occur before their effects can clearly be discerned on the resistivity profiles. Thus it is virtually impossible to estimate in advance the expected accuracy of these curve-fitting processes. This might lead us to ask whether the curve-fitting method is ambiguous. In other words, can two differently layered models have the same apparent resistivity? The answer to this question has been given by Langer (5), who showed that if the conductivity of the ground is a function of depth only, then the potentials about point-current electrodes uniquely determine this function. The "inverse" potential problem for horizontally stratified models, therefore, appears to have a unique solution.

This has led a number of workers to propose schemes for the direct interpretation of apparent resistivity measurements. We might mention in particular the work of Slichter (6), Belluigi (7), Vosoff (8), and Chetayev (9). The problem is not at all simple, however, since Langer's theorem does not imply that substantial changes in the stratification will necessarily lead to *substantial* changes in the potentials. It is, in fact, well known that changing the conductivity and the thickness of a layer in such a way that the product of these quantities remains the same produces very little effect upon apparent resistivities unless the thickness becomes large. Consequently, while it is certainly possible in theory to design methods which will determine the conductivity-depth function from surface potentials, it is very difficult to obtain reasonable results from these methods when the measurements are affected by near-surface heterogeneity or when the ground fails to conform exactly to the theoretical model.

14-7 *The Effect of Aeolotropy*

In aeolotropic media, the situation is further complicated by the fact that in general the equipotential surfaces are no longer normal to the direction of current flow. Recalling (14-1), in an aeolotropic medium we shall have

$$J_i = -\sigma_{ik} \frac{\partial \mathcal{U}}{\partial x_k} \qquad i, k = 1, 2, 3 \qquad (14\text{-}9)$$

using the indicial notation introduced in Chap. 2. Therefore when we apply the continuity condition, we find that

$$\sigma_{ik} \frac{\partial^2 \mathcal{U}}{\partial x_i \, \partial x_k} = 0 \qquad (14\text{-}10)$$

provided that the material is homogeneous. Equation (14-10) is of practical interest when the medium is transversely-isotropic, i.e., when σ is the same in all horizontal directions but has a different value for vertical current flow. This situation is common in shales, schists, and other materials which show a definite foliation pattern or bedding, and it is not difficult to imagine why. When we recall that the processes of electrical conduction in rocks are very largely electrolytic, it is easy to imagine that bedding should play an important role by fixing the routes of easiest migration for the fluids. We might expect, therefore, that conductivities parallel with the bedding plane will be larger on the whole than those perpendicular to it, and in most instances (where there is no extensive fracturing) this is what is observed.

In transversely-isotropic media, (14-10) reduces to the following equation:

$$\sigma_t \left(\frac{\partial^2 \mathcal{V}}{\partial r^2} + \frac{1}{r} \frac{\partial \mathcal{V}}{\partial r} \right) + \sigma_v \frac{\partial^2 \mathcal{V}}{\partial z^2} = 0 \tag{14-11}$$

in which σ_t is the "transverse" conductivity and σ_v is the "vertical" conductivity. The off-diagonal elements in the conductivity tensor are effectively zero in this case. By the usual method of separation of variables, we find that the characteristic functions for this equation are

$$e^{\pm f \lambda z} J_0(\lambda r) \qquad \text{where} \qquad f = \sqrt{\frac{\sigma_t}{\sigma_v}} = \sqrt{\frac{\rho_v}{\rho_t}}$$

We shall call f the *coefficient of aeolotropy* of the medium. The complementary solution of (14-11) is made up of a superposition of these functions.

The potential near the probe itself will be a function of the type

$$\mathcal{V}(r,z) = \frac{C}{\sqrt{r^2 + f^2 z^2}}$$

since this satisfies (14-11) and also satisfies the boundary condition at $z = 0$. To find C, we consider the flow of current outward through the surface of the small hemispheroid based on $z = 0$, whose center is at P and whose semimajor and semiminor axes in the vertical plane are a and a/f, respectively. According to the form we have found for $\mathcal{V}$, this surface will be an equipotential. If we resolve the potential gradient into its radial and vertical components, then according to (14-9) the component of $\mathbf{J}$ which is normal to this surface at Q (see Fig. 14-7) is

$$J_n = - \frac{\sigma_t C}{a^2 \sqrt{\cos^2 \vartheta + f^2 \sin^2 \vartheta}}$$

Since this quantity is uniform over the narrow ribbon around Oz whose area is

$$dS = \frac{2\pi a^2 \cos \vartheta \; d\vartheta}{\cos^2 \vartheta + f^2 \sin^2 \vartheta}$$

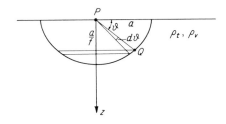

Fig. 14-7 A point source of current at the surface of a transversely-isotropic ground.

we may write the following integral for the total current I:

$$I = \int_S J_n dS = -2\pi\sigma_t C \int_0^{\pi/2} \frac{\cos\vartheta \, d\vartheta}{(\cos^2\vartheta + f^2 \sin^2\vartheta)^{3/2}}$$

from which we get

$$I = \frac{2\pi\sigma_t C}{f}$$

Hence $$C = \frac{I\rho_m}{2\pi}$$ where $$\rho_m = \sqrt{\rho_v \rho_t}$$

and $$\mathcal{V}(r,z) = \frac{I\rho_m}{2\pi \sqrt{r^2 + f^2 z^2}}$$

The general solution of (14-11) therefore becomes

$$\mathcal{V}(r,z) = \frac{I\rho_m}{2\pi \sqrt{r^2 + f^2 z^2}} + \int_0^\infty [A(\lambda)e^{-f\lambda z} + B(\lambda)e^{f\lambda z}] J_0(\lambda r) \, d\lambda \quad (14\text{-}12)$$

which is identical with the solution in an isotropic medium if we replace ρ with ρ_m and z with fz. We find, therefore, that subject to these substitutions the formula (14-7) and the curves derived therefrom will apply equally well to transversely-isotropic media. This leads us to the interesting conclusion that transverse aeolotropy cannot be detected from the form or behavior of the potentials and that if it is present and unaccounted for, it will cause depth calculations to be in error by a factor f. A coefficient of aeolotropy as large as 2 does not appear to be uncommon in some rocks, although we must again stress the very small number of experimental determinations published. It should be realized nonetheless that the uncritical use of type curves where aeolotropy exists may lead to substantial errors in depth interpretations.

14-8 Horizontal Traverses and Resistivity Exploration

When the vertical electrical structure of the ground has been ascertained at one location, direct-current methods may be used for reconnaissance structural mapping in much the same way as the gravity or the seismic refraction methods. The procedure is to select an electrode spacing which is of the order of the depth of the structural interface to be mapped, chosen so that the apparent resistivity measurements fall on the most sensitive portion of

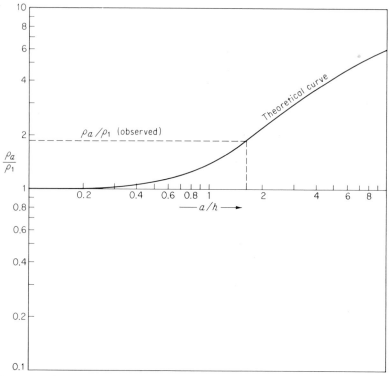

Fig. 14-8   **The use of apparent resistivity curves in horizontal traverses.**

the "expansion" curve.   Keeping a fixed at this value, the center point of the array is moved to a new location nearby.   A change in the value of ρ_a indicates either a change in a/h or in ρ_1, or in both.   Changes in ρ_2 are likely to have less effect.   If surface conditions remain the same, a new value for h may be found by following the appropriate theoretical curve to the new value of ρ_a/ρ_1 (Fig. 14-8) and determining the new value for a/h.   If changes in ρ_1 are suspected, however, the value should be checked by means of a short electrode spread and the value of ρ_a/ρ_1 adjusted upward or downward accordingly.   In principle it makes no difference whether the electrode array is moved in a direction at right angles to or in line with itself; but in practice, especially when dips become appreciable, better results for some reason seem to be found by moving the electrodes in line.

     Mapping by the d-c resistivity method is relevant only to regional investigations.   By its very nature it tends to smooth out lateral changes, and it is quite insensitive to local deformities.   It is also unresponsive to structural changes taking place below the uppermost layer, and so it is generally used for mapping variations in the thickness of that layer.   The method has been used with some success in certain kinds of engineering investigations, and accounts from Europe and the U.S.S.R. indicate that it has given satisfactory results in regional geological reconnaissance.

14-9 Potentials over Sloping Interfaces

In Sec. 14-7 we considered one of the ways in which the ground might fail to conform in its electrical properties to the ideally stratified model. Now we shall consider one of the ways in which it may fail to conform structurally, viz., by supposing that the bedding planes or interfaces are dipping. The question that we now wish to raise is whether or not the effect of dip is recognizable in the shape of resistivity expansion profiles. If it is, it will play some part in the interpretation of horizontal traverses.

The problem of potentials over dipping strata is too complicated to solve mathematically when more than two media are involved, but Maeda (10) has worked out a solution for the special case of two media in contact along a single sloping interface. Referring to Fig. 14-9, we consider a point electrode situated at a point P and delivering current to the ground at the rate of I amp. We now choose a system of cylindrical coordinates (r,ϑ,z) such that the z axis follows the line of outcrop of the contact and P lies in the plane $z = 0$. The contact is thus defined as the plane $\vartheta = \alpha$, and the two media occupy the regions $0 \leq \vartheta < \alpha$, $\alpha < \vartheta \leq \pi$. To determine $\mathcal{U}$ anywhere on the ground, we are required to find solutions of $\nabla^2 \mathcal{U} = 0$, which satisfy the boundary conditions at $\vartheta = \alpha$ and at $\vartheta = 0, \pi$.

We proceed as usual by writing out Laplace's equation

$$\nabla^2 \mathcal{U} = \frac{\partial^2 \mathcal{U}}{\partial r^2} + \frac{1}{r}\frac{\partial \mathcal{U}}{\partial r} + \frac{1}{r^2}\frac{\partial^2 \mathcal{U}}{\partial \vartheta^2} + \frac{\partial^2 \mathcal{U}}{\partial z^2} = 0$$

and solve by the method of separation of variables. The a priori conditions on the solution are that it must converge for large values of z (which may be either positive or negative) and that it must behave properly for small values of r and converge for large ones. The only characteristic functions which can be made to satisfy all these requirements are of the type

$$\mathcal{U}(r,\vartheta,z) \sim \cos \lambda z \cosh s\vartheta\; K_{is}(\lambda r)$$

where K_{is} is the modified Bessel function of the second kind, of imaginary order. The complementary solution will be made up of a superposition of these functions, and to this we must add the particular integral $\mathcal{U} = I\rho_1/2\pi R$

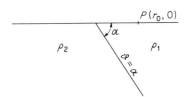

Fig. 14-9 A point source of current near a sloping interface between two uniform media.

in the medium in which the source is located.   Thus in medium 1

$$\mathcal{V}_1(r,\vartheta,z) = \frac{I\rho_1}{2\pi R} + \int_0^\infty \int_0^\infty A(s,\lambda) \cos \lambda z \cosh s\vartheta K_{is}(\lambda r) \, ds \, d\lambda$$

while in medium 2

$$\mathcal{V}_2(r,\vartheta,z) = \int_0^\infty \int_0^\infty B(s,\lambda) \cos \lambda z \cosh s(\pi - \vartheta) K_{is}(\lambda r) \, ds \, d\lambda$$

The boundary conditions which these solutions must satisfy are the following:

$$\frac{\partial \mathcal{V}_1}{\partial \vartheta} = 0 \qquad \vartheta = 0 \qquad \text{or} \qquad \frac{\partial \mathcal{V}_2}{\partial \vartheta} = 0 \qquad \vartheta = \pi$$

$$\mathcal{V}_1 = \mathcal{V}_2 \qquad \vartheta = a$$

$$\frac{1}{\rho_1} \frac{\partial \mathcal{V}_1}{\partial \vartheta} = \frac{1}{\rho_2} \frac{\partial \mathcal{V}_2}{\partial \vartheta} \qquad \vartheta = a$$

Note that the first condition is satisfied automatically by our choice of ϑ dependence.   The conditions at $\vartheta = a$ provide us with two equations with which to determine two unknown functions A and B.   Here we must introduce the modified form of the Lipschitz integral identity

$$\frac{1}{R} = \frac{2}{\pi} \int_0^\infty \cos \lambda z K_0(\lambda P) \, d\lambda$$

where

$$P = \sqrt{r^2 + r_0^2 - 2rr_0 \cos \vartheta}$$

plus the addition theorem for modified Bessel functions

$$K_0(\lambda P) = \frac{2}{\pi} \int_0^\infty \cosh s(\pi \pm \vartheta) K_{is}(\lambda r_0) K_{is}(\lambda r) \, ds$$

to yield

$$\frac{I\rho_1}{2\pi R} = \frac{2I\rho_1}{\pi^3} \int_0^\infty \int_0^\infty \cos \lambda z \cosh s(\pi \pm \vartheta) K_{is}(\lambda r_0) K_{is}(\lambda r) \, ds \, d\lambda$$

in which the $+$ sign is taken in medium 2 and the $-$ sign in medium 1. Substituting this expression together with those for $\mathcal{V}_1$ and $\mathcal{V}_2$ into the boundary conditions and solving, we get

$$A(s,\lambda) = \frac{2I\rho_1}{\pi^3} K_{is}(\lambda r_0) \frac{k \sinh 2s(\pi - a)}{\sinh \pi s - k \sinh s(\pi - 2a)}$$

$$B(s,\lambda) = \frac{2I\rho_1}{\pi^3} K_{is}(\lambda r_0) \frac{(1 + k) \sinh \pi s}{\sinh \pi s - k \sinh s(\pi - 2a)}$$

where $k = \dfrac{\rho_2 - \rho_1}{\rho_2 + \rho_1}$ as before.

The resulting integrals can be reduced to a series of terms involving Legendre functions of the second kind.   These series have been evaluated numerically in a limited number of cases, and it is instructive to study the forms of the apparent resistivity curves for the Wenner electrode configuration. Figure 14-10, taken from Maeda's paper, illustrates the forms of the apparent resistivity curves to be expected from a Wenner expansion taken parallel with the strike of the contact, where $k = \pm 0.6$. When the electrode interval is scaled in units of the perpendicular distance from the survey line to the contact plane, the curves are almost indistinguishable from apparent resistivity curves appropriate to a flat layer. The only change wrought by the dip is a reduction of the resistivity contrast that would be inferred from the single-layer master curves. Thus the presence of dip is ordinarily difficult to detect by the Wenner expansion method and is a source of some ambiguity in the interpretations.

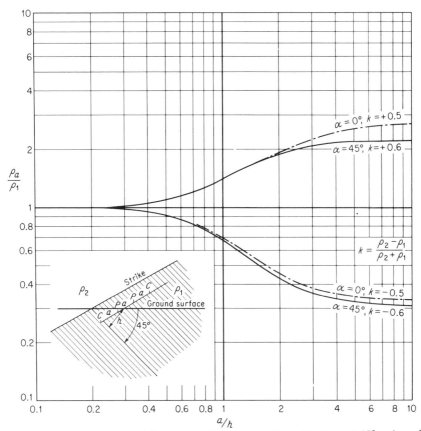

Fig. 14-10 **Apparent resistivity curves for an interface dipping at 45° using the Wenner method, $k = \pm 0.6$.** [**After Maeda (10).**] **Curves for a uniform layer are shown for comparison. The effect of dip is barely evident in the shapes of the curves, but it is almost equivalent to a reduction in the resistivity contrast between the two media.**

In these examples we have worked exclusively with the Wenner electrode array.   Other arrangements of the electrodes are certainly possible and in some special situations may even help to resolve ambiguities of the type just described.   The conversion of curves from one electrode arrangement to another is easily made, once the basic potential functions have been tabulated, and can be further facilitated by constructing special nomograms based upon specific earth models.   A variety of aids of this kind exists for the horizontally stratified ground model, so that conversions may be made between the Wenner and any of several other configurations [for example, see Zavadskaya (11)].

14-10   *Horizontal Exploration and the Interpretation of Electrical Resistivity Anomalies*

Inferring the depth of horizontal or sloping boundaries by "depth-sounding" methods is one aspect of the direct-current interpretation problem.   Another is the analysis of local anomalies in the surface potentials caused by bodies having electrical properties different from those of their surroundings.   The situation for a conducting body placed in a uniform electric field is illustrated in a highly simplified way in Fig. 14-11.   The "spreading" of the equipotential lines at the surface of the ground due to the presence of the conductor

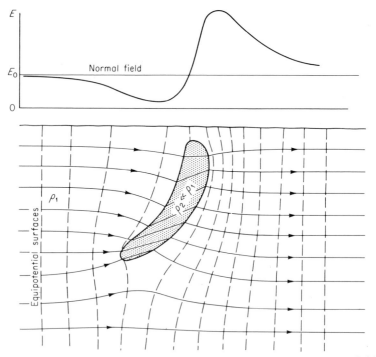

Fig. 14-11   An irregular conducting body in a uniform electric field.

beneath it means that the measured difference in potential across a given distance drops within this region. Interpretation of these anomalies usually consists in trying to solve the "inverse" potential problem with the aid of simple models. The solutions are used to estimate the size, shape, and depth of the disturbing body.

The number of models which can be used for making direct interpretations is severely limited by mathematical considerations. Exact solutions of the boundary-value problem can be found only if a system of coordinates exists in which all boundaries *including the ground surface* are level surfaces and in which Laplace's equation also separates. These requirements are so stringent that we would indeed be limited if no other approach were open. Fortunately, it is often permissible to relax these conditions somewhat and make available a wider variety of models for use in the interpretations. These solutions can be further augmented by the use of approximate numerical methods which can be applied even to nonseparable boundaries. Additional interpretational aids can be compiled from measurements made on laboratory scale models. Thus there should be no need to force the geological interpretations to fit too narrow a set of mathematical requirements.

14-11   The Conducting Sphere

One of the few problems in electrical resistivity theory which is exactly soluble, and which illustrates the difficulties imposed by the use of exact methods, is that of finding the potential distribution on the surface of a uniform half-space in which a conducting sphere lies imbedded. The solution is useful to us, because it indicates the magnitude of the anomaly that can be expected from a conducting mass of a given size. Since it is a common practice in mining geophysics to employ nearly uniform electric fields in exploring for ore bodies (the uniformity is achieved by placing the current probes a large distance apart), we shall first consider how to calculate the anomaly in the electric field.

The geometry is shown in Fig. 14-12. The first task is to find a set of coordinates in which Laplace's equation separates and in which the sphere and the plane are both level surfaces. Such a set is the bispherical system obtained by rotating the coordinates shown in Fig. 14-13 about the z axis. If we define a set of coordinates μ, η, and φ by means of the parametric relations

$$x = \frac{f \sin \eta \cos \varphi}{\cosh \mu - \cos \eta} \qquad y = \frac{f \sin \eta \sin \varphi}{\cosh \mu - \cos \eta} \qquad z = \frac{f \sinh \mu}{\cosh \mu - \cos \eta}$$

then the surfaces $\mu = \text{const}$ are spheres of radius $f/\sinh \mu$ with centers at $(0,0, f \coth \mu)$; the surfaces $\eta = \text{const}$ are fourth-order surfaces which may be either cusped or pointed depending upon whether η is less or greater than $\frac{1}{2}\pi$; and the surfaces $\varphi = \text{const}$ are a family of axial planes. The range of

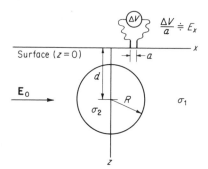

Fig. 14-12 A conducting sphere in a uniform electric field.

μ is from 0 to ∞ if $z \geq 0$, and η goes from 0 to π. The two poles of the system are at $\mu = \pm \infty$, which correspond to the points $(0,0, \pm f)$. Thus the surface of the conducting sphere (whose radius is R) is the surface $\mu = \sinh^{-1} f/R = \mu_0$, say. The ground surface is the surface $\mu = 0$. The center of the sphere lies at a depth $d = f \coth \mu_0$, from which it follows that $f = \sqrt{d^2 - R^2}$. Thus the system of coordinates is defined by the two parameters R and d.

If we put $\mho = \sqrt{\cosh \mu - \cos \eta}\, F$, then Laplace's equation $\nabla^2 \mho = 0$ transforms to

$$\frac{\partial^2 F}{\partial \mu^2} + \frac{1}{\sin \eta} \frac{\partial}{\partial \eta} \left(\sin \eta \frac{\partial F}{\partial \eta} \right) + \frac{1}{\sin^2 \eta} \frac{\partial^2 F}{\partial \varphi^2} - \frac{1}{4} F = 0$$

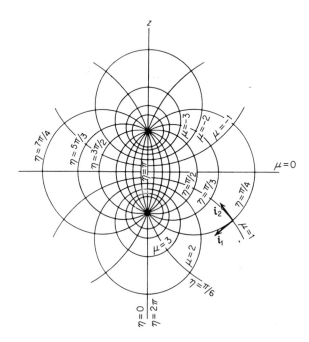

Fig. 14-13 Bispherical coordinates.

If we then set $F(\mu,\eta,\varphi) = M(\mu)H(\eta)\Phi(\varphi)$, this equation in turn separates into the three equations

$$\frac{d^2\Phi}{d\varphi^2} = -m^2\Phi \qquad \frac{d^2M}{d\mu^2} = (n+\tfrac{1}{2})^2 M$$

and

$$\frac{1}{\sin\eta}\frac{d}{d\eta}\left(\sin\eta\,\frac{dH}{d\eta}\right) + \left[n(n+1) - \frac{m^2}{\sin^2\eta}\right]H = 0$$

The first two of these equations are familiar to us, and the third becomes the associated Legendre equation if we change the independent variable to $\cos\eta$. Consequently, the characteristic functions of Laplace's equation in bispherical coordinates are

$$\sqrt{\cosh\mu - \cos\eta}\; e^{\pm(n+\frac{1}{2})\mu} P_n^{|m|}(\cos\eta)\; \frac{\cos}{\sin}\, m\varphi \qquad |m| \le n$$

If in the absence of the sphere there is a uniform field E_0 in the direction Ox, we may write for the undisturbed potential

$$\mathcal{V}_0 = -E_0 x = -\frac{E_0 f \sin\eta \cos\varphi}{\cosh\mu - \cos\eta}$$

By expanding $(\cosh\mu - \cos\eta)^{-\frac{1}{2}}$ by the binomial theorem in powers of $\cos\eta$, it is easy to show that

$$(\cosh\mu - \cos\eta)^{-\frac{1}{2}} = \sqrt{2}\sum_{n=0}^{\infty} e^{-(n+\frac{1}{2})\mu} P_n(\cos\eta) \qquad \mu > 0$$

and consequently

$$\frac{\sin\eta}{(\cosh\mu - \cos\eta)^{\frac{3}{2}}} = 2\frac{\partial}{\partial\eta}\frac{-1}{\sqrt{\cosh\mu - \cos\eta}} = 2\sqrt{2}\sum_{n=1}^{\infty} e^{-(n+\frac{1}{2})\mu} P_n^1(\cos\eta)$$

Thus we have

$$\mathcal{V}_0 = -\sqrt{8}\,E_0 f \cos\varphi \sqrt{\cosh\mu - \cos\eta}\sum_{n=1}^{\infty} e^{-(n+\frac{1}{2})\mu} P_n^1(\cos\eta)$$

The general solution for the potential outside the sphere is

$$\mathcal{V}_1 = \mathcal{V}_0 + \mathcal{V}^{(S)}$$

where $\mathcal{V}^{(S)}$ is the secondary, or anomalous, potential due to the presence of the sphere. In terms of the characteristic functions of Laplace's equation, the full expression is

$$\mathcal{V}_1 = \sqrt{8}\,E_0 f\sqrt{\cosh\mu - \cos\eta}\left\{-\sum_{n=1}^{\infty} e^{-(n+\frac{1}{2})\mu} P_n^1(\cos\eta)\cos\varphi\right.$$
$$+ \sum_{n=0}^{\infty}\sum_{m=-n}^{m=n}\left[A_n^m\cosh(n+\tfrac{1}{2})\mu\right.$$
$$\left.\left. + B_n^m\sinh(n+\tfrac{1}{2})\mu\right]\frac{\cos}{\sin}\,m\varphi\,P_n^{|m|}(\cos\eta)\right\}$$

Inside the sphere, since we must choose a solution that remains finite as $\mu \to \infty$, we put

$$\mathcal{V}_2 = \sqrt{\cosh \mu - \cos \eta} \sum_{n=0}^{\infty} \sum_{m=-n}^{m=n} C_n{}^m e^{-(n+\frac{1}{2})\mu} \begin{matrix} \cos \\ \sin \end{matrix} m\varphi \, P_n{}^{|m|}(\cos \eta)$$

To these general solutions, we must now apply the boundary conditions, viz.,

$$\frac{\partial \mathcal{V}^{(S)}}{\partial \mu} = 0 \qquad \mu = 0$$

$$\mathcal{V}_1 = \mathcal{V}_2 \qquad \mu = \mu_0$$

and

$$\sigma_1 \frac{\partial \mathcal{V}_1}{\partial \mu} = \sigma_2 \frac{\partial \mathcal{V}_2}{\partial \mu} \qquad \mu = \mu_0$$

All terms except those in which $m = 1$ will disappear at once. The expression that results for the anomalous potential field on the plane $z = 0$ $(\mu = 0)$ is

$$\mathcal{V}^{(S)}(0,\eta,\varphi) = \sqrt{8} \, E_0 f \sqrt{1 - \cos \eta} \sum_{n=1}^{\infty} G_n(\eta) P_n{}^1 (\cos \eta) \cos \varphi$$

where G_n is rather a complicated function involving as parameters μ_0, σ_1, and σ_2. If $\sigma_2 \gg \sigma_1$, the expression takes the following simpler form

$$\mathcal{V}^{(S)}(0,\eta,\varphi) \doteq \sqrt{32} \, E_0 f \sqrt{1 - \cos \eta} \sum_{n=1}^{\infty} [e^{(2n+1)\mu_0} + 1]^{-1}$$

$$\times P_n{}^1 (\cos \eta) \cos \varphi \qquad (14\text{-}13)$$

In practice, the difference in potential is measured at two points on the ground surface at a fixed distance a apart. In the absence of anomalous conductors, the difference in potential should remain constant. The effect brought about by the presence of the sphere, however, will be

$$\Delta \mathcal{V}^{(S)} = \mathcal{V}^{(S)}(x + a) - \mathcal{V}^{(S)}(x)$$

which, to a suitable approximation, may be written

$$\Delta \mathcal{V}^{(S)} = a \left(\frac{\partial \mathcal{V}^{(S)}}{\partial x} \right)_{z=0} = - \frac{a(1 - \cos \eta)}{f} \left(\frac{\partial \mathcal{V}^{(S)}}{\partial \eta} \right)_{\mu=0} \qquad (14\text{-}14)$$

The effect of replacing the potential difference with the potential gradient is merely to sharpen the response somewhat. Otherwise it is of little consequence. Thus the electric field anomaly due to the sphere is found by differentiating (14-13).

The complicating factor in the solution of the sphere problem has been the necessity of making the ground surface, as well as the sphere, a surface of separation of the coordinates.   Another way of satisfying the boundary condition at $z = 0$ is to remove the ground surface entirely and to introduce a "mirror-image" sphere above the ground, with its center at $(0,0, -f$ coth $\mu_0)$.   This would not affect the analysis in any way, and we would still get (14-13) as the solution.   If, however, we were to neglect altogether the interaction between the sphere and its image (or, what is the same thing, the interaction between the sphere and the boundary), the problem would then simplify enormously.   In this case we should have two isolated, non-interacting conducting spheres in a uniform electric field E_0 that is transverse to the line joining their centers (Fig. 14-14).   We may then calculate the potential of either one of them on the assumption that the other is absent.   (Note that this still satisfies the boundary condition on $z = 0$, but no longer satisfies the condition on the sphere.)   Thus, the potential on $z = 0$ due to either sphere is found as the solution of the boundary-value problem of a sphere in a uniform electric field, viz.,

$$\mathcal{V}^{(S)} = - \left(\frac{\sigma_2 - \sigma_1}{\sigma_2 + 2\sigma_1} \right) \frac{E_0 R^3 x}{(x^2 + d^2)^{3/2}}$$

where x is measured from the point above the center.   The total anomalous potential will be twice this amount.   And so the electric field anomaly will be, assuming that $\sigma_2 \gg \sigma_1$,

$$\Delta \mathcal{V}^{(S)} \doteq 2 E_0 R^3 a \frac{(2x^2 - d^2)}{(x^2 + d^2)^{5/2}} \tag{14-15}$$

The question now arises of whether and under what conditions (14-15) is a satisfactory approximation to (14-14).   Calculations show that the two formulas agree to within 10 percent when $d \geq 1.3R$ [see, for example, Lipskaya (12)].   Thus the sphere must be close to the ground surface before the interaction between the conductor and its image will have an

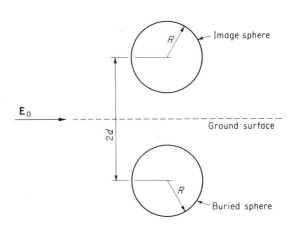

Fig. 14-14   A conducting sphere and its mirror image.

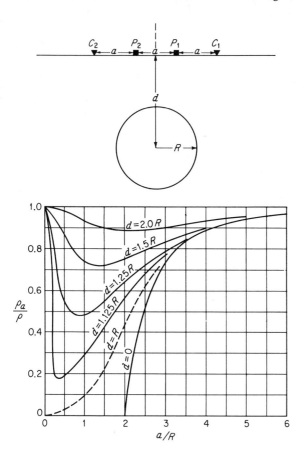

Fig. 14-15 Curves showing the maximum effect of a buried conducting sphere on the apparent resistivity of an otherwise uniform ground as measured by the Wenner method. [After Van Nostrand (13).]

appreciable effect upon the potentials, and it is usually justifiable to ignore it in practice. Thus we replace the ground surface with a noninteracting image of the model. This simply amounts to multiplying the anomalous potential gradient on $z = 0$ by two.

The effect of a highly conducting sphere upon the apparent resistivity of an otherwise uniform ground, as measured by a four-electrode array, has been calculated by Van Nostrand (13). The essential results are summarized in the diagram shown in Fig. 14-15, in which the change in apparent resistivity over the center of the sphere is given as a function of the depth of the sphere for several different radii. The unit of length is in this case the interelectrode spacing. The curves indicate, as formula (14-15) does also, that the sphere becomes scarcely detectable by d-c conduction methods if d becomes much larger than $1.5R$. But this depth limitation is, after all, scarcely more restrictive than that which applies to gravity and magnetic measurements, and it should not be regarded as a shortcoming peculiar to the electrical methods.

14-12   The Conducting Ribbon in a Uniform Field

Tabular conducting bodies having an extended strike, such as mineralized veins or shears, can very effectively be represented by the two-dimensional ribbon model.   To solve the boundary-value problem of a conducting ribbon in a horizontal uniform electric field, we select the system of coordinates defined by

$$x' = f \cosh \mu \cos \nu \qquad y' = f \sinh \mu \sin \nu$$

The lines $\mu = $ const represent a family of confocal ellipses whose foci are at $(\pm f, 0)$.   The range of μ is from 0 to ∞ (Fig. 14-16).   The lines $\nu = $ const are a family of confocal hyperbolas having the same foci as the ellipses, and the range of ν is from 0 to 2π.   The ribbon is represented by the thin ellipse $\mu = \mu_0$, where μ_0 is very small.

Laplace's equation in this system of coordinates is

$$\frac{\partial^2 \mho}{\partial \mu^2} + \frac{\partial^2 \mho}{\partial \nu^2} = 0$$

and the characteristic functions, which we obtain by the method of separation of variables, will be linear combinations of $\cosh n\mu$, $\sinh n\mu$, $\cos n\nu$, and $\sin n\nu$.   If the ribbon dips at an angle d when the direction of E_0 is horizontal and to the right, as shown in Fig. 14-17, then the potential at large distances from the ribbon, ignoring the presence of the ground surface, will be

$$\mho_0 = -E_0 x' \cos d + E_0 y' \sin d$$
$$= -E_0 f (\cos d \cosh \mu \cos \nu - \sin d \sinh \mu \sin \nu)$$

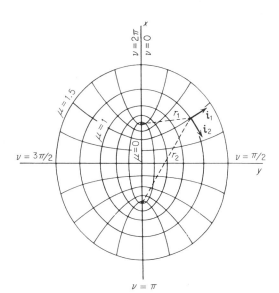

Fig. 14-16   Elliptical coordinates.

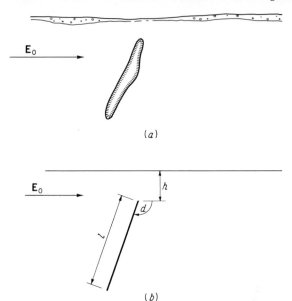

(a)

(b)

Fig. 14-17 A conducting ribbon in a uniform electric field.

It follows that $n = 1$ and therefore that $\mho^{(S)}$ must be some linear combination of cosh μ, sinh μ, cos ν, and sin ν. But $\mho^{(S)}$ must also vanish as $\mu \to \infty$, and the only linear combination of sinh μ and cosh μ that does so is $e^{-\mu} = \cosh \mu - \sinh \mu$. Thus

$$\mho^{(S)} = e^{-\mu}(A \cos \nu - B \sin \nu)$$

The boundary condition to be satisfied by $\mho^{(S)}$ is that in the limit, as $\mu_0 \to 0$ and as σ (the electrical conductivity of the ribbon) becomes large, $\mho_0 + \mho^{(S)} \to 0$ at $\mu = \mu_0$. Consequently

$$A = E_0 f \cos d \, e^{\mu_0} \cosh \mu_0 \quad \text{and} \quad B = E_0 f \sin d \, e^{\mu_0} \sinh \mu_0$$

We may now set sinh $\mu_0 = w/2f$, where w is the thickness of the ribbon. Provided that $w \ll f$, we may also set cosh $\mu_0 \doteq 1$. This allows us to write

$$\mho^{(S)} = \tfrac{1}{2} E_0 e^{-\mu}(l \cos d \cos \nu - w \sin d \sin \nu) \qquad (14\text{-}16)$$

if we replace $2f$ with the ribbon width l. If the ribbon is vertical ($d = 90°$), the potential is proportional to the thickness w, and therefore it vanishes as $w \to 0$. This is as we should expect, because the ribbon will in this case lie parallel with the equipotential surfaces, and being a very good conductor, it should scarcely disturb them. It follows that narrow, steeply dipping veins, no matter how conductive, will be difficult to detect by employing horizontal d-c fields transverse to their strike direction, unless they lie beneath a very shallow cover. This fact is well known in practice.

To draw theoretical profiles of the horizontal gradient of $\mho^{(S)}$ on $z = 0$, it is first necessary to convert from μ and ν to x' and y' in (14-16). The inverse relations are

$$l \cosh \mu = \sqrt{y'^2 + (l/2 + x')^2} + \sqrt{y'^2 + (l/2 - x')^2}$$

and $$l \cos \nu = \sqrt{y'^2 + (l/2 + x')^2} - \sqrt{y'^2 + (l/2 - x')^2}$$

We then apply the transformation

$$x' = x \cos d + y \sin d \qquad y' = -x \sin d + y \cos d$$

which makes the y axis vertical. And finally, if h is the depth of the over-burden, setting $y = h + \frac{1}{2}l \sin d$ puts the x axis along the surface of the ground. The anomalous potential due to the "image ribbon" is found by applying the same set of transformations and substituting $-d$ for d and $-y$ for y. On the ground surface ($y = 0$), the addition of the image term merely has the effect of multiplying $\mathcal{V}^{(S)}$ by 2.

"Type curves," or theoretical profiles, are calculated in this case by taking the horizontal gradient of $2\mathcal{V}^{(S)}(x)$. Two examples of such curves, which correspond to two different values of d, are shown in Fig. 14-18, in which distances are scaled in units of the ribbon-width l. Like the magnetic profiles across a dip-magnetized step, the information contained in these curves is difficult to distil into a pair of characteristic estimators for d and h. (There seems to be, unfortunately, an inverse relationship between the sensitivity of an estimator and its confidence value in curves of this type, and we must be guided more by what can be observed in practice than what we would like to use.) The only interpretable features of d-c anomaly profiles of this kind, as a rule, are the total peak-to-trough change in the electric-field intensity and the ratio of the maximum positive to negative anomalous effect. Horizontal gradients, width ratios, etc. are seldom apt to

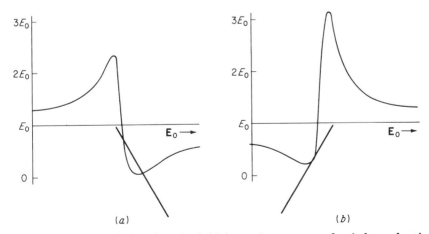

Fig. 14-18 **Profiles of the electric field intensity across a buried conducting ribbon in a uniform primary electric field.**

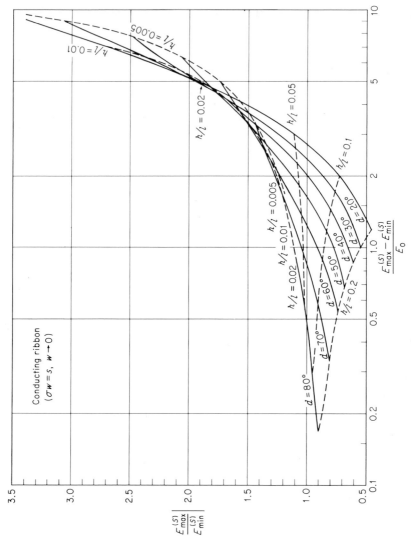

Fig. 14-19 Curves showing the maximum peak-to-trough anomaly versus the ratio of positive-to-negative intensity across a two-dimensional conducting ribbon.

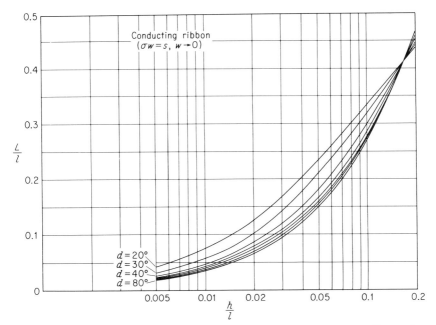

Fig. 14-20 **Curves showing the quantity L/l versus h/l for the two-dimensional conducting ribbon.**

be reliable for a variety of reasons. When we plot the peak-to-trough amplitude against the amplitude ratio of the anomaly, we obtain the set of characteristic curves shown in Fig. 14-19. These show very clearly how difficult it is to separate the effects of depth and dip by any direct method. Consequently, either an indirect or a qualitative approach to interpretation seems to be almost inevitable. This very often consists simply in comparing the anomaly with a few theoretical profiles.

Since the abscissas of type curves are given in units of l, Fig. 14-20 may be useful in establishing the horizontal scale. This is the complementary set to Fig. 14-19 and shows L/l versus h/l with d as a parameter, where L is the horizontal distance between the peak and the trough of $E^{(S)}$.

14-13   *The Conducting Half-plane or Ribbon near a Point Source*

In the event that a three- or four-electrode system is used to traverse the ground, the presence of one or both of the current electrodes near the conducting body means that the primary electric field is not uniform at the interface. In that case, the solution of the field equations is based upon a different approach altogether. In the first place, since the strength of the primary field diminishes as the square of the distance from the current probe, the deeper parts of the conductive body will generally have little

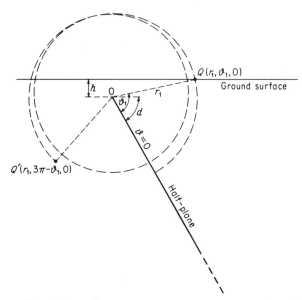

Fig. 14-21   The conducting half-plane model.

effect upon the distribution of the potential near the ground surface.   Therefore extended conductors such as veins or shears may often be represented by the conducting half-plane model.   This is a body that is both infinitely thin and infinitely conductive, having an infinite strike length and an infinite depth extent as indicated in Fig. 14-21.   The parameters of this model are only two: the dip d and the depth of overburden h.   The natural unit of length, in terms of which h may be expressed, is the interelectrode spacing a.

The problem of calculating the potential around a point charge located in the vicinity of a conducting half-plane was solved by Sommerfeld. The solution is based upon the conformal transformation $z = w^2$, which transforms the half-plane into a uniform sheet (see Sec. 17-7).   The potential due to a point source in Riemann (w) space can be found very easily by the method of images and transformed back again into ordinary (z) space; but a difficulty is raised by the fact that one complete revolution in w space corresponds to two revolutions, or windings, in z space, and the image point lies in the second.   It was therefore necessary to find the Green's function in real space which corresponds to the function $1/|\mathbf{r} - \mathbf{r}_0|$ in a space of two windings.   This function was found by Sommerfeld to be

$$G(\mathbf{r},\mathbf{r}_0) = \frac{1}{\pi R} \tan^{-1}\left(-\frac{R}{g} \right) \qquad (14\text{-}17)$$

where

$$g = 2\sqrt{rr_0}\cos\frac{\vartheta - \vartheta_0}{2}$$

and

$$R = [r^2 + r_0{}^2 + (z - z_0)^2 - 2rr_0 \cos(\vartheta - \vartheta_0)]^{1/2}$$

Here r, ϑ, and z are the *cylindrical* coordinates of the point $P(\mathbf{r})$, while r_0, ϑ_0, and z_0 are the cylindrical coordinates of $Q(\mathbf{r}_0)$ in a system whose axis lies along the edge of the half-plane. Thus the total potential in the vicinity of a point electrode delivering current to the ground (whose resistivity is ρ) at the rate of I amp near a conducting half-plane will be

$$\mathcal{V}(r,\vartheta,z) = \frac{I}{2\pi^2 \rho R_1} \tan^{-1}\left(-\frac{R_1}{g_1}\right) + \frac{I}{2\pi^2 \rho R_2} \tan^{-1}\left(-\frac{R_2}{g_2}\right) \quad (14\text{-}18)$$

where the subscript 1 indicates that the quantities $(r_0,\vartheta_0,z_0,)$ in the functions R and g are identified with the coordinates $(r_1, \vartheta_1,0)$ of the source, while the subscript 2 indicates that (r_0,ϑ_0,z_0) are identified with the coordinates of the image. In other words, in R_1 and g_1 we put $r_0 = r_1$, $\vartheta_0 = \vartheta_1$, and $z_0 = 0$. In R_2 and g_2 we put $r_0 = r_1$, $\vartheta_0 = 3\pi - \vartheta_1$, and $z_0 = 0$.

To find the apparent resistivity of the ground in the vicinity of a conducting half-plane, we transform from cylindrical to rectangular coordinates in the expression (14-18). The effect of the surface of the ground is roughly allowed for by multiplying the anomalous potential on the ground surface by 2. Substituting this result into Eq. (14-5), for example, gives the formula for the apparent resistivity ρ_a at an arbitrary position x for the Wenner electrode configuration. This will be in the form

$$\rho_a = f\left(\frac{x}{a}; \frac{h}{a}, d\right)$$

which permits us to calculate "type curves" showing ρ_a/ρ versus x/a for various values of the parameters h/a and d.

Similar sets of curves for depth-limited bodies can also be constructed by using the conducting ribbon model instead of the half-plane. The Green's function for a point source near a conducting ribbon has also been given by Sommerfeld (14)

14-14 The Oblate Spheroid

There are several reasons for introducing the oblate spheroid as a fourth (and final) model for the interpretation of d-c conduction surveys. The oblate hemispheroid is a plausible shape for representing surface features such as sinks, swamps, glacial deposits, or muskeg. Such bodies can produce intense anomalies, and it will be useful to have some rules for their recognition. We may study the effect on electric-field anomalies of limiting the strike length of tabular conductors such as veins by drawing type curves for the conducting disk, which is a degenerate oblate spheroid. In the interpretation of drill-hole surveys, flat-lying conducting zones such as sedimentary ore deposits may be described in terms of oblate spheroidal outlines. Thus the model is useful in many ways.

The system of oblate spheroidal coordinates is obtained by rotating the elliptic coordinates shown in Fig. 14-16 about the line $\nu = \pi/2$. It is defined by the relations

$$x = f \cosh \mu \cos \nu \cos \varphi \qquad y = f \cosh \mu \cos \nu \sin \varphi \qquad z = f \sinh \mu \sin \nu$$

The surfaces $\mu = \text{const}$ are a family of confocal oblate spheroids, ranging in size from the disk of radius f in the (x,y) plane $(\mu = 0)$ to a sphere of infinitely large radius $(\mu \to \infty)$. The surfaces $\nu = \text{const}$ are a family of confocal hyperboloids of one sheet. They range in size from the negative z axis $(\nu = -\pi/2)$ to the entire (x,y) plane $(\nu = 0)$, to the positive z axis $(\nu = \pi/2)$. The third coordinate φ specifies the family of axial planes intersecting along the z axis. The range of φ is from 0 to 2π.

Laplace's equation in oblate spheroidal coordinates [Morse and Feshbach (15), Sec. 10-3] is

$$\frac{1}{\cosh \mu} \frac{\partial}{\partial \mu}\left(\cosh \mu \frac{\partial \mathcal{V}}{\partial \mu}\right) + \frac{1}{\cos \nu} \frac{\partial}{\partial \nu}\left(\cos \nu \frac{\partial \mathcal{V}}{\partial \nu}\right) + \left(\frac{1}{\cos^2 \nu} - \frac{1}{\cosh^2 \mu}\right)\frac{\partial^2 \mathcal{V}}{\partial \varphi^2} = 0$$

and if we make the usual separation of variables, we obtain as the characteristic functions linear combinations of

$$(\cos m\varphi, \ \sin m\varphi) \qquad P_n^{|m|}(\sin \nu)$$

and $\qquad P_n^{|m|}(i \sinh \mu) \qquad Q_n^{|m|}(i \sinh \mu) \qquad |m| \leq n$

$Q_n^{|m|}$ is the second solution of Legendre's associated equation, and its definition is [see, for example, Morse and Feshbach (15), p. 1327]

$$Q_n^{|m|}(z) = (-)^m(z^2 - 1)^{m/2} \frac{d^m}{dz^m} Q_n(z)$$

where

$$Q_n(z) = \frac{1}{2^n n!} \frac{d^n}{dz^n}\left[(z^2 - 1)^n \log_e\left(\frac{z+1}{z-1}\right) - \frac{1}{2} \log_e\left(\frac{z+1}{z-1}\right) P_n(z) \right]$$

The reason why it is necessary to introduce a second solution of Legendre's equation for μ and not for ν is the different ranges of these two variables. Since ν goes from $-\pi/2$ to $\pi/2$, the range of $\sin \nu$ is from -1 to $+1$. Within this interval, the well-behaved (i.e., convergent) solution of Legendre's associated equation is the function $P_n^{|m|}$. Outside of this range, however, $P_n^{|m|}$ diverges. Since μ goes from 0 to ∞, $\sinh \mu$ does also. A second solution of Legendre's equation therefore becomes necessary in order to restore good behavior as $\mu \to \infty$. Since $z = \pm 1$ are singular points of Legendre's equation, the solution which behaves properly for $|z| \geq 1$ will diverge inside the region $0 \leq |z| \leq 1$. Thus it is necessary to construct a linear combination of the two solutions in order to have good behavior everywhere.

1. Bedrock depression. Let us now consider the problem of a hemispheroidal depression $\mu = \mu_0$, which is filled with material having a conductivity σ_2 and which is in a uniform electric field $\mathbf{E}_0$. Since the plan is symmetrical, we may choose the x axis to be in the direction of $\mathbf{E}_0$; so that at large distances from the depression, provided that the ground is otherwise homogeneous,

$$\mathcal{V}_0 = -E_0 x = -E_0 f \cosh \mu \cos \nu \cos \varphi$$

In terms of the characteristic functions of Laplace's equation this may be written as

$$\mathcal{V}_0 = i E_0 f \cos \varphi P_1{}^1(i \sinh \mu) P_1{}^1(\sin \nu)$$

To this must be added a second potential function whose purpose is to satisfy the conditions at the boundary of the depression and which vanishes at large distances from it. This secondary potential may be formed by taking linear combinations of the characteristic functions of Laplace's equation which are well-behaved as $\mu \to \infty$, i.e.,

$$\mathcal{V}^{(S)} = \sum_{n=0}^{\infty} \sum_{m=-n}^{m=n} (a_n{}^m \cos m\varphi + b_n{}^m \sin m\varphi) P_n{}^{|m|}(\sin \nu) Q_n{}^{|m|}(i \sinh \mu)$$

$$\mu \geq \mu_0 \quad (14\text{-}19)$$

Inside the depression, we require a potential that is well-behaved as $\mu \to 0$, i.e.,

$$\mathcal{V}_2 = \sum_{n=0}^{\infty} \sum_{m=-n}^{m=n} (c_n{}^m \cos m\varphi + d_n{}^m \sin m\varphi) P_n{}^{|m|}(\sin \nu) P_n{}^{|m|}(i \sinh \mu)$$

$$\mu \leq \mu_0 \quad (14\text{-}20)$$

while outside, the total potential will be

$$\mathcal{V}_1 = \mathcal{V}_0 + \mathcal{V}^{(S)}$$

The boundary conditions to be satisfied by these potentials are as follows:

$$\mathcal{V}_1 = \mathcal{V}_2 \qquad \mu = \mu_0$$

$$\sigma_1 \frac{\partial \mathcal{V}_1}{\partial \mu} = \sigma_2 \frac{\partial \mathcal{V}_2}{\partial \mu} \qquad \mu = \mu_0$$

and

$$\frac{\partial \mathcal{V}_1}{\partial \nu} = 0 \qquad \nu = 0$$

When all these conditions are satisfied, we get upon substituting for $P_1{}^1$ and $Q_1{}^1$

$$\mathcal{V}_2 = \frac{2\sigma_1 E_0 f \cosh \mu \cos \nu}{(\sigma_2 - \sigma_1) \sinh^2 \mu_0 - 2\sigma_1 - (\sigma_2 - \sigma_1) \sinh \mu_0 \cosh^2 \mu_0 \tan^{-1}(\operatorname{csch} \mu_0)}$$

$$(14\text{-}21)$$

As a limiting case, we might consider the effect of a depression that is very shallow in relation to its diameter. The approximations that may be introduced into (14-21) when μ_0 is small allow us to write

$$\frac{E}{E_0} = \frac{1}{E_0} \frac{\partial \mathcal{V}_2}{\partial x} \doteq \frac{4\sigma_1 f}{4\sigma_1 f + \pi s}$$

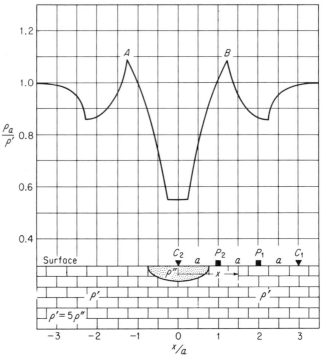

Fig. 14-22 Profile of the apparent resistivity measured by the Wenner system across the center of an oblate hemispheroidal depression. [After Cook and Van Nostrand (16).]

in which $s = \sigma_2 t$ is the "surface conductivity" of the deposit, where t is the depth of the depression and f is its approximate radius.   Both in this formula and in (14-21) we observe that the electric field is constant inside and outside the sink but that it changes abruptly at the boundary.

The effects of hemispheroidal depressions on the profiles of apparent resistivity measured by four-electrode methods have been worked out by Cook and Van Nostrand (16).   One of their profiles, reproduced in Fig. 14-22, shows a Wenner traverse taken diametrically across the center of an oblate depression filled with conducting material.   The most obvious features of this curve are the cusps that occur at one-half of the interelectrode distance beyond the edges of the sink.   The occurrence of cusps in apparent resistivity profiles is usually regarded as strong evidence of surface inhomogeneity.

2. The conducting disk.   By extending the theory of the previous section, we may study the effect produced by a conducting disk upon an electric field.   In mining geophysics, the disk may be used as a model to represent the effect of an ore body having finite dimensions.   If the plane of the disk is inclined at an angle d to the ground surface and if ϑ is the complement of the angle between the direction of $\mathbf{E}_0$ and the trace of the disk, then at large distances from the body the potential will be

$$\mathfrak{v}_0 = -E_0(z' \sin d + x' \cos d \cos \vartheta + y' \cos d \sin \vartheta)$$

where the x' and y' axes are taken in the plane of the disk. This may also be written

$$\mathfrak{v}_0 = iE_0 f[\sin d \, P_1^0(\eta)P_1^0(i\xi) + \cos d \cos (\vartheta - \varphi)P_1^1(\eta)P_1^1(i\xi)]$$

where we have used the abbreviations $\eta = \sin \nu$, $\xi = \sinh \mu$. To this we add the secondary potential (14-19), retaining only those terms in which $n = 1$, $m = 0$, 1. The boundary condition to be satisfied is that in the limit, as the thickness of the disk becomes small and its conductivity becomes large, the total potential must vanish on the disk. Therefore, for the anomalous potential outside the disk, the formula is

$$\mathfrak{v}^{(S)} \doteq - \frac{2sE_0 \sin d}{\sigma_2} \sin \nu \tan^{-1} (\operatorname{csch} \mu)$$

$$- \frac{4fsE_0 \cos d \cos (\vartheta - \varphi)}{\pi s - 4\sigma_1 f} \cos \nu \, [\tanh \mu - \cosh \mu \tan^{-1} (\operatorname{csch} \mu)] \quad (14\text{-}22)$$

where f is the radius of the disk and s is its surface conductivity.

3. Buried oblate spheroid. This model can be used to determine the lateral extent of conducting zones that have been intersected with the drill. When a single intersection is made, the question of how far the mineralized zone extends from the hole often arises. Sometimes an answer is sought with the aid of drill-hole resistivity surveys. In the four-electrode method the standard Wenner array or some variant of it may be used, but in any case both the source and the sink of current will be located close to the points at which potential measurements are to be made. In the three-electrode method the current sink is kept somewhere at the ground surface, so that for practical purposes it may be considered to be at an infinite distance from the potential probes. The potential difference between two

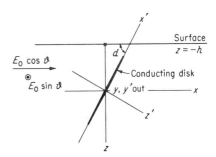

Fig. 14-23 A conducting disk in a uniform electric field.

points in the vicinity of a single current probe will then be measured. A third possible arrangement is to keep one current electrode and one potential probe at ground level. The other two probes are lowered into the hole, and readings of the potential near the current electrode are taken.

In all these methods a new element is introduced into the problem by the proximity of the current source to the body. We must add to the complementary solution of Laplace's equation a particular integral of the form

$$\mathcal{V}_0 = \frac{I\rho}{4\pi R}$$

in order to give the correct form to the behavior of the field close to the source. In terms of the characteristic functions of Laplace's equation in the oblate spheroidal system

$$\frac{I\rho}{4\pi R} = \frac{I\rho}{4\pi} \frac{2}{f} \sum_{n=0}^{\infty} (2n+1) \sum_{m=-n}^{m=n} i^m \left[\frac{(n-|m|)!}{(n+|m|)!} e^{im(\varphi-\varphi_0)} P_n^{|m|}(\eta_0) P_n^{|m|}(\eta) \right]$$

$$\times \begin{cases} P_n^{|m|}(i\xi_0) Q_n^{|m|}(i\xi) & \xi > \xi_0 \\ Q_n^{|m|}(i\xi_0) P_n^{|m|}(i\xi) & \xi \le \xi_0 \end{cases} \quad (14\text{-}23)$$

when R is the distance of the point $P(\xi,\eta,\varphi)$ from the source at $C(\xi_0,\eta_0,\varphi_0)$ [Morse and Feshbach (15), Sec. 10-3]. What must be done, therefore, is to add to (14-23) whichever of the two functions (14-19) or (14-20) is appropriate to the medium containing the source C. The boundary conditions at the spheroidal surface are then introduced in order to determine the con-

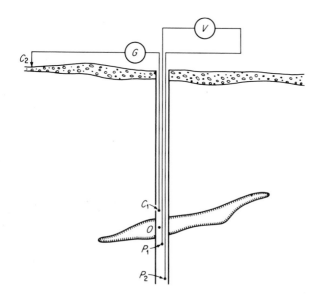

Fig. 14-24 The three-electrode method for making drill-hole resistivity surveys.

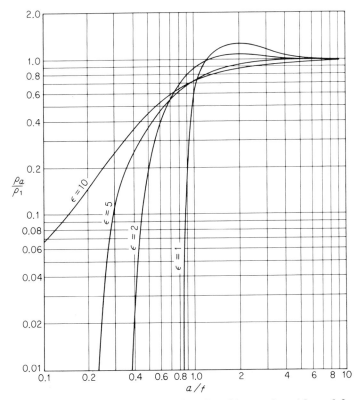

Fig. 14-25 Apparent resistivity curves for the oblate spheroid model using the three-electrode drill-hole method. [After Seigel (17).]

stants in the complementary solution.   The process is straightforward, but the expressions are tedious.   We shall omit the details.

Since the practical interest in this problem centers on determining the lateral extent of the body, it should be sufficient to consider the apparent resistivity curves appropriate to a few special arrangements of the electrodes. In the three-electrode methods, which seem to be best suited to this type of investigation, the basic configuration is illustrated in Fig. 14-24.   The three probes are uniformly spaced, the current electrode and the nearest potential probe straddling the intersection symmetrically.   The interelectrode spacing a may be varied, keeping the point O fixed.   If this geometry is used, the quantities to be determined are the radius and eccentricity of the spheroid, provided it can be assumed that the conductor is flat-lying.   Figure 14-25 shows a set of calculations made by Seigel (17) in which the apparent resistivity is given as a function of a/f for various values of the eccentricity ϵ. One of the interesting features of this set of curves is the great difference between the behavior of bodies that are fairly oblate ($\epsilon = 5$) and that of the disk ($\epsilon = \infty$).

14-15   *The Numerical Calculation of Type Curves*

In order to extend the range of numerical aids available for making quantitative interpretations of d-c conduction anomalies, we wish to make some suggestions about the calculation of profiles over conducting bodies having an arbitrary cross section.   This requires a numerical method for solving boundary-value problems of the Dirichlet type, and in practice we can only attempt such calculations in two dimensions.   Thus we are limited in our choice of problems to two-dimensional bodies in transverse electric fields. For solving problems of this type, one of the most powerful techniques is the relaxation method [Southwell (18)].   Within the context of our interests, the aim of the relaxation method is to calculate the solution of Laplace's equation numerically at a finite number of lattice points instead of determining it mathematically everywhere.   It does this by replacing the partial differential equation with its finite-difference approximation, which in effect reduces the problem to the solution of a very large number of simultaneous linear algebraic equations, each of which involves only a few unknowns.   A four-point formula is generally used.   This means that for the distribution of lattice points shown in Fig. 14-26 (referred to as a "star") the Laplacian of $\mathcal{U}$ at the point 0 is given approximately by

$$(\nabla^2 \mathcal{U})_0 = \left(\frac{\partial^2 \mathcal{U}}{\partial x^2}\right)_0 + \left(\frac{\partial^2 \mathcal{U}}{\partial y^2}\right)_0 \doteq \frac{1}{a^2}(\mathcal{U}_1 + \mathcal{U}_2 + \mathcal{U}_3 + \mathcal{U}_4 - 4\mathcal{U}_0) \quad (14\text{-}24)$$

with an error about equivalent to $a^4(\nabla^4\mathcal{U})_0/12$.   This error may be made as small as we please by choosing a to be small, but the number of algebraic equations will go up accordingly.   Thus we see that a change in $\mathcal{U}$ at 0 affects $\nabla^2\mathcal{U}$ not only there, but also, although to a lesser extent, at the four neighboring points as well.   On account of this interaction, the point values of the function are linked one to another like the threads of a net.

An essential condition for the solution of any problem is that the normal gradient at the surface of the ground must vanish.   If the ground is flat, it is always possible to replace it with a row of lattice points.   If these occupy the position 1 in their respective "stars" (Fig. 14-26), then the boundary condition at the ground surface will be met by making

$$3\mathcal{U}_1 - 4\mathcal{U}_0 + \mathcal{U}_3 = 0 \quad (14\text{-}25)$$

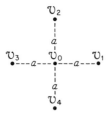

Fig. 14-26   A relaxation "star" for $\nabla^2\mathcal{U} = 0$.

for all "stars" which touch the surface, since this is proportional to the finite difference approximation to $\partial \mho/\partial n$ at the point 1.   If the ground undulates, recourse must be taken to the use of "irregular stars," procedures for which are described in Southwell's book.

The boundary conditions at the interface between two media are dealt with in a like manner.   If the boundary happens to contain one of the lattice points (point 1 of the "star" in medium 1 and point 3 of the adjacent star in medium II, for example), these conditions will both be satisfied by making

$$[\sigma(3\mho_1 - 4\mho_0 + \mho_3)]_I = -[\sigma(3\mho_3' - 4\mho_0' + \mho_1')]_{II} \qquad (14\text{-}26)$$

[If the body happens to be a very good conductor, we set $\mho_1 = 0$ in (14-24) for "stars" which touch the boundary at point 1.]   If on the other hand the boundary cuts through one of the mesh sides (as may often be the case), then the equivalent approximation based on "irregular stars" (or "fictitious nodes") must be used.   If some freedom exists in drawing the boundary, it is helpful to draw it in such a way that it cuts as few mesh sides as possible.

The usual way in which to approach the problem of calculating $\mho^{(S)}$ is to take as an initial solution the function $\mho = -Ex$.   Keeping the values of this function fixed at a suitably large distance from the disturbing body, we force the lattice values to obey (14-25) and (14-26) and at the same time make use of relaxation to ensure that the residuals (14-24) are everywhere small.   When the residuals are reduced to a satisfactory level, the problem is numerically solved.   There are several special techniques which can very substantially reduce the time and labor needed to achieve such solutions, with the aid of which the method becomes more practicable.   As an example of a problem which would be difficult to solve analytically, the relaxation solution for a prism of triangular cross section in a uniform field is described and illustrated in chap. 3 of Southwell's book.

14-16   The Effects of Topography

Probably no other group of geophysical methods is so adversely affected by topography and near-surface conditions as the d-c resistivity methods are. The problems stem partly from physical and partly from geometrical causes. On the one hand, surface deposits often show extreme variability in their electrical conductivities because of heterogeneity, variable moisture conditions, and the effects of weathering and erosion.   We have already considered the effects of surface deposits on electric fields in Sec. 14-14.   On the other hand, the forms of the land surface may have a strong influence on shaping the equipotential surfaces even under uniform conditions, introducing local gradients which may simulate or confuse real anomalies.   It is the second of these problems that we shall now consider.

We must first of all bear in mind that the surface of the ground, whatever its shape, is a "stream line" for the current flow.   Irregularities in the

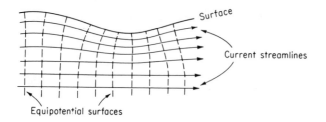

Surface

Current streamlines

Equipotential surfaces

Fig. 14-27 Illustrating the effect of topography on an otherwise uniform electric field.

shape of the land will therefore introduce a local squeezing together or drawing apart of the equipotential surfaces, which instruments measuring potential differences will register as spurious "anomalies" (Fig. 14-27). In areas of moderate-to-strong relief, this topographic effect may very easily distort the shape of an anomaly. Unfortunately, it is easier to diagnose the difficulty than to prescribe for it, although the relaxation principle does suggest a possible remedy. If we can find a solution of Laplace's equation which satisfies the requirements $\partial \mathcal{v}/\partial n = 0$ at the ground surface and which equals the function $\mathcal{v} = -Ex$ at a suitable distance below the ground, the values of this function will give the topographic effect. The calculations will be based upon the use of "irregular stars" (six-pointed ones, since the problem is generally in three dimensions), and the procedure is to begin with lattice values of $\mathcal{v} = -Ex$ and then to perturb them in three dimensions until the condition at the boundary is fulfilled and the residuals are sufficiently small.

References

1. G. N. Watson, "A Treatise on the Theory of Bessel Functions," Cambridge, London, 1944.

2. S. S. Stefanescu and C. and M. Schlumberger, Sur la distribution électrique potentielle autour d'une prise de terre ponctuelle dans un terrain à couches horizontales, homogènes et isotropes, *Jour. Physique et le Radium*, vol. 1, pp. 132–140, 1930.

3. H. Mooney and G. Wetzel, "The Potentials about a Point Probe in a Two-, Three-, and Four-layered Earth," University of Minnesota, Minneapolis, 1957.

4. H. C. Spicer, Investigation of Bedrock Depths by Electrical-resistivity Methods in the Ripon–Fond du Lac Area, Wisconsin, *U.S. Geol. Survey*, Circular 69, 1950 (37 pp).

5. R. E. Langer, An Inverse Problem in Differential Equations, *Am. Math. Soc. Bull.*, ser. 2, vol. 29, pp. 814–820, 1933 (see also vol. 42, pp. 747–754, 1936).

6. L. B. Slichter, Interpretation of Resistivity Prospecting for Horizontal Structure, *Physics*, vol. 4, pp. 307–322 (and p. 407), 1933.

7. A. Belluigi, Asymptotische Formeln Zur Bestimmung des von einer oder von mehreren Elektroden im geschichteten Boden, hervorgerufenen Potentials, *Zeits. Geophysik*, vol. 23, pp. 135–168 and 182–196, 1957.

8. K. Vosoff, Numerical Resistivity Analysis: Horizontal Layers, *Geophysics*, vol. 23, pp. 536–556, 1958.

9. D. N. Chetayev, A Converse Problem in the Theory of Electrical Geophysical Exploration, *Izv. Akad. Nauk S.S.S.R., Ser. Geogr. i Geofiz.*, pp. 141–144, 1957.

10. K. Maeda, Apparent Resistivity over Dipping Beds, *Geophysics*, vol. 20, pp. 123–139, 1955.

11. T. N. Zavadskaya, On the Transformation of Curves Obtained in Electrical Profiling, *Prikladnaya Geofiz.*, vol. 19, pp. 47–56, 1958 (in Russian). See also R. Cassinis, A Practical Criterion for Converting between Apparent Resistivity Diagrams Obtained with Different Arrangements, *Annali Geofisica*, vol. 11, pp. 233–236, 1958.

12. N. V. Lipskaya, The Field of a Point Electrode Observed on the Earth's Surface near a Buried Conducting Sphere, *Izv. Akad. Nauk S.S.S.R., Ser. Geogr. i Geofiz.*, vol. 8, pp. 409–427, 1949.

13. R. G. Van Nostrand, Limitations on Resistivity Methods as Inferred from the Buried Sphere Problem, *Geophysics*, vol. 18, pp. 423–433, 1953.

14. A. Sommerfeld, Über verzweigte Potentiale im Raum, *London Math. Soc. Proc.*, vol. 28, pp. 395–429, 1897. See also H. S. Carslaw, On Multiform Solutions of Partial Differential Equations of Physical Mathematics, and Their Practical Applications, *London Math. Soc. Proc.*, vol. 30, pp. 121–136, 1899.

15. P. Morse and H. Feshbach, "Methods of Theoretical Physics," McGraw-Hill, New York, 1953.

16. K. L. Cook and R. G. Van Nostrand, Interpretation of Resistivity Data over Filled Sinks, *Geophysics*, vol. 19, pp. 761–790, 1954.

17. H. Seigel, Ore Body Size Estimation in Electrical Prospecting, *Geophysics*, vol. 17, pp. 907–914, 1952.

18. R. V. Southwell, "Relaxation Methods in Theoretical Physics," Oxford, London, 1949.

chapter 15   Electromagnetic
Induction Methods

In this chapter we shall describe briefly a variety of electromagnetic prospecting methods. In doing so, we depart from our usual practice of leaving commentaries on practical techniques to other texts, for the simple reason that the literature in this area is scanty and consists mainly of descriptions of obsolete practices. We shall deal in the main with prospecting systems which employ man-made primary fields, since these are the least well served by the existing literature in spite of the fact that they are well known among mining geophysicists. Shorter descriptions of the natural-field methods will also be included, together with references to fuller treatments.

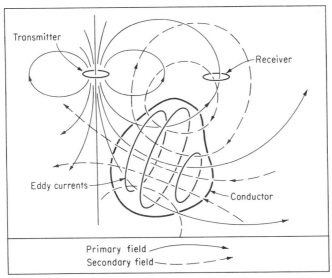

Fig. 15-1   A generalized picture of electromagnetic induction prospecting.

15-1   *Methods of Prospecting*[1]

Electromagnetic-induction prospecting methods (1)(9) which make use of man-made primary fields work in roughly the following way: An alternating magnetic field is established by passing alternating current through a coil or along a very long wire.   This field is measured with a receiver consisting of a coil connected to a sensitive electronic amplifier, meter, or potentiometer bridge.   The frequency of the alternating current is chosen such that an insignificant eddy-current field is induced in the ground if it has an average electrical conductivity.   Ordinarily this sets an upper limit on the operating frequency of about 5,000 cps, although cases are known in which frequencies as high as 50 kc are used.   But if the source and receiver are brought near a more conductive zone, stronger eddy currents may be caused to circulate within it and an appreciable secondary magnetic field will thereby be created. Close to the conductor this secondary or anomalous field may be comparable in magnitude with the primary or normal field (which prevails in the absence of conducting zones), in which case it can very easily be detected by the receiver.   Prospecting for these anomalous zones is carried out by systematically traversing the ground either with the receiver unit alone or with the source and receiver in combination, depending upon the system in use.

In practice, the distance between source and receiver is usually a few

---

[1] Some *methods* of electromagnetic prospecting appear to be patented.   The various techniques here described are not necessarily all to be found "in the public domain." Prospective users are warned to investigate patent rights, not only to the equipment they plan to build or use, but also to the method of using it.

hundred, and seldom more than a few thousands, of feet.   At frequencies smaller than 5,000 cps this is only a very small fraction of a free-space wavelength.   This is a most important point to consider in understanding the principle of the electromagnetic induction methods.   It means that in the regions within which observations are taken radiation is very slight and phase retardations are negligible, so that the effects of propagation can be wholly disregarded.   Electromagnetic induction methods are therefore much more closely related to potential field methods, such as the gravity and magnetic methods, than to the seismic method which depends upon the propagation of waves.   The apparently dissimilar practices used in the field for recording data will be more easily understood if we keep fixed in our mind the notion that the electromagnetic receiving unit is little more than an a-c magnetometer.

15-2   Dip-angle Techniques

In the so-called *dip-angle* or *tilt-angle* techniques, the receiver measures the direction of the total (i.e., primary plus secondary) magnetic field and very little else.   The receiving unit consists of a coil connected through an electronic amplifier of high gain to a set of headphones or (more rarely) to a meter.   The direction of the magnetic field is sensed by rotating the coil about a diameter until a minimum signal is obtained.   When it has been nulled, the total field vector must lie somewhere in the plane of the coil. Two successive observations are usually made in order to determine the true orientation of this vector.   The *strike* is found first by rotating the coil about a vertical axis, and afterward the *dip* is measured by rotating it about a horizontal axis perpendicular to the strike direction.

   The above description is actually a bit oversimplified.   In the absence of any underground conductor, the signal in the receiver will vanish altogether when the coil axis lies perpendicular to the direction of the primary magnetic field.   But if a secondary magnetic field exists, it is generally dis-

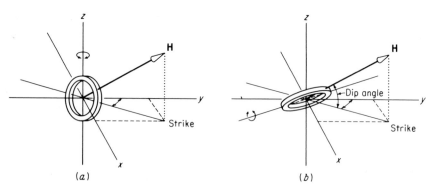

Fig. 15-2   **Finding the strike and dip of an alternating magnetic field H.**

placed from the primary field in phase as well as in direction. Thus the total field may be elliptically polarized and cannot be represented by a simple vector. The receiver cannot then be nulled completely, but it will sense a minimum signal when it contains within its plane the major axis of the polarization ellipse.

Dip-angle techniques which employ man-made primary fields use a loop for a transmitter. The primary field is therefore essentially like that of an alternating magnetic dipole. In some cases the alternating current supplied to the transmitter is obtained from a battery-powered electronic oscillator, and in others, from a portable gasoline-driven a-c generator. The recent trend has been toward the more portable electronic type of apparatus, which allows separations of several hundreds of feet between receiver and transmitter. For distances exceeding about 1,000 ft, however, the more powerful gasoline-driven apparatus seems to be required.

The apparatus can be used in a variety of different ways. To a large extent, the choice depends upon the properties of the conducting zones which are deemed to be favorable, and upon the nature of the terrain in which the work is to be done. The following methods have at various times found popularity in Canada:

1. Parallel-line method. This method requires a portable transmitter unit. It is a vertical loop which is so oriented that the receiver position lies in its plane. The receiver is used to find first the strike and then the dip of the total magnetic field, although the strike measurement serves as a rule only to locate the plane of the dip and is seldom recorded. In the absence of an anomaly the field at the receiver will be parallel to the axis of the transmitting coil, and the dip angle will therefore be zero. Traversing is carried out by moving the transmitter and receiver simultaneously along a pair of parallel picket lines and taking readings at regular intervals. If the strike of the conductors can be anticipated, the picket lines are laid out at right angles to this direction. The spacing between lines will depend somewhat upon the circumstances, but a distance of 400 ft is frequently used. The transmitter is aimed toward the receiver by aligning its axis with the picket line and relying upon the survey control to position the receiver correctly. It can also, when necessary, be aimed roughly by shouting back and forth between the operators. In fairly flat country either method is usually sufficiently accurate.

2. Fixed-transmitter method. The relative position of the transmitter and receiver is exactly the same as that in the parallel-line method. The plan of operations is rather different, however. The transmitter is kept at one position and the receiver is moved along a picket line nearby, making dip-angle measurements at regular intervals. The plane of the transmitter must be rotated with each observation so that it always contains the receiver position. This may be accomplished by calculating the required angles relative to the picket line and then turning the transmitter coil according to

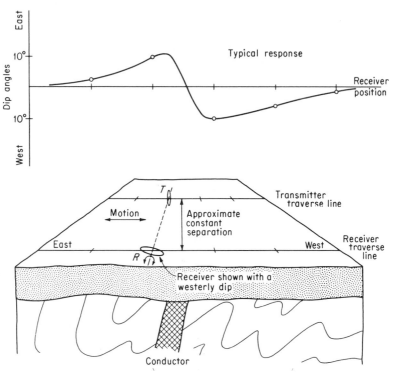

Fig. 15-3   Illustrating the vertical-loop, parallel-line, dip-angle method.

a prearranged time schedule, or the operators may coordinate their movements by shouting if the distance permits.

Either an electronic or a gasoline-powered transmitter may be used, depending upon the separations required between the transmitter and the receiver. It is common practice for several traverses to be made from one transmitter site, along picket lines laid out at regular intervals extending out to the useful limit of sensitivity of the apparatus. Thus the traverses may extend to distances of about 1,200 to 1,600 ft from the transmitter if a gasoline-powered apparatus is used, and to distances of about 600 to 800 ft with battery-powered equipment.

3. "Shoot-back" method. A major difficulty with the two previous systems occurs in rugged terrain. If the receiver lies above or below the axis of the transmitter, any misalignment of the two coils will lead to a fictitious dip angle. However, a prospecting system has been designed specifically to overcome this difficulty and it appears to be excellent even for ordinary work.[1]

This system requires a receiver and a transmitter in each unit. The two units are set at a fixed distance apart (usually 200 ft or so) along a single

---

[1] D. Crone, Crone Geophysics Ltd., Toronto, Canada.

traverse line, and measurements are made at regular intervals along the line. The axis of one coil is pointed toward the other coil but dips at an angle of 15° below horizontal. While this coil transmits, a dip angle is measured at the other by rotating it about a horizontal axis perpendicular to the traverse line. Then the roles of transmitter and receiver are interchanged. The coil that was formerly passive now transmits with its axis pointed toward the other but inclined at an angle of 15° *above* horizontal, while the coil that was formerly the transmitter now becomes the receiver and measures the dip angle. If no conductors are present, this measurement will be nearly identical with the first measurement, regardless of substantial elevation differences. However, an anomalous field will tend to have an opposite effect upon the two dip angles. The two measurements are therefore subtracted and the difference used as an index of the anomaly. Interpretation of the results will be somewhat more complicated than that of the regular dip-angle methods, because each measurement is the result of two independent experiments. With sufficient advance preparation of type curves, however, this obstacle can presumably be overcome.

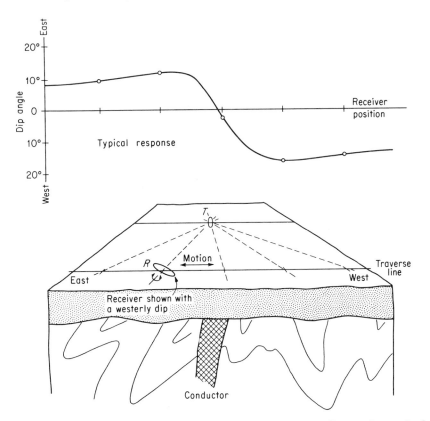

Fig. 15-4 Illustrating the vertical-loop, fixed-transmitter, dip-angle method.

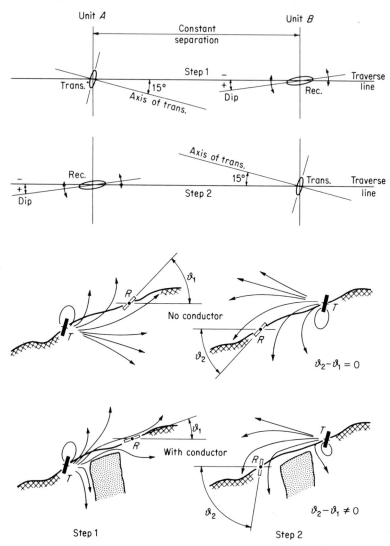

Fig. 15-5 The "shoot-back" dip-angle method. First unit *A* transmits while *B* measures the dip angle. Then unit *B* transmits while *A* measures. A difference in the dip measurements indicates an anomaly.

4. Natural field method. An important dip-angle method which uses natural electromagnetic fields has been described by Ward (2) and is known as AFMAG. The primary field, within the frequency band generally employed, is essentially a series of noise pulses generated by distant lightning flashes, called *sferics*. These disturbances propagate over the entire world, because the atmosphere, which is nonconducting and therefore nonabsorbing and which is situated between a relatively conducting earth and ionosphere, becomes a natural waveguide.   The efficiency of the waveguide is such that flashes occurring anywhere on the globe can make an effective contribution to the total field at any point.   Consequently, the pulse rate is reasonably high, and when detected through a very narrow-band filter, it becomes more or less a continuous signal.   Diurnal and seasonal variations occur in the mean signal level and are related to changes in world-wide thunderstorm activity and ionospheric conditions.   The direction of the magnetic field in most areas is found to be practically horizontal, as a result of the preferred mode of transmission in the atmospheric waveguide.

The AFMAG receiver, although similar in principle to a conventional dip-angle detector, is quite different in detail.   It would be very difficult indeed to find a null by using a single coil when the primary field strength is changing continuously.   Therefore two coils are mounted at right angles to each other, and the difference between their filtered outputs is measured. This gives a much sharper null position than a single coil is able to do. Strike and dip measurements are made at regular intervals in much the same way as with the other dip-angle methods, and near a conductor the field data show a considerable resemblance to those obtained by the fixed-trans-mitter method.   In fact the AFMAG method is sometimes considered to be a fixed-transmitter method, with the transmitter removed to infinity. Because the primary field intensity is substantially uniform over the entire survey area (rather than concentrated within a small region as it is when a dipole transmitter is used), the frequencies commonly used in AFMAG are lower than those in the other dip-angle methods.   The range used in com-mercially manufactured AFMAG sets seems to be between 50 and 500 cps, as against 500 to 5,000 cps with the conventional apparatus.

Each of the dip-angle methods has certain advantages.   The choice of method will depend largely on the particular difficulties which the area to be surveyed presents.   All of them, however, have one problem in common: it is exceedingly difficult on the basis of a single series of measurements to distinguish between a large conductivity and a small depth of burial.   In most cases this ambiguity can be removed by making measurements at two frequencies which are well separated from each other.   Accordingly, many field sets are of the dual-frequency type.

15-3   *Intensity Measurement*

In principle, the distortion of a known magnetic field by a conductor can be determined simply by measuring the field intensity at several locations.   In practice, measurements of this kind often prove to be inadequate because the intensity of the primary field changes so rapidly with position.   Thus even a small error in the relative positioning of the receiver and transmitter may lead to a change in the measured field intensity which is quite indistinguish-able from a good anomaly.   This problem can be overcome to a large extent

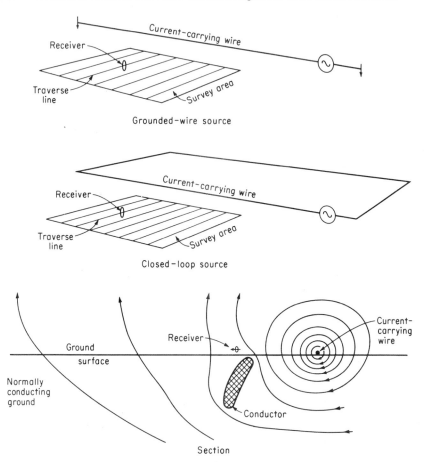

Grounded–wire source

Closed–loop source

Section

Fig. 15-6 A "long-wire" electromagnetic prospecting system, in which the receiver measures the intensity of the horizontal field component.

by providing some sort of connection between the transmitting and receiving units which keeps them a fixed distance apart, and this is very often done. When a mechanical link is used for this purpose, however, it is customary to pass a reference signal along it from the transmitter to the receiver so that a phase comparison between the observed and transmitted fields can be made. This is a system about which we shall have more to say later on.

One method which does make use of straightforward amplitude measurements is the so-called "long-wire" system. A single wire cable, which may be several miles in length, is laid out in a straight line near the region to be surveyed. Current is supplied to the cable by an a-c generator, and the return circuit is made by way of the ground through electrodes at either end. Sometimes the cable is closed on itself by making a large loop remote from the survey area. The technique is to measure the intensity of the horizontal component of the total magnetic field in a direction transverse to the long wire and to look for disturbances in the normal pattern of this component. Such measurements may be carried out even at a considerable distance from the source. The practical limit depends on many factors, but it is usually of the order of one-half or one-fourth of the length of the wire. Because of the two-dimensional nature of the primary field and also because the distance from the wire to the detector is large, small errors in position are not likely to have serious consequences. The physical basis of interpretation is more complicated than it is for a dipole source because of the presence of conduction currents in the ground. However, the method is rather simple to operate and is very sensitive. One difficulty with it is that it is very apt to accentuate large linear features such as faults and shears, because the cable influences such a large volume of the ground. Another is that the sensitivity to different shapes and kinds of conductor changes with the distance of the receiver from the long wire, and so it is difficult to set up quantitative interpretation procedures. Despite these drawbacks, which all long-wire methods have in common, this is a system which can be very useful in reconnaissance because of its superior sensitivity to deeply buried conductors.

15-4 *Phase Component Measuring Systems*

In the theory of electromagnetic-induction prospecting we shall learn that the anomalous magnetic field is not, as a general rule, in phase with the primary field. The phase difference between the two fields, as a matter of fact, is of the first importance in geophysical interpretation. Among other things, it may disclose information about the average conductivity of the anomalous zone. In consequence, many types of field apparatus are designed so that phase comparison can be made. It is usually also advantageous to eliminate the primary field from the observations, since the secondary field, not the total field, is the object of interest. Measurements of this type can be made only if there is a mechanical link between the receiver and the transmitter, which is used for the dual purpose of maintaining an accurate separation between the coils and of obtaining a reference signal from the transmitter for the phase measurement. Usually the receiving unit is designed to balance out the undisturbed primary field and to measure separately the intensities of the inphase and quadrature components of the secondary field as fractions of the undisturbed primary field strength at the receiver position.

    1. Horizontal-loop method. One of the oldest, and perhaps the most universally popular, of all electromagnetic-induction methods uses receiver and transmitter coils which are horizontal and kept at a fixed distance apart. The receiver measures both the inphase and quadrature components of the secondary or anomalous field, usually as a percentage of the primary field intensity. The reference signal passes to the receiver along a cable attached to the transmitter. This cable also controls the separation between the two coils. The coils are moved in line along picket lines, and readings are taken at regular intervals. A reading accuracy of about 1 percent can be expected if the terrain is not rough.

One advantage of the horizontal-loop system is that it is symmetrical. Since the coils are maintained in a fixed relative position, the measurement is equivalent to determining their mutual inductance. The direction of traverse therefore makes no difference, since we are assured by the rule of reciprocity that measurements cannot be affected by interchanging the source with the receiver. For this and other reasons, data obtained by the horizontal-loop method are usually the most easily interpretable.

2. Long-wire, ratio detection method. Phase measurements are also used in one of the "long-wire" methods, in which the transmitter is a long cable that is either grounded at both ends or else (and in this case more frequently) closed by an extremely large loop. The receiving unit consists of two horizontal coils kept at a fixed distance apart by a taut cable. Both the phase difference and the amplitude ratio of the signals induced in the two coils are measured. If no conductors are in the vicinity, the phase difference will be zero and the amplitude ratio will be a smooth function of the distance from the transmitter. The distance between the coils is chosen so that an

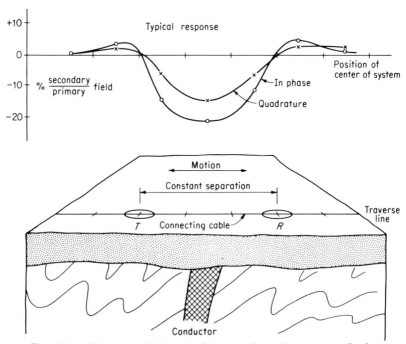

Fig. 15-7   Horizontal-loop, inphase, and quadrature method.

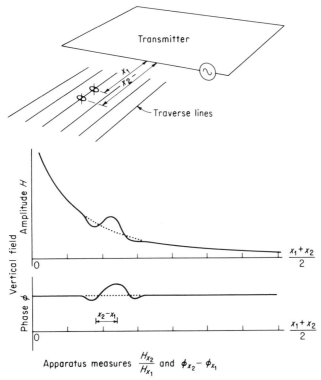

Fig. 15-8 A long-wire electromagnetic system using a differential type of detector.

anomalous field will produce a signal in one coil that differs considerably from that in the other.   The method has about the same sensitivity as the intensity-measuring system, plus the added advantage of being able to measure phase differences which aid greatly in the interpretations.

15-5   *Airborne Electromagnetic Induction Methods*

A great variety of instruments has been developed for airborne electromagnetic prospecting, and it is not possible to describe them all (6).   On the whole, it can be said that in the design of these systems a compromise is struck between measuring what is most useful from the interpretational standpoint and what is most practicable in terms of sensitivity.   The more recently built systems generally employ a dipole type of transmitter and measure the inphase and quadrature components of the secondary magnetic field.   This type of measurement is generally conceded to be the most desirable, in principle.   However, a number of other systems have been found also to be very effective and we shall describe some of them.

1. **Phase component measuring systems.** As with the ground equipment of this same type, a comparatively rigid connection between the receiver and the transmitter is necessary in order to balance out the primary field at the receiver accurately.   The intensities of the inphase and quadrature components of the secondary field are recorded continuously during flight and are given as fractions of the primary field strength.   A sensitivity of a few parts per million is obtainable with most instruments operating under good flying conditions.   Ultimately, sensitivity depends upon how rigidly the spacing and the relative orientation of the two coils can be maintained during flight.   Flexures due to air turbulence will often produce changes in the amount of primary field flux which cuts the receiver coil, thus causing spurious "anomalies" to appear in the record of the inphase component.   The quadrature component will obviously be unaffected.   On the other hand, the metal of the aircraft may give rise to some secondary magnetic field, and in that case any relative motion between the electromagnetic system and the aircraft will produce spurious noise in both phase components.

The systems which are in most common use at present may be considered as essentially two superposed dipoles (3).   By this we mean that the distance between the receiver and the transmitter coils is so small in comparison with the flight height that for practical purposes both coils may be considered to lie at the same point.   Three different coil arrangements are illustrated in Fig. 15-9.   These three well-proved systems all have the axes of both coils directed along the line of flight.   The data obtained by these apparently different arrangements are exactly comparable if the secondary field strength $H^{(S)}$ is measured in terms of the dipole moment m of the transmitter.   Note the dimensions of the ratio of these two quantities

$$\frac{H^{(S)}}{m} = \frac{\text{amp-turns/m}}{\text{amp-turns m}^2} = \text{m}^{-3} \text{ in the mks system}$$

This ratio can be determined from measurements of $H^{(S)}/H^{(P)}$ in parts per million by applying the formula for the field intensity of a small, current-carrying loop.   According to the formula the primary field strength $H^{(P)}$ at the receiver is

$$H^{(P)} = \frac{m}{4\pi} \frac{k}{l^3}$$

where $k = -1$ if the coils are coplanar
      $= 2$ if they are coaxial
and $l =$ the distance between coils

A measurement in parts per million can therefore be converted to a measurement of $H^{(S)}/m$ by applying the appropriate factor

$$\frac{H^{(S)}}{m} = \frac{10^{-6}k}{4\pi l^3} \frac{H^{(S)}}{H^{(P)}}$$

when $H^{(S)}/H^{(P)}$ is in ppm.

The height at which these instruments should be flown depends very largely upon the sensitivity of the apparatus and, of course, upon the terrain. Heights of 60 to 100 ft have been flown with helicopters, while about 300 ft is more common for a fixed-wing aircraft. Height is a critical factor in the interpretation of data and must be measured continuously during flight with a radio altimeter. The operating frequency of the instrument, which is fixed beforehand, depends to some extent upon the average height at

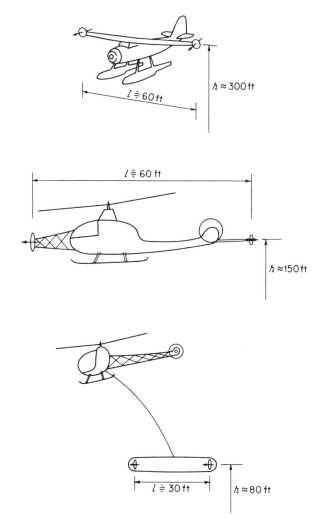

Fig. 15-9   Airborne electromagnetic systems of the "double-dipole" type.

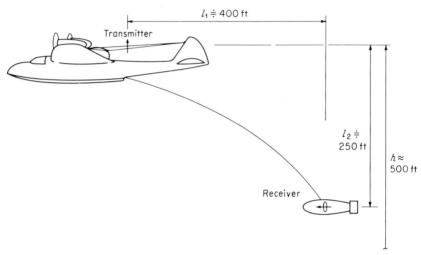

Fig. 15-10   Disposition of receiver and transmitter in a quadrature measuring airborne electromagnetic system.

which the system is to be flown.   The higher-flying systems generally use frequencies in the range 300 to 600 cps and the lower-flying systems in the range 1,000 to 4,000 cps.   The horizontal position of the aircraft is customarily determined from continuous-strip photography, and flight lines are generally laid out at one-quarter- to one-eighth-mile intervals.

2. Quadrature system.   One of the earliest successful airborne electromagnetic systems employs what is essentially a measurement of the quadrature component (4).   Because the inphase component is not detected, the stability requirements in the relative positioning of the receiver and transmitter are very much relaxed.   The coil system used is shown in Fig. 15-10.   To remove the ambiguity in the interpretation created by the lack of an inphase measurement, the measurements are made continuously at two different frequencies, one at about 400 cps and the other at about 2,000 cps.

3. Long-wire method.   The long-wire, intensity-measuring, ground electromagnetic system can also be adapted for use with airborne receiving equipment.   The airborne survey is conducted essentially like a ground survey except that the detector coil is mounted in a light aircraft and the intensity measurement is recorded continuously along the flight lines.   The method has a particular advantage in that the same transmitter can be used in following up the airborne measurements on the ground.   Correlation between the ground and air surveys should therefore be very close.

4. Two-plane method.   A method which seems to be very popular among Scandinavian and European mining geophysicists is the two-plane, rotary field method (5).   Figure 15-11 illustrates the system.   Two aircraft are used, one flying behind the other, keeping as nearly as possible at a constant separation and at the same height above ground. The transmitter and receiver both consist of a horizontal and a vertical coil whose axes are perpendicular to the flight direction. The two coils of the transmitter are supplied with equal amounts of alternating current displaced in phase by 90°, making the source equivalent to a rotating dipole. At the receiver, the measurement is made by shifting the phase of the output of one of the two coils by 90° (ahead or behind, as is appropriate) and then taking the difference between the two signals. The two phase-components of the differential signal are then measured using the signal in the vertical coil as a reference. In the absence of a secondary field the difference will be very small in spite of the comparatively large changes in relative positioning which will inevitably occur between the two aircraft. If steeply dipping conductors are traversed in a direction nearly perpendicular to their strike, the signal detected by the horizontal receiving coil will be much larger than that picked up by the vertical coil. The system then responds in much the same way as the horizontal-loop ground apparatus. In other circumstances, however, the response is much more complicated.

5. Transient system. A novel electromagnetic induction system which is quite different from all the others in its operating principle is the transient system (6). The transmitter coil is energized with what is essentially a step current. In the absence of conductors a sharp transient pulse proportional to the time derivative of the magnetic field is induced in the receiver. When a conductor is present, however, a sudden change in magnetic field intensity will induce in it a flow of current which will tend to slow the decay of the field. Figure 15-12 illustrates this situation. The switching is repeated several times a second as the aircraft follows its flight line, so that the signal is virtually continuous.

A major advantage of the transient system is that the receiver "listens" only while the transmitter is "quiet," so that the problems arising out of relative motion between transmitter and receiver are virtually eliminated. Moreover, if the entire decay of the secondary field can be observed, the response is equivalent to a-c measurements made over the whole of the frequency spectrum. It is important to note in this connection, however, that not the decay function itself but only its time derivative can be recorded, if a coil is used as the detector. This means that the anomalous fields which

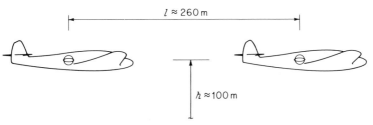

Fig. 15-11   Two-plane airborne electromagnetic system.

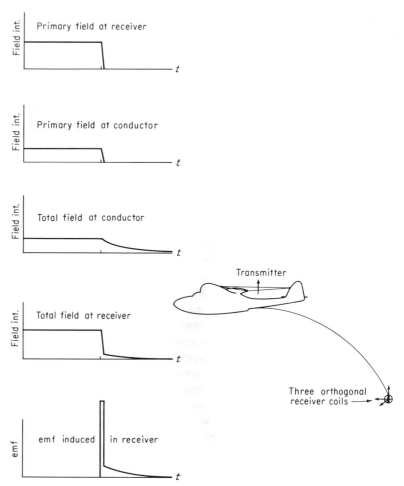

Fig. 15-12   Transient airborne electromagnetic system.

decay very slowly are suppressed in amplitude more than the others, and since these are the very ones generally associated with good conductors, there would seem to be an inherent weakness in the system similar to that in an alternating field apparatus in which only the quadrature component is measured.   On the other hand, it may be that future developments in high-frequency resonance magnetometers will make it possible to measure the field itself, rather than its time derivative, and so provide a remedy.

15-6   Methods of "Depth Sounding"

"Depth sounding" can be performed by the electromagnetic induction method in a manner almost exactly analogous to the way it is done by d-c conduction.   A receiving and a transmitting coil are set up in a given relative orientation, and a series of measurements is made as the separation between them is changed.   It is also possible (at least in theory) to achieve virtually similar results by keeping the coils a fixed distance apart and by varying the frequency of the current.   For horizontally stratified media, a theorem analogous to Langer's for the resistivity method has been proved by Slichter (7).   It demonstrates that if an alternating magnetic dipole lies at the surface of a medium whose electrical conductivity is a function of depth only, the conductivity of the medium everywhere at depth may be completely determined by measurements of the field intensity at the surface. The theorem has not led to a practical method for direct interpretation, but it is valuable from the standpoint of establishing the uniqueness of solutions to depth-sounding problems.

In actual fact, depth sounding by electromagnetic induction methods of the type just described appears to be very little used.   Two simple reasons probably explain this.   Firstly, theoretical curves for more than one surface layer overlying a half-space are not generally available; and secondly, the apparatus required is more complicated and more expensive—and the results are probably not more certain—than for the d-c conduction method.   In any case the relative merits of the two methods have not been well determined.

The one electromagnetic method of depth sounding which *has* assumed a great importance is the so-called magnetotelluric method described by Cagniard (8).   This method utilizes natural electromagnetic disturbances generated by fluctuating ionospheric currents with periods ranging from about 0.1 sec up to many minutes.   With modification, it has made possible very deep sounding (to depths of the order of 100 km) without the vast electrode arrays spread over huge distances on the ground surface that would be required by d-c methods in order to achieve a similar penetration. In fact its useful range seems to begin at just about the depths where resistivity methods become impracticable.   The magnetotelluric method consists in measuring the geoelectric field strength at the ground surface in one horizontal direction and the geomagnetic field intensity in the horizontal direction at right angles to the first.   Generally, a wide-band recording of the two signals is made on magnetic tape and replayed later for analysis into harmonic components.   According to theory, if the ground is uniform its resistivity ρ will be given by

$$\rho = \frac{1}{\omega\mu} \left[\frac{(E_\omega)_x}{(H_\omega)_y} \right]^2$$

where ω is the angular frequency of the signal, μ is the magnetic permeability of the ground ($= \mu_0$, usually), and $(E_\omega)_x$ and $(H_\omega)_y$ are the intensities of the harmonic components of the signals $E_x(t)$ and $H_y(t)$ at the angular frequency ω.   For horizontally stratified media this formula defines an apparent resistivity $\rho_a(\omega)$ from which the depths to the various boundaries can be determined by fitting the observations to theoretical curves.

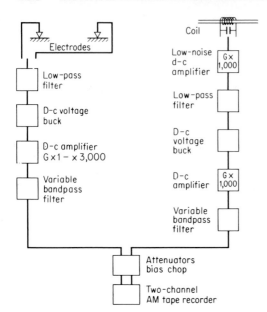

Fig. 15-13 **Instrumentation for magnetotelluric field measurements.** [**Courtesy** *Jour. Geophys. Research*]

The apparatus required for this type of work is costly. A block diagram for a recorder is shown in Fig. 15-13. The basic detector unit for the electric field measurement is just a pair of electrodes set at a considerable distance apart (separations of a few kilometers are not uncommon). For measuring the magnetic field, either a very sensitive coil or a high-frequency resonance magnetometer can be used [although in the former case it is $\partial H(t)/\partial t$ which is actually recorded, not $H(t)$]. The recorder, however, is not the heart of the system, for harmonic analyses of the two signals and cross-correlation tests are both essential to processing the data. So far, the most practical method for doing this seems to be by digitizing the records and performing the harmonic analysis on an electronic computer.

The magnetotelluric technique is still in a stage of very rapid development and will no doubt pass through several modifications before it becomes standardized. Certainly the incentive is high, on account of the great depths to which it seems capable of yielding information. The electrical conductivity in the mantle, for instance, is expected to be strongly influenced by temperature; and deep electrical sounding may be able to discern the shapes and depths of "isothermal surfaces" underneath the crust. Preliminary surveys seem already to have discovered the predicted rapid rise in conductivity at a depth of about 70 km. In petroleum prospecting, the area of application is close to that of the aeromagnetic method. Because a strong resistivity contrast between the sedimentary and crystalline rocks usually occurs at the base of the sedimentary section, the method has been useful as an additional reconnaissance tool for estimating the thickness of sedimentary basins.

15-7 *Interpretation of Electromagnetic Surveys*

As with all other geophysical methods, it is necessary to have both a proper understanding of the physical processes and an adequate supply of mathematical aids in order to make any advance in the quantitative interpretation of electromagnetic induction surveys. With the electromagnetic induction method, obtaining a correct grasp of the physical process seems to be a little more difficult than with the gravity, magnetic, or d-c conduction methods. We are therefore required in the beginning to spend some time on the study of a few very simple and purely heuristic models before going on to investigate those which are actually used in the quantitative interpretation of anomalies. Among the latter, only very simple shapes such as the sphere, the infinite sheet, and the layered half-space offer general solutions of sufficient tractability, although as we shall later see, if the body is extremely conductive, wider possibilities are open.

Luckily, we are not bound to use analytical solutions exclusively in compiling aids to electromagnetic interpretation. In fact, it is usually possible to measure the response of any field system to an *idealized* conductor by making small-scale measurements in the laboratory. The "modeling" method for obtaining theoretical responses works so well in practice that most major users of electromagnetic induction prospecting methods make use of such apparatus for compiling interpretational aids. The measurements obtained with model apparatus can be used either directly by comparing them with the field measurements or indirectly by furnishing the data necessary to produce systematized interpretational aids such as characteristic curves. The latter approach is particularly useful in interpreting airborne surveys, because the large quantities of data make individual curve matching impracticable.

It is important to clarify one apparently deceptive feature of model measurements. This is brought about very largely by their name. Measurements made in the laboratory are in no way different from calculations made by analytical techniques, insofar as their application to geophysical interpretation is concerned. In either case, the models employed as a basis for calculation or for measurement are a highly abstracted representation of any geological situation. If the modeling system is used routinely as an aid, it is used for the most part as an analogue device for finding solutions which are unobtainable mathematically at a comparable cost. Seldom is a scale model of a real geological conductor built for such a purpose, because rarely ever is enough known about its electrical properties or about its shape in order to make the construction possible.

In view of many practical considerations, nearly all of the theory appropriate to electromagnetic induction methods of prospecting is based upon uniform conductors having very simple shapes. These are concepts which, as we know, are exceedingly idealized in the realm of geology. The conductors with which we have to deal in practice are zones of sulfide mineralization which are notorious for their irregularity and inhomogeneity. We cannot therefore expect a high degree of accuracy in the interpretations, except perhaps under unusually favorable circumstances. Nonetheless, by a judicious use of interpretational aids limits can often be placed on such properties of the anomalous body as its conductivity, depth to top, dip and strike extent, and width; and these limits can be extremely useful in guiding further exploration and development, so much so, in fact, that electromagnetic induction methods have earned a place of unique importance in mining geophysics.

References

1. "Methods and Case Histories in Mining Geophysics," Canadian Inst. of Mining and Metallurgy (Sixth Commonwealth Mining and Metallurg. Cong.), Mercury, Montreal, 1957.

2. S. H. Ward, AFMAG—Airborne and Ground, *Geophysics*, vol. 24, pp. 761–789, 1959.

3. D. Boyd and B. C. Roberts, Model Experiments and Survey Results from a Wing-tip-mounted Electromagnetic Prospecting System, *Geophys. Prosp.*, vol. 9, pp. 411–420, 1961.

4. N. R. Paterson, Experimental and Field Data for Dual Frequency Phase-shift Method of Airborne Electromagnetic Prospecting, *Geophysics*, vol. 26, pp. 601–617, 1961.

5. G. Tornqvist, Some Practical Results of Airborne Electromagnetic Prospecting in Sweden, *Geophys. Prosp.*, vol. 6, pp. 112–126, 1959.

6. R. H. Pemberton, Airborne Electromagnetics in Review, *Geophysics*, vol. 27, pp. 691–713, 1962.

7. L. B. Slichter, An Inverse Boundary Value Problem in Electrodynamics, *Physics*, vol. 4, p. 418, 1933.

8. L. Cagniard, Basic Theory of the Magneto-telluric Method of Geophysical Prospecting, *Geophysics*, vol. 18, pp. 605–635, 1953.

9. D. S. Parasnis, "Principles of Applied Geophysics," Methuen, London, 1962.

chapter 16   *Electromagnetic Theory*

To calculate the magnetic fields produced by electromagnetic induction in voluminous media is neither a simple nor straightforward matter. The principles of electromagnetic induction have been well understood for over a century and are concisely codified in Maxwell's equations, but in the form in which they are usually presented these equations are not very helpful in solving problems. We shall find it useful to review at this time the points in electromagnetic theory which we shall need in Chap. 17, where we shall concern ourselves with the interpretation of geophysical measurements.

Let us assume that the ground is made up of a number of clearly defined zones or regions within which the electrical conductivity, the magnetic permeability, and the dielectric capacitivity are all isotropic and constant. We shall try to calculate the magnetic fields due to sources of certain simple types placed on or above the ground surface in the vicinity of these zones. In most cases we shall use boundary-value methods to solve formally for the magnetic field everywhere, including the regions of anomalous conductivity.   However, we need only make calculations of the field intensity at points where the receiver is located.

16-1   *Equations for the Electromagnetic Field*

Five field vectors are needed to determine the electric and magnetic fields within a given region.   These are the four vectors **B**, **H**, **E**, and **D**, which describe the electromagnetic field; and the electric current density **J**, which describes the motion of free charge.   Their names, together with their mks units and dimensions, are

B = the magnetic induction in webers per square meter, M/TQ

H = the magnetic field intensity in ampere-turns per meter, Q/LT

E = the electric field intensity in volts per meter, ML/QT^2

D = the electric displacement in coulombs per square meter, Q/L^2

J = the electric current density in amperes per square meter, Q/L^2T

The five vectors are related, in part, by Maxwell's equations

$$\nabla \times \mathbf{E} = -\frac{\partial \mathbf{B}}{\partial t} \tag{16-1}$$

and
$$\nabla \times \mathbf{H} = \mathbf{J} + \frac{\partial \mathbf{D}}{\partial t} \tag{16-2}$$

The first of these equations is the mathematical formulation of Faraday's law, viz., that "circulating" about any time-varying magnetic field there exists an electric field such that the total emf generated around any closed path C $\left(\oint_c \mathbf{E} \cdot d\mathbf{s}\right)$ is proportional to the negative rate of change of the magnetic flux passing through C.   The second equation states that around any current field there "circulates" a magnetic field such that $\oint_c \mathbf{H} \cdot d\mathbf{s}$ around any closed path C is proportional to the *total* current, both conduction and displacement, which flows through C.   (The quantity $\partial \mathbf{D}/\partial t$ is given the name *displacement current density*.)

From (16-1) and (16-2) we can derive conditions on the vectors **B** and **D** themselves rather than on their time derivatives.   By taking the divergence

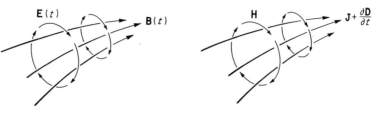

Fig. 16-1   Fig. 16-2

of (16-1), we obtain

$$\boldsymbol{\nabla} \cdot \boldsymbol{\nabla} \times \mathbf{E} \equiv 0 = -\boldsymbol{\nabla} \cdot \frac{\partial \mathbf{B}}{\partial t} = -\frac{\partial}{\partial t} \boldsymbol{\nabla} \cdot \mathbf{B}$$

The divergence of **B** is therefore time-independent, and since **B** is generally time-varying, it follows that

$$\boldsymbol{\nabla} \cdot \mathbf{B} = 0 \qquad (16\text{-}3)$$

By taking the divergence of (16-2) we obtain

$$\boldsymbol{\nabla} \cdot \boldsymbol{\nabla} \times \mathbf{H} \equiv 0 = \boldsymbol{\nabla} \cdot \mathbf{J} + \boldsymbol{\nabla} \cdot \frac{\partial \mathbf{D}}{\partial t} = \boldsymbol{\nabla} \cdot \mathbf{J} + \frac{\partial}{\partial t} \boldsymbol{\nabla} \cdot \mathbf{D}$$

To reduce this further, we must employ a relationship involving the electric-charge density ρ. The equation of continuity which follows from the definition of current as the rate of flow of indestructable charge is

$$\boldsymbol{\nabla} \cdot \mathbf{J} = -\frac{\partial \rho}{\partial t} \qquad (16\text{-}4)$$

and we therefore obtain

$$\frac{\partial}{\partial t} (\boldsymbol{\nabla} \cdot \mathbf{D} - \rho) = 0$$

Since both **D** and ρ may be time-varying, this suggests that

$$\boldsymbol{\nabla} \cdot \mathbf{D} = \rho \qquad (16\text{-}5)$$

However, Eq. (16-5) is more general than necessary, for it is easy to show that in any region of nonvanishing conductivity the charge density will reach its equilibrium (steady-state) value in an extremely short time $(t \approx \epsilon/\sigma)$. This implies that charge does not accumulate appreciably during the flow of current, so that in fact

$$\boldsymbol{\nabla} \cdot \mathbf{J} = 0 \qquad (16\text{-}6)$$

and therefore also

$$\boldsymbol{\nabla} \cdot \mathbf{E} = \boldsymbol{\nabla} \cdot \mathbf{D} = 0 \qquad (16\text{-}7)$$

Besides Ohm's law, we have two other empirical relationships among the four electromagnetic-field vectors in isotropic, continuous media. In free space we can set

$$\mathbf{B} = \mu_0 \mathbf{H} \qquad (16\text{-}8)$$

where $\mu_0 = 4\pi \times 10^{-7}$ henry/m (dimensions ML/Q^2), and then we find that

$$\mathbf{D} = \epsilon_0 \mathbf{E} \qquad (16\text{-}9)$$

where $\epsilon_0 = 8.854 \times 10^{-12}$ farad/m (dimensions $Q^2 T^2 / ML^3$)
$\doteq 1/36\pi \times 10^{-9}$ farad/m

Within most isotropic media similar proportionalities are found to hold, and thus we can write

$$\mathbf{B} = \mu\mathbf{H}$$

$$\mathbf{D} = \epsilon\mathbf{E} \tag{16-10}$$

and also

$$\mathbf{J} = \sigma\mathbf{E}$$

The ratios μ/μ_0 and ϵ/ϵ_0 are called, respectively, the relative magnetic permeability and the dielectric capacitivity of the medium, and in the mks system they are dimensionless. The magnetic permeability is closely related to the magnetic susceptibility k introduced in Chap. 11, since the flux density $\mathbf{B}$ can always be expressed in terms of the field intensity $\mathbf{H}$ and a magnetization vector $\mathbf{M}$, viz.,

$$\mathbf{B} \equiv \mu_0(\mathbf{H} + \mathbf{M})$$

But in most materials the magnetization $\mathbf{M}$ is proportional to $\mathbf{H}$

$$\mathbf{M} = \kappa\mathbf{H}$$

where κ is the mks magnetic susceptibility, and so

$$\mathbf{B} = \mu_0(1 + \kappa)\mathbf{H} \qquad \text{as well as} \qquad \mathbf{B} = \mu\mathbf{H}$$

In cgs electromagnetic units (primed symbols) the corresponding formula is

$$\mathbf{B'} = \mu_0'(\mathbf{H'} + 4\pi\mathbf{M'}) = \mu_0'(1 + 4\pi k)\mathbf{H'}$$

where $\mu_0' = 1$ gauss/oersted, and thus we obtain the interrelationships

$$\frac{\mu'}{\mu_0'} = \frac{\mu}{\mu_0} = 1 + \kappa = 1 + 4\pi k \tag{16-11}$$

By using the relations (16-10), we can eliminate three of the five variables from Maxwell's equations and reduce them to the following set:

$$\boldsymbol{\nabla} \times \mathbf{E} = -\mu\frac{\partial\mathbf{H}}{\partial t} \tag{16-12}$$

$$\boldsymbol{\nabla} \cdot \mathbf{H} = 0 \tag{16-13}$$

$$\boldsymbol{\nabla} \times \mathbf{H} = \sigma\mathbf{E} + \epsilon\frac{\partial\mathbf{E}}{\partial t} \tag{16-14}$$

$$\boldsymbol{\nabla} \cdot \mathbf{E} = 0 \tag{16-15}$$

In cases where there exists a current density $\mathbf{J}_0$ due to sources independent of the electromagnetic field, we must write instead of Eq. (16-14)

$$\boldsymbol{\nabla} \times \mathbf{H} = \sigma\mathbf{E} + \epsilon\frac{\partial\mathbf{E}}{\partial t} + \mathbf{J}_0 \tag{16-16}$$

The four equations can be reduced still further by taking the curl of (16-12) and (16-14) and substituting each into the other. Then by making use of the vector identity $\nabla \times (\nabla \times A) = \nabla(\nabla \cdot A) - \nabla \cdot \nabla A = \nabla(\nabla \cdot A) - \nabla^2 A$, where $\nabla^2 A$ is to be interpreted as the Laplacian operator acting on the *rectangular* components[1] of A, we obtain

$$\nabla^2 E - \sigma\mu \frac{\partial E}{\partial t} - \epsilon\mu \frac{\partial^2 E}{\partial t^2} = 0 \tag{16-17}$$

and

$$\nabla^2 H - \sigma\mu \frac{\partial H}{\partial t} - \epsilon\mu \frac{\partial^2 H}{\partial t^2} = 0 \tag{16-18}$$

From these two necessary (but not sufficient) relationships we observe that both E and H must propagate as a dissipative wave motion.

In most cases we shall be dealing with alternating fields, and we may therefore assume for H and E a time dependence which is of the form $H(r,t) = \text{Re } H(r,\omega)e^{i\omega t}$ where ω is the angular frequency of the field (i.e., $\omega = 2\pi f$). Equations (16-17) and (16-18) then become

$$\nabla^2 E = i\sigma\mu\omega E - \epsilon\mu\omega^2 E \tag{16-19}$$

and

$$\nabla^2 H = i\sigma\mu\omega H - \epsilon\mu\omega^2 H \tag{16-20}$$

16-2 The Long-wavelength Approximation

Now let us pause briefly to consider the relative magnitudes of the terms appearing in Eqs. (16-19) and (16-20) in order to ascertain which are significant. Among the coefficients appearing on the right-hand sides of these equations are the physical properties ϵ, μ, and σ. With the exception of water $(\epsilon/\epsilon_0 \doteq 80)$ dielectric capacitivities seldom vary by more than an order of magnitude, and typical for most rocks and minerals is the value $\epsilon \doteq 9\epsilon_0 \doteq 8 \times 10^{-11}$ farad/m. Even among ferromagnetic minerals the relative magnetic permeability remains less than 3, and in the vast majority of cases it is very close to unity; thus we may set $\mu \doteq \mu_0 \doteq 1.3 \times 10^{-6}$ henry/m. About electrical conductivity, however, we can say nothing at all; for this property may take on values ranging from practically zero in air, through values of the order of 10^{-3} mho/m in relatively nonconducting rocks, to values as great as 10^4 mhos/m in certain semimetallic minerals. The magnitude of the remaining parameter ω depends upon the apparatus. Typical frequencies are about 1,000 cps (6,000 radians/sec). This figure

---

[1] In rectangular coordinates the expression $\nabla \cdot \nabla A$ reduces to $i_x \nabla^2 A_x + i_y \nabla^2 A_y + i_z \nabla^2 A_z$. In curvilinear systems, this is not the case, because the components of A transform with the base vectors. Hence we can only interpret the divergence of the gradient of a vector as $\nabla \cdot \nabla A = \nabla(\nabla \cdot A) - \nabla \times (\nabla \times A)$. In practical cases, however, it is often easier to find the cartesian components of A, perform the operation ∇^2, and then to reconstruct the resulting vector in the original coordinate system. We therefore regard it as permissible to use the concise notation $\nabla^2 A$ from this point onward, always keeping in mind that it applies only to the rectangular components of A.

is rarely exceeded in large-scale prospecting systems whose overall dimensions are as large as 10^3 m.

If we now compare the magnitudes of the terms in (16-19) and (16-20) at distances of, let us say, less than 10^3 m from the source, we find that above the ground, where $\epsilon = \epsilon_0$, $\mu = \mu_0$, and $\sigma = 0$,

$$\nabla^2 \frac{\mathbf{H}}{\mathbf{E}} \doteq 0$$

since the receiver lies well within the quasi-static zone[1] of the transmitter. On the other hand, in normally conducting rocks where $\epsilon \doteq 9\epsilon_0$, $\mu \doteq \mu_0$, and $\sigma \approx 10^{-3}$ mho/m

$$\nabla^2 \frac{\mathbf{H}}{\mathbf{E}} \doteq [(-4 \times 10^{-9} + 7 \times 10^{-6}i) \ m^{-2}] \frac{\mathbf{H}}{\mathbf{E}}$$

and in highly conducting zones where σ may run as high as 10^4 mhos/m

$$\nabla^2 \frac{\mathbf{H}}{\mathbf{E}} \doteq [(-4 \times 10^{-9} + 70i) \ m^{-2}] \frac{\mathbf{H}}{\mathbf{E}}$$

In both of the latter instances the real part of the coefficient of $\mathbf{H}$ or $\mathbf{E}$ is so small in relation to the imaginary part that we may neglect it. In normally conducting rocks even the imaginary term has a very small effect in most cases, but in highly conducting zones it definitely does not. Thus we conclude that in nonconducting regions the fields obey the equation

$$\nabla^2 \frac{\mathbf{H}}{\mathbf{E}} = 0 \tag{16-21}$$

while in zones having an appreciable conductivity they satisfy

$$\nabla^2 \frac{\mathbf{H}}{\mathbf{E}} = i\sigma\mu\omega \frac{\mathbf{H}}{\mathbf{E}} \tag{16-22}$$

The interesting feature of these equations is that neither one of them represents the propagation of steady waves. Equation (16-21) is obeyed by static electric or magnetic fields, and (16-22) may be written as

$$\nabla^2 \frac{\mathbf{H}}{\mathbf{E}} = \sigma\mu \frac{\partial}{\partial t} \left(\frac{\mathbf{H}}{\mathbf{E}} \right) \tag{16-23}$$

which is the vector diffusion equation.

To learn something about the fundamental nature of the solutions of (16-23), we may limit ourselves to a situation in which the fields depend

---

[1] At distances which are small in relation to the wavelength of radiation from an antenna, the fields are everywhere almost in phase with the antenna current. This close-in region is accordingly called the *quasi-static zone.*

upon only one variable and vary sinusoidally with time. If the magnetic field is plane-polarized in the y direction and propagates in the x direction, the solutions of (16-23) are

$$H_y(x,t) = H_0 \exp\left[i\omega t \pm (i\sigma\mu\omega)^{\frac{1}{2}}x\right]$$

and

$$J_z(x,t) = \sigma E_z(x,t) = J_0 \exp\left[i\omega t \pm (i\sigma\mu\omega)^{\frac{1}{2}}x\right]$$

where

$$J_0 = (i\sigma\mu\omega)^{\frac{1}{2}}H_0$$

according to (16-2). Separating the real and imaginary terms in the exponentials and retaining only those terms which remain finite as $x \to \infty$, we

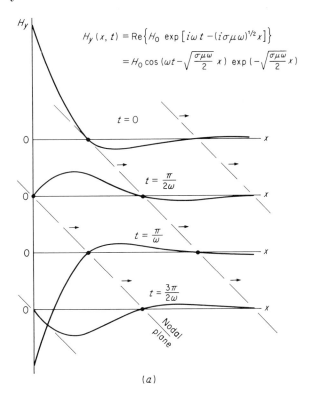

$$H_y(x,t) = \mathrm{Re}\left\{H_0 \exp\left[i\omega t - (i\sigma\mu\omega)^{1/2}x\right]\right\}$$

$$= H_0 \cos\left(\omega t - \sqrt{\frac{\sigma\mu\omega}{2}}\,x\right) \exp\left(-\sqrt{\frac{\sigma\mu\omega}{2}}\,x\right)$$

$t = 0$

$t = \dfrac{\pi}{2\omega}$

$t = \dfrac{\pi}{\omega}$

$t = \dfrac{3\pi}{2\omega}$

Nodal plane

(a)

Fig. 16-3 Diffusion (in one dimension) of a magnetic field into a conductor. (a) Showing the damped wave nature of $H_y(x,t)$; (b) showing the effect of different values of the parameter $\sigma\mu\omega$.

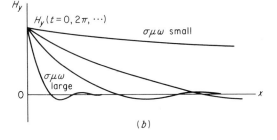

$H_y(t = 0, 2\pi, \cdots)$

$\sigma\mu\omega$ small

$\sigma\mu\omega$ large

(b)

may write

$$H_y = H_0 \exp\left[-\left(\frac{\sigma\mu\omega}{2}\right)^{\!\frac{1}{2}} x\right] \exp\left[i\left(\omega t - \left(\frac{\sigma\mu\omega}{2}\right)^{\!\frac{1}{2}} x\right)\right]$$

and

$$J_z = (i\sigma\mu\omega)^{\frac{1}{2}} H_0 \exp\left[-\left(\frac{\sigma\mu\omega}{2}\right)^{\!\frac{1}{2}} x\right] \exp\left[i\left(\omega t - \left(\frac{\sigma\mu\omega}{2}\right)^{\!\frac{1}{2}} x\right)\right]$$

These formulas represent a highly damped, dispersive wave. Figure 16-3 shows a series of graphs of the real part of H_y which illustrate this. We see that if the product $(\sigma\mu\omega/2)^{-\frac{1}{2}}$ (which has the dimensions of a length) is very small compared with the thickness of the conductor, the magnetic field will barely penetrate the conductor before it is damped out. The induced current field responsible for the decay of H_y from its initial value will also be concentrated near the surface, and in the limit, as $\sigma\mu\omega \to \infty$, it will become a surface current. On the other hand, for very small values of $\sigma\mu\omega$ we find that the magnetic field penetrates the conductor almost without attenuation, while the current field, although distributed throughout the conductor, has a vanishingly small intensity. In the intermediate range we may expect to find a moderately strong current field induced in the conductor, which will in turn produce an appreciable alteration in the magnetic field. These currents will not be in phase with the magnetic field or uniformly distributed, but will in general tend to crowd toward the outside of the conductor.

We cannot, of course, expect a one-dimensional analysis to apply in detail to three-dimensional fields. However, the main features of the phenomenon ought to be similar and we should expect, if l is a typical dimension of the conductive zone, that when $\sigma\mu\omega l^2 \gg 1$, the magnetic field will effectively vanish in the interior of the conductor. A secondary field sufficient to produce complete cancellation of the internal primary field must thus be created. Conversely, if $\sigma\mu\omega l^2 \ll 1$, the conductor should have very little effect upon the field, and its presence will be undetectable. Between these extremes we should expect the conductor to perturb the magnetic field somewhat, by generating some secondary field.

16-3 Boundary Conditions

In addition to satisfying the field equations within the several regions into which we have imagined the ground to be divided, $\mathbf{E}$ and $\mathbf{H}$ must also satisfy certain boundary conditions at the interfaces. The field vectors and their derivatives must be continuous within the various regions, but where discontinuities occur in σ or μ, discontinuities in $\mathbf{E}$ and $\mathbf{H}$ will also exist. To derive the boundary conditions, we must go back to Maxwell's equations (16-1), (16-2), and (16-3) and to the conservation law (16-6), none of which makes any assumptions about the uniformity of the medium. Working in much the same way as we did in Sec. 14-2, where we derived the boundary conditions for direct-current fields, we find from (16-1), $\nabla \times \mathbf{E} = -\partial\mathbf{B}/\partial t$,

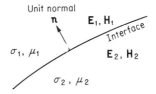

Fig. 16-4

that

$$\mathbf{n} \times (\mathbf{E_1} - \mathbf{E_2}) = 0 \qquad (16\text{-}24)$$

i.e., that the electric field tangential to the interface is continuous. From (16-6), $\nabla \cdot \mathbf{J} = 0$, we deduce that

$$\mathbf{n} \cdot (\sigma_1 \mathbf{E_1} - \sigma_2 \mathbf{E_2}) = 0 \qquad (16\text{-}25)$$

i.e., that the current density normal to the interface is continuous; and from (16-2), $\nabla \times \mathbf{H} = \mathbf{J} + \partial \mathbf{D}/\partial t$, that

$$\mathbf{n} \times (\mathbf{H_1} - \mathbf{H_2}) = 0 \qquad (16\text{-}26)$$

i.e., that the magnetic field tangential to the interface is continuous. This last equation does not hold when either σ_1 or σ_2 becomes infinite, for then a finite current can flow even within an infinitesimal distance from the boundary. Denoting the density of such a surface current with the symbol $\mathbf{K}$ amp/m, we then obtain in place of (16-26) the relation

$$\mathbf{n} \times (\mathbf{H_1} - \mathbf{H_2}) = \mathbf{K} \qquad (16\text{-}27)$$

Finally, from (16-3), $\nabla \cdot \mathbf{B} = 0$, we obtain

$$\mathbf{n} \cdot (\mu_1 \mathbf{H_1} - \mu_2 \mathbf{H_2}) = 0 \qquad (16\text{-}28)$$

in other words, that the flux density normal to the interface is continuous.

In connection with (16-27), which applies when the conductivity of one region is made very large, we can refer back to our argument in Sec. 16-3 where we deduced that when $\sigma\omega \to \infty$ within a region, $\mathbf{H}$ (and also $\mathbf{E}$) must vanish there. The boundary conditions on $\mathbf{H}$ therefore reduce to the simpler forms

$$\mathbf{n} \cdot \mathbf{H_1} = 0 \quad \text{and} \quad \mathbf{n} \times \mathbf{H_1} = \mathbf{K} \quad \text{when } \omega\sigma_2 \to \infty \quad (16\text{-}29)$$

when it is clear that only time-varying fields are being considered.

16-4   *The Electromagnetic Potential*

We have progressed about as far as we can usefully go with the two field vectors $\mathbf{E}$ and $\mathbf{H}$. In solving boundary-value problems in electromagnetic theory, it is extremely inconvenient to deal with a pair of vectors which, as well as satisfying Eqs. (16-17) and (16-18), must also be interrelated by either one of Maxwell's equations (16-1) or (16-2). We shall now introduce a vector potential from which **E** and **H** can both be derived. This will reduce the number of unknowns to a minimum and simplify the finding of solutions.

Actually there are many different vector potentials from which to choose, and we try to select whichever one effects the greatest reduction in the number of variables, according to the special circumstances of the problem under consideration. One of the most widely used is the magnetic vector potential **A**, from which we derive the magnetic induction **B** as follows:

$$\mathbf{B} = \nabla \times \mathbf{A} \tag{16-30}$$

thereby satisfying automatically the condition that $\nabla \cdot \mathbf{B} = 0$. In the mks system **A** is measured in webers per meter and has the dimensions ML/QT.

Now according to Faraday's law (16-1), we observe that

$$\nabla \times \mathbf{E} = -\frac{\partial \mathbf{B}}{\partial t} = -\frac{\partial}{\partial t}(\nabla \times \mathbf{A})$$

and this is satisfied by letting

$$\mathbf{E} = -\frac{\partial \mathbf{A}}{\partial t} \tag{16-31}$$

A further condition on the vector potential derives from (16-6) and (16-7), viz., $\nabla \cdot \mathbf{D} = 0 = \nabla \cdot \mathbf{E} = \nabla \cdot \mathbf{J}$. According to (16-31), it then follows that $\nabla \cdot \mathbf{A}$ is time-independent. The only circumstance in which this can be so when **A** itself is time-varying is that the divergence must vanish. As a third equation, therefore, we may now add

$$\nabla \cdot \mathbf{A} = 0 \tag{16-32}$$

If we now insert (16-30) and (16-31) into Eq. (16-14), we obtain the equation

$$-\nabla \times (\nabla \times \mathbf{A}) = \mu\sigma\frac{\partial \mathbf{A}}{\partial t} + \epsilon\mu\frac{\partial^2 \mathbf{A}}{\partial t^2}$$

which must be satisfied everywhere within a homogeneous medium, and by virtue of (16-32) this reduces to the wave equation

$$\nabla^2 \mathbf{A} = \mu\sigma\frac{\partial \mathbf{A}}{\partial t} + \epsilon\mu\frac{\partial^2 \mathbf{A}}{\partial t^2} \tag{16-33}$$

Thus we have shown that if a vector field **A** can be found which satisfies (16-32) and (16-33), we can derive from it a set of electromagnetic field vectors **H** and **E** which satisfy Maxwell's equations. We have not, however, shown that every solution of Maxwell's equations has a vector potential **A**. Since we shall not be concerned with any cases in which the vector potential does not exist, we need not pause to consider this question.

Since we shall assume sinusoidally time-varying fields and the long wavelength approximation in most cases, we can immediately simplify (16-33) to

$$\nabla^2 \mathbf{A} = 0 \qquad (16\text{-}34)$$

in nonconducting regions, and

$$\nabla^2 \mathbf{A} - i\mu\sigma\omega \mathbf{A} = 0 \qquad (16\text{-}35)$$

in conducting zones. Or, if a current field exists due to external sources which are independent of the electromagnetic field, we must use (16-16) instead of (16-14) in deriving (16-33) and add $-\mu\mathbf{J}_0$ to the right-hand sides of (16-33), (16-34), and (16-35).

The vector potential can be very useful in calculating the magnetic field of known current distributions as well as in solving boundary-value problems. In the absence of displacement current (16-33) can be written as

$$\nabla^2 \mathbf{A} = -\mu \mathbf{J} \qquad (16\text{-}36)$$

which is a Poisson equation for each of the cartesian components of $\mathbf{A}$. The general solution of these equations at points which are external to the regions in which electric currents are flowing is, in rectangular coordinates,

$$\mathbf{A}(\mathbf{r}) = \frac{\mu}{4\pi} \int_V \frac{\mathbf{J}(\mathbf{r}_0)}{|\mathbf{r} - \mathbf{r}_0|} \, d^3 r_0 \qquad (16\text{-}37)$$

where $\mathbf{J} = 0$ outside the volume V. This expression can be regarded as a set of formulas for the three rectangular components of $\mathbf{A}$. It is not generally valid in a curvilinear coordinate system.

16-5 The Magnetic Field of a Loop

We shall now consider the use of the magnetic vector potential in calculating the magnetic field around a current-carrying loop in free space. First let us choose a system of circular cylindrical coordinates (ρ, φ, z) whose axis coincides with that of the loop and whose origin lies at its center. On account of the symmetry of $\mathbf{J}$ about the coordinate axis, $\mathbf{A}$ will have no component in any

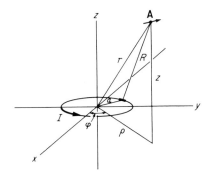

Fig. 16-5 Vector potential of a current-carrying loop situated with its center at the origin and having its axis in the Oz direction.

axial plane.   Thus the only cylindrical component of **A** which remains is the φ component, and (16-37) therefore reduces to the scalar equation

$$A_\varphi = \frac{\mu_0}{4\pi} \int_V \frac{\cos(\varphi_0 - \varphi) J_\varphi(\mathbf{r}_0)}{|\mathbf{r} - \mathbf{r}_0|} d^3 r_0 \tag{16-38}$$

where the integral is to be evaluated around the circumference of the wire loop, since this is the only region in which electric currents are flowing.   If I amp of current flow through the loop and if its winding cross-sectional area is very small compared with its radius a, then the integral (16-38) reduces to

$$A_\varphi = \frac{\mu_0 I}{4\pi} \int_0^{2\pi} \frac{a \cos(\varphi_0 - \varphi) \, d\varphi_0}{[\rho^2 + a^2 + z^2 - 2a\rho \cos(\varphi_0 - \varphi)]^{1/2}}$$

Now we can, without loss of generality, set $\varphi = 0$.   If we then expand the denominator of the integrand by the binomial theorem, we may write

$$A_\varphi = \frac{\mu_0 I}{2\pi} \int_0^\pi \frac{a \cos \varphi_0}{(\rho^2 + z^2)^{1/2}} \left[1 + \frac{a\rho \cos \varphi_0 - \frac{1}{2}a^2}{(\rho^2 + z^2)} + \cdots \right] d\varphi_0$$

$$\doteq \frac{\mu_0 I a^2 \rho}{4(\rho^2 + z^2)^{3/2}} \qquad \text{if } a^2 \ll \rho^2 + z^2 \tag{16-39}$$

Since we now know **A**, the magnetic field is given by (16-30) and (16-8). Thus

$$\mathbf{H} = \frac{1}{\mu_0} \nabla \times \mathbf{A}$$

$$= \frac{1}{\mu_0} \left[-\frac{\partial A_\varphi}{\partial z} \mathbf{i}_\rho + \frac{1}{r} \frac{\partial}{\partial r}(r A_\varphi) \mathbf{i}_z \right]$$

$$= \frac{I a^2}{4} \left[\frac{3\rho z}{(\rho^2 + z^2)^{5/2}} \mathbf{i}_\rho + \frac{2z^2 - \rho^2}{(\rho^2 + z^2)^{5/2}} \mathbf{i}_z \right] \tag{16-40}$$

This is identical with the expression for the magnetic field due to a dipole situated at the center of the coil, whose magnetic moment has a magnitude

$$m = \pi a^2 I \tag{16-41}$$

and is directed along Oz.   We find therefore that a loop transmitter can be represented by an oscillating magnetic dipole as long as we measure its field at distances greater than a few loop diameters from its center.   At the same time we note that the expression for the vector potential due to an axial dipole having a moment m is

$$\mathbf{A} = \frac{\mu_0 m}{4\pi} \frac{\rho}{(\rho^2 + z^2)^{3/2}} \mathbf{i}_\varphi \tag{16-42}$$

and we may verify also that the general expression for a dipole in any other direction is

$$\mathbf{A}(\mathbf{r}) = \frac{\mu_0}{4\pi} \mathbf{m} \times \nabla_0 \left(\frac{1}{|\mathbf{r} - \mathbf{r}_0|} \right) \tag{16-43}$$

16-6   *Mutual Inductance*

Mutual inductance is a term used to describe the interaction at a distance between electrical circuits, due to electromagnetic induction. It arises in electromagnetic induction prospecting because in the majority of cases the transmitter and the receiver are both simple coils, and we are concerned with the interaction between them. Accordingly, we shall consider only the problem of mutual inductance between two coils. A more general treatment of the subject may be found elsewhere (1).

We shall assume the transmitter and receiver are simple loops of small winding cross section held in fixed positions and orientations with respect to one another. The emf induced in the receiver is proportional to the time rate of change of the current in the transmitter, and the constant of proportionality is called the *coefficient of mutual inductance*. Thus if the current flowing through the transmitter loop is I_1 amp and if it is sinusoidally time-varying with an angular frequency ω, we may write for the emf in the receiver

$$\mathcal{E}_2 = -i\omega M_{12} I_1 \qquad \text{volts} \tag{16-44}$$

where M_{12} is the coefficient of mutual inductance measured in henrys. By virtue of the Helmholtz reciprocity law, the receiving and transmitting functions of the two coils may be interchanged without affecting the value of M_{12}. Thus

$$M_{12} = M_{21}$$

If the two loops are in free space, the magnetic potential of the transmitter at a point $\mathbf{r}_2$ within the receiver loop (see Fig. 16-6) is, according to (16-37),

$$\mathbf{A}(\mathbf{r}_2) = \frac{\mu_0 I_1}{4\pi} \oint_1 \frac{d\mathbf{l}_1}{|\mathbf{r}_2 - \mathbf{r}_1|}$$

where the subscript 1 applies to quantities measured at the transmitter position and the subscript 2, to quantities measured at the receiver position.

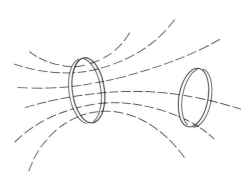

Fig. 16-6   The mutual coupling between two loops.

The magnetic flux intercepted by the receiver will therefore be

$$\Phi_2 = \int_2 [\nabla \times \mathbf{A}(\mathbf{r}_2)] \cdot d\mathbf{S}_2$$

and since

$$\mathcal{E}_2 = -\frac{\partial \Phi_2}{\partial t} = -i\omega\Phi_2$$

it follows that

$$M_{12} = \frac{\mu_0}{4\pi} \int_2 d\mathbf{S}_2 \cdot \left(\nabla_2 \times \oint_1 \frac{d\mathbf{l}_1}{|\mathbf{r}_2 - \mathbf{r}_1|} \right) \tag{16-45}$$

If the distance between loops is a good deal larger than the diameters of the loops themselves, the expression (16-45) simplifies very considerably. To begin with, the magnetic potential of the transmitter at the receiver position can be written, approximately, as

$$\mathbf{A}(\mathbf{r}_2) = \frac{\mu_0 a_1{}^2 I_1}{4} \left(\mathbf{n}_1 \times \nabla_1 \frac{1}{|\mathbf{r}_2 - \mathbf{r}_1|} \right)$$

where $\mathbf{n}_1$ is a unit vector in the direction of the transmitter axis. The emf induced in the receiver will be, to the same approximation,

$$\mathcal{E}_2 = -i\omega\Phi_2$$

$$= -\frac{i\omega\mu_0\pi a_1{}^2 a_2{}^2 I_1}{4} \nabla_2 \times \left(\mathbf{n}_1 \times \nabla_1 \frac{1}{|\mathbf{r}_2 - \mathbf{r}_1|} \right) \tag{16-46}$$

Two cases of special importance in electromagnetic induction prospecting are the following:

1. The loops are coaxial and the distance between their centers is l. We then find from (16-46) that

$$M_{12} = \frac{\mu_0\pi a_1{}^2 a_2{}^2 I_1}{2l^3} \qquad l \gg a_1, a_2 \tag{16-47}$$

2. The loops are coplanar and the distance between their axes is l. In this case it turns out that

$$M_{12} = \frac{-\mu_0\pi a_1{}^2 a_2{}^2 I_1}{4l^3} \qquad l \gg a_1, a_2 \tag{16-48}$$

In both cases the unit normal vectors $\mathbf{n}_1$ and $\mathbf{n}_2$ are assumed to be parallel. A third arrangement also often used in practice is one in which the axes of the two loops intersect at right angles. This is the coil configuration most commonly used with the dip-angle techniques (Sec. 15-2), and in this case the mutual inductance between the two coils in free space is zero.

If the transmitter and receiver are kept in fixed relative positions, as is the case in some prospecting systems, the only circumstance that will cause their mutual inductance to change is the presence of a conducting and/or permeable zone in the vicinity. Since the strength of the secondary field produced by such a body must be proportional to the primary field intensity, which is itself proportional to the electric current flowing in the transmitter, and since the additional emf induced in the receiver must be proportional to the time rate of change of the secondary field which penetrates it, we may write in a manner similar to (16-44)

$$\Delta \mathcal{E}_2 = -i\omega \, \Delta M_{12} I_1$$

The *electromagnetic anomaly* is just the dimensionless quotient

$$\frac{\Delta \mathcal{E}_2}{\mathcal{E}_2} = \frac{\Delta M_{12}}{M_{12}}$$

and is usually measured in parts per million or as a percentage. In a very real sense, therefore, electromagnetic anomalies are simply those changes in the coefficient of mutual inductance between a pair of coils that are brought about by their proximity to a conducting region. However, it should be noted from remarks made in Sec. 16-2 that these changes are not always in phase with the transmitter current I_1 and that ΔM_{12} is therefore usually a complex quantity.

16-7 Scaling

Next we shall investigate the conditions under which similitude can exist between different electromagnetic systems. By this we mean that each system is a perfect replica of the other except for a change of scale. First of all we imagine two systems having identical geometries, but one having linear dimensions l times smaller than those of the other. We can then derive relationships between the physical parameters σ, μ, ϵ and the time t in the two systems, if electromagnetic responses in the one system are to be identical with those in the other.

If we describe the electromagnetic field in the smaller system (to which we shall henceforth refer as the "model" system) by the vectors $\mathbf{H}_1$ and $\mathbf{E}_1$ and in the larger system (which we shall call the "natural" system) by $\mathbf{H}_2$ and $\mathbf{E}_2$, then we know that both sets of vectors must satisfy (16-17) and (16-18),

$$\left(\nabla_1{}^2 - \sigma_1\mu_1 \frac{\partial}{\partial t_1} - \epsilon_1\mu_1 \frac{\partial^2}{\partial t_1{}^2} \right) \frac{\mathbf{H}_1}{\mathbf{E}_1} = 0 \qquad (16\text{-}49)$$

$$\text{and} \qquad \left(\nabla_2{}^2 - \sigma_2\mu_2 \frac{\partial}{\partial t_2} - \epsilon_2\mu_2 \frac{\partial^2}{\partial t_2{}^2} \right) \frac{\mathbf{H}_2}{\mathbf{E}_2} = 0 \qquad (16\text{-}50)$$

At the same time, if the first system is to be a model of the second, the vectors $\mathbf{H}_1$ and $\mathbf{E}_1$ must also satisfy Eq. (16-50) when distances and times

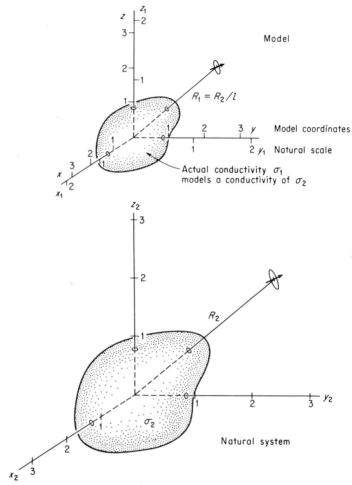

Fig. 16-7   Relationship between a full-scale system and its model.

are scaled appropriately. Thus, putting $l = x_2/x_1 = y_2/y_1 = z_2/z_1$ and $k = t_2/t_1$, we shall have

$$\left(\nabla_2{}^2 - \sigma_2\mu_2\frac{\partial}{\partial t_2} - \epsilon_2\mu_2\frac{\partial^2}{\partial t_2{}^2}\right)\begin{matrix}\mathbf{H}_1\\\mathbf{E}_1\end{matrix} = 0 \qquad (16\text{-}51)$$

where

$$\nabla_2{}^2 = \frac{\partial^2}{\partial(lx_1)^2} + \frac{\partial^2}{\partial(ly_1)^2} + \frac{\partial^2}{\partial(lz_1)^2} = \frac{1}{l^2}\nabla_1{}^2$$

$$\frac{\partial}{\partial t_2} = \frac{\partial}{\partial(kt_1)} = \frac{1}{k}\frac{\partial}{\partial t_1}$$

and

$$\frac{\partial^2}{\partial t_2{}^2} = \frac{\partial^2}{\partial(kt_1)^2} = \frac{1}{k^2}\frac{\partial^2}{\partial t_1{}^2}$$

Accordingly

$$\frac{\nabla_2^2}{\nabla_1^2} = \frac{\sigma_2\mu_2\,\partial/\partial t_2}{\sigma_1\mu_1\,\partial/\partial t_1} = \frac{\epsilon_2\mu_2\,\partial^2/\partial t_2^2}{\epsilon_1\mu_1\,\partial^2/\partial t_1^2}$$

i.e.,
$$l^2 = \frac{\sigma_1\mu_1 k}{\sigma_2\mu_2} = \frac{\epsilon_1\mu_1 k^2}{\epsilon_2\mu_2} \tag{16-52}$$

We now have three relations among the five scaling factors l, k, σ_2/σ_1, μ_2/μ_1, and ϵ_2/ϵ_1, which would seem to allow us considerable freedom in choosing a model system. However it is not possible to scale μ and ϵ arbitrarily, for while the dielectric constant and the magnetic permeability of materials can be controlled to some extent, the free-space values cannot be altered. Consequently we must have

$$\frac{\mu_2}{\mu_1} = \frac{\epsilon_2}{\epsilon_1} = 1$$

and the scaling ratios are thus completely determined by the relations

$$l^2 = \frac{k\sigma_1}{\sigma_2} = k^2 \tag{16-53}$$

Of course, if the real phenomenon is quasi-static, it is unnecessary to scale the terms of (16-17, 16-18) involving ϵ, and the scaling equation has again one surplus degree of freedom

$$l^2 = \frac{k\sigma_1}{\sigma_2} \tag{16-54}$$

However, we must be careful that the displacement current in the model does not become so large that its effect is no longer negligible. Fortunately, in modeling electromagnetic prospecting systems, this term is usually decreased rather than increased, and so it raises no difficulty.

As an example, we may consider a field system operating at a frequency of 1,000 cps, which is to be modeled at a linear scale of 1 to 500. To duplicate the response of the field system at the smaller scale, we would require that

$$l^2 = 500^2 = \frac{k\sigma_1}{\sigma_2} = k^2$$

i.e.,
$$k = 500 \qquad \frac{\sigma_1}{\sigma_2} = 500$$

Since the relationship between time in the model system (t_1) and in the natural system (t_2) is $t_2 = kt_1$, the operating frequency of the model will therefore have to be 500,000 cps and all conductivities will have to be increased 500 times. This is often inconvenient experimentally, since there is a gap in conductivity between the sulfide minerals, whose conductivities range up to about 10^4 mhos/m, and the metals, most of which have conductivities above 10^6 mhos/m.   To avoid this difficulty we may use the freedom allowed by the quasi-stationary scaling equation (16-55) and choose a more convenient value for σ_1/σ_2, say 10^4, for which an operating frequency of 25,000 cps is required.   In some cases the same electronic apparatus can be employed both in the laboratory and in the field if the operating frequency is not scaled.   We then have $k = l$ and $\sigma_1/\sigma_2 = l^2$.

16-8   Elliptical Polarization

In virtually every type of electromagnetic induction prospecting system an alternating magnetic field is employed.   In Sec. 16-2 we learned that in the long-wavelength approximation the primary field in free space obeys a Laplace type of equation and suffers no change in phase from point to point. When conductors are present and a secondary field exists, however, we generally find that there is a phase difference between the primary and secondary fields.   In some prospecting systems, particularly those of the "dip-angle" type, the receiver responds to the total field produced by the transmitter.   We shall find it instructive to consider what happens when two space vectors, which alternate at the same frequency but which are out of phase with each other, are superimposed.

The two vectors $\mathbf{A} \cos \omega t$ and $\mathbf{B} \cos (\omega t + \varphi)$ differ in direction by an arbitrary space angle α and in phase by a phase angle φ.   To find their sum we must first choose a set of cartesian coordinates and resolve the two vectors into their rectangular space components.   This is a routine matter, and we shall not actually perform the operation since the trigonometry would obscure our final result.   If we take x,y coordinates in the plane of the two vectors, however, we can always resolve them into pairs of rectangular components $(A_x \cos \omega t, A_y \cos \omega t)$ and $(B_x \cos (\omega t + \varphi), B_y \cos (\omega t + \varphi))$. The two components in each direction can then be summed to yield a pair of orthogonal quantities $X \cos (\omega t + \varphi_1)$ and $Y \cos (\omega t + \varphi_2)$.   Let us now

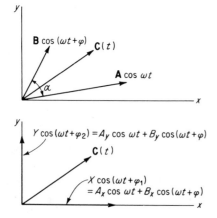

Fig. 16-8   Addition of two alternating vector quantities.

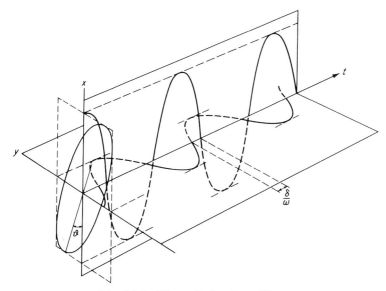

Fig. 16-9   The polarization ellipse.

add these vectorially and consider the behavior of their sum as t advances. If we write

$$\mathbf{C}(t) = \mathbf{i}_x C_x(t) + \mathbf{i}_y C_y(t) \tag{16-55}$$

where

$$C_x(t) = X \cos(\omega t + \varphi_1)$$

and

$$C_y(t) = Y \cos(\omega t + \varphi_2)$$

then by eliminating t we obtain

$$\frac{C_x^2}{X^2} + \frac{C_y^2}{Y^2} - \frac{2 C_x C_y}{XY} \cos \delta = \sin 2\delta \tag{16-56}$$

where

$$\delta = \varphi_2 - \varphi_1$$

This is the equation of an ellipse whose major axis is inclined at an angle ϑ with respect to the x axis, given by

$$\tan 2\vartheta = \frac{XY \cos \delta}{X^2 - Y^2} \tag{16-57}$$

It becomes plain that $\mathbf{C}(t)$ never vanishes but rotates in space, changing continuously in magnitude as it does so, so that it describes an "ellipse of polarization."

The following special cases are important. When $\delta = 0$, Eq. (16-56) reduces to

$$\frac{C_x^2}{X^2} + \frac{C_y^2}{Y^2} - \frac{2 C_x C_y}{XY} = \left(\frac{C_x}{X} - \frac{C_y}{Y} \right)^2 = 0$$

so that

$$Y C_x = X C_y$$

This is a straight line through the origin having a slope Y/X, indicating that **C** is a simple alternating vector. This situation obtains whenever $\varphi = 0$.

In the case $\delta = \pi/2$, we have

$$\frac{C_x{}^2}{X^2} + \frac{C_y{}^2}{Y^2} = 1$$

which is an ellipse whose axes are in the Ox and Oy directions. The tilt angle ϑ is therefore zero or $n\pi/2$. This happens when **A** and **B** are at right angles and $\varphi = \pi/2$. In addition, in the special case $X = Y$ and $\delta = \pi/2$ we find that

$$C_x{}^2 + C_y{}^2 = X^2 = Y^2$$

which is the equation of a circle.

We can now see why it may not be possible to find a perfect null by rotating the axis of a receiving coil in the x,y plane. The amplitude of the signal is proportional to the component of **C** lying in the direction of the coil axis. It will therefore be a maximum when the axis of the coil is parallel with the major axis of the polarization ellipse and a minimum when the coil is aligned with the minor axis. The ratio of maximum to minimum signal amplitude measures the ratio of the elliptic axes, which gives a rough indication of the phase difference between the two vectors if some assumption can be made about their relative magnitudes. Thus, for example, if we identify the primary field intensity **H** with the vector **A** whose direction we define to be Ox and the secondary field intensity Δ**H** with the vector **B** and then if we can suppose that $\Delta H \ll H$, it follows from (16-56) that the ratio of the minor to the major axis of the polarization ellipse is approximately

$$\frac{\Delta H \sin \alpha \sin \varphi}{H}$$

while
$$\tan 2\vartheta \doteq \frac{\Delta H}{H} \sin \alpha \cos \varphi + \left(\frac{\Delta H}{H}\right)^2 \frac{\sin 2\alpha \sin \varphi}{2} \tag{16-58}$$

A secondary magnetic field which is nearly in phase with the primary field produces a very narrow polarization ellipse and consequently a rather sharp null. As the secondary field shifts out of phase, so the null will broaden. Observations of this type can sometimes be useful as a means of gaining a rough idea of whether the electrical conductivity of the anomalous zone is large or small.

References

1. W. R. Smythe, "Static and Dynamic Electricity," 2d ed., McGraw-Hill, New York, 1950.

chapter 17   Theory of Electromagnetic Induction

In this chapter we shall work out in some detail the responses of conducting bodies having known shapes, sizes, conductivities, and permeabilities to magnetic fields from various sources. The approach will be essentially formal, our purpose being to keep the necessary mathematical preliminaries to geophysical interpretation collectively in view. The models we choose will in some cases be applicable in a direct sense to geophysical interpretation at a later time, but on balance they are selected mainly for a heuristic purpose. At this early stage, a thorough understanding of the induction process is the key to our further advancement with the theory. Virtually all discussion of practical applications will be deferred until Chap. 18.

The order in which we shall proceed is as follows: We shall begin by investigating the response of a horizontal-loop system to a simple closed circuit in order to set the scene for more advanced models. Then we shall study the response of a conducting permeable sphere to a uniform (but alternating) magnetic field. In spite of their evident simplicity, these two problems in

485

themselves reveal the essential nature of induction effects. After completing this introduction to the subject, we shall take up the study of more advanced models which have some interpretational value, restricting our attention mainly to dipolar sources. This limitation permits us, as a rule, to ignore the very small amount of induction which takes place in ordinary rocks, or at least to disregard its interaction with the induction fields surrounding conductive bodies. Thus we aver that the true response would be little changed if the conductive body were removed from its geological environment and were to be seated in empty space. Even with this simplification, few problems involving dipolar sources are analytically soluble, and we shall find our choice of models limited.

In fact, the models for which tractable mathematical solutions can be found for a dipolar source are easily enumerated. They are the sphere, the thin horizontal sheet, the infinitely conductive half-plane, and the horizontally stratified half-space. We can find a use for all these solutions in geophysical interpretation. We shall discuss the responses of these different models to both alternating and, where possible, transient magnetic fields. We shall also examine the effects of dissemination of the conducting material and of interactions between nearby conductors, in order to try to bring the geological overtones back into focus. And having provided this background, we shall then turn our attention, in Chap. 18, to the practical problems of interpretation.

17-1   Conducting Loop

The first problem to which we turn our attention is that of determining the response of a fixed-coil, inphase and quadrature prospecting system to a single, closed, conductive circuit. Figure 17-1 illustrates the problem under consideration. It is clearly very far removed from any recognizable geological situation, but we shall find that this simple model demonstrates many of the essential features of eddy-current induction. Moreover, we can derive the solution from simple circuit theory, without need for the ponderous machinery of electromagnetic field theory.

Let us suppose that an alternating current $I_0 e^{i\omega t}$ is made to flow in the transmitting coil. This current generates an alternating magnetic field in

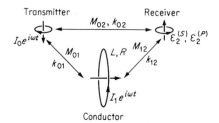

Fig. 17-1 **A circuit representation of an electromagnetic prospecting system.**

the surrounding environment, which in turn induces an emf both in the underground circuit and in the receiving coil. These emfs are governed by Faraday's law

$$\mathcal{E}_j = -M_{ij}\frac{dI_i}{dt}$$

where $\mathcal{E}_j$ is the emf induced in one circuit by a current I_i flowing in another, if M_{ij} is their mutual inductance. The emf induced in the receiver by the primary field is therefore

$$\mathcal{E}_2{}^{(P)} = -M_{02}\frac{d}{dt}I_0 e^{i\omega t} = -i\omega M_{02}I_0 e^{i\omega t}$$

and the emf induced in the underground circuit is

$$\mathcal{E}_1 = -i\omega M_{01}I_0 e^{i\omega t}$$

To the latter we must add $\mathcal{E}_1^\dagger$, the sum of the voltage drop across the resistance of the circuit and the back emf generated by the self-inductance, when a current $I_1 e^{i\omega t}$ flows around the loop. Thus

$$\mathcal{E}_1^\dagger = -RI_1 e^{i\omega t} - L\frac{d}{dt}I_1 e^{i\omega t} = -(R + i\omega L)I_1 e^{i\omega t}$$

To find the current I_1, we observe that around any closed circuit the total emf must vanish, i.e.,

$$\mathcal{E}_1 + \mathcal{E}_1^\dagger = 0$$

and therefore

$$I_1 e^{i\omega t} = -\frac{i\omega M_{01}}{R + i\omega L}I_0 e^{i\omega t}$$

$$= -\frac{M_{01}}{L}\left[\frac{i\omega L(R - i\omega L)}{R^2 + \omega^2 L^2}\right]I_0 e^{i\omega t}$$

This is the solution for the eddy current induced in the underground circuit. However, we are interested only in the secondary magnetic field which this current produces, and particularly in the emf which the field induces in the receiving coil. The latter will be given by

$$\mathcal{E}_2{}^{(S)} = -i\omega M_{12}I_1 e^{i\omega t}$$

where M_{12} is the mutual inductance between the underground circuit and the receiver. In most cases the apparatus measures this anomalous voltage by comparing it with the emf induced by the primary field in the absence of the circuit; in other words, it measures $\mathcal{E}_2{}^{(S)}/\mathcal{E}_2{}^{(P)}$. Thus the electromagnetic anomaly, or the *response* of the prospecting system to the buried loop, is given by

$$\frac{\mathcal{E}_2{}^{(S)}}{\mathcal{E}_2{}^{(P)}} = \frac{-i\omega M_{12} I_1 e^{i\omega t}}{-i\omega M_{02} I_0 e^{i\omega t}} = -\frac{M_{01} M_{12}}{M_{02} L} \left[\frac{i(\omega L/R)(1 - i\omega L/R)}{1 + (\omega L/R)^2} \right]$$

$$= -\frac{M_{01} M_{12}}{M_{02} L} \left(\frac{\alpha^2 + i\alpha}{1 + \alpha^2} \right) \quad (17\text{-}1)$$

where $\alpha = \omega L/R$.

To examine the nature of this expression, we note that $-M_{01} M_{12}/M_{02} L$ depends only on the relative sizes and positions of the three circuits. The remainder of (17-1) is a complex function of a dimensionless quantity $\alpha = \omega L/R$ which depends upon the frequency of the field as well as on the electrical properties of the buried loop. To the first quantity we give the name *coupling coefficient*, and we can demonstrate its significance a little more clearly by rewriting the mutual inductance coefficients in terms of the self-inductances and of the individual coupling coefficients of the three circuits. Thus if

$$M_{ij} = k_{ij} \sqrt{L_i L_j} \qquad |k| \leq 1$$

it follows that

$$-\frac{M_{01} M_{12}}{M_{02} L} = -\frac{k_{01} k_{12}}{k_{02}}.$$

We find accordingly that the *coupling coefficient* measures the amount of flux that couples the receiver to the transmitter through the underground circuit, in relation to that which couples the receiver to the transmitter directly. Its value changes with the position of the system but not with the electrical properties of the buried circuit nor with the frequency of the alternating current.

To the complex function $f(\alpha) = (\alpha^2 + i\alpha)/(1 + \alpha^2)$ we give the name *response function*, and to the dimensionless variable α, the name *response parameter*. This part of the response has to do strictly with the electrical properties of the underground loop and with the frequency of the alternating current. In fact, it must already be apparent that only the position and size of the loop will affect the coupling coefficient, whereas only the resistance and size will affect the response function. Figure 17-2 is a graph of the real and imaginary parts of $f(\alpha)$.

There are two limiting features of the response function which are immediately evident. The first is that for large values of the response parameter it tends to an upper (real) limit of one, i.e., $\lim_{\alpha \to \infty} [f(\alpha)] = 1$. The maximum response which the buried (ground) circuit can possibly produce is therefore

$$\left[\frac{\mathcal{E}_2{}^{(S)}}{\mathcal{E}_2{}^{(P)}} \right]_L = -\frac{M_{01} M_{12}}{M_{02} L} = -\frac{k_{01} k_{12}}{k_{02}} \quad {}^* \quad (17\text{-}2)$$

* The subscript L will be used to denote an expression valid at large values of the response parameter, when the self-inductance L is the controlling factor in determining the induced currents.

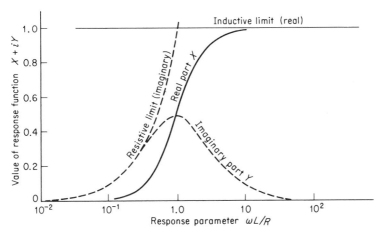

Fig. 17-2 **Response function** $\mathbf{f}(\alpha) = \mathbf{X}(\alpha) + \mathbf{iY}(\alpha)$ **of a loop in an alternating magnetic field.**

Since this is real, we learn that the response at the inductive limit is purely inphase and has no quadrature component.

At the other extreme, when α is very small, $f(\alpha) \to i\alpha$ and the response becomes

$$\left[\frac{\mathcal{E}_2^{(S)}}{\mathcal{E}_2^{(P)}}\right]_R = -\frac{i\omega L}{R}\frac{k_{01}k_{12}}{k_{02}}\text{*} \tag{17-3}$$

which is small when compared with the inductive limit, is in quadrature with the primary field, and is proportional to the frequency of the field. Between these two limits the response undergoes a smooth transition both in phase and amplitude. Starting with small values of the response parameter, the amplitude of response increases (linearly at first but less and less rapidly afterward) until the inductive limit is reached. Meanwhile the phase angle of the response function, which began at 90°, shifts toward zero, passing through 45° as the amplitude of the response equals $\sqrt{2}/2$ times the maximum value.

The changes in response are easy enough to understand. When $\omega L/R$ is small, the amount of current induced in the underground circuit will also be small, and the secondary magnetic field will be everywhere very much smaller than the primary field. Therefore each process of induction (receiver from transmitter, underground circuit from transmitter, and receiver from underground circuit) can be considered quite independently. Since each involves a phase shift of 90°, the secondary emf produced by two induction processes will be in quadrature with the primary emf produced by only one.

---

* The subscript R will be used to denote an expression valid at very low values of the response parameter, when the resistance of the underground circuit is the controlling factor in determining the response.

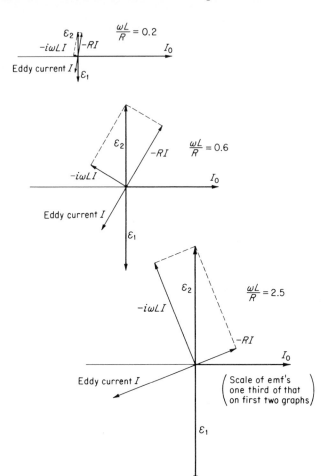

Fig. 17-3   Phasor graphs showing the evolution of the induced current I in a circuit as the frequency of the inducing field is increased.   The emf's in the circuit are $\mathcal{E}_1$ due to the transmitter current I_0, and $\mathcal{E}_2$ due to I.

As α becomes larger, the secondary magnetic field induces an emf in the underground circuit which begins to become appreciable in relation to that induced by the primary field.   The phase angle of the current in the buried loop, and therefore the phase angle of the secondary magnetic field, must shift in order that the net induced emf and the resistive loss should exactly balance.   At the inductive limit this balance virtually becomes an equality between the emfs induced by the primary and by the secondary magnetic fields in the buried conductor.   The induced current and the secondary magnetic field must therefore be in phase with, but in opposition to, the primary field.

The inductive limit for a circuit corresponds, in fact, to the case of perfect conductivity discussed in Sec. 16-3.   There we noted that the total magnetic field cannot possibly penetrate into a perfectly conducting medium. In our simple circuit the equivalent situation occurs when the total primary and secondary magnetic fluxes which cut the circuit become equal and opposite. The total flux therefore falls to zero, although the total field at any point does not necessarily vanish.

We may also examine the coupling coefficient in a little more detail and so justify qualitatively the shape of the profile of response of a horizontal loop system to a more or less vertical sheetlike conductor, which was shown in Fig. 15-7. In many cases the signs of k_{01}, k_{12}, and k_{02} can be determined simply by sketching a flux diagram. An example is shown in Fig. 17-4. Arbitrary positive directions are assigned to each element so that k_{02} is always positive. It also remains constant because the separation between

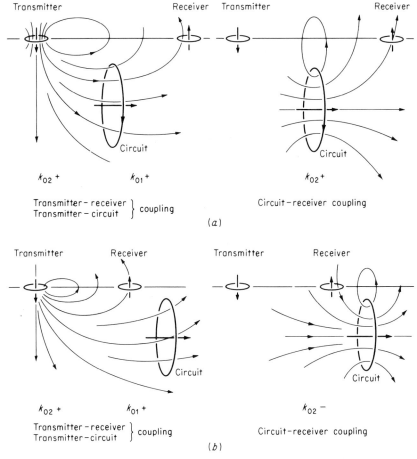

Fig. 17-4 Coupling of a horizontal-loop prospecting system with a simple circuit. (*a*) **Prospecting system astride the conductor;** (*b*) **prospecting system to one side of the conductor.**

the coils is not changed during a traverse. When receiver and transmitter straddle the buried circuit, both k_{01} and k_{12} will have the same sign. When one of the coils is directly over the edge of the circuit, the coupling coefficient appropriate to that coil will vanish, and when both are on the same side of it, the signs will be opposite. A profile of the response will therefore begin by being positive when both coils lie on one side of the body; then it will change sign as the leading coil crosses the edge of the conductor and will remain negative as long as the two coils continue to straddle the edge. It reverts to positive after the trailing coil has passed over the top of the conductor. It requires little insight to perceive that $|k_{01}k_{12}/k_{02}|$ will usually be greater when the coils straddle the conductor than when they both lie on one side of it. And, since the coupling coefficient is independent of which of the coils carries the current, the profile will be the same regardless of whether the leading coil receives or transmits. A profile over a vertical conductor should therefore be symmetrical.

17-2   Conducting Sphere in a Uniform Field

We shall next consider the response of a conducting, permeable sphere in a uniform, alternating magnetic field (10, 12, 16a). Except for some special applications related to natural-field methods, this model is, like the loop, useful in a strictly heuristic sense. Nonetheless, we can use it to demonstrate that the significant features of eddy-current induction revealed with the loop model hold also for a solid body. We can also investigate the effect of the magnetic permeability on the response.

First we choose a set of spherical polar coordinates (r,ϑ,φ) with the polar axis in the direction of the primary field $\mathbf{H}_0$. In terms of the base vectors $\mathbf{i}_r$, $\mathbf{i}_\vartheta$, $\mathbf{i}_\varphi$, we can write

$$\mathbf{H}_0 = H_0 \cos \vartheta \mathbf{i}_r - H_0 \sin \vartheta \mathbf{i}_\vartheta = \nabla \times \tfrac{1}{2}H_0 r \sin \vartheta \mathbf{i}_\varphi$$

A uniform, alternating primary field can therefore be obtained from a vector

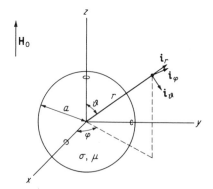

Fig. 17-5 Conducting sphere in a uniform magnetic field.

potential

$$\mathbf{A}_0 = \tfrac{1}{2}\mu_0 H_0 r \sin \vartheta e^{i\omega t}\mathbf{i}_\varphi \qquad (17\text{-}4)$$

and if the source is actually a dipole situated at some considerable distance from the sphere, the primary field vector $\mathbf{H}_0$ can be calculated from (16-40).

From the axial symmetry of the problem and by the use of a little physical intuition, we can see that $\mathbf{E}$, $\mathbf{J}$, and therefore $\mathbf{A}$ will have only φ components and that $\mathbf{H}$ and $\mathbf{B}$ can therefore have only ϑ and r components.

In Sec. 16-4 we showed that $\mathbf{A}$ must satisfy the equations

$$\nabla^2 \mathbf{A} - \sigma\mu \frac{\partial \mathbf{A}}{\partial t} = 0 \qquad \nabla \cdot \mathbf{A} = 0 \qquad (17\text{-}5)$$

where $\nabla^2 \mathbf{A}$ indicates that the Laplacian operator ∇^2 operates separately on each of the *rectangular* components of $\mathbf{A}$. For our problem we may write

$$\mathbf{A} = A(r,\vartheta)\mathbf{i}_\varphi$$

which in rectangular components becomes

$$\mathbf{A} = -A \sin \varphi \mathbf{i}_x + A \cos \varphi \mathbf{i}_y$$

If we now operate upon the rectangular components with the operator ∇^2 and then recombine them, we find that

$$\nabla^2 \mathbf{A} = \left(\nabla^2 A - \frac{A}{r^2 \sin^2 \vartheta}\right) \mathbf{i}_\varphi$$

and (17-5) therefore becomes

$$\nabla^2 A - \frac{A}{r^2 \sin^2 \vartheta} = \sigma\mu \frac{\partial A}{\partial t} \qquad (17\text{-}6)$$

This is a linear scalar equation which we can solve by the usual method of separation of variables. We begin by supposing that there exists a solution of the type

$$A(r,\vartheta,t) = R(r)\Theta(\vartheta)e^{i\omega t}$$

Equation (17-6) then separates into a modified Bessel equation and an associated Legendre equation, viz.,

$$(1 - u^2)\frac{d^2\Theta}{du^2} - 2u\frac{d\Theta}{du} - \left[\frac{1}{(1 - u^2)} - n(n + 1)\right]\Theta = 0 \qquad (17\text{-}7)$$

and $$\frac{d^2R}{dr^2} + \frac{2}{r}\frac{dR}{dr} - \left[k^2 + \frac{n(n + 1)}{r^2}\right]R = 0 \qquad (17\text{-}8)$$

where $u = \cos \vartheta$, $k^2 = i\sigma\mu\omega$, and n is the separation constant. Equation (17-7) has a real solution $P_n{}^1(u)$, and (17-8) has two solutions $r^{-\frac{1}{2}}I_{n+\frac{1}{2}}(kr)$ and $r^{-\frac{1}{2}}I_{-n-\frac{1}{2}}(kr)$, provided that $k \neq 0$. In an insulating region where

$k = 0$, Eq. (17-8) reduces to

$$\frac{d}{dr}\left(r^2 \frac{dR}{dr}\right) - n(n+1)R = 0$$

which has the solution

$$R(r) = Cr^n + Dr^{-(n+1)} \tag{17-9}$$

We can now write the general solutions of (17-5) for the regions outside and inside the sphere. Outside, the field must be the sum of the primary field A_0 plus a complementary term which vanishes at infinity. Since

$$A_0 = \tfrac{1}{2}\mu_0 H_0 r \sin \vartheta e^{i\omega t} = \tfrac{1}{2}\mu_0 H_0 r P_1{}^1(u)e^{i\omega t}$$

we shall require a complementary term in which the separation constant $n = 1$. As a result, the expression for the external potential is

$$A_e = \tfrac{1}{2}\mu_0 H_0 r P_1{}^1(u)e^{i\omega t} + Dr^{-2}P_1{}^1(u)e^{i\omega t} \tag{17-10}$$

while inside the sphere, since there can be no singularities, the potential will be

$$A_i = F P_1{}^1(u)r^{-\frac{1}{2}}I_{3\!/\!2}(kr)e^{i\omega t} \tag{17-11}$$

Since the boundary conditions are homogeneous, no other terms are required.

The conditions to be applied at the boundary $r = a$ are given by (16-28) and (16-26), which in terms of the vector potential are

$$\frac{\partial A_e}{\partial u} = \frac{\partial A_i}{\partial u}$$

and

$$\frac{1}{\mu_0}\frac{\partial}{\partial r}(rA_e) = \frac{1}{\mu}\frac{\partial}{\partial r}(rA_i) \tag{17-12}$$

Substituting for A_e and A_i, performing the differentiations, and solving for D, we obtain

$$D = -\frac{a^3\mu_0 H_0}{2}\left[\frac{(\mu_0/2 - 2\mu)I_{3\!/\!2}(ka) + \mu_0 ka I'_{3\!/\!2}(ka)}{(\mu_0/2 + \mu)I_{3\!/\!2}(ka) + \mu_0 ka I'_{3\!/\!2}(ka)}\right]$$

If we now make use of the following relationship for the derivative of the Bessel function,

$$I'_n(x) = I_{n-1}(x) - \frac{n}{x}I_n(x) = I_{n+1}(x) + \frac{n}{x}I_n(x)$$

and introduce the expressions

$$I_{1\!/\!2}(ka) = \sqrt{\frac{2}{\pi ka}}\sinh(ka)$$

and

$$I_{-1\!/\!2}(ka) = \sqrt{\frac{2}{\pi ka}}\cosh(ka)$$

we obtain for the anomalous vector potential

$$\mathbf{A}^{(S)} = -\mathbf{i}_\varphi \frac{\mu_0 \sin \vartheta}{4\pi r^2} 2\pi a^3 H_0 e^{i\omega t}$$

$$\times \left\{ \frac{[\mu_0(1 + k^2 a^2) + 2\mu] \sinh (ka) - (2\mu + \mu_0)ka \cosh (ka)}{[\mu_0(1 + k^2 a^2) - \mu] \sinh (ka) + (\mu - \mu_0)ka \cosh (ka)} \right\} \quad (17\text{-}13)$$

This we recognize from (16-39) and (16-42) to be the field of a magnetic dipole oriented in the direction of the polar axis and having a moment

$$m = -2\pi a^3 H_0 e^{i\omega t}(X + iY) \quad (17\text{-}14)$$

where $X + iY$ is the complex quantity in brackets in (17-13). The anomalous field intensity at the receiver can of course be calculated by inserting this m into (16-40). The behavior of the complex function $X + iY$ is shown in the graphs of Fig. 17-6, where we have chosen as the independent

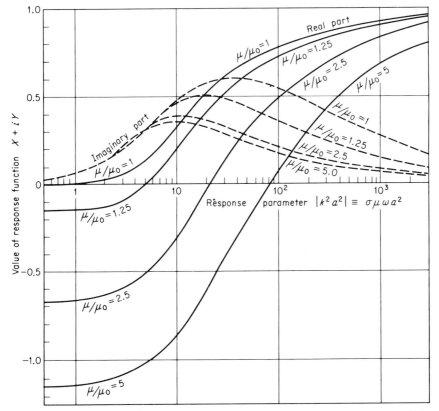

Fig. 17-6 **Response function of a sphere in a uniform alternating field.** [After Wait (16*b*).]

variable the quantity $|k^2a^2| = \sigma\mu\omega a^2$. When $\mu = \mu_0$, the function simplifies to

$$X + iY = 1 - \frac{3}{ka}\coth(ka) + \frac{3}{k^2a^2}$$

From the graphs we see that $X + iY$ has a real limit of 1 for large values of ka. Moreover, it is the only term in (17-13) which involves either the frequency of the alternating field or the conductivity of the sphere. It is, therefore, the response function of a sphere in a uniform field and is in all respects analogous to the response function of a loop obtained in the previous section. The response parameter in this case is $\alpha = |k^2a^2| = \sigma\mu\omega a^2$. When $\mu = \mu_0$, the response function is very similar to that of the simple loop, the only difference being that the transition from the resistive limit to the inductive limit of response takes place more slowly in the sphere. In all likelihood, this is because the induced current system has greater freedom to change its distribution in a massive conductor than in a circuit.

The effect of a magnetic permeability which is greater than μ_0 is twofold. First, it increases the value of the response parameter so that the transition from quadrature to inphase response takes place at slightly lower values of the parameters σ, ω, and a. Secondly, at low values of the response parameter an additional moment is introduced which is inphase and in the same direction as the primary field. This is just the induced magnetization, which occurs exactly as it does in a static field. We note, however, that when the response parameter is large, the conductive response completely overwhelms this moment.

It is worth our while to examine the limiting magnetic moment of the sphere at very high and at very low frequencies. By inserting the first few terms in the series expansions for $\sinh(ka)$ and $\cosh(ka)$ into (17-14), we find that as $|k^2a^2| \to 0$

$$m_R = \left[\underbrace{4\pi a^3\left(\frac{\mu - \mu_0}{\mu + 2\mu_0}\right)}_{\text{magnetization}} - \underbrace{2\pi a^3\left(\frac{i\sigma\mu_0\omega a^2}{15}\right)\left(\frac{3\mu_0}{\mu + 2\mu_0}\right)}_{\text{initial induction}}\right]H_0e^{i\omega t} \quad (17\text{-}15)$$

$$= -2\pi a^3\left(\frac{i\sigma\mu_0\omega a^2}{15}\right)H_0e^{i\omega t} \qquad \text{when } \mu = \mu_0$$

On the other hand, as $|k^2a^2| \to \infty$

$$m_L = -2\pi a^3 H_0 e^{i\omega t} \qquad (17\text{-}16)$$

These two limiting expressions could have been obtained without solving the boundary-value problem. Since the boundary-value method is not always tractable for more complex models than the sphere, it will be instructive to examine alternative procedures for deriving limits.

Starting with m_R, we shall assume for the moment that $\mu = \mu_0$. At the resistive limit only the primary magnetic field $H_0 e^{i\omega t}$ is of any consequence in the induction process.   If we divide the sphere into annular rings, each having a radius $r \sin \vartheta$ and a cross-sectional area $r \, d\vartheta \, dr$, we can calculate the emf ε induced in each ring, viz.,

$$\varepsilon = -i\omega\mu_0\pi r^2 \sin^2 \vartheta H_0 e^{i\omega t}$$

Equating this to the resistive potential drop in the ring and noting that the resistance around the circuit is $2\pi \sin \vartheta / \sigma \, d\vartheta \, dr$, we obtain for the current flow

$$d^2 I = -\frac{i\sigma\mu_0\omega r^2 \sin \vartheta \, dr \, d\vartheta \, H_0 e^{i\omega t}}{2}$$

The contribution to the dipole moment from this ring will be

$$d^2 m_R = \pi(r \sin \vartheta)^2 \, d^2 I = -\frac{i\sigma\mu_0\omega\pi r^4 \sin^3 \vartheta \, dr \, d\vartheta \, H_0 e^{i\omega t}}{2}$$

and if we then integrate over the entire cross section of the sphere, we find that

$$m_R = 2 \int_0^a dr \int_0^\pi d\vartheta \left(-\frac{i\sigma\mu_0\omega\pi r^4 \sin^3 \vartheta}{2} \right) H_0 e^{i\omega t}$$

$$= -2\pi a^3 \left(\frac{i\sigma\mu_0\omega a^2}{15} \right) H_0 e^{i\omega t}$$

which is identical with (17-15) when $\mu = \mu_0$.

When $\mu \neq \mu_0$, we have only to replace $\mu_0 H_0 e^{i\omega t}$ with the actual flux density inside the sphere and to add a term to account for the induced magnetization.   The latter has been discussed in the case of static fields (in cgs units) in Chap. 11.   The field $\mathbf{H}_i$ inside a permeable sphere placed in a uniform external field $\mathbf{H}$ is (in mks units)

$$\mathbf{H}_i = \mathbf{H} - \tfrac{1}{3}\mathbf{M} \tag{17-17}$$

where $\mathbf{M}$ is the magnetization given by

$$\mathbf{M} = \kappa\mathbf{H}_i = \left(\frac{\mu}{\mu_0} - 1 \right)\mathbf{H}_i \tag{17-18}$$

κ being the mks magnetic susceptibility.   Now the flux density inside the sphere is

$$\mathbf{B}_i = \mu\mathbf{H}_i = \left(\frac{3\mu_0}{\mu + 2\mu_0} \right)\mu H_0 e^{i\omega t} \qquad \text{since} \qquad \mathbf{H} = \mathbf{H}_0 e^{i\omega t}$$

and the additional dipole moment due to the magnetization is

$$m = \tfrac{4}{3}\pi a^3 M = \tfrac{4}{3}\pi a^3 \frac{(\mu_0 - \mu)}{(\mu + 2\mu_0)} H_0 e^{i\omega t}$$

This gives for the total dipole moment

$$m_R = -2\pi a^3 \left(\frac{i\mu_0\omega a^2}{15}\right)\left(\frac{3\mu_0}{\mu + 2\mu_0}\right) H_0 e^{i\omega t} + 4\pi a^3 \left(\frac{\mu - \mu_0}{\mu + 2\mu_0}\right) H_0 e^{i\omega t}$$

which is identical with (17-15).

Similarly, the inductive, or high-frequency, limit m_L is obtained by finding the *effective* uniform magnetization $\mathbf{M}'$, which would make the flux density vanish everywhere inside the sphere. Thus

$$0 = \mathbf{B}_i = \mu_0(\mathbf{H}_i + \mathbf{M}') = \mu_0\left(\mathbf{H}_0 e^{i\omega t} + \frac{2}{3}\mathbf{M}'\right) \tag{17-19}$$

hence

$$\mathbf{M}' = -\frac{3}{2}\mathbf{H}_0 e^{i\omega t}$$

and consequently

$$m_L = -\left(\frac{4}{3}\pi a^3\right)\frac{3}{2}H_0 e^{i\omega t} = -2\pi a^3 H_0 e^{i\omega t}$$

just as we obtained in (17-16).

17-3 Conducting Sheet in a Dipole Field

Another model for which we can obtain a reasonably simple solution and one which we can later use in interpretation is that of an alternating dipole source in the vicinity of a thin, infinitely extensive conducting sheet. This model can be applied quantitatively to the study of the response of an airborne system to a layer of conducting overburden.

To proceed with the analysis of this problem, we must develop a special boundary equation which holds for thin conductive sheets. Such a sheet is one whose thickness s is very small yet whose conductivity σ is so large that the product σs remains finite even as s vanishes. The sheet may be infinite in extent or bounded, as desired. For convenience we shall consider it to be flat and lying in the plane $z = 0$ of a cartesian coordinate system. Since the conductivity may be infinite, the current density $\mathbf{J}$ may also be infinite inside the sheet. We therefore define a surface current $\mathbf{K} = s\mathbf{J} = \sigma s\mathbf{E}$, which will remain finite. Obviously $\mathbf{K}$ will have only x and y components.

A time-varying magnetic field which penetrates the sheet will generally suffer a discontinuity; and if we divide the total field $\mathbf{H}$ into a secondary component $\mathbf{H}^{(S)}$, generated by the surface currents $\mathbf{K}$, and a primary component $\mathbf{H}^{(P)}$, generated by currents external to the sheet, symmetry tells us that $\mathbf{H}^{(P)}$ and the normal component of $\mathbf{H}^{(S)}$ will be continuous through the sheet and that the tangential components of $\mathbf{H}^{(S)}$ will change in sign (Fig. 17-7). Written analytically, these conditions are

$$\mathbf{H}_+^{(P)} - \mathbf{H}_-^{(P)} = 0 \qquad \mathbf{n} \cdot [\mathbf{H}_+^{(S)} - \mathbf{H}_-^{(S)}] = 0 \qquad \mathbf{n} \times [\mathbf{H}_+^{(S)} + \mathbf{H}_-^{(S)}] = 0$$

$$\tag{17-20}$$

Maxwell's equations (16-1) and (16-2), modified for the quasi-stationary state, can now be used to find the necessary relationships between $\mathbf{E}$ and $\mathbf{H}$ and between $\mathbf{K}$ and $\mathbf{H}^{(S)}$ inside the sheet; then by employing Ohm's law $\mathbf{K} = \sigma s \mathbf{E}$, a relationship between $\mathbf{H}$ and $\mathbf{H}^{(S)}$ can be derived (7), (12). This turns out to be

$$\left[\frac{\partial H_x^{(S)}}{\partial x} + \frac{\partial H_y^{(S)}}{\partial y} \right]_{z=0^+} = - \left[\frac{\partial H_x^{(S)}}{\partial x} + \frac{\partial H_y^{(S)}}{\partial y} \right]_{z=0^-} = - \frac{\sigma s \mu_0}{2} \frac{\partial H_z}{\partial t}$$

Then, by making use of the fact that $\nabla \cdot \mathbf{H} = 0$ everywhere outside of the sheet, we can reduce these equations to

$$\left[\frac{\partial H_z^{(S)}}{\partial z} \right]_{z=0^+} = - \left[\frac{\partial H_z^{(S)}}{\partial z} \right]_{z=0^-} = \frac{\sigma s \mu_0}{2} \frac{\partial H_z}{\partial t} \qquad (17\text{-}21)$$

This boundary equation can be used in the following way: In the normal routine for solving boundary-value problems, the solutions of the equation $\nabla^2 \mathbf{A} = 0$ or of $\nabla^2 \mathbf{A} - \sigma \mu\, \partial \mathbf{A}/\partial t = 0$ within each homogeneous region of space have finally to be determined by applying the boundary conditions (16-26) and (16-28). There is no need, however, to find a solution within the interior of a *thin* sheet conductor. All that is necessary is to find a solution of the equation $\nabla^2 \mathbf{A} = 0$ which is valid *outside* of the sheet and which satisfies the special boundary condition (17-21), the continuity conditions (17-20), and whatever particular conditions are required at the source and at infinity.

We can solve very simply by this method for the field of a source which varies *stepwise* in time near an infinite, flat conducting sheet. Let us suppose that the primary field has the following time dependence:

$$\mathbf{H}^{(P)}(\mathbf{r},t) = 0 \qquad t < 0 \qquad \mathbf{H}^{(P)}(\mathbf{r},t) = \mathbf{H}^{(P)}(\mathbf{r}) \qquad t \geq 0$$

Since the rate of change in the field is infinitely great at the moment when

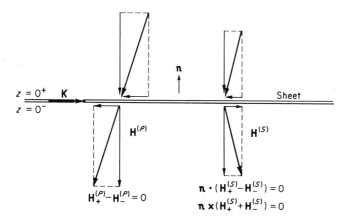

Fig. 17-7 Continuity conditions through a thin sheet conductor.

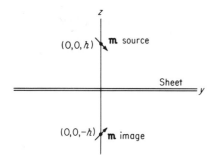

Fig. 17-8   Image solution for the field produced by an external dipole source above a perfectly conducting sheet.

the step occurs, it is physically obvious that no component of **H** can possibly penetrate the sheet during this instant. [This can be more rigorously derived from (17-21).]   We may therefore make an appeal to simple image theory to give us the total field when $t = 0$.   If the source is a magnetic dipole of moment **m** situated at $(0,0,h)$, then when $t = 0$, the primary field is

$$\mathbf{H}^{(P)} = -\frac{1}{4\pi} \left(\boldsymbol{\nabla}\{\mathbf{m} \cdot \boldsymbol{\nabla}_0[(x - x_0)^2 + (y - y_0)^2 + (z - z_0)^2]^{-\frac{1}{2}}\}\right)_{x_0=y_0=0,z=h}$$

$$(17\text{-}22)$$

The secondary field when $t = 0$, measured on the same side of the sheet as the source, will be the field of an image dipole situated at the mirror point below the sheet (Fig. 17-8), i.e.,

$$\mathbf{H}^{(S)} = -\frac{1}{4\pi} \left(\boldsymbol{\nabla}\{\overline{\mathbf{m}} \cdot \boldsymbol{\nabla}_0[(x - x_0)^2 + (y - y_0)^2 + (z - z_0)^2]^{-\frac{1}{2}}\}\right)_{x_0=y_0=0,z_0=-h}$$

where $\overline{\mathbf{m}} = m_x\mathbf{i}_x + m_y\mathbf{i}_y - m_z\mathbf{i}_z$

and thus

$$\mathbf{H}^{(S)} = -\frac{1}{4\pi} \left(\boldsymbol{\nabla}\{\mathbf{m} \cdot \boldsymbol{\nabla}_0[(x - x_0)^2 + (y - y_0)^2 + (z + z_0)^2]^{-\frac{1}{2}}\}\right)_{x_0=y_0=0,z_0=h}$$

$$(17\text{-}23)$$

$$t = 0 \qquad z > 0$$

When $t > 0$ on the other hand, the current system established in the sheet during the instant of the switching-on of the primary magnetic field will begin to decay, and along with it will also decay the secondary field $\mathbf{H}^{(S)}$. The solution for $\mathbf{H}^{(S)}(t)$ is simple, however, for if $\mathbf{H}^{(S)}(x,y,z)$ is a solution of the boundary condition (17-21) at $t = 0$, then $\mathbf{H}^{(S)}\left(x,y,z \pm \frac{2t}{\sigma\mu_0 s}\right)$ is a solution at time t.   The plus sign must be used when z is positive and the minus sign when z is negative.   Thus the solution for the secondary field due to a dipole source varying stepwise in time is

$$\mathbf{H}^{(S)} = -\frac{1}{4\pi}\left(\boldsymbol{\nabla}\left\{\mathbf{m} \cdot \boldsymbol{\nabla}_0\left[(x - x_0)^2 + (y - y_0)^2 \right.\right.\right.$$
$$\left.\left.\left. + \left(z + z_0 + \frac{2t}{\sigma\mu_0 s}\right)^2\right]^{-\frac{1}{2}}\right\}\right)_{x_0=y_0=0,z_0=h} \quad (17\text{-}24)$$

To find the response due to an alternating magnetic dipole we must take the Fourier transform of (17-24). From Fourier's integral theorem we know that any transient can be decomposed into a spectrum of harmonic components and that the spectral amplitude of $\mathbf{H}^{(S)}(t)$ relative to that of a unit step is given by

$$\check{\mathbf{H}}^{(S)}(\omega) = i\omega \int_0^\infty e^{-i\omega t}\mathbf{H}^{(S)}(t)\,dt \qquad (17\text{-}25)$$

The response due to a primary magnetic field $\mathbf{H}^{(P)}e^{i\omega t}$ is therefore just $\check{\mathbf{H}}^{(S)}(\omega)e^{i\omega t}$. This result is derived in a little more detail in Sec. 17-10.

To work out an example, we shall take the case of a *double-dipole* electromagnetic system. The dipole moment of the transmitter and the direction of the receiver axis are taken to be in the Ox direction at $(0,0,h)$. The solution for a step source is then

$$
\begin{aligned}
H_x^{(S)}(t) &= -\frac{m}{4\pi}\left\{\frac{\partial^2}{\partial x\,\partial x_0}\left[(x-x_0)^2 + \left(2h + \frac{2t}{\sigma\mu_0 s}\right)^2\right]^{-\frac{1}{2}}\right\}_{x_0=0} \\
&= -\frac{m}{4\pi}\left(2h + \frac{2t}{\sigma\mu_0\omega}\right)^{-3} \qquad t \geq 0 \qquad (17\text{-}26)
\end{aligned}
$$

For an alternating source this becomes

$$
\begin{aligned}
\check{H}_x^{(S)}(\omega) &= \frac{m}{4\pi}\, i\omega \int_0^\infty e^{-i\omega t}\left(2h + \frac{2t}{\sigma\mu_0 s}\right)^{-3}dt \\
&= \frac{m}{4\pi}\frac{1}{(2h)^3}\,(i\alpha)^3 e^{-i\alpha}\int_\alpha^\infty \frac{e^{-i\beta}}{\beta^3}\,d\beta
\end{aligned}
$$

where α is an abbreviation for the quantity $\sigma\mu_0\omega sh$. By separating the integrand into real and imaginary components and integrating twice by parts, this expression can be written in terms of the tabulated sine and cosine integrals. The final result is

$$\check{H}_x^{(S)} = -\frac{m}{4\pi(2h)^3}[X(\alpha) + iY(\alpha)]e^{i\omega t} \qquad (17\text{-}27)$$

where $X(\alpha) = \frac{1}{2}(\alpha^2 - \alpha^3\{\cos\alpha[\pi/2 - Si(\alpha)] + \sin\alpha Ci(\alpha)\})$

$Y(\alpha) = \frac{1}{2}(\alpha - \alpha^3\{\sin\alpha[\pi/2 - Si(\alpha)] - \cos\alpha Ci(\alpha)\})$

A graph of the response function $X + iY$ is given in Fig. 17-9, and we recognize the same general form as we have found previously. When we examine (17-27), we find that the inductive limit of $H_x^{(S)}$ is just the field of the image of the transmitter in the sheet, which obtains from (17-26) when $t = 0$. This is to be expected since the boundary condition on $\mathbf{H}(t)$ when $t = 0$ is exactly the same as that for $\check{\mathbf{H}}(\omega)$ when $\alpha \to \infty$; viz., no part of the field may penetrate the sheet. Moreover, we note that at the resistive limit the response is proportional to α.

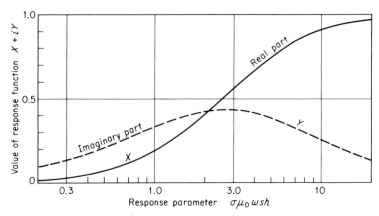

Fig. 17-9 **Response function for a double-dipole prospecting system in the presence of a thin sheet conductor.**

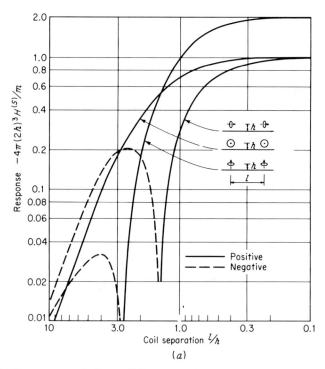

Fig. 17-10 **Responses of three different coil systems to a highly conductive fraction of the primary field strength.**

An interesting feature of this solution is that the response parameter α contains h, a parameter of the prospecting system rather than of the body. This is because the conductor is so greatly extended that the dimensions of the current system are determined by the proximity of the source to the sheet rather than by a dimension of the sheet itself.

To find the sensitivity of some electromagnetic prospecting systems to an extended uniform sheet, it is instructive to calculate the inductive limit of response for a variety of coil configurations. The results are shown in Fig. 17-10a. The calculations are simple, since they involve only the formula for the field of a dipole. For the three systems presented we find a variety of responses. Two of them exhibit a null and a change of sign. All the responses, however, become proportional to h^{-3} for small values of l/h. For practical purposes this limit is reached at about $l/h = \frac{1}{3}$ or $\frac{1}{4}$, and for smaller values the systems could be considered as being "double dipoles." If the practical range of l/h becomes greater than unity, it is more revealing to calculate $H^{(S)}/H^{(P)}$ than $H^{(S)}/m$. This is done in Fig. 17-10b. We see that in all three cases the limiting response for small values of h/l is unity.

To find the responses of these same coil configurations to a sheet of finite surface conductivity, it would be necessary to revise Eq. (17-26). The

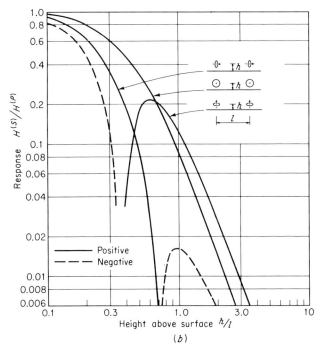

(b)

region with a flat extended surface. (a) **In terms of dipole moment;** (b) **as a**

response functions will, in general, be more complicated than the one shown in Fig. 17-9, especially for the two systems which exhibit a null in their response at the inductive limit.

17-4   Uniformly Conducting Ground in a Dipole Field

Another model, similar in its application to the thin sheet, is a uniformly conducting half-space.   To simplify matters, we shall take as the source a simple magnetic pole of strength p.   This is in all ways analogous to a point electric pole, except of course that it cannot have a real, separate physical existence.   From the response of the model to a polar field, however, we can find the response to a dipolar field simply by differentiation with respect to the coordinates of the source.   Since this can be done as a final step, the magnetic pole is an invention of convenience which saves a great deal of cumbersome algebra.

Because the problem now possesses cylindrical symmetry, it is convenient to work in cylindrical coordinates ρ, φ, z.   If the magnetic source lies on the axis at $\rho = 0$, $z = z_0$, we shall have

$$\mathbf{H}^{(P)} = -\frac{p}{4\pi} \nabla(\rho^2 + z_1{}^2)^{-\frac{1}{2}} e^{i\omega t} \qquad z_1 = z - z_0$$

By again introducing the Lipschitz integral, which we have already used in Sec. 14-5, this can be written

$$\mathbf{H}^{(P)} = -\frac{p e^{i\omega t}}{4\pi} \nabla \int_0^\infty J_0(\lambda\rho) e^{-\lambda|z_1|} \, d\lambda$$

i.e.,

$$H_\rho{}^{(P)} = \frac{p e^{i\omega t}}{4\pi} \int_0^\infty J_1(\lambda\rho)\lambda e^{\lambda z_1} \, d\lambda \left.\begin{array}{l} \\ \\ \end{array}\right\} z_1 \le 0 \qquad (17\text{-}28)$$

and

$$H_z{}^{(P)} = -\frac{p e^{i\omega t}}{4\pi} \int_0^\infty J_0(\lambda\rho)\lambda e^{\lambda z_1} \, d\lambda$$

Now let us place the surface of the ground at $z = 0$.   Above this surface the conductivity vanishes, while below it is a constant σ.   The problem is now to find a solution of $\nabla^2 \mathbf{A} = 0$ in $z > 0$ which at $z = 0$ is matched to a cor-

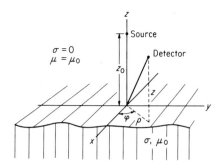

Fig. 17-11   Coordinates for a conducting half-space.

responding solution of $\nabla^2 \mathbf{A} - i\sigma\mu\omega\mathbf{A} = 0$ in $z < 0$ through the boundary conditions.

To determine the magnetic field in $z < 0$, we follow the same procedure that we used with the conducting sphere (Sec. 17-2). From axial symmetry we realize that the vector potential $\mathbf{A}$ will have only a φ component $A(\rho, z)$. (Note that this will also satisfy automatically the condition $\nabla \cdot \mathbf{A} = 0$.) To be a solution of Maxwell's equations, this must satisfy

$$\nabla^2 A - \frac{A}{\rho^2} = \sigma\mu \frac{\partial A}{\partial t}$$

which we obtain by writing ρ^2 for $r^2 \sin^2 \vartheta$ in (17-6). By assuming the existence of a solution in the form $A = R(\rho)Z(z)e^{i\omega t}$, we can separate this equation into two parts, viz.,

$$\frac{d^2 R}{d\rho^2} + \frac{1}{\rho} \frac{dR}{d\rho} + \left(\lambda^2 - \frac{1}{\rho^2} \right) R = 0$$

and

$$\frac{d^2 Z}{dz^2} - (\lambda^2 + k^2)Z = 0 \qquad \lambda \geq 0$$

where $k^2 = +i\sigma\mu\omega$.

Solutions of these two equations are $R(\rho) = J_1(\lambda\rho)$, $Y_1(\lambda\rho)$, and $Z(z) = \exp \pm (\lambda^2 + k^2)^{1/2}z$. Since a singularity at $\rho = 0$ is not permissible, $Y_1(\lambda\rho)$ must be discarded, and since A must vanish as $z \to -\infty$, the positive sign will be used in the exponential term. Thus the magnetic potential in $z < 0$ is, by superposition,

$$A = \int_0^\infty f(\lambda)J_1(\lambda\rho)e^{\gamma z}\, d\lambda \qquad \text{where} \qquad \gamma^2 = \lambda^2 + k^2$$

The field to which this corresponds is $\mathbf{H} = \dfrac{1}{\mu} \nabla \times A \mathbf{i}_\varphi$, and it has the following components:

$$\left. \begin{aligned} H_\rho &= -\frac{e^{i\omega t}}{\mu} \int_0^\infty \gamma f(\lambda)J_1(\lambda\rho)e^{\gamma z}\, d\lambda \\[2mm] H_z &= \frac{e^{i\omega t}}{\mu} \int_0^\infty \lambda f(\lambda)J_0(\lambda\rho)e^{\gamma z}\, d\lambda \end{aligned} \right\} z \leq 0 \qquad (17\text{-}29)$$

and

In $z > 0$ the same arguments hold except that k is zero and the secondary field must vanish as $z \to +\infty$. For the sum of the primary and secondary fields we therefore write the expressions (absorbing λ and $4\pi/p$ into the unknown part of the integrand)

$$\left. \begin{aligned} H_\rho &= \frac{pe^{i\omega t}}{4\pi} \int_0^\infty [e^{\lambda z_1} - g(\lambda)e^{-\lambda z}]\lambda J_1(\lambda\rho)\, d\lambda \\[2mm] H_z &= -\frac{pe^{i\omega t}}{4\pi} \int_0^\infty [e^{\lambda z_1} + g(\lambda)e^{-\lambda z}]\lambda J_0(\lambda\rho)\, d\lambda \end{aligned} \right\} 0 \leq z \leq z_0 \qquad (17\text{-}30)$$

The boundary conditions (16-26) and (16-28) tell us that

$$\frac{p\lambda}{4\pi}\left[e^{-\lambda z_0} - g(\lambda)\right] = -\frac{\gamma}{\mu}\,f(\lambda)$$

and

$$-\mu_0\left(\frac{p\lambda}{4\pi}\right)\left[e^{-\lambda z_0} + g(\lambda)\right] = \lambda f(\lambda)$$

Therefore, eliminating $f(\lambda)$, we obtain

$$g(\lambda) = -\left(\frac{\gamma - \lambda}{\gamma + \lambda}\right)e^{-\lambda z_0}$$

Consequently, we arrive at the solution for the anomalous magnetic field in $0 < z \le z_0$

$$H_\rho{}^{(S)} = \frac{pe^{i\omega t}}{4\pi}\int_0^\infty \left(\frac{\gamma - \lambda}{\gamma + \lambda}\right)\lambda J_1(\lambda\rho)e^{-\lambda(z_0 + z)}\,d\lambda$$

and

$$H_z{}^{(S)} = \frac{pe^{i\omega t}}{4\pi}\int_0^\infty \left(\frac{\gamma - \lambda}{\gamma + \lambda}\right)\lambda J_0(\lambda\rho)e^{-\lambda(z_0 + z)}\,d\lambda$$

(17-31)

If the source is a magnetic dipole, these expressions must be differentiated at $(0,0,z_0)$ in the direction of the dipole moment, i.e.,

$$H_{dipole} = dl\,\frac{\partial H_{pole}}{\partial l} = \frac{m}{p}\,\frac{\partial H_{pole}}{\partial l}$$

when the dipole moment is $m = p\,dl$. When the dipole moment is in the Oz direction, we obtain immediately

$$H_\rho{}^{(S)} = -\frac{m}{4\pi}\,e^{i\omega t}\int_0^\infty \left(\frac{\gamma - \lambda}{\gamma + \lambda}\right)\lambda^2 J_1(\lambda\rho)e^{-\lambda(z_0 + z)}\,d\lambda$$

$$H_\varphi{}^{(S)} = 0$$

(17-32)

and

$$H_z{}^{(S)} = -\frac{m}{4\pi}\,e^{i\omega t}\int_0^\infty \left(\frac{\gamma - \lambda}{\gamma + \lambda}\right)\lambda^2 J_0(\lambda\rho)e^{-\lambda(z_0 + z)}\,d\lambda$$

For a horizontal dipole we must consider two possibilities, viz., a moment directed either toward the receiver position or else at right angles to this line. To find the components of the field in either case, we must first resolve $H_\rho{}^{(S)}$ in (17-31) into its x and y components

$$H_x{}^{(S)} = \frac{pe^{i\omega t}}{4\pi}\,\frac{x}{\sqrt{x^2 + y^2}}\int_0^\infty \left(\frac{\gamma - \lambda}{\gamma + \lambda}\right)J_1(\lambda\sqrt{x^2 + y^2})e^{-\lambda(z + z_0)}\,d\lambda$$

and

$$H_y{}^{(S)} = \frac{pe^{i\omega t}}{4\pi}\,\frac{y}{\sqrt{x^2 + y^2}}\int_0^\infty \left(\frac{\gamma - \lambda}{\gamma + \lambda}\right)J_1(\lambda\sqrt{x^2 + y^2})e^{-\lambda(z + z_0)}\,d\lambda$$

The first case can then be worked out by applying to each of these formulas the operation $\dfrac{m}{p}\left(\dfrac{\partial}{\partial x_0}\right)_{y=0}$. The second, which represents a pair of vertical, coplanar coils, may be solved by applying the operation $\dfrac{m}{p}\left(\dfrac{\partial}{\partial y_0}\right)_{y=0}$. In either case it is assumed that the receiver lies somewhere in the plane $y = 0$. Since x_0 and y_0 have not been retained explicitly in the equations, however, it is expedient to replace the operations in x_0 and y_0 with the equivalent operations in x and y, i.e.,

$$\left(\frac{\partial}{\partial x_0}\right)_{y=0} = -\left(\frac{\partial}{\partial x}\right)_{y=0}$$

$$\left(\frac{\partial}{\partial y_0}\right)_{y=0} = -\left(\frac{\partial}{\partial y}\right)_{y=0}$$

Thus for the *in-line* dipole ($\mathbf{m} = m\mathbf{i}_x$) we find that

$$(H_x{}^{(S)})_{y=0} = -\frac{m}{4\pi}e^{i\omega t}\int_0^\infty \left(\frac{\gamma - \lambda}{\gamma + \lambda}\right)\left[\lambda^2 J_0(\lambda x) - \frac{\lambda}{x}J_1(\lambda x)\right]e^{-\lambda(z+z_0)}\,d\lambda$$

$$(H_y{}^{(S)})_{y=0} = 0 \tag{17-33}$$

and $\quad (H_z{}^{(S)})_{y=0} = \dfrac{m}{4\pi}e^{i\omega t}\displaystyle\int_0^\infty \left(\dfrac{\gamma - \lambda}{\gamma + \lambda}\right)\lambda^2 J_1(\lambda x)e^{-\lambda(z+z_0)}\,d\lambda$

while for the *transverse* dipole ($\mathbf{m} = m\mathbf{i}_y$)

$$(H_x{}^{(S)})_{y=0} = 0 = (H_z{}^{(S)})_{y=0}$$

and $\quad (H_y{}^{(S)})_{y=0} = -\dfrac{m}{4\pi}e^{i\omega t}\displaystyle\int_0^\infty \left(\dfrac{\gamma - \lambda}{\gamma + \lambda}\right)\dfrac{\lambda}{x}J_1(\lambda x)e^{-\lambda(z+z_0)}\,d\lambda \tag{17-34}$

Having obtained formal solutions to the problem, we must now turn our attention to the evaluation of the integrals. In the special case in which both source and receiver lie on the ground surface ($z = z_0 = 0$) it is possible to obtain fairly simple closed expressions for them. The procedure to follow is to rewrite

$$\frac{\gamma - \lambda}{\gamma + \lambda} = \frac{(\gamma - \lambda)^2}{k^2} = 1 + 2\frac{\lambda^2}{k^2} - 2\frac{\gamma\lambda}{k^2} \qquad (\gamma^2 \equiv k^2 + \lambda^2)$$

Each solution then takes the form of a sum of several integrals of the type

$$\int_0^\infty \lambda^m \gamma^n \frac{J_0(\lambda\rho)}{J_1(\lambda\rho)}\,d\lambda \qquad \begin{array}{l} m = 1, 2, 3, 4 \\ n = 1, 0 \end{array}$$

Integrals of this kind may be evaluated by setting $z = 0$ in the Lipschitz and Sommerfeld integral identities

$$\int_0^\infty J_0(\lambda\rho)e^{-\lambda z}\,d\lambda = \frac{1}{(\rho^2 + z^2)^{1/2}}$$

$$\int_0^\infty \frac{\lambda}{\gamma}J_0(\lambda\rho)e^{-\gamma z}\,d\lambda = \frac{\exp - k(\rho^2 + z^2)^{1/2}}{(\rho^2 + z^2)^{1/2}}$$

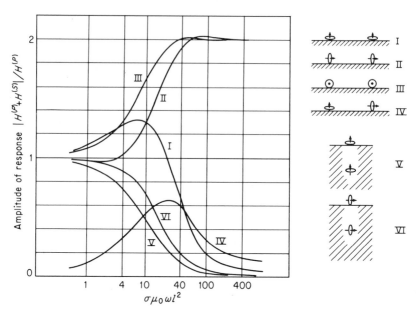

Fig. 17-12   Response of a variety of prospecting systems situated on the surface of a conducting half-space [After Wait (16c).]

and in similar formulas derived from them by differentiation with respect to ρ and z. The matter has been examined in considerable detail by Wait (16d,e). As an example we give his result for a pair of widely spaced loops lying on the ground surface

$$\frac{\mathcal{E}^{(S)} + \mathcal{E}^{(P)}}{\mathcal{E}^{(P)}} = \frac{2}{k^2\rho^2}\,[9 - (9 + 9k\rho + 4k^2\rho^2 + k^3\rho^3)e^{-k\rho}] \qquad k^2 \equiv i\sigma\mu_0\omega$$

and in Fig. 17-12 show his numerical data for this and for other coil systems.

In the general case $(z, z_0 > 0)$, however, the integrals must be evaluated numerically, and this can be a major undertaking. Because the integrands oscillate, convergence becomes a problem, and it is difficult to sustain much accuracy in the calculations. The fact that $(\gamma - \lambda)/(\gamma + \lambda)$ is complex adds an additional complication. When we examine the integrand in detail, we see that it vanishes for large values of λ, the criterion being $\lambda \gg \rho^{-1}$ or $\lambda \gg (z + z_0)^{-1}$. In the event that $|k|\rho$ or $|k|(z + z_0) \gg 1$, the factor $(\gamma - \lambda)/(\gamma + \lambda)$ will remain very close to unity throughout the important range of integration. Then, in the limit, we can evaluate the integrals in closed form with the help of the Lipschitz identity. When we do so, we arrive at the same result that we obtain for a thin, perfectly conducting sheet by the method of images (Sec. 17-5), thus indicating that we have reached the inductive limit of the response. Otherwise numerical evaluation is necessary. It is reasonably straightforward, provided that the Bessel function makes few oscillations within that range of λ in which the integrand has a significant magnitude. This condition is fulfilled when $|k|\rho$ or ρ/z are small (<1). Wait (16e) has published a tabulation of the integrals within this range, and from these tables we show in Fig. 17-13 the response to a uniformly conducting ground of a double dipole electromagnetic system situated at a height h above the surface. It is instructive to compare this with the response of the same system to a thin conducting sheet. An important point to notice in this example is that the response parameter for a uniform ground is $|k^2h^2| = \sigma\mu_0\omega h^2$, while it was $\sigma\mu_0\omega sh$ for a thin sheet.

For large values of $\rho|k|$, particularly when $z + z_0$ is small but not negligible, the difficulties in computing become severe. Slichter and Knopoff (11) have given some results for the vertical dipole as part of a more general problem, and their curves are shown in Fig. 17-14.

The problem of a conducting layer on top of a conducting half-space—or, for that matter, of any number of layers on a half-space—can be solved formally by a straightforward extension of the procedure used above. The solutions will have exactly the same form as that for the simple half-space except for the complex term $(\gamma - \lambda)/(\gamma + \lambda)$, which will be replaced by a more complicated function. Numerical evaluation will thus be even more lengthy. Treatments of the layer and half-space problem, some numerical data, and an example of its use are given in references (16f), (11), and (5).

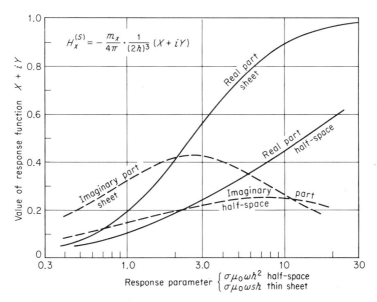

Fig. 17-13   **Response of a "double-dipole" prospecting system to a conducting half-space, with the response to a thin sheet reproduced from Fig. 17-9 for comparison.**

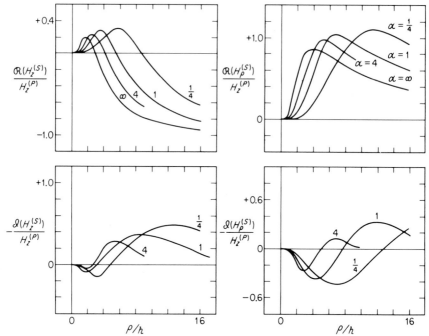

Fig. 17-14   A vertical dipole over a conducting half-space.   Source and detector at height h above the surface.   Curves are given for $\sigma\mu\omega\, h^2 = \frac{1}{4}$, 1, 4, and ∞. [Slichter and Knopoff, (11).]

17-5   *Magnetotelluric Depth Sounding*

It is now appropriate to consider the model which is used in vertical exploration, or "depth sounding," by means of micropulsation measurements [Cagniard (2)].   The source of the magnetic and electric fields is visualized as a vast but virtually uniform, time-varying current sheet that flows through the ionosphere parallel with the earth's surface.   The time variations which concern us have harmonic components whose periods lie in the range from 0.01 to 1,000 sec, while the linear dimensions of the system may be as large as several hundreds of kilometers.

To construct a mathematical model of this system, we take a rectangular frame of coordinates with the origin at the ground surface and Oz directed vertically downward.   The ground is assumed to consist of a number of homogeneous layers having different conductivities overlying a conducting half-space.   The ionospheric current sheet is considered to be infinite in extent and to be at a height h above the ground.   It is then immediately clear that the problem is one-dimensional, since the electromagnetic field vectors will change in the z direction only.

The scale of this model is so entirely different from that of models used in conventional electromagnetic prospecting that we cannot immediately take it for granted that the long wavelength approximation, which we have heretofore employed, is valid. Actually, the fields usually are quasi-static, but because this fact does not really simplify the analysis, we shall derive an unrestricted solution for the problem.

The general solution of (16-20) in one dimension (assuming a harmonic time dependence of the fields) is

$$\mathbf{H}(t) = (H_x \mathbf{i}_x + H_y \mathbf{i}_y) e^{i(\omega t \pm kz)} \qquad k = (i\sigma\mu\omega - \epsilon\mu\omega^2)^{1/2}$$

By (16-14), the corresponding electric field is

$$\mathbf{E}(t) = (\pm H_y \mathbf{i}_x \mp H_x \mathbf{i}_y) \frac{\mu\omega}{k} e^{i(\omega t \pm kz)}$$

Now $\mathbf{H}$ and k are both complex, and they will differ within each region. The conditions to be satisfied at the several interfaces are (16-24) and (16-26), viz.,

$$\mathbf{i}_z \times (\mathbf{E}_1 - \mathbf{E}_2) = 0 \qquad \text{and} \qquad \mathbf{i}_z \times (\mathbf{H}_1 - \mathbf{H}_2) = 0$$

except across the current sheet, where the condition on the magnetic field is

$$\mathbf{i}_z \times (\mathbf{H}_1 - \mathbf{H}_2) = \mathbf{K} \qquad \mathbf{K} \cdot \mathbf{i}_z = 0 \qquad (17\text{-}35)$$

In addition to these requirements, at large values of $|z|$ the fields must vary as

$$e^{ikz} \qquad \text{when} \qquad z \to +\infty \qquad \text{and} \qquad e^{-ikz} \qquad \text{when} \qquad z \to -\infty$$

in order that the intensities shall eventually attenuate and propagate away from the source.

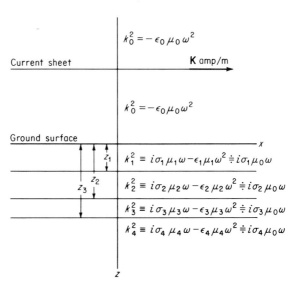

Fig. 17-15 Coordinates for the magnetotelluric "depth-sounding" problem.

The simplest situation occurs when the ground has a uniform conductivity. In this case we may assume that $\mathbf{H}$ and $\mathbf{E}$ are given by

$$(H_x^{(1)}\mathbf{i}_x + H_y^{(1)}\mathbf{i}_y)e^{i(\omega t - k_0 z)} \qquad \text{and} \qquad -(H_y^{(1)}\mathbf{i}_x - H_x^{(1)}\mathbf{i}_y)\frac{\mu_0\omega}{k_0}e^{i(\omega t - k_0 z)}$$

respectively, above the current sheet; by

$$[(H_x^{(2)}\mathbf{i}_x + H_y^{(2)}\mathbf{i}_y)e^{-ik_0 z} + (H_x^{(3)}\mathbf{i}_x + H_y^{(3)}\mathbf{i}_y)e^{ik_0 z}]e^{i\omega t}$$

$$\text{and} \qquad -[(H_y^{(2)}\mathbf{i}_x - H_x^{(2)}\mathbf{i}_y)e^{-ik_0 z} - (H_y^{(3)}\mathbf{i}_x - H_x^{(3)}\mathbf{i}_y)e^{ik_0 z}]\frac{\mu_0\omega}{k_0}e^{i\omega t}$$

between the current sheet and the ground; and by

$$(H_x^{(4)}\mathbf{i}_x + H_y^{(4)}\mathbf{i}_y)e^{i(\omega t + k_1 z)} \qquad \text{and} \qquad (H_y^{(4)}\mathbf{i}_x - H_x^{(4)}\mathbf{i}_y)\frac{\mu_1\omega}{k_1}e^{i(\omega t + k_1 z)}$$

$$(17\text{-}36)$$

beneath the ground surface. By applying the boundary conditions at both interfaces we could, if we wished, solve for the four H_y's in terms of $\mathbf{K}$. (If $\mathbf{K}$ is assumed to be in the x direction, it is obvious that all the H_x's must vanish.) However it is not really necessary to carry through these calculations, because the form of the equations alone shows that the conductivity of the ground can be calculated from measurements made at the ground surface. The quantities required are the amplitude of the electric field in one direction and the amplitude of the magnetic field in the perpendicular direction, since the ratio of these two quantities, according to (17-36), is given by

$$\frac{E_x^{(4)}}{H_y^{(4)}} = \frac{\mu_1\omega}{k_1} = \left(\frac{\mu_0\omega}{\sigma_1}\right)^{1/2}e^{-\pi/4}$$

i.e.,

$$\rho_1 = \frac{(|E_x|/|H_y|)^2}{\mu_0\omega} \qquad (17\text{-}37)$$

In practice the measurement of H_y is made just above the ground surface and that of E_x just beneath it. However the boundary conditions assure us that it makes no difference on which side of the interface the measurements are made.

If we next assume that the ground consists of a simple homogeneous layer resting on a half-space, we may write

$$\mathbf{H} = \mathbf{i}_y(H_y^{(4)}e^{-ik_1 z} + H_y^{(5)}e^{ik_1 z})e^{i\omega t}$$

$$\mathbf{E} = -\mathbf{i}_x(H_y^{(4)}e^{-ik_1 z} - H_y^{(5)}e^{ik_1 z})\frac{\mu_1\omega}{k_1}e^{i\omega t}$$

$$\text{and} \qquad \mathbf{H} = \mathbf{i}_y H_y^{(6)}e^{i(\omega t + k_2 z)} \qquad (17\text{-}38)$$

$$\mathbf{E} = \mathbf{i}_x H_y^{(6)}\frac{\mu_2\omega}{k_2}e^{i(\omega t + k_2 z)}$$

respectively, for the fields in the layer and in the half-space beneath it.

Application of the boundary conditions at the subsurface interface $z = z_1$ yields

$$H_y^{(4)} = -H_y^{(5)}e^{i2k_1z_1}\left(\frac{k_1\mu_2 - k_2\mu_1}{k_1\mu_2 + k_2\mu_1}\right) = -H_y^{(5)}e^{i2k_1z_1}\left(\frac{\sigma_1^{1/2} - \sigma_2^{1/2}}{\sigma_1^{1/2} + \sigma_2^{1/2}}\right)$$

and from this we find that the ratio of field strengths at $z = 0$ is given by

$$\frac{E_x}{H_y} = (\mu_0\omega\rho_1)^{1/2}e^{-i\pi/4}\left[\frac{1 + e^{i2z_1(i\sigma_1\mu_0\omega)^{1/2}}(\sigma_1^{1/2} - \sigma_2^{1/2})/(\sigma_1^{1/2} + \sigma_2^{1/2})}{1 - e^{i2z_1(i\sigma_1\mu_0\omega)^{1/2}}(\sigma_1^{1/2} - \sigma_2^{1/2})/(\sigma_1^{1/2} + \sigma_2^{1/2})}\right]$$

If the quantity $\rho_a = (|E_x|/|H_y|)^2/\mu_0\omega$ is called the *apparent resistivity* of the ground (as in depth-sounding by d-c methods) then

$$\rho_a = \rho_1 \frac{|1 + e^{i2z_1(i\sigma_1\mu_0\omega)^{1/2}}(\sigma_1^{1/2} - \sigma_2^{1/2})/(\sigma_1^{1/2} + \sigma_2^{1/2})|^2}{|1 - e^{i2z_1(i\sigma_1\mu_0\omega)^{1/2}}(\sigma_1^{1/2} - \sigma_2^{1/2})/(\sigma_1^{1/2} + \sigma_2^{1/2})|^2} \tag{17-39}$$

Figure 17-16 shows a set of master curves for the complete range of resistivity contrasts. The resemblance between these graphs and the master curves for the Wenner method (Fig. 14-5) is remarkable, and they are used in a similar way. Harmonic components of $H_y(t)$ and $E_x(t)$ are measured by special apparatus, and values of the apparent resistivity are calculated for a

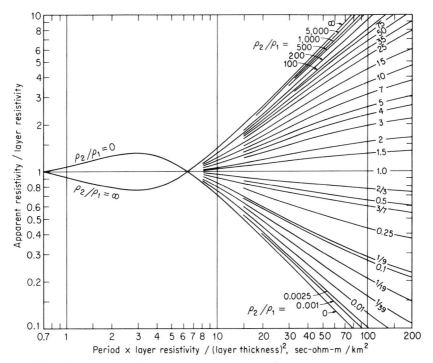

Fig. 17-16 **Master curves for the apparent resistivity of a single-layered ground measured by the magnetotelluric method.** **[Yungel (18).]**

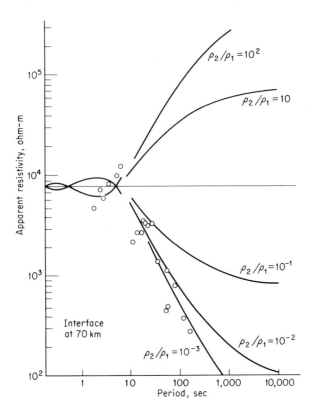

Fig. 17-17 Apparent resistivity versus period at Littleton, Massachusetts. The resistivity of the layer was assigned the value 8000 ohm-m according to earlier d-c conduction measurements. [Cantwell and Madden, *Jour. Geophys. Research*, vol. 65, pp. 4,202–4,205, 1960.]

wide range of frequencies, plotted on logarithmic paper, and fitted to the master curves. The relative displacement of the ordinate scales of master and field curves determines ρ_1, while the displacement of the two abscissa scales gives $\rho_1 z_1^{-2}$. The best-fitting curve also yields the conductivity contrast. An example of such an interpretation is shown in Fig. 17-17.

Curves for one and two layers are readily available (2), (18), and others are not too difficult to compute. The mathematical solutions can be obtained by a simple extension of the solution for a single layer. Of course, the expression for the apparent resistivity of a multilayered ground is more complicated than (17-39) but is not necessarily intractable.

A remarkable feature of the magnetotelluric method is that the analysis does not depend upon the amplitude, position, or frequency spectrum of the source. In fact it can also be shown that it is practically independent of the angle of incidence of the primary field (2). It does, however, depend upon the uniformity of the primary-field intensity over horizontal distances which are *very* large in comparison with the effective depth of penetration of the field into the ground—a condition that obviously becomes more difficult to satisfy as the frequency decreases. The requirement that the field be uniform horizontally over large distances places a restriction on the general applicability of the Cagniard method of magnetotelluric depth sounding. Price (8) [and also Wait (16c)] has examined the restriction in considerable detail. To resolve the difficulty, he has put forward an alternative method in which several magnetic recorders are to be operated simultaneously at different locations on the ground surface. Harmonic analysis of the records must then be performed for both time and space coordinates. The theoretical model to which these data are fitted consists of a uniformly stratified ground on the surface of which the magnetic field varies sinusoidally with time and also with horizontal distance.

17-6   Conducting Sphere in a Dipole Field

A great deal of practical significance attaches to the problem of predicting the response of a compact mineralized zone to a ground or airborne prospecting system. From a practical point of view, the most suitable model to choose in order to represent the mineral body is the conducting sphere, which we have already introduced in Sec. 17-2. This time, however, we are required to determine the response in a dipolar field, but for reasons of economy we shall start with a magnetic pole source and then obtain our dipole result by differentiation.

Taking a set of spherical coordinates (r, ϑ, φ) and placing the magnetic pole (which has a strength p) at $(r_0, 0, 0)$, we may write for the primary field

$$
\begin{aligned}
\mathbf{H}^{(P)} &= -\frac{pe^{i\omega t}}{4\pi} \nabla \frac{1}{R} \\
&= -\frac{pe^{i\omega t}}{4\pi} \nabla \frac{1}{r_0} \sum_{n=0}^{\infty} \left(\frac{r}{r_0}\right)^n P_n(\cos \vartheta) \qquad r < r_0 \qquad (17\text{-}40) \\
&= -\frac{pe^{i\omega t}}{4\pi} \nabla \frac{1}{r} \sum_{n=0}^{\infty} \left(\frac{r_0}{r}\right)^n P_n(\cos \vartheta) \qquad r_0 < r
\end{aligned}
$$

If the surface of the spherical conductor lies at $r = a$, the first form will be the appropriate one to use at the boundary, since $a < r_0$. Taking the gradient in (17-40), we obtain for the components of $\mathbf{H}^{(P)}$

$$
H_r^{(P)} = -\frac{pe^{i\omega t}}{4\pi} \frac{1}{rr_0} \sum_{n=1}^{\infty} n \left(\frac{r}{r_0}\right)^n P_n(\cos \vartheta)
$$

$$
H_\vartheta^{(P)} = \frac{pe^{i\omega t}}{4\pi} \frac{1}{rr_0} \sum_{n=1}^{\infty} \left(\frac{r}{r_0}\right)^n P_n{}^1(\cos \vartheta) \qquad (17\text{-}41)
$$

$$
H_\varphi^{(P)} = 0
$$

To represent the field inside the sphere, we shall require a vector potential which has only a φ component, just as in Sec. 17-2. Thus the general expression for the interior potential A_i in spherical coordinates is

$$A_i = r^{-\frac{1}{2}}e^{i\omega t} \sum_{n=1}^{\infty} [B_n P_n{}^1(\cos \vartheta) + C_n Q_n{}^1(\cos \vartheta)][I_{n+\frac{1}{2}}(kr) + D_n I_{-n-\frac{1}{2}}(kr)]$$

where $k^2 \equiv i\sigma\mu\omega$. Inside the sphere, however, there can be no singularities, and so this immediately reduces the solution to

$$A_i = r^{-\frac{1}{2}}e^{i\omega t} \sum_{n=1}^{\infty} B_n P_n{}^1(\cos \vartheta) I_{n+\frac{1}{2}}(kr) \qquad r < a$$

The field components to which this potential corresponds are

$$H_r = \frac{1}{\mu} r^{-\frac{3}{2}}e^{i\omega t} \sum_{n=1}^{\infty} B_n I_{n+\frac{1}{2}}(kr) n(n+1) P_n(\cos \vartheta)$$

$$H_\vartheta = -\frac{1}{\mu} r^{-\frac{3}{2}}e^{i\omega t} \sum_{n=1}^{\infty} B_n P_n{}^1(\cos \vartheta)[\tfrac{1}{2} I_{n+\frac{1}{2}}(kr) + kr I'_{n+\frac{1}{2}}(kr)] \quad (17\text{-}42)$$

and $H_\varphi = 0$

Outside of the sphere, the general solution for the anomalous magnetic potential is

$$A^{(S)} = e^{i\omega t} \sum_{n=1}^{\infty} [E_n P_n{}^1(\cos \vartheta) + F_n Q_n{}^1(\cos \vartheta)](G_n r^n + r^{-(n+1)})$$

Again $Q_n{}^1$ can be dropped immediately, and since $A^{(S)} \to 0$ as $r \to \infty$, so also can the r^n term. We are then left with the expression

$$A^{(S)} = e^{i\omega t} \sum_{n=1}^{\infty} E_n P_n{}^1(\cos \vartheta) r^{-(n+1)} \qquad r \geq a$$

At the risk of elaborating the analysis beyond its proper economical length, we may set down the expressions for the components of the total magnetic field outside of the sphere (the factors μ and $p/4\pi$ have been removed from E_n) thus

$$H_r = -\frac{p e^{i\omega t}}{4\pi} \sum_{n=1}^{\infty} \left[\frac{n}{rr_0} \left(\frac{r}{r_0}\right)^n - E_n n(n+1) r^{-(n+2)} \right] P_n(\cos \vartheta)$$

$$(17\text{-}43)$$

$$H_\vartheta = \frac{p e^{i\omega t}}{4\pi} \sum_{n=1}^{\infty} \left[\frac{1}{rr_0} \left(\frac{r}{r_0}\right)^n + E_n n r^{-(n+2)} \right] P_n{}^1(\cos \vartheta)$$

If we now apply to (17-42) and (17-43) the continuity conditions on μH_r and on H_ϑ at $r = a$, we can solve for E_n and find that

$$E_n = \frac{a^{2n+1}}{(n+1) r_0{}^{n+1}} \left\{ \frac{[\mu_0/2 - (n+1)\mu] I_{n+\frac{1}{2}}(ka) + \mu_0 ka I'_{n+\frac{1}{2}}(ka)}{(\mu_0/2 + n\mu) I_{n+\frac{1}{2}}(ka) + \mu_0 ka I'_{n+\frac{1}{2}}(ka)} \right\} \quad (17\text{-}44)$$

If we abbreviate this expression by writing

$$E_n = \frac{a^{2n+1}}{(n+1)r_0^{n+1}}(X_n + iY_n)$$

we obtain as final solutions for the components of the anomalous magnetic field produced by a pole source the following expressions:

$$H_r^{(S)} = \frac{pe^{i\omega t}}{4\pi} \sum_{n=1}^{\infty} (X_n + iY_n) \frac{r_0 a^{2n+1}}{(rr_0)^{n+2}} nP_n(\cos\vartheta)$$

$$H_\vartheta^{(S)} = \frac{pe^{i\omega t}}{4\pi} \sum_{n=1}^{\infty} (X_n + iY_n) \frac{r_0 a^{2n+1}}{(rr_0)^{n+2}} \frac{n}{n+1} P_n^1(\cos\vartheta)$$

and $\qquad H_\varphi^{(S)} = 0$

For a radial dipole source $(\mathbf{m} = m_r \mathbf{i}_r)$

$$H_{dipole}^{(S)} = \frac{m}{p} \frac{\partial}{\partial r_0} H_{pole}^{(S)}$$

which yields for the field components

$$H_r^{(S)} = -\frac{m_r}{4\pi} e^{i\omega t} \sum_{n=1}^{\infty} (X_n + iY_n) \frac{a^{2n+1}}{(rr_0)^{n+2}} n(n+1) P_n(\cos\vartheta)$$

$$H_\vartheta^{(S)} = -\frac{m_r}{4\pi} e^{i\omega t} \sum_{n=1}^{\infty} (X_n + iY_n) \frac{a^{2n+1}}{(rr_0)^{n+2}} nP_n^1(\cos\vartheta) \qquad (17\text{-}45)$$

$$H_\varphi^{(S)} = 0$$

For a transverse dipole the operation to be performed is equivalent to a

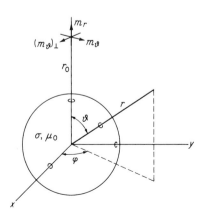

Fig. 17-18 Spherical conductor in a dipole field.

rotation of the polar axis.   Thus

$$\frac{\partial \mathbf{H}^{(S)}}{\partial l}\, \delta l = \frac{m}{pr_0}\left(\frac{\partial \mathbf{H}^{(S)}}{\partial \vartheta_0}\right)_{\varphi - \varphi_0 = 0,\, \pi/2}$$

Two cases must be distinguished: (1) a dipole which lies in the same axial plane as the receiver ($\varphi - \varphi_0 = 0$) or (2) one which lies in the plane perpendicular to this direction ($\varphi - \varphi_0 = \pi/2$).   Since the rotation at the source will rotate the base vectors at the receiver position, we must be careful in taking the derivative.   Noting that

$$\frac{\partial \mathbf{H}^{(S)}}{\partial \vartheta_0} = \frac{\partial H_r^{(S)}}{\partial \vartheta_0}\, \mathbf{i}_r + \frac{\partial \mathbf{i}_r}{\partial \vartheta_0}\, H_r^{(S)} + \frac{\partial H_\vartheta^{(S)}}{\partial \vartheta_0}\, \mathbf{i}_\vartheta + \frac{\partial \mathbf{i}_\vartheta}{\partial \vartheta_0}\, H_\vartheta^{(S)}$$

we find that

$$\frac{m_\vartheta}{pr_0}\left(\frac{\partial \mathbf{H}^{(S)}}{\partial \vartheta_0}\right)_{\varphi - \varphi_0 = 0} = \frac{m_\vartheta}{pr_0}\left[\frac{\partial H_r^{(S)}}{\partial \vartheta_0}\, \mathbf{i}_r + \frac{\partial H_\vartheta^{(S)}}{\partial \vartheta_0}\, \mathbf{i}_\vartheta\right]_{\varphi - \varphi_0 = 0}$$

$$= -\frac{m_\vartheta}{pr_(}\left[\frac{\partial H_r^{(S)}}{\partial \vartheta}\, \mathbf{i}_r + \frac{\partial H_\vartheta^{(S)}}{\partial \vartheta}\, \mathbf{i}_\vartheta\right]_{\varphi - \varphi_0 = 0} \qquad (17\text{-}46)$$

and $$\frac{(m_\vartheta)_\perp}{pr_0}\left(\frac{\partial \mathbf{H}^{(S)}}{\partial \vartheta_0}\right)_{\varphi - \varphi_0 = \frac{\pi}{2}} = \frac{(m_\vartheta)_\perp}{pr_0}\left[\frac{\partial \mathbf{i}_\vartheta}{\partial \vartheta_0}\, H_\vartheta^{(S)}\right]_{\varphi - \varphi_0 = \frac{\pi}{2}}$$

$$= -\frac{(m_\vartheta)_\perp}{pr_0}\left[\csc \vartheta\, H_\vartheta^{(S)} \mathbf{i}_\varphi\right]_{\varphi - \varphi_0 = \frac{\pi}{2}} \qquad (17\text{-}47)$$

Finally, we arrive at the following expressions for the anomalous field components: For case 1, as described above,

$$H_r^{(S)} = \frac{m_\vartheta}{4\pi}\, e^{i\omega t} \sum_{n=1}^{\infty} (X_n + iY_n)\, \frac{a^{2n+1}}{(rr_0)^{n+2}}\, nP_n^1(\cos \vartheta)$$

$$H_\vartheta^{(S)} = -\frac{m_\vartheta}{4\pi}\, e^{i\omega t} \sum_{n=1}^{\infty} (X_n + iY_n)\, \frac{a^{2n+1}}{(rr_0)^{n+2}}\left[n^2 P_n(\cos \vartheta)\right. \qquad (17\text{-}48)$$

$$\left. - \frac{n}{n+1}\, \cot \vartheta\, P_n^1(\cos \vartheta)\right]$$

and $H_\varphi^{(S)} = 0$

For case 2

$$H_r^{(S)} = H_\vartheta^{(S)} = 0 \qquad (17\text{-}49)$$

and $$H_\varphi^{(S)} = -\frac{(m_\vartheta)_\perp}{4\pi}\, e^{i\omega t} \sum_{n=1}^{\infty} (X_n + iY_n)\, \frac{a^{2n+1}}{(rr_0)^{n+2}}\left[\frac{n}{n+1}\, \csc \vartheta\right.$$

$$\left. \times P_n^1(\cos \vartheta)\right]$$

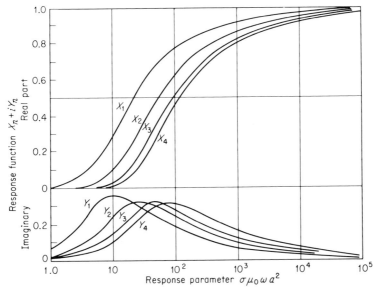

Fig. 17-19 Response functions of the induced multipole moments of order 1 to 4, inclusive, for a spherical conductor in a dipole field. [Wait (16*b*).]

In examining the solutions, we first note that if the dipole is far from the sphere, only the first term is significant and the solution degenerates to that for a uniform field which we obtained in Sec. 17-2. In that event, the secondary field of the sphere will be identical with the field of a simple magnetic dipole. As the source is brought closer to the sphere, higher terms in the series become effective, and the secondary field of the sphere is enhanced by the excitation of multipole moments of higher orders. These need not, in general, all have the same phase, since the response function of each moment is different from all the others. Wait (16*b*) has published graphs of the response functions for $n = 1, 2, 3$, and 4, which are reproduced in Fig. 17-19. These reveal a rather important fact, viz., that the multipole moments of higher orders are less easily excited than those of the lower orders, as evidenced by the observation that they reach the inductive limit at higher values of the response parameter $\sigma\mu\omega a^2$.

A computer is practically a necessity if good use is to be made of this solution in producing type curves. Unless the source is several radii distant from the sphere, more than four terms of the series are necessary to achieve satisfactory accuracy; and tables or graphs of the response functions corresponding to moments higher than the fourth have not yet been published. On the other hand, the solution at the inductive limit, as always, presents us with some comparatively simple computations because the response functions are not involved. Figure 17-20 illustrates some typical results.

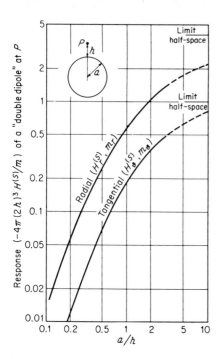

Fig. 17-20 Response of a double-dipole prospecting system to a perfectly conducting sphere.

17-7 Conducting Half-plane in a Dipole Field

The most useful model so far developed for the interpretation of electromagnetic anomalies over mineralized zones is a thin, conducting sheet of infinite dip and strike extent, termed a *half-plane*. Unfortunately, no exact solution has yet been discovered for the field due to a sheet of finite surface conductivity in the vicinity of a dipole-source,[1] but the solution for infinite conductivity (corresponding to the inductive limit) is available (17). The problem was first solved by Sommerfeld (13) in 1897, and the following paragraphs are a direct adaptation of his method.

If we wish to determine the inductive limit of the response, we need concern ourselves only with the magnetic field outside of the conductor. By virtue of (16-2), the curl of this field must vanish everywhere, and we may represent a sinusoidally time-varying field with a scalar potential ψ as follows:

$$\mathbf{H} = -\nabla \psi e^{i\omega t} \qquad (17\text{-}50)$$

But in free space, $\nabla \cdot \mathbf{H} = 0$, and thus the potential must satisfy Laplace's

---

[1] Jeans (4) has presented a solution of this problem, the proof of which unfortunately contains a fundamental error that invalidates the result. Wesley (17b) has given an approximate solution which is open to controversy. For a criticism of Wesley's result, we refer to Wait (16g).

equation

$$\nabla^2 \psi = 0$$

At the inductive limit, the boundary conditions state that the magnetic field does not penetrate the conductor [see (16-29)], so that

$$\mathbf{n} \cdot \mathbf{H} = \frac{\partial \psi}{\partial n} = 0$$

at the surface of the sheet. Problems of this type (called Neumann's problem) have a common occurrence in potential theory. If the problem were in two dimensions only—as would be the case if the source were a line dipole—it could be solved very simply by the use of the conformal transformation

$$z = \bar{z}^2$$

i.e.,
$$\rho = \bar{\rho}^2 \qquad \varphi = 2\bar{\varphi}$$

which transforms the half-plane into an infinite sheet (Fig. 17-21). The problem of the sheet is easily solved by the method of images, and if we transform these functions back into x,y coordinates, we immediately have the solution for the half-plane problem. The total field is the sum of a source and an image term, but the potentials do not have the familiar form

$$\psi = \frac{\mathbf{m}}{4\pi} \cdot \nabla_0(-2 \ln R)$$

$$= -\frac{\mathbf{m}}{4\pi} \cdot \nabla_0 \ln [\rho_0{}^2 + \rho^2 - 2\rho_0\rho \cos (\varphi_0 - \varphi)]^{\frac{1}{2}} \qquad (17\text{-}51)$$

but are given by the transformation of this expression onto the z plane

$$\psi = -\frac{\mathbf{m}}{2\pi} \cdot \nabla_0 \ln \left[\rho_0 + \rho - 2 \sqrt{\rho_0\rho} \cos \left(\frac{\varphi_0 - \varphi}{2} \right) \right]^{\frac{1}{2}} \qquad (17\text{-}52)$$

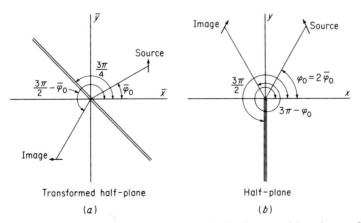

Fig. 17-21 Solution of the two-dimensional half-plane problem by application of the conformal transformation $z = \bar{z}^{\frac{1}{2}}$.

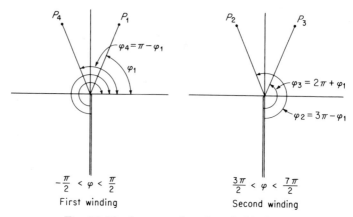

Fig. 17-22   Image points in a half-plane.

The function

$$-2 \ln \left[\rho_0 + \rho - 2 \sqrt{\rho_0 \rho} \cos \left(\frac{\varphi_0 - \varphi}{2} \right) \right]^{\frac{1}{2}}$$

is referred to as "Green's function in two dimensions for a space of two windings," and it corresponds to the ordinary two-dimensional Green's function $-2 \ln R$. The three-dimensional form of Green's function in a space of two windings corresponding to the ordinary Green's function $1/R$ (in $\bar{z}$ space) was found by Sommerfeld to be

$$G(\mathbf{r}, \mathbf{r}_0) = \frac{1}{\pi R} \tan^{-1} \left(-\frac{R}{g} \right) \tag{17-53}$$

where

$$g = 2 \sqrt{\rho_0 \rho} \cos \left(\frac{\varphi_0 - \varphi}{2} \right)$$

and

$$R = \left[\rho_0^2 + \rho^2 + (z_0 - z)^2 - 2\rho_0 \rho \cos \left(\frac{\varphi_0 - \varphi}{2} \right) \right]^{\frac{1}{2}}$$

The derivation of this result is rather involved and will not be given here. With the help of this expression, however, the method of images can be extended to the three-dimensional problem. Thus the potential of the total field due to a magnetic dipole situated at (ρ_1, φ_1, z_1) in the vicinity of a conducting half-plane is

$$\psi = \frac{\mathbf{m}}{4\pi} \cdot \boldsymbol{\nabla}_0 \frac{1}{\pi R_1} \tan^{-1} \left(-\frac{R_1}{g_1} \right) + \frac{\bar{\mathbf{m}}}{4\pi} \cdot \boldsymbol{\nabla}_0 \frac{1}{\pi R_2} \tan^{-1} \left(-\frac{R_2}{g_2} \right) \tag{17-54}$$

where $\bar{\mathbf{m}}$ is the reflection of $\mathbf{m}$ in the $x = 0$ plane. The subscript 1 indicates that the coordinates (ρ_0, φ_0, z_0) are identified with those of the source, viz.,

$$\rho_0 = \rho_1 \qquad \varphi_0 = \varphi_1 \qquad z_0 = z_1$$

after the gradient has been taken; while the subscript 2 indicates that (ρ_0, φ_0, z_0) are similarly identified with the image coordinates (ρ_2, φ_2, z_2), viz.,

$$\rho_0 = \rho_2 = \rho_1 \qquad \varphi_0 = \varphi_2 = 3\pi - \varphi_1 \qquad z_2 = z_1$$

To obtain the potential $\psi^{(S)}$ of the secondary field, we subtract from ψ the primary field potential

$$\psi^{(P)} = \frac{\mathbf{m}}{4\pi} \cdot \nabla_0 \frac{1}{R_1}$$

and make use of the identity

$$\frac{1}{R_1} = \frac{1}{\pi R_1} \tan^{-1}\left(-\frac{R_1}{g_1}\right) + \frac{1}{\pi R_1} \tan^{-1}\left(\frac{R_1}{g_1}\right)$$

$$= \frac{1}{\pi R_1} \tan^{-1}\left(-\frac{R_1}{g_1}\right) + \frac{1}{\pi R_1} \tan^{-1}\left(-\frac{R_3}{g_3}\right)$$

where $\qquad \rho_3 = \rho_1 \qquad \varphi_3 = \varphi_1 + 2\pi \qquad z_3 = z_1$

and therefore $\qquad R_3 = R_1 \qquad g_3 = -g_1$

The result is

$$\psi^{(S)} = -\frac{\mathbf{m}}{4\pi} \cdot \nabla_0 \frac{1}{\pi R_3} \tan^{-1}\left(-\frac{R_3}{g_3}\right) + \frac{\bar{\mathbf{m}}}{4\pi} \cdot \nabla_0 \frac{1}{\pi R_2} \tan^{-1}\left(-\frac{R_2}{g_2}\right)$$

or $\qquad \psi^{(S)} = -\frac{\mathbf{m}}{4\pi} \cdot \nabla_0 \frac{1}{\pi R_1} \tan^{-1}\left(\frac{R_1}{g_1}\right) + \frac{\bar{\mathbf{m}}}{4\pi} \cdot \nabla_0 \frac{1}{\pi R_4} \tan^{-1}\left(\frac{R_4}{g_4}\right) \qquad (17\text{-}55)$

where $\qquad \rho_4 = \rho_2 = \rho_1 \qquad \varphi_4 = \varphi_2 - 2\pi = \pi - \varphi_1 \qquad z_4 = z_1$

The last form has the advantage that both the source and its image point lie in the first winding, so that there is no difficulty in transforming from cylindrical to rectangular coordinates.

The two remaining steps are to perform the directional differentiation $\mathbf{m} \cdot \nabla_0$ and to calculate the gradient $-\nabla\psi$, in order to obtain the components of the magnetic field vector. This leads to some formidable algebra. The equations are much more wieldy if rectangular coordinates are employed

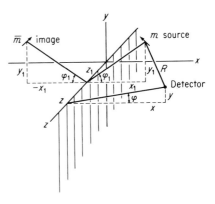

Fig. 17-23 Coordinates for the half-plane solution.

in the R subfunctions and cylindrical coordinates in the g subfunctions. The field corresponding to the first term of (17-55) is

$$\mathbf{u}_1 = \frac{1}{4\pi^2 R_1{}^2} \left[\frac{2\,\nabla R_1(\partial R_1/\partial l)}{R_1} - \frac{\partial}{\partial l}(\nabla R_1) \right] \tan^{-1}\frac{R_1}{g_1}$$

$$+ \frac{1}{4\pi^2(R_1{}^2 + g_1{}^2)} \left\{ \left[\frac{g_1\dfrac{\partial}{\partial l}(\nabla R_1) - R_1\dfrac{\partial}{\partial l}(\nabla g_1)}{R_1} \right] \right.$$

$$+ \frac{2}{(R_1{}^2 + g_1{}^2)} \left[R_1\left(\nabla R_1\frac{\partial g_1}{\partial l} + \nabla g_1\frac{\partial R_1}{\partial l}\right) \right.$$

$$\left. \left. - g_1\left(2\,\nabla R_1\frac{\partial R_1}{\partial l} - \nabla g_1\frac{\partial g_1}{\partial l} + \frac{g_1{}^2}{R_1{}^2}\nabla R_1\frac{\partial R_1}{\partial l}\right) \right] \right\} \quad (17\text{-}56)$$

where $\qquad\qquad \dfrac{\partial}{\partial l} = \mathbf{m}\cdot\nabla_0 \qquad$ and where

$$R_1 = [(x_1 - x)^2 + (y_1 - y)^2 + (z_1 - z)^2]^{1/2}$$

$$g_1 = 2\sqrt{\rho_1\rho}\,\cos\left(\frac{\varphi_1 - \varphi}{2}\right)$$

and where the derivatives appearing in (17-56) are the following:

$$\frac{\partial R_1}{\partial l} = \frac{1}{R_1}[m_x(x_1 - x) + m_y(y_1 - y) + m_z(z_1 - z)]$$

$$\frac{\partial g_1}{\partial l} = \sqrt{\frac{\rho}{\rho_1}}\left[m_\rho\cos\left(\frac{\varphi_1 - \varphi}{2}\right) - m_\varphi\sin\left(\frac{\varphi_1 - \varphi}{2}\right)\right]$$

$$\nabla_x R_1 = -\frac{(x_1 - x)}{R_1} \qquad \nabla_y R_1 = -\frac{(y_1 - y)}{R_1} \qquad \nabla_z R_1 = -\frac{(z_1 - z)}{R_1}$$

$$\nabla_\rho g_1 = \sqrt{\frac{\rho_1}{\rho}}\cos\left(\frac{\varphi_1 - \varphi}{2}\right) \qquad \nabla_\varphi g_1 = \sqrt{\frac{\rho_1}{\rho}}\sin\left(\frac{\varphi_1 - \varphi}{2}\right) \qquad \nabla_z g_1 = 0$$

$$\frac{\partial}{\partial l}\nabla_x R_1 = \left[\frac{(x_1 - x)}{R_1{}^2}\left(\frac{\partial R_1}{\partial l}\right) - \frac{m_x}{R_1}\right]$$

$$\frac{\partial}{\partial l}\nabla_y R_1 = \left[\frac{(y_1 - y)}{R_1{}^2}\left(\frac{\partial R_1}{\partial l}\right) - \frac{m_y}{R_1}\right]$$

$$\frac{\partial}{\partial l}\nabla_z R_1 = \left[\frac{(z_1 - z)}{R_1{}^2}\left(\frac{\partial R_1}{\partial l}\right) - \frac{m_z}{R_1}\right]$$

$$\frac{\partial}{\partial l}\nabla_\rho g_1 = \frac{1}{2\sqrt{\rho_1\rho}}\left[m_\rho\cos\left(\frac{\varphi_1 - \varphi}{2}\right) - m_\varphi\sin\left(\frac{\varphi_1 - \varphi}{2}\right)\right]$$

$$\frac{\partial}{\partial l}\nabla_\varphi g_1 = \frac{1}{2\sqrt{\rho_1\rho}}\left[m_\rho\sin\left(\frac{\varphi_1 - \varphi}{2}\right) + m_\varphi\cos\left(\frac{\varphi_1 - \varphi}{2}\right)\right]$$

$$\frac{\partial}{\partial l}\nabla_z g_1 = 0$$

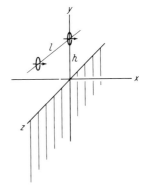

Fig. 17-24 A situation in which the half-plane solution simplifies greatly.

The second term is obtained in exactly the same way as the first term after substituting the coordinates of P_4 for P_1 and the components of $\bar{m}$ for those of m, i.e.,

$$x_4 = -x_1 \qquad y_4 = y_1 \qquad z_4 = z_1$$

$$\rho_4 = \rho_1 \qquad \varphi_4 = \pi - \varphi_1$$

$$\bar{m}_x = -m_x \qquad \bar{m}_y = m_y \qquad \bar{m}_z = m_z$$

The secondary magnetic field is then

$$\mathbf{H}^{(S)} = (\mathbf{u}_1 - \mathbf{u}_4)e^{i\omega t} \tag{17-57}$$

A computer is required to calculate curves from this massive expression. A dramatic simplification takes place, however, when a pair of coplanar vertical loops lie directly above the edge of the sheet (Fig. 17-24). The

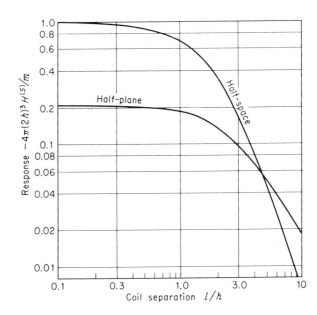

Fig. 17-25 Response obtained by the configuration shown in Fig. 17-24.

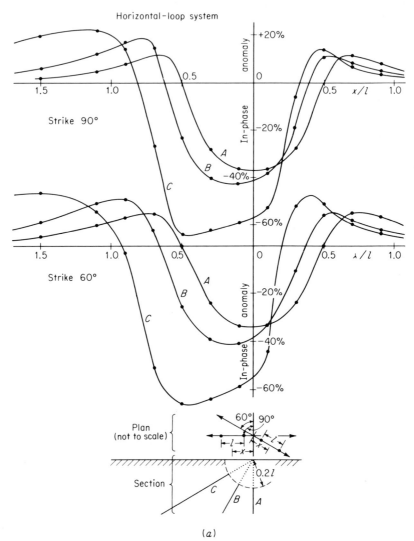

(a)

Fig. 17-26 **Some type curves computed from the solution for a perfectly conducting half-plane.** (a) **For the horizontal-loop method;** (b) **for the vertical-loop parallel-line method;** (c) **for the vertical-loop fixed-transmitter method.**

anomaly then becomes

$$\frac{H_x^{(S)}}{m_x} = \frac{2}{4\pi^2 l^3}\left[\tan^{-1}\left(\frac{l}{2h}\right) - \frac{l}{2h}\right]$$

$$= \frac{1}{4\pi(2h)^3}\left(\frac{2}{\pi}\right)\left[\frac{1}{3} - \frac{1}{5}\left(\frac{l}{2h}\right)^2 + \frac{1}{7}\left(\frac{l}{2h}\right)^4 \cdots\right] \qquad l \leq 2h$$

This is plotted in Fig. 17-25 along with the equivalent result for an infinitely conductive half-space taken from Fig. 17-10a. In Fig. 17-26 a number of profiles are shown which relate to three of the ground prospecting systems described in Chap. 15. They were computed from the half-plane solution (17-57), and illustrate the sort of interpretational aids that such a solution is able to provide.

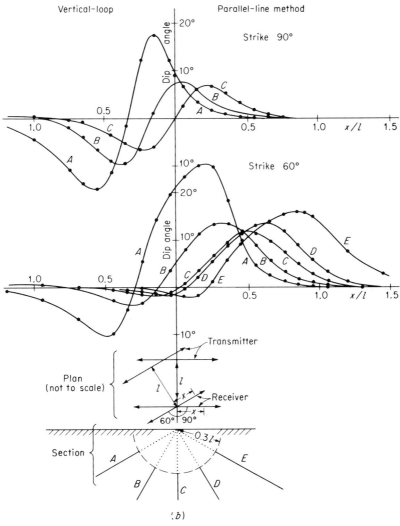

Fig. 17-26 (*continued*)

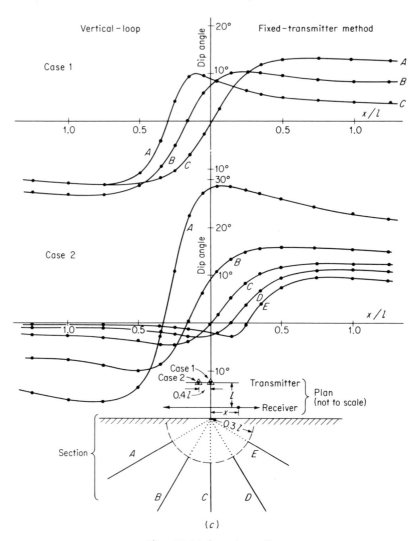

Fig. 17-26 (*continued*)

17-8 The Ribbon and the Disk Models

Two other models for which the secondary magnetic fields can be found in the inductive limit are the circular disk and the ribbon. When the radius of the disk or the width of the ribbon becomes very large, they both effectively become half-planes; and it is in showing how the finite extent of a conductor diminishes the response that they are particularly useful.

Both of these problems can be solved either by the method of images (13), (3) or by the method of separation of variables. The first leads to closed expressions of the kind (17-55), and the second, to infinite series of eigenfunctions such as we obtained in Sec. 17-6 for the sphere.   The eigenfunctions for the disk are the associated Legendre functions in an oblate spheroidal coordinate system, and for the ribbon they are the Mathieu functions in an elliptic cylindrical system.

The second method has been employed by Douloff (3) to calculate some profiles for the disk model, and the first method by Martin (6), for the ribbon.   Inevitably, the series solution converges very slowly when the source and detector are near the conductor, while the image solution for the ribbon suffers from a special restriction that both the source and detector must lie in a single plane perpendicular to the ribbon.

Figure 17-27 summarizes a few of the calculations for the disk and ribbon models.   It shows the maximum anomaly measured by a horizontal-loop prospecting system when it traverses a vertical conductor.   From these graphs we note that the disk must have a diameter several times greater than the coil separation l in order that the maximum effect may approach that of a half-plane, while the ribbon must have a width only slightly greater than l in order to accomplish the same result.

Although solutions at the inductive limit yield a good deal of interesting

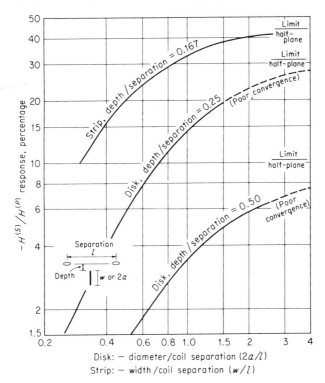

Fig. 17-27   Peak response of a horizontal-loop prospecting system to a vertical disk and to a ribbon (strip) of infinite conductivity.

information about the models they represent, we would also like to know how the responses behave below this limit.   For the disk we can find an approximate answer to this question without becoming involved in a major tour de force of mathematical analysis.   For, by analogy with the solution for a sphere, we would expect that the series solution for a disk of finite surface conductivity will have the same form as the series solution at the inductive limit, but contain a response function in each term.   The first term, which will be the most important one unless the disk is so large that it approximates a half-plane, will be the response to a uniform axial field. This problem by itself would be tractable, but it is not necessary for us to attack it.   We shall be quite content to find the resistive and inductive limits of this term, since these will tell us almost all we need to know about the solution.

Figure 17-28 shows the disk in cross section.   The half-thickness z is related to the radius ρ by the formula

$$z = \frac{s}{2a} (a^2 - \rho^2)^{\frac{1}{2}}$$

We may either consider the conductivity to be a constant σ, or suppose it to be a function of the radius

$$\sigma(\rho) = \sigma \frac{a}{s} (a^2 - \rho^2)^{-\frac{1}{2}}$$

so that the surface conductivity $2z\sigma(\rho)$ is a constant σs.   In fact, we shall do both.

Following the method which we used previously to find the resistive limit for the response of a sphere in a uniform field (Sec. 17-2), we begin with the contribution to the dipole moment from an annulus of radius ρ and width

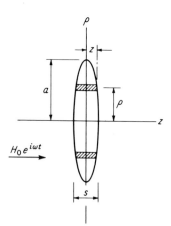

Fig. 17-28   Cross section of an oblate spheroidal disk in a uniform axial field.

$d\rho$

$$dm_R = -i\pi\sigma\mu_0\omega H_0 e^{i\omega t}\frac{s}{2a}(a^2 - \rho^2)^{\frac{1}{2}}\rho^3\,d\rho \qquad \sigma(\rho) = \sigma$$

or $\qquad dm_R = -i\pi\sigma\mu_0\omega H_0 e^{i\omega t}s\rho^3\,d\rho \qquad 2z\sigma(\rho) = \sigma s$

Integrating from $\rho = 0$ to $\rho = a$, these expressions give

$$m_R = -\frac{\pi}{15}i\sigma\mu_0\omega s a^4 H_0 e^{i\omega t} \qquad \sigma(\rho) = \sigma$$

or $\qquad m_R = -\frac{\pi}{8}i\sigma\mu_0\omega s a^4 H_0 e^{i\omega t} \qquad 2z\sigma(\rho) = \sigma s$

for the resistive limits of the leading term. The inductive limit of the response can be found from the solution of the problem of a permeable spheroid in a uniform, axial magnetostatic field. This can be found in many texts on electromagnetic theory, and it is shown that within the spheroid there is a uniform magnetization $\mathbf{M}$ which produces a uniform magnetic field $N\mathbf{M}$. Thus

$$\mathbf{H}_i = \mathbf{H} - N\mathbf{M}$$

where $\mathbf{H}_i$ = the internal magnetic field, $\mathbf{H}$ = the primary axial magnetic field, and N = the mks demagnetization factor, whose value depends on the ellipticity of the spheroid but which is given approximately by the formula

$$N = 1 - \frac{\pi}{4}\frac{s}{a} \qquad \text{when} \qquad \frac{s}{a} \ll 1$$

At the inductive limit, the eddy currents produce an *effective* magnetization $\mathbf{M}$ which is just sufficient to cause the flux density inside the disk to vanish; i.e.,

$$\mathbf{B}_i = \mu_0(\mathbf{H}_i + \mathbf{M})$$

$$= \mu_0[\mathbf{H} + (1 - N)\mathbf{M}] = \mu_0\left(\mathbf{H} + \frac{\pi}{4}\frac{s}{a}\mathbf{M}\right) = 0$$

Therefore, if $\mathbf{H} = \mathbf{H}_0 e^{i\omega t}$,

$$\mathbf{M} = -\frac{4}{\pi}\frac{a}{s}\mathbf{H}_0 e^{i\omega t}$$

and $\qquad m_L = M \times \text{volume} = -\frac{8}{3}a^3 H_0 e^{i\omega t}$

We have now gained enough information roughly to sketch in the response between the two limits. The induced magnetic moment must be

$$m = -\frac{8}{3}a^3 H_0 e^{i\omega t}[X(\alpha) + iY(\alpha)]$$

where $\qquad\qquad \alpha = \sigma\mu_0\omega s a \qquad\qquad\qquad\qquad (17\text{-}58)$

and although we have not determined the response function $X(\alpha) + iY(\alpha)$ explicitly, we can guess at its form from past experience. We also know its limits at large and small values of the response parameter, viz.,

$$(X + iY) = \frac{i\pi\alpha}{40} \qquad \sigma(\rho) = \sigma \qquad \alpha \to 0$$

$$= \frac{i3\pi\alpha}{64} \qquad 2z\sigma(\rho) = \sigma s \qquad \alpha \to 0$$

and $\qquad \lim_{\alpha \to \infty} (X + iY) = 1$

Figure 17-29 shows a conjecture drawn from the available facts. Some data that were obtained by Ashour (1) for the disk of uniform surface conductivity by using an approximate method are also included.

17-9 Multiple Conductors

There are many situations in which a prospecting system will respond to two or more conductors at once. Unlike gravity and magnetic measurements, the total effect will not, in general, be a simple sum of the individual responses, since inductive interaction between the bodies will play a part. The multiconductor problem must therefore be studied as a whole. The variety of possibilities is of course unlimited, but we may perceive the nature of the effect by considering two models, each of which represents an extreme situation. On the one hand, there is the problem of only two distinctly separate conductors situated close to and parallel with one another, and on the other, the problem of a zone of disseminated mineralization in which the conducting material is fragmented into countless particles of small but varying sizes.

In neither case is it possible to find a model that is both representative

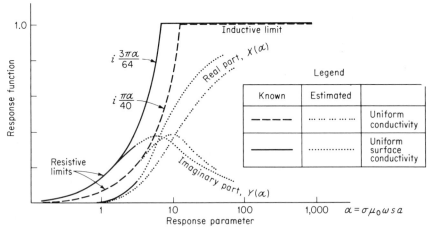

Fig. 17-29   Limits of the response function of an oblate spheroidal disk in a uniform axial field.

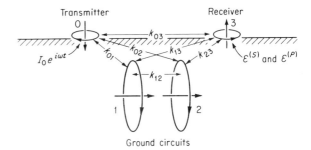

Fig. 17-30 A horizontal loop prospecting system energizing a pair of interacting circuits.

and tractable, but with the aid of simpler, heuristic models we can shed some light on the physical consequences of interaction. For the first, we shall choose a pair of interacting loops; and for the second, a packing of small conducting spheres.

The first problem is solved by an extension of the analysis given in Sec. 17-1. The situation under study is illustrated in Fig. 17-30, together with the notation to be used. If a current $I_0 e^{i\omega t}$ flows in the transmitter, we can determine the currents $I_1 e^{i\omega t}$ and $I_2 e^{i\omega t}$ in the two underground circuits by solving a pair of linear equations. We can then proceed to calculate the anomalous signal $\mathcal{E}^{(S)}$ which the two circuits induce in the receiver and which, expressed in terms of the undisturbed signal from the transmitter, constitutes the electromagnetic anomaly. The analysis leads to some rather cumbersome algebraic expressions, which give as a final result

$$\frac{\mathcal{E}^{(S)}}{\mathcal{E}^{(P)}} = -\frac{k_{01}k_{13}}{k_{03}}\left[\left(\frac{Q_1 A - Q_1 Q_2 B}{A^2 + B^2}\right) + i\left(\frac{Q_1 Q_2 A + Q_1 B}{A^2 + B^2}\right)\right]$$
$$- \frac{k_{02}k_{23}}{k_{03}}\left[\left(\frac{Q_2 A - Q_1 Q_2 B}{A^2 + B^2}\right) + i\left(\frac{Q_1 Q_2 A + Q_2 B}{A^2 + B^2}\right)\right]$$
$$- k_{12}\left(\frac{k_{01}k_{23}}{k_{03}} + \frac{k_{02}k_{13}}{k_{03}}\right)\left[\left(\frac{Q_1 Q_2 B}{A^2 + B^2}\right) - i\left(\frac{Q_1 Q_2 A}{A^2 + B^2}\right)\right] \quad (17\text{-}59)$$

where
$$A = Q_1 + Q_2$$
$$B = 1 + (k_{12}{}^2 - 1)Q_1 Q_2$$
$$Q_1 = \frac{\omega L_1}{R_1} \qquad Q_2 = \frac{\omega L_2}{R_2}$$

Although we have found the formal solution of the problem, a certain amount of effort is required in order to see what this expression really signifies. For a start, we note that if the mutual coupling between the two underground circuits vanishes (i.e., if $k_{12} = 0$), (17-59) simplifies to

$$\frac{\mathcal{E}^{(S)}}{\mathcal{E}^{(P)}} = -\frac{k_{01}k_{13}}{k_{03}}\left[\frac{Q_1{}^2 + iQ_1}{(1 + Q_1{}^2)}\right] - \frac{k_{02}k_{23}}{k_{03}}\left[\frac{Q_2{}^2 + iQ_2}{(1 + Q_2{}^2)}\right] \qquad k_{12} = 0$$

which is just the addition of the responses to the two circuits separately. This is as it should be, of course, since we have specifically excluded any

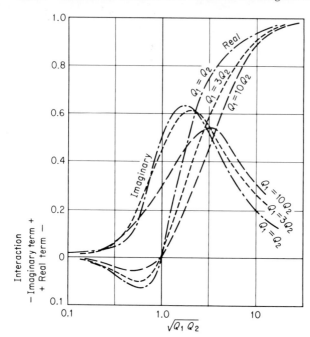

interaction between them.   This condition will apply only in the event that the underground loops are widely separated.

If on the other hand the two circuits merge together so that

$$k_{12} = 1 \qquad k_{01} = k_{02} \qquad k_{13} = k_{23}$$

it follows that

$$\frac{\mathcal{E}^{(S)}}{\mathcal{E}^{(P)}} = -\frac{k_{01}k_{13}}{k_{03}}\left[\frac{(Q_1 + Q_2)^2 + i(Q_1 + Q_2)}{1 + (Q_1 + Q_2)^2}\right]$$

which is the response to a single loop whose response parameter is $Q_1 + Q_2$.

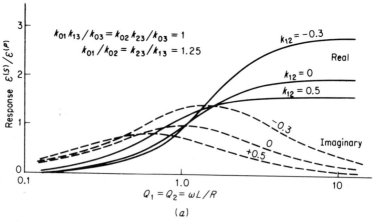

Fig. 17-32   (a) Response of two interacting circuits calculated from (17-59); model.   [Ranasinghe (9).]

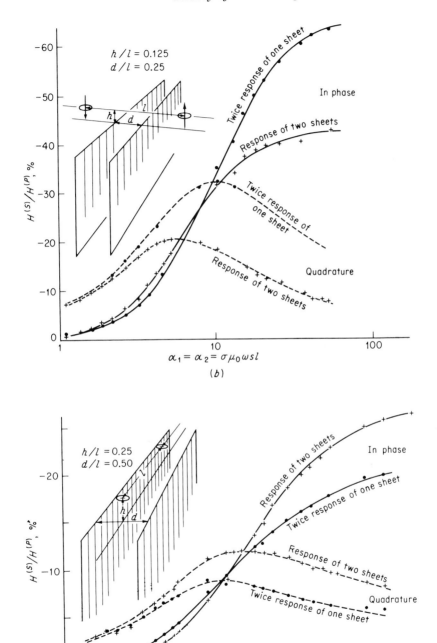

(b) and (c) response of two interacting conducting sheets measured on a scale

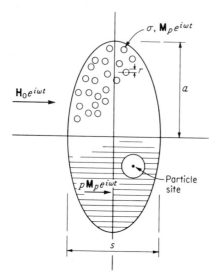

Fig. 17-33 A spheroidal envelope packed with small conducting spheres.

Between these two extremes we can obtain a good indication of the result to be expected by writing the approximation to (17-59) when k_{12} is small. Then k_{12}^2 can be neglected and we obtain

$$\frac{\mathcal{E}^{(S)}}{\mathcal{E}^{(P)}} = - \frac{k_{01}k_{13}}{k_{03}}\left[\frac{Q_1^2 + iQ_1}{(1 + Q_1^2)}\right] - \frac{k_{02}k_{23}}{k_{03}}\left[\frac{Q_2^2 + iQ_2}{(1 + Q_2^2)}\right]$$
$$- k_{12}\left(\frac{k_{01}k_{23}}{k_{03}} + \frac{k_{02}k_{13}}{k_{03}}\right)\left[\frac{Q_1Q_2(1 - Q_1Q_2) - iQ_1Q_2(Q_1 + Q_2)}{(1 + Q_1^2)(1 + Q_2^2)}\right] \quad (17\text{-}60)$$

The first two terms of this expression are just the responses to the two circuits taken individually, and the third obviously represents the interaction between them. The form of the interaction is independent of the coupling coefficients and is shown in Fig. 17-31.

Two simple examples of this interaction are given in Fig. 17-32a. In both cases the coupling coefficients between the conductors and the prospecting system are equal and of the same sign. For the sake of simplicity the response parameters are assumed to be equal also. However, the interaction between the conductors has been taken as positive in one case and negative in the other. Numerical values have been assigned to the various coupling coefficients and the responses are calculated from (17-59).

In the case in which k_{12} is positive, the net response is intermediate between the result for noninteracting and the result for merging circuits. Interaction decreases the magnitude of the total response at the inductive limit, but the limit is attained at somewhat lower values of Q_1 and Q_2.

In the case in which k_{12} is negative, exactly the opposite occurs. The inductive limit of total response is increased, but the response reaches this limit more slowly.

In order to demonstrate that this rather crude theory actually does apply to continuous conductors, some scale-model measurements using a horizontal-loop prospecting system and rectangular sheet conductors are also included in Fig. 17-32. The resemblance between the model results and the calculations for the loops is striking.

We now turn our attention to a model of a disseminated conductor. We shall consider a spheroidal envelope which is uniformly filled with very small spherical conductors. Figure 17-33 illustrates the situation. The packing fraction p is assumed to be small, so that each particle will be isolated from its neighbors. The primary field $\mathbf{H}_0 e^{i\omega t}$ is uniform and directed along the axis of the spheroid.

The method of analysis which we shall use is similar to that used by Tesche (15) and is analogous to the Lorentz method of calculating the dielectric constants of a crystal. The field acting on a given particle is composed of two parts, viz., the primary field $\mathbf{H}_0 e^{i\omega t}$ and the net field produced by all the other particles. As we have noted previously, an important property of a uniformly magnetized body enclosed by a second-degree surface is that it has a uniform internal field. The strength of the field produced by an axial magnetization is given by

$$\mathbf{H}_M = -N\mathbf{M}$$

where the demagnetization factor N ranges between 0 and 1 depending on the ellipticity, as indicated in Fig. 17-34. It is therefore not unreasonable to suppose that sample averages of the moments induced in the particles will be the same throughout the spheroid and that in an average sense the field at any particle site will be independent of position. The word "average" is employed because it is obvious that variations in the field intensity *must* occur in the neighborhood of every particle and that local variations

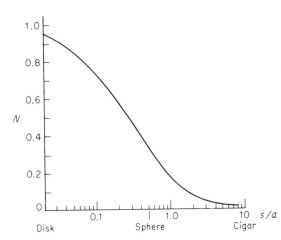

Fig. 17-34 Axial demagnetization factor of a uniform spheroid.

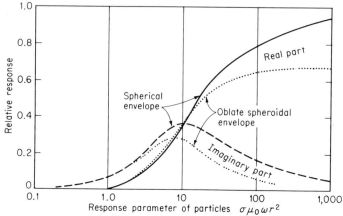

Fig. 17-35   Relative responses of a spherical and an oblate spheroidal envelope packed with small spheres.   The volume of the envelope is the same in both cases and the packing fraction is 0.5.

in the strength of the induced moments will thus arise whenever two or more particles happen to be closer together or farther apart than normal.

If the average value of the induced magnetizations of the spherical particles is $\mathbf{M}_p e^{i\omega t}$, we can evaluate the interaction field at any given site. To a good approximation, provided p is small, each particle behaves as if it were at the center of a spherical hole within a uniformly magnetized spheroid. This produces a field

$$\mathbf{H}_M = (-N_{spheroid} + N_{sphere})\mathbf{M}$$
$$= -(N_{spheroid} - N_{sphere})p\mathbf{M}_p e^{i\omega t}$$

where $N_{sphere} = \frac{1}{3}$.
The *total* field at the particle site is therefore

$$\mathbf{H}e^{i\omega t} = \mathbf{H}_0 e^{i\omega t} - (N - \frac{1}{3})p\mathbf{M}_p e^{i\omega t}$$

We derived the relationship between $\mathbf{M}_p$ and $\mathbf{H}$ in Sec. 17-2.   It is given by (17-14), viz.,

$$\mathbf{M}_p e^{i\omega t} = -\frac{3}{2}\mathbf{H}e^{i\omega t}(X + iY)$$

Therefore

$$\mathbf{M}_p e^{i\omega t} = -\frac{\frac{3}{2}(X + iY)\mathbf{H}_0 e^{i\omega t}}{1 + p(N - \frac{1}{3})(X + iY)}$$

If the total volume of the spheroid is V, the total induced moment will be

$$\mathbf{m} = p V \mathbf{M}_p e^{i\omega t} = -\frac{\frac{3}{2}pV(X + iY)\mathbf{H}_0 e^{i\omega t}}{1 + p(N - \frac{1}{3})(X + iY)} \tag{17-61}$$

We arrive accordingly at an expression which is just the sum of the induced magnetic moments of the particles as if each were completely isolated from its neighbors and modified by a factor (the square bracket in the denominator) which expresses the effect of interactions. It is noteworthy that the effect of the interactions disappears as the shape of the envelope approaches a sphere. This fact has been verified experimentally by Tesche for a packing fraction approaching $\frac{2}{3}$, which is far larger than the greatest value for which we might expect our theory to be valid. In the event that the envelope approaches a disk in shape, the denominator of (17-61) becomes

$$1 + \tfrac{2}{3}p(X + iY)$$

The effect of interactions between grains is shown in Fig. 17-35 when the packing fraction is 0.5. We observe that it does not greatly alter the basic form of the response, although it does change it in magnitude.

Another interesting corollary follows from (17-61) if we assume that the spherical particles are not identical but have different radii. Figure 17-36 shows the response function for a spherical packing of particles of which the volume fraction is normally distributed according to the logarithm of the

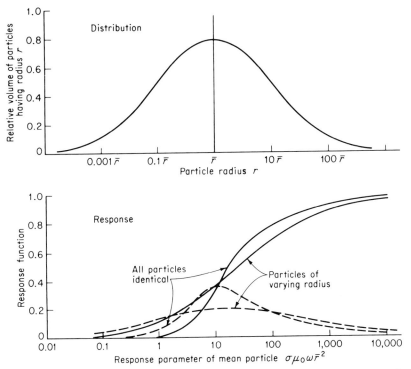

Fig. 17-36 Response function of a spherical packing of small spheres whose radii are not identical.

radius. The most noticeable effect is that the transition from imaginary to real response takes place over a much wider range of the response parameter.

It follows from these results that a finely disseminated conductor is capable of a very substantial response—of the order of p times that of a continuous body having similar dimensions and conductivity. However, we note in this respect that the transition point of the response depends upon the response parameter of the individual particles, so that, other factors being equal, a very much higher frequency must be used on a disseminated than on a continuous conductor in order to reach equivalent points on their respective response functions.

17-10 Transient Response

There is at present at least one type of electromagnetic prospecting apparatus which utilizes a transient primary field rather than an alternating one. It also happens, as we saw in Sec. 17-3, that certain problems can be solved more simply if the source varies stepwise in time than if it varies harmonically. It is thus worth our while to consider the theoretical connection between the two types of response.

The Fourier integral theorem states (with certain reservations) that any single-valued function can be synthesized from, or decomposed into, a continuous spectrum of harmonic components. Thus a unit-step function can be written in spectral form as follows:

$$u(t) = \frac{1}{2\pi} \int_{-\infty}^{+\infty} v(\omega)e^{i\omega t}\, d\omega$$

where $v(\omega)$ is the spectrum of $u(t)$ and is given by

$$v(\omega) = \int_{-\infty}^{\infty} u(t)e^{-i\omega t}\, dt$$

$$= \int_{0}^{\infty} e^{-i\omega t}\, dt$$

This latter integral is divergent, but it can be evaluated by introducing a factor $e^{-pt}(p > 0)$ and taking a limit

$$v(\omega) = \lim_{p\to 0} \int_{0}^{\infty} e^{-pt}e^{-i\omega t}\, dt = -\frac{i}{\omega}$$

If the Fourier transform of $u(t)$ is $-i/\omega$, then it is clear that the transform of the secondary field $\mathbf{H}^{(S)}(t)$ produced by a stepwise varying source $\mathbf{m}u(t)$ is related to the response $\check{\mathbf{H}}^{(S)}(\omega)e^{i\omega t}$ due to a steady-state source $\mathbf{m}e^{i\omega t}$ as follows:

$$\text{Spectrum of } \mathbf{H}^{(S)}(t) = \check{\mathbf{H}}^{(S)}(\omega)\left(-\frac{i}{\omega}\right)$$

i.e.,

$$\mathbf{H}^{(S)}(t) = -\frac{1}{2\pi} \int_{-\infty}^{\infty} \check{\mathbf{H}}^{(S)}(\omega)\frac{i}{\omega} e^{i\omega t}\, d\omega \tag{17-62}$$

It is important to note that Fourier transforms must be defined in the range $\infty > \omega > -\infty$, although real harmonic responses exist only for positive frequencies. This seemingly paradoxical situation arises from the use of complex variable notation in which a real, steady-state field is represented by

$$\mathbf{H}(t) = \text{Re } [\check{\mathbf{H}}(\omega)e^{i\omega t}]$$

In order that (17-62) should always yield a real function for $\mathbf{H}^{(S)}(t)$ we must have

$$\check{\mathbf{H}}^{(S)}(-\omega) = [\check{\mathbf{H}}^{(S)}(\omega)]^*$$

where the asterisk indicates the complex conjugate.

Equation (17-62) shows that if we have observed or calculated $\check{\mathbf{H}}^{(S)}(\omega)$ over the whole range of frequencies, we can always work out the transient response $\mathbf{H}^{(S)}(t)$. Conversely, we know from the Fourier theorem that

$$-\frac{i}{\omega} \check{\mathbf{H}}^{(S)}(\omega) = \int_{-\infty}^{\infty} \mathbf{H}^{(S)}(t)u(t)e^{-i\omega t} dt$$

and therefore

$$\check{\mathbf{H}}^{(S)}(\omega) = i\omega \int_{0}^{\infty} \mathbf{H}^{(S)}(t)e^{-i\omega t} dt \qquad (17\text{-}63)$$

The harmonic response is therefore calculable if the transient response $\mathbf{H}^{(S)}(t)$ is known for $t > 0$. This is in fact how we obtained the steady-state solution for the flat conducting sheet in the presence of a magnetic dipole in Sec. 17-3.

Let us now work out the step response of a simple circuit and compare it with the harmonic response. In Sec. 17-1 we ascertained that if an alternating current $I_0 e^{i\omega t}$ flows through the transmitting coil, the receiver detects primary and secondary induced emfs given by

$$\mathcal{E}_2{}^{(P)} = -i\omega M_{02} I_0 e^{i\omega t}$$

and

$$\mathcal{E}_2{}^{(S)} = i\omega \frac{M_{01}M_{12}}{L} \left(\frac{Q^2 + iQ}{1 + Q^2} \right) I_0 e^{i\omega t} \qquad (17\text{-}64)$$

where $Q = \omega L/R$.

The corresponding result for a step source $I_0 u(t)$ could be found simply by applying (17-62) to these equations. However, it is no more difficult to start afresh from first principles, and this has the additional advantage of giving us a better physical insight into the problem.

Following the same path by which we reached the harmonic solution in Sec. 17-1, we introduce separate expressions for the emfs $\mathcal{E}_1(t)$ and $\mathcal{E}_1^{\dagger}(t)$ induced in the underground circuit by the current $I_0 u(t)$ in the transmitter and by the current $I(t)$ in the circuit itself, i.e.,

$$\mathcal{E}_1(t) = -M_{01} \frac{d}{dt} [I_0 u(t)]$$

$$\mathcal{E}_1^{\dagger}(t) = -RI(t) - L \frac{d}{dt}[I(t)]$$

Since the sum of these potentials must vanish, we have

$$-\left(L\frac{d}{dt} + R\right)I(t) = M_{01}I_0\frac{d}{dt}[u(t)] \tag{17-65}$$

To solve this expression for $I(t)$, we define the step function $u(t)$ as follows:

$$u(t) = 0 \qquad t < 0$$
$$= \frac{t}{\tau} \qquad 0 \leq t < \tau$$
$$= 1 \qquad t > \tau$$

where τ vanishes in the limit.

Before switching on the current in the transmitter

$$I(t) = 0 \qquad t < 0 \tag{17-66}$$

while during the interval from $t = 0$ to $t = \tau$

$$-\left(L\frac{d}{dt} + R\right)I(t) = \frac{M_{01}I_0}{\tau}$$

which has the solution

$$I(t) = -\frac{M_{01}I_0}{R\tau} + Ae^{-Rt/L}$$

In order that $I(t)$ should vanish when $t = 0$, the constant A must be $= M_{01}I_0/R\tau$. Therefore

$$I(t) = -\frac{M_{01}I_0}{R\tau}(1 - e^{-Rt/L}) \tag{17-67}$$

For small values of $t(t \ll L/R)$ this expression reduces to

$$I(t) \doteq -\frac{M_{01}I_0}{R\tau}\frac{Rt}{L}$$

and in particular

$$I(\tau) \doteq -\frac{M_{01}I_0}{L}$$

When $t > \tau$, Eq. (17-65) becomes

$$\left(L\frac{d}{dt} + R\right)I(t) = 0$$

which has the solution $I(t) = Be^{-Rt/L}$. The constant B is determined by applying the condition

$$I(\tau) \doteq -\frac{M_{01}I_0}{L}$$

and since τ is extremely small $(\tau \ll L/R)$, it follows that

$$I(t) \doteq -\frac{M_{01}I_0}{L}e^{-Rt/L} \qquad t > 0 \tag{17-68}$$

The result tells us that a step in the transmitter current induces a similar sharp rise in current in the underground circuit. Instead of remaining constant, however, the induced current decays exponentially with a time constant $T = L/R$.

The emfs induced in the receiving coil by the transmitter and the buried circuit will be

$$\mathcal{E}_2^{(P)} = - M_{02} \frac{d}{dt} [I_0 u(t)] = - M_{02} I_0 \delta(t)$$

$$\mathcal{E}_2^{(S)} = - M_{12} \frac{d}{dt} I(t) = \frac{M_{01} M_{12}}{L} I_0 \left[\delta(t) - \frac{R}{L} e^{-Rt/L} \right]$$

$$(17\text{-}69)$$

where

$$\delta(t) = 0 \qquad t < 0$$

$$= \frac{1}{\tau} \qquad 0 \le t \le \tau$$

$$= 0 \qquad t > \tau$$

Figure 17-37 shows the form of the transients.

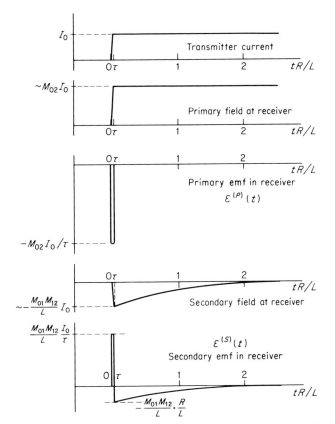

Fig. 17-37 Transient response of a simple circuit. Compare with the harmonic response shown in Fig. 17-2.

There is an evident relationship between the harmonic and transient responses. The reciprocal time constant of the exponential decay of the secondary field corresponds to the angular frequency Ω at which the harmonic response has a phase angle of $45°$, since

$$\frac{\Omega L}{R} = 1 \qquad \text{therefore} \qquad \Omega = \frac{R}{L} = \frac{1}{T}$$

Furthermore, the amplitude of the secondary magnetic field strength at the beginning of the decay is equivalent to the inductive limit of response to a harmonic source. However, there is no such correspondence between the induced emfs in the two cases, since the latter are proportional to the time derivatives of the magnetic flux passing through the receiver and not to the flux itself.

17-11 Transient Response to a Sphere

Now we shall take a brief look at the transient problem as applied to the simplest of the solid models, the uniform sphere. The steps to be followed have been clearly marked out in previous sections, and they are not unduly difficult to execute. The response to a conducting sphere in the presence of a dipolar harmonic source is given by the three formulas (17-45), (17-48), and (17-49), the choice of which will depend upon the orientation of the dipole. These formulas are of such a length that we shall not reproduce any of them here. The essential processes in the solution for a dipolar step source consist in substituting these expressions into the Fourier inversion formula (17-62) and in evaluating the integrals. Since only the time domain will be affected by the substitutions, the response function (17-44) is the only part of these expressions which is involved in the integrations.

The evaluation of the integrals, while not presenting any unusual difficulties, is certainly rendered more tractable if we put $\mu = \mu_0$. In that case we get, for $n = 1$,

$$X_1 + iY_1 = 1 + \frac{3}{k^2 a^2} - \frac{3 \coth ka}{ka} \qquad k^2 = i\mu_0\sigma\omega$$

The transient response for $n = 1$ is therefore found by evaluating the integral

$$R_1(t) = -\frac{1}{2\pi} \int_{-\infty}^{\infty} \left(\frac{i}{\omega} + \frac{3}{\beta\omega^2} - \frac{i^{1/2} \, 3 \coth \sqrt{i\beta\omega}}{\sqrt{\beta} \, \omega^{3/2}} \right) e^{i\omega t} \, d\omega \qquad \beta = \mu_0\sigma a^2$$

$$= \lim_{p \to 0} \frac{1}{2\pi i} \int_{p-i\infty}^{p+i\infty} \left(\frac{1}{s} + \frac{3}{\beta s^2} - \frac{3 \coth \sqrt{\beta s}}{\sqrt{\beta} \, s^{3/2}} \right) e^{st} \, ds$$

This is most easily done by using the Cauchy residue theorem. The result,

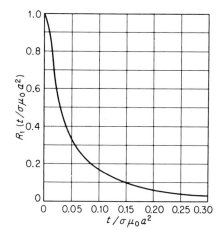

Fig. 17-38 The transient response of a conducting sphere in a uniform magnetic field that varies stepwise in time. [After Wait (16a).]

given by Wait (16a), is

$$R_1(t) = \frac{3t}{\beta} + 1 - \frac{2}{\sqrt{\pi}} \sqrt{\frac{t}{\beta}} - \sum_{n=1}^{\infty} 2n \left[\frac{1}{n} \sqrt{\frac{t}{\beta}} \, \Phi' \left(\frac{n}{\sqrt{t/\beta}} \right) \right.$$
$$\left. - 2 \left(1 - \Phi \left(\frac{n}{\sqrt{t/\beta}} \right) \right) \right]$$

where $\Phi(x) = \sqrt{2/\pi} \int_0^x e^{-\xi^2} \, d\xi$ is a tabulated function.

The decay of the dipole moment of the sphere following excitation by a primary field that varies stepwise with time is shown in Fig. 17-38 as a function of the dimensionless quantity $t/\mu_0\sigma a^2$. Transient effects due to the higher moments are more difficult to calculate; however, we can surmise from Fig. 17-9 that they will decay away more rapidly than the dipole moment effect because the response functions are all displaced toward higher frequencies. Thus the excitation of multipole moments of orders higher than the first will have the effect of speeding the decay of the secondary magnetic field.

References

1. A. A. Ashour, Induction of Electric Currents in a Uniform Circular Disk, *Mechanics and Applied Mathematics Quart. Jour.*, vol. 3, pp. 119–128, 1950.

2. L. Cagniard, Basic Theory of the Magnetotelluric Method of Geophysical Prospecting, *Geophysics*, vol. 18, pp. 605–635, 1953.

3. A. A. Douloff, The Response of a Disk in a Dipole Field, *Geophysics*, vol. 26, pp. 452–464.

4. J. H. Jeans, Finite Current Sheets, *London Math. Soc. Proc.*, vol. 31, pp. 151–169, 1899.

5. G. V. Keller and F. C. Frischknecht, Electrical Resistivity Surveys on the Athabaska Glacier, Alberta, Canada, *Natl. Bur. Standards Jour. Research*, vol. 64D, pp. 439–448, 1960.

6. L. Martin, "Field outside a Conducting Strip in the Presence of a Magnetic Dipole," Unpub. M.A. thesis, University of Toronto, 1960.

7. J. C. Maxwell, "A Treatise on Electricity and Magnetism," 3d ed. (2 vol.), Clarendon, Oxford, 1892.

8. A. T. Price, The Theory of Magnetotelluric Methods When the Source Field Is Considered, *Jour. Geophys. Research*, vol. 67, pp. 1,907–1,918, 1962.

9. V. V. C. Ranasinghe, "Inductive Interaction between Nearby Conducting Bodies in a Time Varying Magnetic Field," Unpub. M.A. thesis, University of Toronto, 1962.

10. L. B. Slichter, The Observed and Theoretical Response of Conducting Spheres, *Am. Inst. Mining Metall. Petroleum Engineers Trans.*, vol. 97, pp. 443–459, 1932.

11. L. B. Slichter and L. Knopoff, Field of an Alternating Magnetic Dipole on the Surface of a Layered Earth, *Geophysics*, vol. 24, pp. 77–88, 1959.

12. W. R. Smythe, "Static and Dynamic Electricity," 2d ed., p. 397, McGraw-Hill, New York, 1950.

13. A. Sommerfeld, Über verzweigte Potentiale im Raum, *London Math. Soc. Proc.*, vol. 28, pp 395–429, 1897.

14. E. C. Stoner, The Demagnetization Factor for Ellipsoids, *Philos. Mag.*, vol. 36 (7th ser.), pp. 803–820, 1945.

15. F. R. Tesche, "Instrumentation of Electromagnetic Modelling and Applications to Electromagnetic Prospecting," Unpub. Ph.D. thesis, University of California, Los Angeles, 1951.

16. J. R. Wait, (a) A Conducting Sphere in a Time-varying Magnetic Field, *Geophysics*, vol. 16, pp. 666–672, 1951.

(b) A Conducting Permeable Sphere in the Presence of a Coil Carrying an Oscillating Current, *Canadian Jour. Physics*, vol. 31, pp. 670–678, 1953.

(c) On the Relation between Telluric Currents and the Earth's Magnetic Field, *Geophysics*, vol. 19, pp. 281–289, 1954.

(d) Mutual Coupling of Wire Loops Lying on a Homogeneous Ground, *Geophysics*, vol. 19, pp. 290–296, 1954.

(e) Mutual Electromagnetic Coupling of Loops over a Homogeneous Ground, *Geophysics*, vol. 20, pp. 630–637, 1955.   Additional note, *Geophysics*, vol. 21, pp. 479–484, 1956.

(f) Induction by an Oscillating Magnetic Dipole over a Two-layer Ground, *Appl. Science Research-Sec. B*, vol. 7, pp. 73–80, 1958.

(g) On the Electromagnetic Response of an Imperfectly Conducting Thin Dyke, *Geophysics*, vol. 24, pp. 167–171, 1959.

17. J. P. Wesley, (a) Response of a Dyke to an Oscillating Dipole, *Geophysics*, vol. 23, pp. 128–133, 1958.

(b) Response of a Thin Dyke to an Oscillating Dipole, *Geophysics*, vol. 23, pp. 134–143, 1958.

18. S. H. Yungel, Magnetotelluric Sounding Three Layer Interpretation Curves, *Geophysics*, vol. 26, pp. 465–473, 1961.

chapter 18 Interpretation of Electromagnetic Surveys

In practice, the interpretation of electromagnetic anomalies is an exercise in model fitting. In common with the interpretation of potential field data of other kinds, the procedure most often followed is to compare observed anomalies with the calculated or measured responses of the apparatus to conductors of various simple shapes, sizes, orientations, and conductivities. If the anomaly pattern can be matched satisfactorily with one of these theoretical responses, we conclude that the appropriate model may resemble the geological body in its more important particulars. If a fit cannot be obtained, we surmise that significant differences must exist between the geological conductor and the interpretational models we have tried to use. Often the nature of the discrepancies between the observed and theoretical responses will give some indication of where these differences lie.

Any fitting process ought to involve the whole anomaly pattern as far as possible. The indirect approach to quantitative interpretation is to compare one or two specially selected profiles of the anomaly with a compilation of theoretical profiles, called *type curves.* Since a substantial effort is required to calculate or measure the data for these curves, it is uneconomical to carry through this part of the work in each interpretation. The theoretical responses should be obtained beforehand for a variety of models and model parameters. These may be used at a later time for making indirect interpretations.

18-1   Type Curves

By "type curve," we mean a particular profile of the response of a prospecting system to one of the interpretational models. Since these profiles are essential to quantitative electromagnetic interpretations, it is worth our while to give some space to describing how portfolios of such curves can best be set up. The data used for their construction may be obtained either by computing theoretical solutions of the kind developed in Chap. 17 or by taking measurements on a laboratory scale model (Sec. 16-6). The greatest problem that confronts us in compiling a suite of type curves is in selecting the values to use for the model parameters, for it would seem that a very large number of curves will be needed in order to handle routine problems in interpretation effectively. Even models characterized by a small number of parameters require an unwieldy number of curves in order to display all the various combinations of these parameters. It is therefore vitally important to eliminate any redundancy. We can help ourselves a great deal in this respect by the use of a little dimensional analysis.

First let us write out the expression for a profile of the theoretical response of an electromagnetic prospecting system to a given model as follows:

$$R = f(x;\sigma,\mu,\omega,\ l_1,\ l_2,\ \dots\ ,\ \Phi_1,\ \Phi_2,\ \dots) \tag{18-1}$$

where R is the quantity measured by the prospecting system (which is almost always a dimensionless measure of the electromagnetic anomaly); x measures the position of the prospecting system along the traverse; σ, μ are the electrical conductivity and the magnetic permeability, respectively, of the conductor; ω is the angular frequency of the current in the transmitter; $l_1, l_2, \dots$ are linear dimensions of the model and of the prospecting system; and $\Phi_1, \Phi_2, \dots$ are angles or dimensionless ratios describing the configuration of the model and of the prospecting system. Now let us try to group these variables in such a way that the smallest possible number will affect the response independently. First, we remember that the three parameters σ, μ, ω always occur in combination in the equations for an alternating electromagnetic field in the quasi-stationary state. They should therefore be similarly grouped in (18-1). Because μ enters alone into the boundary conditions on magnetic fields, however, it must appear separately in (18-1) as well as in the triple product. The principle of dimensional analysis states that the most economical arrangement of variables will be in characteristic, dimensionless groups. Therefore we should rewrite (18-1) in the following form:

$$R = g\left(\frac{x}{l_i}; \sigma\mu\omega l_j l_k, \frac{\mu}{\mu_0}, \frac{l_1}{l_i}, \frac{l_2}{l_i}, \ldots, \Phi_1, \Phi_2, \ldots\right) \tag{18-2}$$

which has two less variables than (18-1).

The choice of which dimensions to choose as l_i, l_j, and l_k will depend, of course, on the system and on the model. However, it is obvious that the scaling unit l_i must be a length which is accessible to direct measurement, and accordingly it must be characteristic of the prospecting system rather than of the model. For ground systems there is often only one characteristic length, viz., the separation of the receiver and transmitter coils. For airborne systems, the height of the aircraft may be used instead.

The rules for the selection of l_j and l_k are less well defined. In general, they must appear in the product $\alpha = \sigma\mu\omega l_j l_k$ as well as in ratios of the kind l_j/l_i and l_k/l_i, so that the number of variables is not affected. However, should one of the dimensions of the conductor happen to be very small in relation to the characteristic dimensions of the system, the conductor will behave essentially as a thin sheet. In that case, we know that the thickness and the conductivity must combine as a product. By choosing l_j as the thickness of the conductor, we can satisfy this condition, and l_j/l_i need not enter into R at all. The number of variables is thus reduced by one more.

At large values of α, the response of any system must clearly reach an inductive limit, so that we may write

$$R_L = h\left(\frac{x}{l_i}; \frac{l_1}{l_i}, \frac{l_2}{l_i}, \ldots, \Phi_1, \Phi_2, \ldots\right) \tag{18-3}$$

which is independent of α and μ/μ_0. In certain simple cases the relative response R/R_L below this limit is controlled *only* by α and μ/μ_0. Examples of this are the sphere in a uniform field (Sec. 17-2) and the infinite sheet or the half-space near a double-dipole prospecting system (Secs. 17-3, 17-4). In each of these three cases the response separates explicitly into a coupling coefficient and a response function just as we found in Sec. 17-1 for the simple circuit. The form of the response parameter itself comes directly from the analysis. For the sphere it is $\sigma\mu\omega a^2$ (a = radius), for the sheet it is $\sigma\mu\omega sh$ (s = thickness, h = height of prospecting system), and for the half-space it is $\sigma\mu\omega h^2$. These three forms suggest a general rule that should be followed in choosing l_j and l_k: They should be the two dimensions which most effectively control the volume of the eddy-current circulation, particularly that part of it closest to the receiver and therefore the most effective in determining the response. If this rule is followed, even in cases more complicated than those we have considered, the amplitude and phase of the secondary field relative to the inductive limit will remain more highly de-

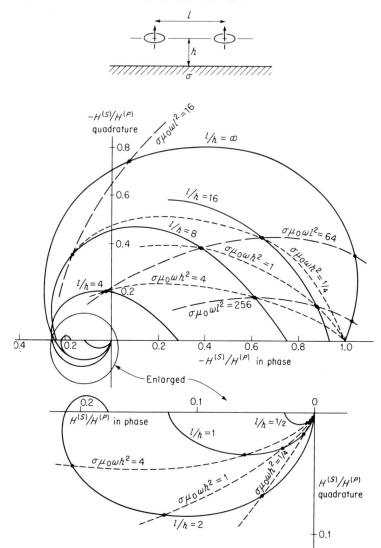

Fig. 18-1   Response of a uniform half-space to a horizontal-loop prospecting system.   If $l/h < 4$, $\sigma\mu_0\omega h^2$ is a satisfactory response parameter.   If it is used much beyond this range, however, it becomes confounded with l/h and ceases to be an effective measure of the response transition.   When l/h is large (> 4), the response parameter should be chosen as $\sigma\mu_0\omega l^2$.   [Data from Slichter and Knopoff, and Wait, Chap. 17, (11) and (16d), respectively.]

pendent upon the value of α than upon any other parameters of the model or of the prospecting system.

Very often, however, there may be several different parameters which will have an appreciable effect upon the size of the induced current system. Within a certain range of distances, one of these might be the dominant factor, and in another range, another. The choice of which one to use in the response parameter then becomes much less important. It is necessary only to avoid using any parameter which is without effect.

As an example, the response of a horizontal-loop electromagnetic system to a conducting half-space is shown in Fig. 18-1. Instead of showing the response function as a function of the response parameter, this diagram shows the secondary field as a complex variable when α and l/h are given constant values. Diagrams of this type are called "phasor diagrams." It will be observed that if $l > 4h$, the response parameter whose effects are most clearly distinguished from those of l/h is $\alpha = \sigma\mu\omega l^2$. This is because when $l \gg h$, the parameter l is clearly more effective than h in determining the size of the induced current vortex that is "seen" from the receiver position. On the other hand, as l becomes smaller, h gains in importance. A better response parameter to use for $l < 4h$ is $\alpha = \sigma\mu\omega h^2$, since there is less ambiguity in the phasor curves between lines of constant l/h and constant α when the response parameter is changed to this new form.

As another example, we may consider the response of a horizontal-loop system to a half-plane conductor. There the coil separation l is usually large compared with the depth to the edge of the conductor, and the best choices for l_j and l_k are obviously l and the sheet thickness s. That the phase of the response is substantially independent of other parameters can be verified from the examples in the following section (see Fig. 18-3). If we consider the response of the same model to a "double-dipole" airborne system, however, the height h of the system above the edge of the conductor, being much greater than l, will more effectively control the size of the eddy-current system. As seen by the receiver, the current system will have dimensions of the order of $h, \cdot h,$ and s. The response parameter should therefore be $\alpha = \sigma\mu\omega sh$.

18-2 Some Examples of Type Curves

Figures 18-2, 18-5, and 18-6 are examples of type curves for a conducting half-plane model. Each set displays the response measured by a different type of ground prospecting apparatus. Figure 18-2 shows the response of a horizontal-loop inphase and quadrature measuring system to the half-plane. Figures 18-5 and 18-6 demonstrate the responses of vertical-loop dip-angle measuring systems of the parallel line and of the fixed transmitter type, respectively. The suites are not meant to be comprehensive. They are

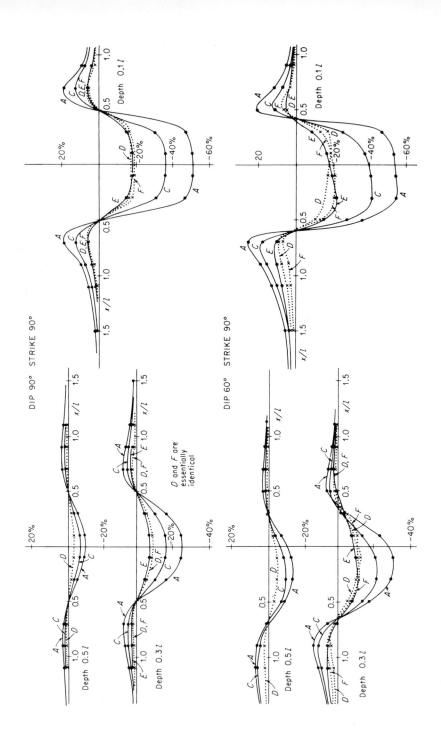

DIP 90° STRIKE 90°

DIP 60° STRIKE 90°

D and F are
essentially
identical

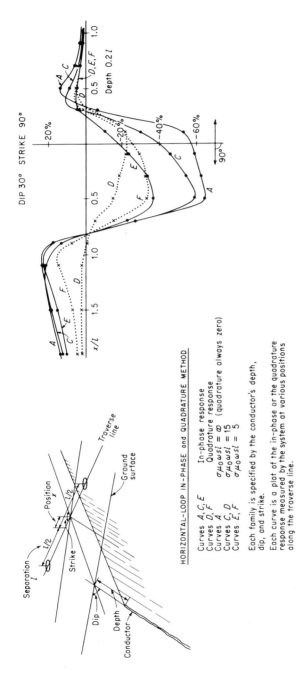

DIP 30° STRIKE 90°

HORIZONTAL–LOOP IN–PHASE and QUADRATURE METHOD

Curves *A,C,E*   In-phase response
Curves *D,F*   Quadrature response
Curves *A* $\sigma\mu_0\omega sl = \infty$ (quadrature always zero)
Curves *C,D* $\sigma\mu_0\omega sl = 15$
Curves *E,F* $\sigma\mu_0\omega sl = 5$

Each family is specified by the conductor's depth,
dip, and strike.

Each curve is a plot of the in-phase or the quadrature
response measured by the system at various positions
along the traverse line.

Fig. 18-2 Some type curves for a half-plane conductor and a horizontal-loop prospecting system.   Curves for $\sigma\mu_0\omega sl = \infty$ were obtained by computing the solution of Sec. 17-7.   The remainder were obtained by scale-model measurements.

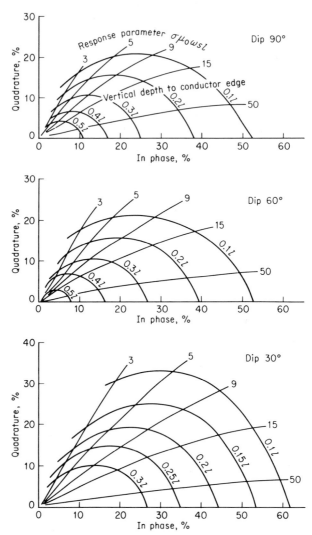

Fig. 18-3   Response (midway between zeros) of a horizontal-loop prospecting system to a half-plane conductor (traverse perpendicular to conductor strike).

intended only to illustrate the more obvious effects of the several parameters upon the forms of the "anomalies."

Horizontal-loop system.   Two curves are given for each example, one showing the percentage of inphase secondary field and the other the percentage of quadrature.   The responses are all essentially similar in form, showing a deep negative value when the coils straddle the upper edge of the conductor and weaker positive values when both coils lie on the same side of it.   Because the measurement is essentially one of mutual inductance, the shape of the profile is unaffected by interchanging the two coils. The examples in Fig. 18-2 all relate to the case in which the traverse line is perpendicular to the strike of the conductor; but the shape of the response is not critically dependent upon this, and profiles taken at 60° to the strike are very similar in appearance to those taken at 90°.

Of the three examples given, data for the horizontal-loop system are the easiest to interpret, because the effect of the depth to the top of the half-plane, the dip, and the value of the response parameter $\alpha = \sigma\mu\omega sl$ all produce very different effects on the anomaly pattern. To a first approximation the response parameter has practically no effect on the shape of the profile. Put in another way, the effect of α is independent of the position of the prospecting system along the traverse. If the inductive limit ($\alpha \to \infty$) is taken as a standard, a finite value of α just reduces the amplitude and shifts the phase of the anomaly in accordance with the usual form of response functions.

The most easily measured properties of the inphase and quadrature profiles are their amplitudes. Figure 18-3 reviews how the amplitudes of the anomaly midway between the two zeros change with depth, dip, and response parameter. A supplementary graph is given in Fig. 18-4 showing the horizontal displacement of the midpoint of the profile from the top edge of the conductor. Phasor characteristic curves of this type have been published by Hedstrom and Parasnis (2) and are in common use in the mining industry. They are particularly useful for making quick interpretations. In theory, one should estimate the dip of the conductor beforehand, since a different set of curves is appropriate to each value of d.

The influence of the dip on the anomaly profile is of secondary importance. It leads to a certain asymmetry, which is not very marked. Two particular effects, however, are most remarkable. Firstly, the area under the positive part of the profile is always greater on the downdip side. Secondly, the ratio of inphase to quadrature response on the negative part of the profile is greater on the downdip side. Since neither of these effects is strongly evident at dips much exceeding 30° and since either can easily be masked by irregularities in the conductor or by proximity to other conducting bodies, it is not worthwhile devising quantitative procedures for estimating d. The best results are obtained simply by direct comparison of the data with the type curves.

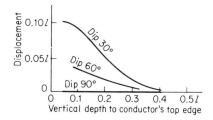

Fig. 18-4 Displacement (in terms of coil separation *l*) of the top edge of the conductor from the midpoint between zeros of the anomaly. The midpoint is always on the up-dip side.

Although the charts of Fig. 18-3 may be used to obtain an estimate of depth and response parameter directly, the whole of the anomaly profile should be compared with the type curves in order to verify that these values do in fact lead to a satisfactory fit. If they do, then by substituting the known values of ω, μ_0, and l into the estimated response parameter an estimate of the surface conductivity σs can be made. Typical values for good conductors lie in the range 10 to 100 mhos, while for some very good conductors such as pyrrhotite veins they may be even higher [Bosschart, (1)].

Conductivity itself cannot be estimated without an independent estimate of s. One frequently used empirical method starts from the premise that on the theoretical profile over a half-plane of moderately steep dip the distance between the two zeros is closely equal to l. This distance is therefore determined from the field profile, and the excess over the value of l is used as an estimate of s. Rather good results are often obtained by this method if the width of the conductor is as great or greater than the depth to its top. If conductivity is estimated by using this value of s, it should be regarded as an apparent conductivity of the entire zone that lies within the conductor's outer envelope. With sulfide conductors, it will rarely reach a value which even approaches the conductivity of the pure minerals. Apparent conductivity values of the order of 1 mho/m are more common. The reason is simply that large volumes of barren nonconducting rock usually exist within the outer envelope of the mineralized zone, and even within the heavily mineralized regions the concentration is rarely 100 per cent.

Dip-angle systems. The profiles shown in Figs. 18-5 and 18-6 are not quite so easy to summarize as the type curves for the horizontal-loop system, and interpretation requires a little more experience. The main characteristic of the curves is that the magnitude of the dip angle reaches a maximum when the receiver is on one side of the conductor, falls to zero approximately over the edge, and then goes through a maximum of opposite sign on the other side. The sense of the *crossover* in dip angle is the same in all cases. The field always dips away from the edge of the conductor and never toward it.

The shape of the curves is highly variable. In a general way the amplitude of the profile is determined by the response parameter and by the depth of the conductor. To a very rough approximation, the curves for different values of α are similar to those for a perfect conductor but are attenuated. Besides having an effect on amplitude, depth also affects the steepness of the crossover from one maximum to the other, but this is a risky feature on which to base an interpretation because it is also very sensitive to the finite width of a real conductor.

One very important feature of these two methods is that the angle which the traverse makes with the strike of the conductor in the parallel-line method, and the position of the transmitter with respect to the conductor trace in the fixed transmitter method, both have a profound influence on the

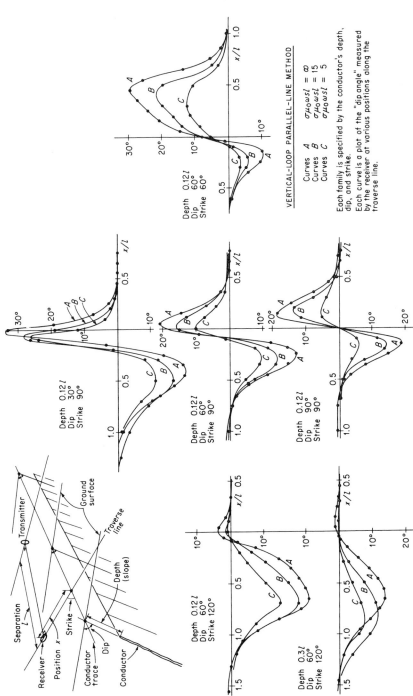

Fig. 18-5 Some type curves for a half-plane model and a vertical-loop, parallel-line, dip-angle measuring prospecting system. Curves A for $\sigma\mu_0\omega sl = \infty$ were obtained from the mathematical solution of Sec. 17-7; the others, by scale modeling.

VERTICAL-LOOP PARALLEL-LINE METHOD

Curves A $\sigma\mu_0\omega sl = \infty$
Curves B $\sigma\mu_0\omega sl = 15$
Curves C $\sigma\mu_0\omega sl = 5$

Each family is specified by the conductor's depth, dip, and strike.

Each curve is a plot of the "dip angle" measured by the receiver at various positions along the traverse line.

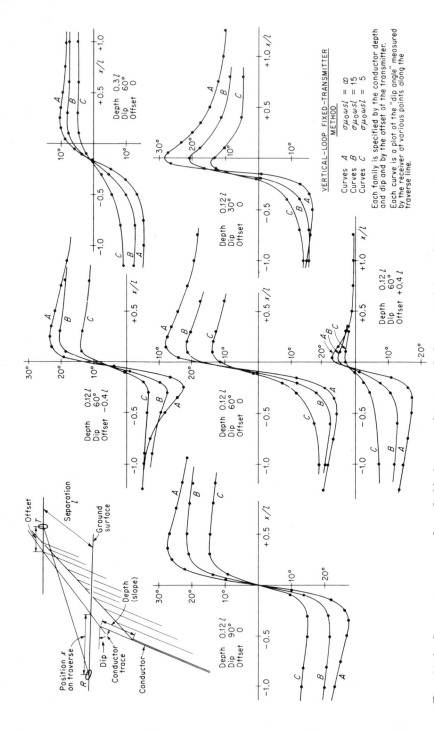

Fig. 18-6 Some type curves for a half-plane conductor and the vertical-loop, fixed-transmitter, dip-angle measuring prospecting system. The curves for $\sigma\mu_0\omega sl = \infty$ are obtained from the mathematical solution of Sec. 17-7; the others, by scale modeling.

VERTICAL-LOOP FIXED-TRANSMITTER
METHOD

Curves A $\sigma\mu_0\omega sl = \infty$
Curves B $\sigma\mu_0\omega sl = 15$
Curves C $\sigma\mu_0\omega sl = 5$

Each family is specified by the conductor depth and dip and by the offset of the transmitter.

Each curve is a plot of the "dip angle" measured by the receiver at various points along the traverse line.

shapes of the type curves.   It is necessary to determine the strike direction before ever attempting to interpret a profile.   This is generally done by correlating crossovers from one traverse line to the next, although it can be reasonably well determined from a single profile of the parallel-line type (Fig. 18-7).   This by itself can be a useful application of the method in certain kinds of reconnaissance work.

Once the strike of the conductor has been determined, the interpreter can proceed to estimate the other parameters by curve matching.   It is usually unprofitable to try to separate the effects of response parameter and depth on a single profile.   The difficulty is best resolved by making measurements at two frequencies, and facilities for doing this are often provided in the apparatus.   The ratio of the amplitudes of the two profiles can be used to estimate α, and the amplitude itself, to measure depth.

If a dual frequency apparatus is not available, somewhat similar results can be achieved by changing l.   A change in the response parameter can be produced in this way, but other extraneous effects may be introduced at the same time because of the change in the transmitter position.

As with the horizontal-loop method, the dip of the half-plane affects only the shape of the profile.   Although its effects are somewhat larger with the dip-angle method than with the horizontal-loop method, they are still difficult to distinguish from either a change in the strike direction or spurious irregularities.   It has been suggested that dip interpretation can be made considerably easier by taking a series of short profiles of the fixed transmitter type, where the transmitter is put first on one side of the conductor's trace, then over it, and finally on the other.

For the fixed transmitter method and in the special circumstance that the interpreter has assured himself that the transmitter was closely on strike and the response of the conductor was very near the inductive limit, estimates of depth and dip can be obtained from two characteristic measures. These are the total "throw" of the anomaly (peak-to-peak amplitude) and

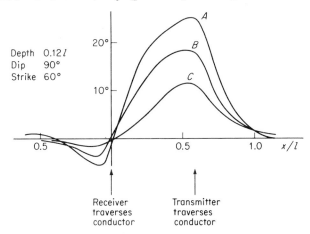

Fig. 18-7 Vertical-loop, parallel-line method (see Fig. 18-5). The transmitter's traversal of the conductor's edge produces a noticeable inflection in the profile, thereby providing a method of estimating the strike of a conductor from a single profile.

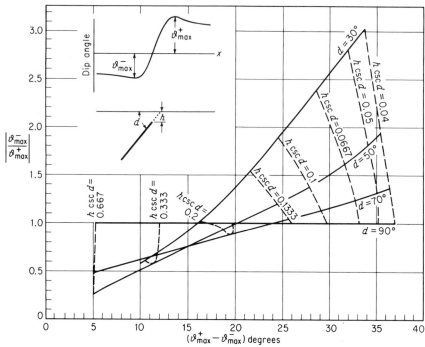

Fig. 18-8 "Characteristic curves" for the conducting half-plane model, using the vertical-loop, fixed-transmitter method. The unit of h is the distance between the transmitter and the picket line along which the receiver moves. If the conductor is deeply buried, $h > 0.2$, the effects of dip become difficult to distinguish in these curves from those of depth.

the ratio of the maximum amplitudes observed on each side of the crossover. The necessary characteristic curves are shown in Fig. 18-8.

A special word of caution is needed in connection with the use of the half-plane model in the interpretation of data taken by the fixed-transmitter method. The apparatus often permits values of l of 1,200 ft or more to be used. To approximate a half-plane, the conductor must then be extremely large, at the very least 2,000 ft long and 1,500 ft deep. If much accuracy is expected, especially in estimating the response parameter, it should be even larger. While conductors of this size are certainly not altogether unknown, the interpreter should expect to see the influence of finite dip and strike extent occurring much more frequently in the anomaly patterns when these large values of l are used.

Figures 18-9, 18-10, and 18-11 show some examples of type curves for the half-plane fitted to field data. The results are by no means perfect and in many ways are typical of what may be expected in the ordinary routine of interpretation. The all-too-common difficulty of having too few measurements along the profile is evident in one of the examples.

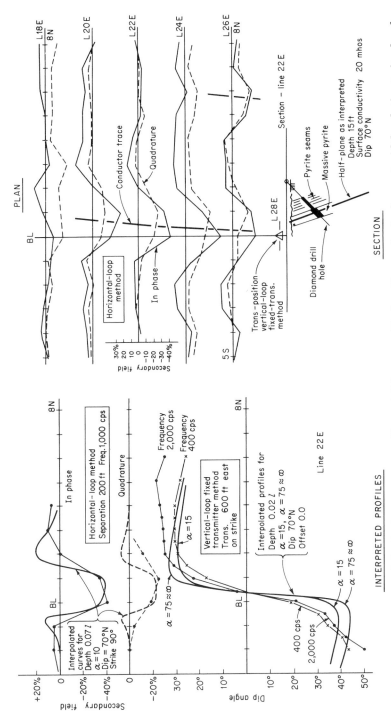

Fig. 18-9 The results of the fitting are reasonably satisfying, considering the conductor's great width relative to its depth of burial. Most of the divergence between the field and theoretical curves can be attributed to this.

561

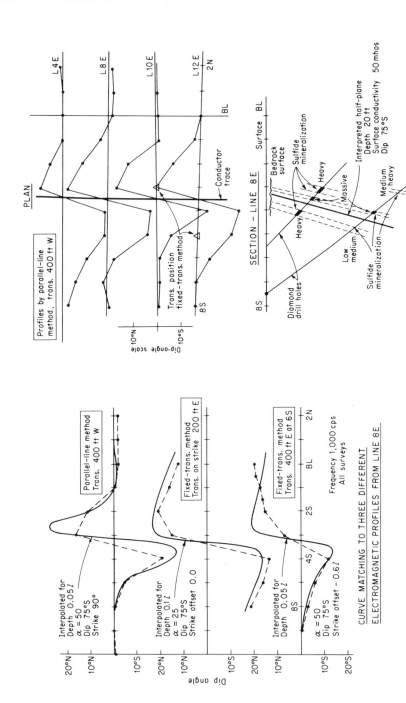

Fig. 18-10 The half-plane model gives profiles which fit the field data reasonably well. However there is considerable ambiguity in each fitting, and if only a single profile were interpreted, very significant errors could be made in positioning the conductor. The interpreted value of surface conductivity is really a lower limit, since the body appears almost as a perfect conductor.

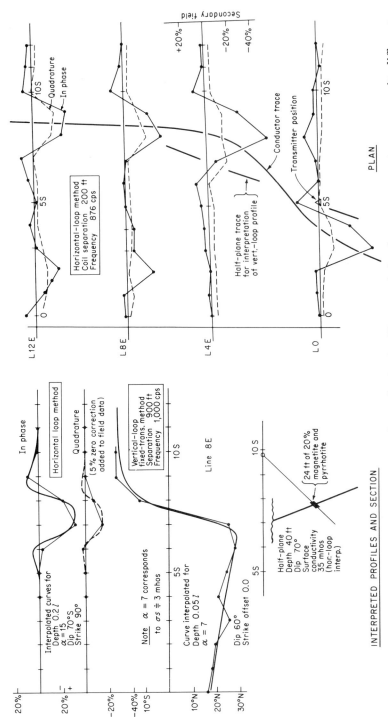

Fig. 18-11 Rather good fits to the field data have been obtained. However, it has been necessary to assume quite different values of surface conductivity in the two different types of survey in order to achieve this. Very probably the vertical-loop data are showing the effect of the conductor's finite depth extent. If the bottom approaches to within about 1,000 ft of the surface, its effect would be enough to account for the discrepancy.

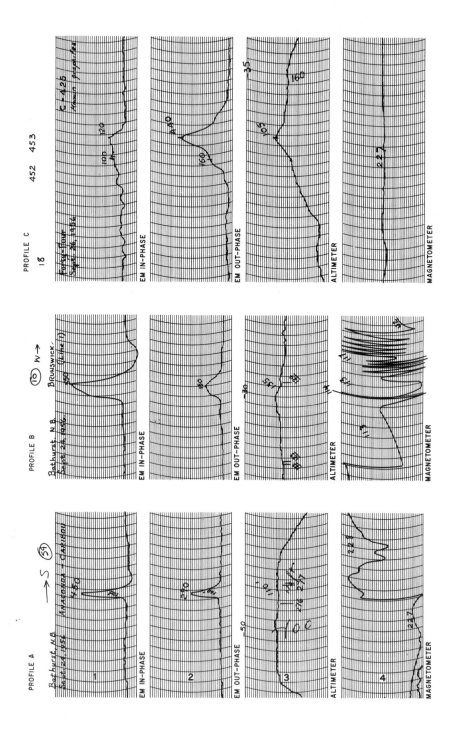

564

18-3 The Interpretation of Airborne Surveys

The objectives of an airborne electromagnetic survey are somewhat more limited than, or are at least different from, those of a ground survey. The primary objective in both cases is to discover zones of high conductivity. Beyond this, a ground survey is usually required to provide some information on the shape of the conductive body and to pinpoint its position sufficiently well that a drill may be accurately aimed toward it. (Additional information will generally be required to decide *if* drilling should be done.) This second objective is never required of an airborne survey. What is desirable is that the electromagnetic system give some indication of the conductor's general characteristics so that only those which offer a reasonable promise of being mineralized zones need be investigated on the ground.

Because the noise level of an airborne electromagnetic apparatus is usually rather high compared with the amplitudes of the anomalies obtained even from the largest and best conductors, a less rigorous approach is taken to interpretation. The common procedure is to compare only the more obvious anomaly characteristics with the values for a variety of theoretical models, with the hope that some indication may be found therefrom about whether the conducting body resembles one model more than the others. Is the conductor more likely to be compact in shape or is it dikelike? Is it greatly extended in one direction or has it a limited strike length? If the answers to such simple questions can be found, they will often provide a real basis for deciding whether, in the circumstances, a follow-up survey on the ground ought to be initiated.

It is easier, and also probably clearer, to give examples of airborne electromagnetic interpretations rather than to discuss the subject in general

Fig. 18-12 Reproduction of inphase and quadrature airborne electromagnetic profiles taken with a horizontal "double-dipole" type of apparatus over three different conductors. The operating frequency was in all cases about 400 cps. Accompanying the electromagnetic profiles are records taken simultaneously with a radio altimeter and an airborne magnetometer. General comments on these profiles: (*a*) Profile *A*: The conductor is a massive sulfide replacement zone which occurs in chlorite schist lying on a graywacke slate series. The absence of any distinct magnetic anomaly indicates that the sulfide is of a nonmagnetic type. The ratio of inphase to quadrature response amplitude (= 1.59) indicates that the body is a moderately good conductor. (*b*) Profile *B*: The conductor is again a sulfide replacement which occurs under very shallow overburden. Notice the intense magnetic anomaly, indicating a very substantial amount of iron. The high ratio of inphase to quadrature response amplitude (= 3.66) suggests that the body is a very good conductor. (*c*) Profile *C*: This set of profiles was flown across a band of graphitic sediments. Note the low ratio of inphase to quadrature response amplitude (= 0.27). The complete absence of any magnetic anomaly suggests that the graphite band is barren of massive sulfides. [Figure and comments courtesy Canadian Aero Service Ltd.]

terms. We shall therefore describe how an interpreter might analyze the data obtained by a system of the "double-dipole" type [Sec. 15-5].

The two most significant and certainly the most easily measurable properties of anomalies obtained with apparatus of this type will be the peak amplitudes of the two phase components. Beyond that, we can usually say whether the profiles are symmetrical or asymmetrical, simple or complicated (i.e., whether there are few or many inflection points), and make some sort of measurement of their widths. All of these qualities are descriptive of the shape of the anomaly profile, and since both phase components are measured continuously during flight, there will be two profiles by which to determine them. Since the phase of the secondary field due to a single conductor will not vary a great deal as the aircraft passes over it, we would expect the two profiles to have similar shapes. The total amount of information is therefore not as great as it at first might seem. Where we require characteristic measures of the anomaly shape, an average taken from the two profiles can be used.

If the anomaly pattern is sufficiently simple and intense, quantitative interpretation methods can be tried. If it is complicated, it is possibly caused by more than one conductor, and the value of these methods is doubtful. Small irregularities are often accounted for by changes in flight height and by instrumental noise.

The first step in interpreting an anomaly is to make a rough estimate of the horizontal extent of the conductor by examining the profile width and by searching for correlations between adjacent flight lines. If the pattern is simple, we can generally put the conductor into one of three categories: (1) compact, (2) broad extent, or (3) linear. As a sort of standard of comparison, the calculated anomaly pattern due to a moderately sized and highly conducting spherical body is shown in Fig. 18-13.

We may then select within each category a simple interpretational model to use for comparison. For example, we may choose the sphere for category 1, the infinite horizontal sheet for category 2, and the half-plane for category 3. The next step is to see if the theoretical response could possibly duplicate the principal features of the observed anomaly without assuming geologically unreasonable values of the model parameters. If it can, the values of the parameters will help us to assess the character of the geological conductor. If not, it may be possible to guess in what way the conductor must differ from the model. Details of how this might be carried out are given in the following paragraphs:

Case 1. To an anomaly of limited dimensions we could fit a spherical conductor by measuring the amplitudes of the two phase components and the width of the anomaly pattern. These three measures would in principle be just sufficient to determine uniquely the values of the response parameter α, the radius a, and the distance h of closest approach to the top of the sphere. We would then have to decide whether or not the sphere so determined had an intelligible geological interpretation. However, the width of the anomaly profile is difficult to measure with great precision, and so strongly does it influence the calculation of h and a that it is better on the whole to work in the opposite direction. A plausible value is assumed for h, and the amplitudes of the two phase components of the anomalous field are used to find values for α and a. The width of the anomaly pattern can then be checked against its theoretical value in order to ascertain whether a reasonable fit has actually been obtained.

The height of the aircraft above the ground is often used as a trial value for h. If the theoretical value for the width of the anomaly pattern exceeds the observed value, we may conclude that the body is not equidimensional but is probably more tabular in form. If the calculated width is too narrow,

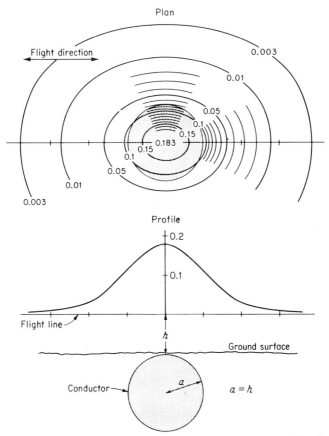

Fig. 18-13 Inphase anomaly measured by a horizontal "double-dipole" prospecting system over a perfectly conducting sphere of radius equal to the flight height. The response H/m is given in units of $-(4\pi)^{-1}\,(2h)^{-3}$. Thus for a coplanar coil system flown at a height of 300 ft and having a coil separation of 60 ft, the peak anomaly would be $+183$ ppm.

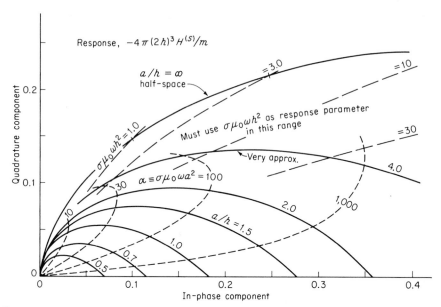

Fig. 18-14   Peak response of a sphere to a horizontal "double-dipole" system (curves are not very accurate beyond $a/h = 1.0$).

the indication is that a larger h should be used.   In other words, the body is either larger and deeper than we have supposed (or possibly displaced to one side of the flight line), or else it is shallow but flat-lying.

The actual determination of α, a, and h may be made by the use of a phasor diagram constructed in exactly the manner described in Sec. 18-2, Fig. 18-3.   The diagram will in this case be constructed from Eqs. (17-48) or (17-49) for the response of a conducting sphere in a dipole field.   Its form is outlined in Fig. 18-14.

Case 2.   In fitting a horizontal conducting sheet to an anomaly pattern, we assume that the lateral extent of the conductor is many times the height of the aircraft.   If this is actually the case, the anomaly should have a flat top.   The properties of the sheet can then be determined solely from the amplitudes of the two phase components of the anomalous field.   The response to a flat conducting sheet of a "double-dipole" system was given in Sec. 17-3 as Eq. (17-27) (see also Fig. 17-9).   For the purpose of making interpretations of airborne data, it is better to translate this result into a phasor diagram such as that shown in Fig. 18-15.   Height and response parameter can then be found directly from the amplitudes of the phase components of the anomalous field.

If we obtain a value for h which is close to the flight height, we can accept the sheet interpretation with few reservations, especially if surface conductors such as swamps or thick clay overburden are found in the area.

If the value of h is significantly larger than the flight height, we may accept it provisionally if it has a reasonable geological interpretation, or we could ascribe the anomaly to other causes, such as a zone of closely spaced vertical conductors lying near the ground surface. It is altogether unlikely that a value of h which is much less than the flight height will be obtained from the calculations, because at any given height the horizontal sheet causes a stronger anomaly than any other shape of conducting body.

Case. 3 Four parameters must be determined in order to fit a half-plane conductor to an electromagnetic anomaly. These are (1) the response parameter $\alpha = \sigma\mu_0\omega sh$, (2) the height h of the system above the edge of the sheet, (3) the dip, and (4) the strike direction. There is no difficulty in determining the strike direction if the anomaly is observed on more than one flight line. In that event, the fitting of the other three parameters proceeds exactly as outlined for the horizontal-loop ground prospecting system (Sec. 18-2). The asymmetry of the anomaly profile can be used roughly to estimate the dip of the half-plane, and the amplitudes of the inphase and quadrature effects may be used to estimate the height and the response parameter. Phasor diagrams similar in form to those of Figs. 18-3, 18-14, and 18-15 are needed to do this, a different one for each combination of strike and dip. The interpretation routine might then run in the following sequence: (1) Estimate the strike direction from inter-flightline correlation, (2) estimate the dip by comparing the record with a few theoretical profiles calculated at a similar strike angle, and (3) select the appropriate phasor graph and use it to estimate h and α from the inphase and quadrature anomaly amplitudes.

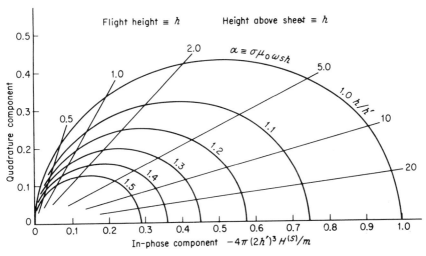

Fig. 18-15 Phasor diagram for interpreting the depth and response factor of a thin horizontal sheet from the response of a horizontal "double-dipole" system.

After obtaining values for h and α, it is possible to compare the expected width of the anomaly pattern against observations if sufficient theoretical data are available to do so.   However, the comparison will not have much significance if the conductor is not nearly vertically dipping, because the width, unhappily, is strongly affected by dip and the dip can be estimated only roughly from the profiles.   Should the value of h turn out to be less than the height of the aircraft above the ground, we may infer that the conductor has a substantial thickness.   Values larger than the flight height may indicate either a certain depth of burial or a limited depth extent of the conducting material, or both.

Quantitative interpretation is not worthwhile when the flight lines run almost parallel to the strike of the conductor.   This is because the response is too sensitive to the exact position of the aircraft and to irregularities in the conductor itself.   Neither of these factors can be ascertained to the degree of precision necessary to allow the use of phasor diagrams.

The foregoing discussion has been somewhat detailed not because these details are important in themselves, but because they demonstrate how interpretation may be done and how the theoretical models can be put to use.   Such details of procedure as we have described will necessarily undergo alterations if a different prospecting system is employed or with any marked differences in the nature of the conductors which are expected.   However, a scheme of interpretation along similar lines to the one just described can usually be developed for each system with a relatively modest amount of preparatory effort [see (2), (3), and (4)].

18-4   Discrimination in Electromagnetic Prospecting Systems

In many instances an airborne electromagnetic apparatus is used solely as a detector of conductive zones which might be caused by massive sulfide mineralization.   Thus its job is to point out localized, highly conducting masses against a background of variable but more extensive and more poorly conducting rocks.   Put another way, the system should be able to discriminate among conductors of various kinds rather than register all zones of anomalous conductivity with equal intensity.   In the ascending scale of mineral conductivities, the point at which discrimination becomes effective will depend very largely upon the "scale" of the system and upon its operating frequency.   Since these vary considerably from one apparatus to another, it is not surprising to find that surveys made by different prospecting systems over the same area often emphasize altogether different geological units.

The "scale" of an electromagnetic prospecting system is rather a loose term which signifies the overall or largest dimension.   In a crude way it is this factor which determines the radius within which a conductor must lie if it is to produce an anomaly whose amplitude is an appreciable fraction of the strongest anomalies which the system commonly detects. Except that it refers to horizontal as well as vertical distances and that it takes no note of the level of instrumental sensitivity achieved by the apparatus, it is quite akin to the much discussed (and misused) term "depth penetration" of the system. In a two-plane airborne electromagnetic system in which the receiver-to-transmitter distance is much larger than the flight height, it will be the former which sets the scale. On the other hand, in a double-dipole system the scale will be determined by the flight height, since this will be the distance of closest possible approach to any conductor, and it will still be very much larger than the coil separation. There are other airborne systems in which both the coil separation and the flight height must be taken into consideration.

In general, electromagnetic systems will respond to different conductors according to the value of the product of conductivity times the square of the average linear dimension of the body, provided that the average dimension is small in relation to the scale value of the system. When the size of the body exceeds the scale of the system, conductivity alone becomes the controlling factor. Thus, when a high power of discrimination is required of the system, certain limitations must be imposed upon its scale value.

To crystallize the above discussion, let us consider a theoretical example. We shall represent the small, good conductor with a sphere and the larger but poorer conductor with a half-space. The prospecting system that we shall choose is a horizontal double dipole. All the data that are necessary to our argument are then available from Fig. 18-14.

Let us contrast the performances of two different prospecting systems, the first having a flight height of 125 ft and the second a flight height of 500 ft. We shall assume that the sphere has a radius of 250 ft so that $a/h = 2$ in the first case and 0.5 in the second. Thus the sphere is a comparatively small conductor to the high-flying system and a fairly large one for the other. If we assign to it a conductivity of 0.37 mhos/m, it will produce an inphase response three times as large as the quadrature response when the frequency is 450 cps in the high-flying system and 1,500 cps in the low-flying system. The difference is due entirely to the fact that the primary magnetic field over the volume of the sphere is comparatively uniform when the system having the larger scale value is used.

A value of 3 for the ratio of inphase to quadrature amplitude would almost certainly be construed as the mark of a "good conductor." However, we must keep in mind that, in order to be discerned among the background effects originating from the larger but less conductive zones, the sphere must produce a signal that has an amplitude which is conspicuously greater than these effects. If we allow the half-space to serve as a model of other much larger but less conductive zones, we may calculate the maximum conductivity that it can possess without masking the effect of the sphere. Let us arbitrarily set as an upper limit on the inphase response to the half-space a value equal to one-half of that to the sphere. The quadrature response may be allowed to take values considerably larger than this. These are roughly the minimal conditions under which the anomaly will be clearly resolved.

According to Fig. 18-14, the appropriate maximum values of the response parameter $\sigma\mu_0\omega h^2$ are 0.3 in the system having the larger scale value and 1.2 in the case of the smaller. These values correspond to conductivities of 0.003 and 0.06 mhos/m, respectively, at the frequencies previously specified.

We therefore arrive at the conclusion that the low-flying system not only can detect the smaller conductor but can also distinguish it from anomalies caused by larger zones whose conductivity reaches up to about one-sixth that of the sphere, while the high-flying system can resolve the local anomaly only if the two conductivities differ by a factor of at least 100. This can be extremely important under certain circumstances. However, we must be careful not to carry the argument too far. Against the advantages in discrimination which obtain at the lower flight elevations, we must consider the decreased ability of the system to detect deeply buried conductors. Low flying also increases the need for a finer line spacing in order to avoid missing important anomalies.

It should also be pointed out that strong discrimination will be decidedly a mixed blessing if there is a significant probability that target conductors may be larger in size and lower in conductivity than the cutoff values for which the system is designed. It is urgently necessary to take considerable care in this matter, and careful thought should be given ahead of time to deciding what the probable size and conductivity of the target bodies will be. This will provide a rational basis for selecting the frequency and the system scale value and will undoubtedly enhance the probability of success.

References

1. R. A. Bosschart, On the Occurrence of Low Resistivity Geological Conductors, *Geophys. Prosp.*, vol. 9, pp. 203–212, 1961.

2. H. Hedstrom and D. S. Parasnis, Some Model Experiments Relating to Electro-Magnetic Prospecting with Special Reference to Airborne Work, *Geophys. Prosp.*, vol. 6, pp. 322–341, 1959.

3. N. R. Paterson, Experimental and Field Data for Dual-frequency Phase-shift Method of Airborne Electromagnetic Prospecting, *Geophysics*, vol. 26, pp. 601–617, 1961.

4. W. G. Wieduwilt, Interpretation Techniques for a Single-frequency Airborne Electromagnetic Device, *Geophysics*, vol. 27, pp. 493–506, 1962.

appendix *Coefficients for Downward Continuation*

The Gauss-Laguerre quadrature formula for evaluating the downward continuation integral (see Sec. 9-9) is

$$\Delta g_z(0) \doteq \sum_{j=1}^{m} M_j(\gamma) \, \Delta g_0(r_j) \qquad \text{(A-1)}$$

where $r_j = 2 \sqrt{\gamma w_j}$, the w_j's being the m roots of the Laguerre polynomial $L_m(w_j) = 0$. $\Delta g_0(r_j)$ is the ring average of Δg around a circle of radius r_j on the plane $z = 0$. The weighting coefficients $M_j(\gamma)$ are defined in Sec. 9-9.

The values of $\Delta g_0(r_j)$ are to be found by interpolation, either from the observations or from contours. In practice it is usually more convenient to use radii which pass through the points of a regular grid (see Henderson, 1960) in lieu of the values defined by (A-1). Thus we convert the formula (A-1) into the equivalent expression

$$\Delta g_z(0) \doteq \sum_{k=0}^{n} B_k(\gamma) \, \Delta g_0(r_k) \qquad \text{(A-2)}$$

in which the radii r_k have values that are fixed so that the circles will pass through certain grid points. The conversion is made by applying a four-point Laplace interpolation formula to yield an expression for each

$$\Delta g_0(2 \sqrt{\gamma w_j})$$

in terms of the values of $\Delta g_0(r_k)$ on the four nearest circles. Thus by combining the interpolation coefficients with the computed values of $M_j(\gamma)$, we arrive at the following table of downward continuation coefficients:

Downward continuation coefficients

k	r_k	n_k	$z = 0.75$		$z = 1.00$		$z = 1.25$	
			$\gamma = \frac{1}{9}$	$\gamma = \frac{1}{6}$	$\gamma = \frac{1}{6}$	$\gamma = \frac{1}{4}$	$\gamma = \frac{1}{4}$	$\gamma = \frac{1}{3}$
0	0	1	+2.1652	+1.2361	+2.5038	+1.2024	+2.3168	+1.2609
1	1	4	+0.3844	+1.8444	+1.4621	+2.6644	+3.4369	+3.5492
2	$\sqrt{2}$	4	−1.2647	−1.7333	−2.6018	−2.2727	−4.0842	−2.8342
3	$\sqrt{5}$	8	−0.1081	−0.1712	−0.1531	−0.3683	−0.4216	−0.7242
4	$\sqrt{8}$	4	−0.0481	−0.0356	−0.0455	−0.0298	−0.0174	−0.0134
5	$\sqrt{10}$	8	−0.0232	−0.0418	−0.0341	−0.0310	−0.0323	−0.0407
6	$\sqrt{13}$	8	−0.0267	−0.0279	−0.0330	−0.0389	−0.0443	−0.0437
7	$\sqrt{17}$	8	−0.0220	−0.0230	−0.0330	−0.0278	−0.0327	−0.0354
8	$\sqrt{20}$	8	−0.0167	−0.0135	−0.0142	−0.0198	−0.0241	−0.0253
9	$\sqrt{25}$	12	−0.0041	−0.0215	−0.0262	−0.0292	−0.0343	−0.0335
10	$\sqrt{34}$	8		−0.0145	−0.0237	−0.0216	−0.0267	−0.0297
11	$\sqrt{40.5}$	16		+0.0018	−0.0013	−0.0151	−0.0196	−0.0164
12	$\sqrt{50}$	12				−0.0126	−0.0164	−0.0199
13	$\sqrt{64.5}$	12						−0.0206

In the table, the third column gives the number of grid points lying on the circle. $\Delta g_0(r_k)$ is taken to be a simple arithmetic average of the n values (weighting for the non-evenness of the distribution of points on the circles can be incorporated into the averages but is usually disregarded), or in more refined calculations it is obtained by numerical quadrature. The larger of the two values of γ given for each z corresponds with "heavy" smoothing, and the smaller with "light" smoothing. The choice between them will depend a good deal upon the amount of roughness present in the data. The use of a regular grid usually involves making a set of interpolations from the original data (sometimes referred to as "digitizing" the map), and is often done with a view to carrying out the subsequent arithmetical operations on an automatic computer.

The Gauss-Laguerre coefficients and the values of w_j were taken from the tables given by Salzer and Zucker (1949).

References

Henderson, R. G., A Comprehensive System of Automatic Computation in Magnetic and Gravity Interpretation, *Geophysics*, vol. 25, pp. 569–585, 1960.

Salzer, H. E., and Ruth Zucker, Tables of the Zeros and Weight Factors of the First Fifteen Laguerre Polynomials, *Bull. Amer. Math. Soc.*, vol. 55, pp. 1004–1012, 1949.

Index